Properties and Applications of Thermoelectric Materials

NATO Science for Peace and Security Series

This Series presents the results of scientific meetings supported under the NATO Programme: Science for Peace and Security (SPS).

The NATO SPS Programme supports meetings in the following Key Priority areas: (1) Defence Against Terrorism; (2) Countering other Threats to Security and (3) NATO, Partner and Mediterranean Dialogue Country Priorities. The types of meeting supported are generally "Advanced Study Institutes" and "Advanced Research Workshops". The NATO SPS Series collects together the results of these meetings. The meetings are co-organized by scientists from NATO countries and scientists from NATO's "Partner" or "Mediterranean Dialogue" countries. The observations and recommendations made at the meetings, as well as the contents of the volumes in the Series, reflect those of participants and contributors only; they should not necessarily be regarded as reflecting NATO views or policy.

Advanced Study Institutes (ASI) are high-level tutorial courses intended to convey the latest developments in a subject to an advanced-level audience

Advanced Research Workshops (ARW) are expert meetings where an intense but informal exchange of views at the frontiers of a subject aims at identifying directions for future action

Following a transformation of the programme in 2006 the Series has been re-named and re-organised. Recent volumes on topics not related to security, which result from meetings supported under the programme earlier, may be found in the NATO Science Series.

The Series is published by IOS Press, Amsterdam, and Springer, Dordrecht, in conjunction with the NATO Public Diplomacy Division.

Sub-Series

A.	Chemistry and Biology	Springer
B.	Physics and Biophysics	Springer
C.	Environmental Security	Springer
D.	Information and Communication Security	IOS Press
E.	Human and Societal Dynamics	IOS Press

http://www.nato.int/science
http://www.springer.com
http://www.iospress.nl

Series B: Physics and Biophysics

Properties and Applications of Thermoelectric Materials

The Search for New Materials for Thermoelectric Devices

edited by

Veljko Zlatić
Institute of Physics, Zagreb, Croatia

and

Alex C. Hewson
Imperial College, Department of Mathematics, London, United Kingdom

Published in cooperation with NATO Public Diplomacy Division

Proceedings of the NATO Advanced Research Workshop on
Properties and Applications of Thermoelectric Materials
Hvar, Croatia
21–26 September 2008

Library of Congress Control Number: 2009927448

ISBN 978-90-481-2891-4 (PB)
ISBN 978-90-481-2890-7 (HB)
ISBN 978-90-481-2892-1 (e-book)

Published by Springer,
P.O. Box 17, 3300 AA Dordrecht, The Netherlands.

www.springer.com

Printed on acid-free paper

Preface

The NATO sponsored Advanced Research Workshop on "Properties and Applications of Thermoelectric Materials" took place on the Croatian island of Hvar over the period 20-26 September, 2008. This subject has attracted renewed interest as concerns with the efficient use of energy resources, and the minimization of environmental damage, have become important current issues. There has been the recognition that thermoelectric devices could play a role in generating electricity from waste heat, enabling cooling via refrigerators with no moving parts, and many other more specialized applications. The main problem in realizing this ambition is the rather low efficiency of such devices for general applications. The workshop addressed that problem by reviewing the latest experimental and theoretical work in this field and by exploring various strategies that might increase the efficiency of thermoelectric devices.

A measure of the potential of a particular material for an efficient thermoelectric device is the figure of merit ZT. This dimensionless ratio depends on the electronic conductivity, the total thermal conductivity and the thermopower (Seebeck coefficient) of the material. For practical applications in a given temperature range a figure of merit should exceed a value of 1, and the search, therefore, is to find or fabricate materials that satisfy this criterion. The difficulty that arises is the conflicting effects of the different factors. The high electrical conductivity of good metals enhances ZT, but this coexists with poor thermopower and high electric thermal conductivity which reduces ZT, so these materials are not suitable. Materials with a high thermopower and low thermal conductivity, on the other hand, tend to have a poor electrical conductivity, so some compromise has to be found between these conflicting factors in optimizing ZT. One existing class of materials which are metallic, but also have a high thermopower, are the compounds and alloys which have strong electron correlation. These materials predominantly contain rare earth or actinide ions, and the enhanced thermopower arises from narrow asymmetric renormalized bands near the Fermi level, resulting from the hybridization of the 4f or 5f states with the conduction electrons. The thermoelectric anomalies due to the proximity of the chemical potential to the Mott–Hubbard gap have also attracted a lot of attention recently. Typical examples are provided by high temperature superconductors and other oxides. The thermoelectric properties of these classes of materials were the main focus of the workshop.

The papers in this volume cover the broad range of approaches, from the experimental work of fabricating, and characterising, the properties of new compounds to enhance ZT, through to theoretical work on renormalized band structure calculations and model Hamiltonians to obtain a deeper understanding of the thermoelectric properties of these materials. The effects of disorder, the proximity to metal-insulator transitions, the properties of layered composite materials, and the introduction of voids or cages into the structure to reduce the lattice thermal conductivity are all explored. Other ways of enhancing ZT by exploiting glassy materials or using the thermoelectric properties of metal-excitonic insulator interfaces are also covered. Several papers dealing with the general physical properties of strongly correlated materials are included as well.

We hope the papers presented here give a guide to the current activity in the field, and convey the fact that real progress is being made. Much work still requires to be done – this is more of an interim report – but we hope it will encourage and stimulate further developments in the field, to bring forward the day when the general production of high efficiency thermoelectric devices can be realized.

Zagreb, London
March 2009

Veljko Zlatić
Alex Hewson

Acknowledgements

The Advanced Research Workshop on "Properties and Applications of Thermoelectric Materials" was made possible through the financial support from the NATO Science for Peace Program. Additional financial support from the European COST-ECOM (P16) Action, The International Institute for Complex and Adaptive Matter (I2CAM), and the Ministry of Science of Croatia is also gratefully acknowledged. We also wish to thank Dr. I. Aviani of the Institute of Physics in Zagreb for his generous and continued help in the organization of the workshop.

Contents

Contributors

P. Adelmann Forschungszentrum Karlsruhe, Institut für Festkörperphysik, PO Box 3640, 76021 Karlsruhe, email: peter.Adelmann@ifp.fzk.de

E. Alves Dep. Física, Instituto Tecnológico e Nuclear/CFN-UL, P-2686-953 Sacavém, Portugal, email: ealves@itn.pt

F. B. Anders Institute for Theoretical Physics, University of Bremen, D-28334 Bremen, Germany
Lehrstuhl für Theoretische Physik II, Technische Universität Dortmund, Otto-Hahn Straße 4, 44221 Dortmund, Germany, email: frithjof.anders@tu-dortmund.de

V. I. Anisimov Institute of Metal Physics, Russian Academy of Science-Ural division, 620219 Yekaterinburg, Russia, email: via@imp.uran.ru

R. Arita Department of Applied Physics, University of Tokyo, Tokyo 113-8656, Japan, email: arita@ap.t.u-tokyo.ac.jp

M. C. Aronson Department of Physics and Astronomy, Stony Brook University, Stony Brook, NY 11974 and Brookhaven National Laboratory, Upton, NY 11973, email: maronson@bnl.gov

N. P. Barradas Dep. Física, Instituto Tecnológico e Nuclear/CFN-UL, P-2686-953, Sacavém, Portugal, email: nunoni@itn.pt

R. E. Baumbach Department of Physics and Institute for Pure and Applied Physical Sciences, University of California at San Diego, La Jolla, CA 92093, USA, email: baumbach@physics.ucsd.edu

M. C. Bennett Department of Physics and Astronomy, Stony Brook University, Stony Brook, NY 11974, email: mcbennett@umich.edu

S. Burdin Institut für Theoretische Physik, Universität zu Köln, Zülpicher Str. 77, 50937 Köln, Germany
Max Planck Institut für Physik Komplexer Systeme, Nöthnitzer Str. 38, 01187 Dresden, Germany, email: burdin@thp.uni-koeln.de

B. Chevalier I.C.M.C.B., CNRS, Université Bordeaux I, 33608-Pessac, France, email: chevalie@icmcb-bordeaux.cnrs.fr

T. Cichorek Institute of Low Temperature Structure Research, Polish Academy of Sciences, 50-950 Wrocław, Poland, email: T.Cichorek@int.pan.wroc.pl

B. Coqblin L.P.S., CNRS UMR 8502, Université Paris-Sud, 91405-Orsay, France, email: coqblin@lps.u-psud.fr

G. Czycholl Institute for Theoretical Physics, University of Bremen, D-28334 Bremen, Germany, email: czycholl@itp.uni-bremen.de

M. Deppe Max Planck Institut für Chemische Physik fester Stoffe, Nöthnitzer Str. 40, 01187 Dresden, Germany, email: deppe@cpfs.mpg.de

A. M. Finkel'stein Department of Condensed Matter Physics, The Weizmann Institute of Science, Rehovot 76100, Israel
Department of Physics, Texas A&M University, College Station, TX 77843-4242, USA, email: alexander.finkelstein@weizmann.ac.il

N. Franco Dep. Física, Instituto Tecnológico e Nuclear/CFN-UL, P-2686-953, Sacavém, Portugal, email: nfranco@itn.pt

C. Godart CNRS, ICMPE, CMTR, 2/8 rue Henri Dunant, 94320 Thiais, France, email: godart@icmpe.cnrs.fr

A. P. Gonçalves Dep. Química, Instituto Tecnológico e Nuclear/CFMC-UL, P-2686-953, Sacavém, Portugal, email: apg@itn.pt

C. Grenzebach Institute for Theoretical Physics, University of Bremen, D-28334 Bremen, Germany, email: cgrenzebach@itp.uni-bremen.de

K. Grube Forschungszentrum Karlsruhe, Institut für Festkörperphysik, PO Box 3640, 76021 Karlsruhe, email: Kai.Grube@ifp.fzk.de

J. J. Hamlin Department of Physics and Institute for Pure and Applied Physical Sciences, University of California at San Diego, La Jolla, CA 92093, USA, email: jhamlin@physics.ucsd.edu

F. Hardy Forschungszentrum Karlsruhe, Institut für Festkörperphysik, PO Box 3640, 76021 Karlsruhe, email: frederic.hardy@ifp.fzk.de

S. Hartmann Max Planck Institut für Chemische Physik fester Stoffe, Nöthnitzer Str. 40, 01187 Dresden, Germany, email: Hartmann@cpfs.mpg.de

K. Haule Physics Department and Center for Materials Theory, Rutgers University, 136 Frelinghuysen Road, Piscataway, NJ, email: haule@physics.rutgers.edu

K. Held Institute for Solid State Physics, Vienna University of Technology, A-1040 Vienna, Austria, http://www.ifp.tuwien.ac.at/cms

Z. Henkie Institute of Low Temperature Structure Research, Polish Academy of Sciences, 50-950 Wrocław, Poland, email: Z.Henkie@int.pan.wroc.pl

P. C. Ho Department of Physics and Institute for Pure and Applied Physical Sciences, University of California at San Diego, La Jolla, CA 92093, USA, email: pcho@csufresno.edu

P. Horsch Max-Planck-Institut für Festkörperforschung, Heisenbergstrasse 1, 70569 Stuttgart, Germany, email: P.Horsch@fkf.mpg.de

Y. Janssen Brookhaven National Laboratory, Upton, NY 11973, email: yjanssen@bnl.gov

U. Köhler Max Planck Institut für Chemische Physik fester Stoffe, Nöthnitzer Str. 40, 01187 Dresden, Germany
Leibniz-Institut für Festkörper- und Werkstoffforschung Dresden, Helmholtzstr. 20, 01069, Dresden, Germany, email: u.koehler@ifw-dresden.de

M. S. Kim Brookhaven National Laboratory, Upton, NY 11973, email: mskim@bnl.gov

W. Knafo Forschungszentrum Karlsruhe, Institut für Festkörperphysik, PO Box 3640, 76021 Karlsruhe
Physikalisches Institut, Universität Karlsruhe, 76128 Karlsruhe, email: knafo@lncmp.org

W. Koshibae Cross-Correlated Materials Research Group (CMRG), RIKEN, Saitama 351-0198, Japan, email: wataru@riken.jp

G. Kotliar Physics Department and Center for Materials Theory, Rutgers University, 136 Frelinghuysen Road, Piscataway, NJ, email: kotliar@physics.rutgers.edu

K. Kuroki University of Electro-Communications 1-5-1 Chofugaoka, Chofu-shi Tokyo 182-8585, Japan, email: kuroki@vivac.e-one.uec.ac.jp

E. B. Lopes Dep. Química, Instituto Technológico e Nuclear/CFMS-UL, P-2686-953 Sacavém, Portugal, email: eblopes@itn.pt

H. v. Löhneysen Forschungszentrum Karlsruhe, Institut für Festkörperphysik, PO Box 3640, 76021 Karlsruhe
Physikalisches Institut, Universität Karlsruhe, 76128 Karlsruhe, email: h.vL@phys.uni-karlsruhe.de

D. E. MacLaughlin Department of Physics, University of California, Riverside, CA 92521-0413, USA, email: macl@physics.ucr.edu

S. Maekawa Institute for Materials Research, Tohoku University, Sendai 980-8577, Japan
CREST, Japan Science and Technology Agency, Sanbancho, Tokyo 102-0075, Japan, email: maekawa@imr.tohoku.ac.jp

K. Maiti Department of Condensed Matter Physics and Materials Science, Tata Institute of Fundamental Research, Homi Bhabha Road, Colaba, Mumbai - 400005, India, email: kbmaiti@tifr.res.in

M. B. Maple Department of Physics and Institute for Pure and Applied Physical Sciences, University of California at San Diego, La Jolla, CA 92093, USA, email: mbmaple@ucsd.edu

C. Meingast Forschungszentrum Karlsruhe, Institut für Festkörperphysik, PO Box 3640, 76021 Karlsruhe, Germany, email: Christoph.Meingast@ifp.fzk.de

K. Michaeli Department of Condensed Matter Physics, The Weizmann Institute of Science, Rehovot 76100, Israel, email: karen.michaeli@weizmann.ac.il

N. Oeschler Max Planck Institut für Chemische Physik fester Stoffe, Nöthnitzer Str. 40, 01187 Dresden, Germany, email: oeschler@cpfs.mpg.de

A. M. Oleś M. Smoluchowski Institute of Physics, Jagellonian University, Reymonta 4, 30059 Kraków, Poland
Max-Planck-Institut für Festkörperforschung, Heisenbergstrasse 1, 70569 Stuttgart, Germany, email: A.M.Oles@fkf.mpg.de

A. Pietraszko Institute of Low Temperature Structure Research, Polish Academy of Sciences, 50-950 Wrocław, Poland, email: A.Pietraszko@int.pan.wroc.pl

A. Rüegg Theoretische Physik, ETH Zürich, 8093 Zürich, Switzerland, email: rueegga@phys.ethz.ch

M. Rontani CNR-INFM Research Center S3, Via Campi 213/A, 41100 Modena MO, Italy, email: rontani@unimore.it

O. Rouleau CNRS, ICMPE, CMTR, 2/8 rue Henri Dunant, 94320 Thiais, France, email: rouleau@icmpe.cnrs.fr

P. Schweiss Forschungszentrum Karlsruhe, Institut für Festkörperphysik, PO Box 3640, 76021 Karlsruhe, email: peter.schweiss@ifp.fzk.de

L. J. Sham Department of Physics, University of California San Diego, Gilman Drive 9500, La Jolla 92093-0319, California, email: lsham@ucsd.edu

L. Shu Department of Physics and Institute for Pure and Applied Physical Sciences, University of California at San Diego, La Jolla, CA 92093, USA, email: lshu@physics.ucsd.edu

M. Sigrist Theoretische Physik, ETH Zürich, 8093 Zürich, Switzerland, email: sigrist@phys.ethz.ch

D. A. Sokolov Brookhaven National Laboratory, Upton, NY 11973, email: fermiliquid@gmail.com

F. Steglich Max Planck Institut für Chemische Physik fester Stoffe, Nöthnitzer Str. 40, 01187 Dresden, Germany, email: Steglich@cpfs.mpg.de

P. Sun Max Planck Institut für Chemische Physik fester Stoffe, Nöthnitzer Str. 40, 01187 Dresden, Germany, email: sun@cpfs.mpg.de

R. Wawryk Institute of Low Temperature Structure Research, Polish Academy of Sciences, 50-950 Wrocław, Poland, email: R. Wawryk@int.pan.wroc.pl

T. Wolf Forschungszentrum Karlsruhe, Institut für Festkörperphysik,
PO Box 3640, 76021 Karlsruhe, email: Thomas.Wolf@ifp.fzk.de

Q. Zhang Forschungszentrum Karlsruhe, Institut für Festkörperphysik,
PO Box 3640, 76021 Karlsruhe, email: qinzhang788@yahoo.com.cn

V. Zlatić Institute of Physics, Bijenicka cesta 46, P.O. Box 304, HR-10001 Zagreb,
Croatia, email: zlatic@ifs.hr

G. Zwicknagl Institut fuer Mathematische Physik, Technische Universitaet
Braunschweig, Mendelssohnstr. 3, 38116 Braunschweig, Germany,
email: g.zwicknagl@tu-bs.de

Strongly Correlated Electron Phenomena in the Filled Skutterudites

M. B. Maple, R. E. Baumbach, J. J. Hamlin, P. C. Ho, L. Shu, D. E. MacLaughlin, Z. Henkie, R. Wawryk, T. Cichorek, and A. Pietraszko

Abstract In the field of modern condensed matter physics, a major effort has been devoted to the study of correlated electron phenomena in solids, particularly those associated with d- and f-electrons of transition metal, rare earth, or actinide ions. One class of materials that has contributed to this enterprise has been the filled skutterudites, which have provided a rich reservoir of new materials including $PrOs_4Sb_{12}$: the first Pr-based heavy fermion superconductor. In this class of compounds, the delicate interplay between several types of physical phenomena can be tuned by means of various "knobs", such as pressure and chemical substitution, to reveal the underlying mechanisms of the correlated electron behavior.

In this paper, we review recent experiments on filled skutterudite compounds based on Ce and Pr "filler ions", with particular attention to the arsenides and antimonides.

1 Introduction

Historically, mineral skutterudites were mined as a source of transition metals (such as Fe and Ni) in Skutterud, Norway, and their crystalline structure was first characterized in 1928 by I. Oftedal [1]. In general, mineral skutterudites have the chemical formula TX_3 where the transition metal T = Co, Rh, or Ir and the

M. B. Maple, R. E. Baumbach, J. J. Hamlin, P. C. Ho, and L. Shu
Department of Physics and Institute for Pure and Applied Physical Sciences, University of California at San Diego, La Jolla, CA 92093, USA
e-mail: mbmaple@ucsd.edu, baumbach@physics.ucsd.edu, jhamlin@physics.ucsd.edu, pcho@csufresno.edu, lshu@physics.ucsd.edu

D. E. MacLaughlin
Department of Physics, University of California, Riverside, CA 92521-0413, USA
e-mail: macl@physics.ucr.edu

Z. Henkie, R. Wawryk, T. Cichorek, and A. Pietraszko
Institute of Low Temperature Structure Research, Polish Academy of Sciences, 50-950 Wrocław, Poland
e-mail: Z.Henkie@int.pan.wroc.pl, R.Wawryk@int.pan.wroc.pl, T.Cichorek@int.pan.wroc.pl, A.Pietraszko@int.pan.wroc.pl

V. Zlatić and A. C. Hewson (eds.), *Properties and Applications of Thermoelectric Materials*, NATO Science for Peace and Security Series B: Physics and Biophysics,

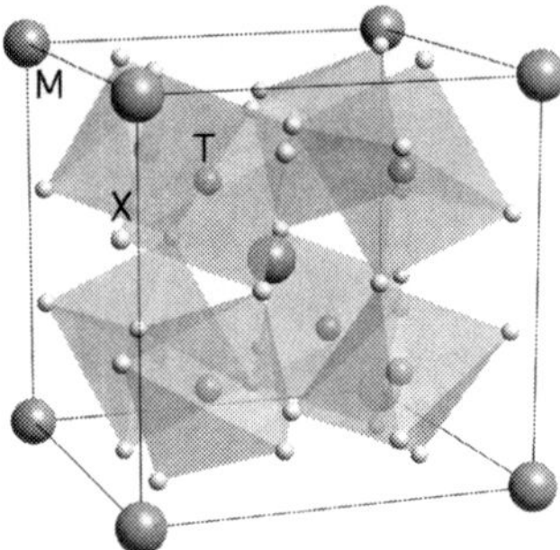

Fig. 1 Crystal structure of the filled skutterudites. The filled skutterudites have the chemical formula MT_4X_{12} and crystallize in the cubic space group $Im\bar{3}$ with two formula units per unit cell. The lattice constant a is large $\sim$ 7.8–9.3 Å. The rare earth position is (0, 0, 0), the transition metal position is (0.25, 0.25, 0.25) and the pnictogen position is (0, y, z) where the parameters $y \sim 0.35$ and $z \sim 0.16$ are variable depending on the particular chemical composition. The transition metal ions are located in distorted canted octahedra that are composed of pnictide ions. The M ions are located in large cages composed of pnictide ions.

pnictogen X = P, As, or Sb. The crystal structure of these compounds has two large voids per unit cell which were exploited in 1977 by W. Jeitschko and D. Braun to accommodate a variety of atoms, thereby introducing the "filled" skutterudite structure and consequently initiating modern interest in the filled skutterudites [2–4]. The skutterudite family of compounds has since attracted concentrated interest due to the wide variety of interesting electronic and magnetic phenomena that they exhibit including: conventional BCS (Bardeen–Cooper–Schrieffer) superconductivity, unconventional superconductivity, heavy fermion behavior, non-Fermi-liquid (NFL) behavior, antiferroquadrupolar order, localized moment magnetic order, spin fluctuations, itinerant electron ferromagnetism, hybridization gap semiconducting behavior, intermediate valence behavior, etc. [5–9]. Filled skutterudites crystallize in the cubic space group $Im\bar{3}$ with two formula units per unit cell (Fig.1) and are characterized by large lattice constants a ~7.8–9.3 Å. The rare earth position is (0, 0, 0), the transition metal position is (0.25, 0.25, 0.25), and the pnictogen position is (0, y, z) where the parameters y $\sim$ 0.35 and $z \sim 0.16$ are variable depending on the particular chemical composition. It is important to note that the variation in the values of y and z is due to the fact that the pnictogens do not form perfect squares and the MX_6 canted octahedra are distorted. This property has important implications for the phenomena that emerge in these systems.

In addition to the wealth of correlated electron phenomena that are observed in these compounds, some of the filled skutterudites show promising thermoelectric properties at elevated temperatures (600–900 K) which generated new interest in the topic of the filled skutterudites in the mid-1990s. The thermoelectric potential of a material is quantified by the figure of merit $Z = \sigma S^2/\kappa$ where σ is the electrical conductivity, S is the Seebeck coefficient or the thermopower, and κ is the thermal conductivity, which is the sum of an electronic (κ_e) and a lattice (κ_l) component. For the filled skutterudites, κ_l is reduced by the motion or "rattling" of the M filler ions which incoherently scatter phonons as the result of the filler ion's large amplitude oscillations, in addition to tunneling between equivalent off-center sites within the X cage. The reader is referred to one of several reviews for more information [5,10,11].

Until the late 1990s, it was generally accepted that the filled skutterudites have the chemical formula MT_4X_{12} where the "filler" atom M = La – Nd, Sm, Eu, Yb, U, or Th, the transition metal T = Fe, Ru, or Os, and the pnictogen X = P, As, or Sb. The element T can be replaced with transition metals from adjacent rows, although this action usually results in a strong reduction of the M filling fraction. More recently, it was shown that filled skutterudites with M = alkali metal, alkaline earth, Hf, or Tl, T = Fe, Ru, or Os, and X = Sb can be formed [12, 13]. Around the same time, filled skutterudites with the small atomic radius rare earth ions M = Y, or Gd-Lu, T = Fe, Ru, or Os and X = P were also synthesized using a high pressure technique [14–16]. Interestingly, the same technique can be used to synthesize the compound $La_xRh_4P_{12}$, which has an unusually high superconducting transition temperature, $T_c = 17$K [17]. Following these developments, the filled skutterudite family was expanded to include M = Sr, Ba, La-Nd, Eu, or Th, T = Pt, and X = Ge [18, 19]. This result is of particular interest because it shows that by maintaining the appropriate charge balance, it is possible to deviate from the "classical" filled skutterudites which include elements from the Fe-transition metal and pnictogen columns. It is also worth noting that the metal oxide compounds $MM'_3T_4O_{12}$, where M can be any cation with a large enough radius, irrespective of the charge state, M' is a Jahn-Teller cation Cu^{2+} or Mn^{2+}, and T can be either a transition metal or a non-transition metal, have recently attracted intense interest [20]. For these compounds, the crystal structure is derived from the simple perovskite ABO_3 structure and is essentially that of the filled skutterudite where an additional crystallographic site is occupied, leading to a quaternary chemical composition.

In this paper, we summarize recent results for the filled skutterudites MT_4X_{12} where M = Ce or Pr, T = Fe, Ru, or Os, and X = P, As, or Sb. In particular, these compounds have been shown to exhibit several types of behavior that are characterized by strong hybridization between the localized f-electron and conduction electron states. Moreover, the delicate interplay between several types of interactions can be tuned by means of several "knobs" such as pressure, chemical composition, and disorder to reveal the underlying mechanisms for the phenomena that are observed. For these reasons, the filled skutterudites continue to play an important role in the development and understanding of correlated electron materials and, as new subsets are synthesized, they likely hold more exciting phenomena yet to be discovered. Finally, due to the ability of this structure to incorporate many different types of atoms, there exists a large phase space of uninvestigated filled skutterudites which remain promising candidates for thermoelectric applications.

2 Ce-based Filled Skutterudites

A particularly interesting subclass of the filled skutterudites consists of the compounds with M = Ce. These compounds are characterized by strong hybridization between the localized $4f$-electron and conduction electron states. The resultant correlated electron behavior is a delicate function of various parameters, the most prominent being the lattice constant a and the electronic configuration of the $T-X$ subsystem.

As the pnictogen changes in the direction X = P, As, and Sb, the lattice constant increases [2–4] and the strength of the hybridization apparently decreases for the compounds CeT_4X_{12} (T = Fe, Ru, or Os). This progression causes the behavior of the Ce atoms to evolve towards that of localized Ce^{3+} ions, although it should be noted that even for the Sb-based compounds the effects of hybridization are not negligible [21]. Conversely, the importance of the electronic configuration is most apparent in the differences in behavior for the series T = Fe, Ru, and Os where the arsenides and phosphides appear to traverse quantum critical points (QCPs) that are centered near the Ru members, both of which show non-Fermi-liquid behavior [22, 23]. Currently, the order parameters for the possible QCPs remain under investigation. The phenomena that emerge as a result of these competing effects are summarized in Fig. 2. The presence of impurities, chemical disorder, and vacancy concentration also are observed to influence the physical phenomena.

For the Ce-based X = P and As examples, the effects of hybridization are first seen as reductions in a from the values expected for typical lanthanide contractions for ions with a valence of 3+ [2–4]. The reduced lattice constants are accompanied by unusual magnetic susceptibilities that are strongly suppressed below the values expected for Ce^{3+} ions according to Hund's rules exhibiting a Curie law T dependence (Fig. 3) [23–26]. In contrast to the phosphides and arsenides, the antimonides show magnetic behavior that is similar to that expected for Ce^{3+} ions at

Ce	*Fe*	*Ru*	*Os*
P	Semicond. $\Delta \sim 1500$ K	Semicond. $\Delta \sim 900$ K	Semicond. $\Delta \sim 400$ K
As	Weakly magnetic semimetal	NFL	Semicond. $\Delta \sim 50$ K
Sb	Heavy fermion $T_{coh} \sim 150$ K	NFL	Semicond. $\Delta \sim 10$ K

Fig. 2 Grid summarizing the physical behavior observed for CeT_4X_{12} compounds. More than half of these compounds are weakly magnetic semiconductors where the gap size is correlated with the lattice constant. The remaining members all show strongly correlated electron behavior with the As-based members exhibiting weak magnetism and the Sb-based compounds displaying magnetic behavior similar to that expected for Ce^{3+} ions. The *dark borders* denote compounds in which Ce appears to have an intermediate valence.

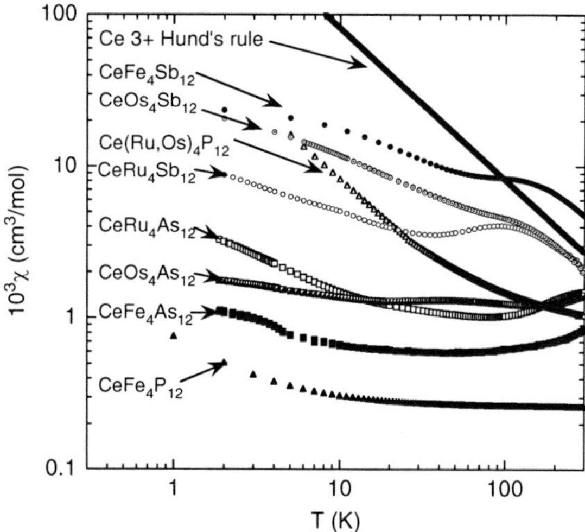

Fig. 3 Magnetic susceptibility $\chi(T)$ for the Ce-based filled skutterudites. The $\chi(T)$ data reveal that for the P and As-based compounds, strong hybridization between the Ce ions and the conduction electron ions results in weakly magnetic or nonmagnetic behavior below room temperature. For the Sb-based compounds, the magnetic susceptibility is similar to that expected for Ce^{3+} ions for $T > 100$ K, below which the development of a coherent state appears to suppress $\chi(T)$. The straight line represents the Curie law behavior of $\chi(T)$ for Ce^{3+} ions according to Hund's rules (i.e., without crystalline electric field splitting). Data are from refs. [22–28].

temperatures above 100 K [22, 27, 28]. However, around 100 K, the magnetic susceptibilities of the antimonides go through a broad hump and, with decreasing T, begin to resemble those seen for the phosphides and arsenides. At low T, these compounds also show a weak increase in $\chi(T)$ with decreasing T, the origin of which remains uncertain but could either be intrinsic, as for NFL behavior [29–32] or T-dependent intermediate valence behavior [33, 34], or extrinsic, due to the presence of a low concentration of magnetic impurities.

For more than half of the Ce-based filled skutterudites, the suppressed magnetic behavior is accompanied by small energy gap semiconducting behavior ($\Delta/k_B =$ 10–1,500 K) as shown in Fig. 4, where the gap size is inversely proportional to the lattice constant [21]. The members of this family that are not semiconductors are particularly interesting, showing non-Fermi-liquid behavior for $CeRu_4As_{12}$ [24] and $CeRu_4Sb_{12}$ [22], and moderately heavy fermion behavior for $CeFe_4Sb_{12}$ [35]. It is also apparent that sample dependence, i.e., vacancies or chemical disorder, plays an important role in the correlated electron behavior seen in these compounds. This result makes it necessary to synthesize high quality single crystal specimens in order to study their intrinsic properties.

As mentioned above, the compounds $CeRu_4As_{12}$ and $CeRu_4Sb_{12}$ display NFL behavior. In typical discussions of NFL physics in d- or f-electron materials, the phenomena are described in terms of interactions between the itinerant electron and localized d- or f-electron states of transition metal or rare earth ions [29–32].

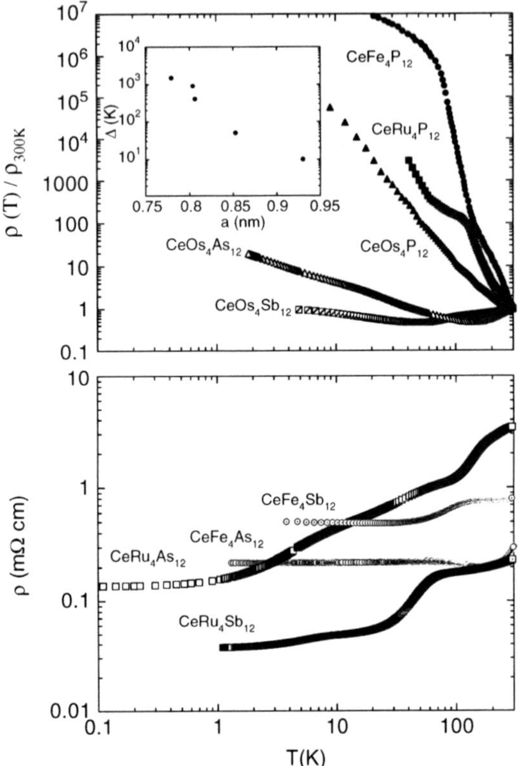

Fig. 4 Electrical resistivity $\rho(T)$ for the Ce-based filled skutterudites. The $\rho(T)$ data reveal that, for more than half of these compounds, the transport behavior is semiconducting with the size of the energy gap decreasing with increasing lattice constant. The remaining members of the series exhibit poor metallic behavior, with non-Fermi liquid behavior appearing at low T for $CeRu_4As_{12}$ and $CeRu_4Sb_{12}$ and a Kondo coherent state emerging at low T for $CeFe_4Sb_{12}$. Data are from Refs. [21–24, 35].

Examples include (1) proximity to a QCP where a second-order phase transition (usually magnetic) is suppressed to 0 K and quantum fluctuations govern physical properties, (2) Kondo disorder where a range of Kondo temperatures T_K are allowed, including $T_K = 0$ K, (3) the Griffiths phase model, and (4) the quadrupolar Kondo model. Since $CeRu_4Sb_{12}$ exhibits magnetic behavior that may indicate that the Ce ions remain in the 3+ valence state to low T, one of these models may be appropriate to describe its behavior. On the other hand, if the Ce ions do not remain in the 3+ valence state, as may be the case for $CeRu_4As_{12}$, then Kondo scenarios are unlikely to describe the physical behavior. As such, the possibility of a QCP where valence fluctuations on the Ce ions give rise to the incipient NFL state has been considered for $CeRu_4As_{12}$. It is also of interest to note that both $CeRu_4Sb_{12}$ and $CeRu_4As_{12}$ are near semiconducting states, as evinced by the behavior of their neighboring compounds, $CeOs_4Sb_{12}$ [21, 28] and $CeOs_4As_{12}$ [23], both of which are small gap semiconductors. Additionally, a recent study of polycrystalline $CeRu_4As_{12}$ revealed small gap semiconducting behavior, in contrast to results for single crystals [24]. Thus, proximity to a 0 K transition to a semiconducting state is also an interesting possible cause for the NFL behavior in these compounds. A related phase diagram has been developed for the series $CeRhSb_{1-x}Sn_x$, where a transition from a hybridization gap insulating ground state to a NFL ground state with increasing x has been discovered [36].

3 Pr-based Filled Skutterudites

The Pr-based filled skutterudites are mainly correlated electron metals which exhibit an unusually wide variety of ground states and phenomena, as summarized in Fig. 5, for which heavy quasiparticle masses are a common feature. The abundance of complicated novel ground states (often magnetic) are found in the vicinity of superconductivity, indicating that the interplay between a number of competing interactions generates the electronic ground states.

Several of the PrT_4X_{12} filled skutterudites show complicated magnetic behavior. For $PrFe_4As_{12}$, pronounced features in magnetic and transport data indicate the onset of long range ferromagnetic order below $\Theta_C = 18\,\mathrm{K}$. A gap-like reduction of $\gamma \sim 340$ mJ/mol K^2 and several other features point to strongly correlated electron behavior likely coupled to a change in magnetic and/or structural order near $T^* \sim 12$ K [37]. $PrOs_4As_{12}$ has a complex phase diagram with two (and possibly more) distinct ordered states for $T < 2.3$K and for magnetic fields $H < 3$ T. Neutron diffraction measurements in zero field show that the ground state is antiferromagnetic with $T_N \sim 2.3$ K, while the nature of the higher T and H ordered state remains to be determined. Specific heat measurements also indicate heavy fermion behavior with $\gamma \sim 1$ J/mol K^2. In the paramagnetic state, the Kondo effect is inferred from the field dependence of a broad peak in the specific heat versus T data and a local minimum in the electrical resistivity versus T data [38]. For $PrFe_4Sb_{12}$, long-range magnetic ordering occurs at $T_M < 4.1$ K and is related to both the Pr and Fe sites, indicating a complex, possibly ferrimagnetic, structure. The magnitude of the specific

Pr	Fe	Ru	Os
P	AFQ order $T_{AFQ} = 6.5$ K	Insulator M-I transition $T_{M\text{-}I} = 62$ K	No order > 70 mK
As	Complex FM order $T_M = 18$ K	BCS SC $T_c = 2.4$ K	AFM order $T_{AFM} = 2.3$ K Unidentified order
Sb	Complex magnetic order $T_M \sim 4.1$ K	BCS SC $T_c = 1.1$ K	Unconv. SC $T_c = 1.85$ K AFQ order

Fig. 5 A tabular summary of the behavior observed for the PrT_4X_{12} (T = Fe, Ru, or Os; X = P, As, or Sb) compounds. The Pr-based filled skutterudite compounds are primarily metals with correlated electron ground states. The *dark borders* indicate compounds that exhibit heavy fermion behavior.

heat at low temperatures indicates electronic mass enhancement, although it has not been possible to isolate the electronic contribution [39].

The phosphorus-based analogues, PrT_4P_{12} also show anomalous magnetic and transport behaviors that are worth noting [5]. The electrical resistivity of $PrFe_4P_{12}$ shows a weak increase with decreasing T for 300–10 K which is followed by a large peak near 5 K, below which $\rho(T)$ decreases strongly with decreasing T [40]. Below the phase transition at 5 K, each Pr ion has a low magnetic moment, and it has been argued that the low T state is antiferroquadrupolar in nature. The application of a magnetic field induces a transition to a heavy-fermion state at low T [41, 42]. In contrast, the compound $PrRu_4P_{12}$ shows a metal to insulator transition near 60 K which is not accompanied by a magnetic transition [43]. Surprisingly, the compound $PrOs_4P_{12}$ is a metal and shows no evidence for any ordered states down to $T \sim$ 100 mK. However, $PrOs_4P_{12}$ does exhibit an abrupt drop in $\rho(T)$ near 50 K and a small kink near 7 K, the origins of which are unknown [43, 44].

Of the Pr-based filled skutterudites, the compound $PrOs_4Sb_{12}$ has received the most attention for several reasons: (1) it exhibits superconductivity with $T_c = 1.85$ K that apparently develops out of a heavy Fermi liquid with a quasiparticle effective mass $m^* \approx 50\ m_e$, (2) it is the first example of a heavy fermion superconductor based on Pr (all of the other known heavy fermion superconductors are compounds of Ce or U); (3) it displays some type of unconventional strong coupling superconductivity that breaks time reversal symmetry, apparently consists of several distinct superconducting phases, some of which appear to have point nodes in the energy gap, and may involve triplet spin pairing of electrons; (4) there is a high field ordered phase between 4.5 and 16 T and below ~1 K that has been identified with antiferroquadrupolar order, indicating that the superconductivity occurs in the proximity of a quadrupolar QCP, similar to the situation with certain Ce and U compounds where superconductivity is found in the vicinity of an antiferromagnetic QCP; and (5) the pairing of superconducting electrons may be mediated by electric quadrupole fluctuations, rather than magnetic dipole fluctuations, which are believed to be responsible for pairing in the Ce and U based heavy fermion superconductors [45]. The extraordinary normal and superconducting state physical properties of $PrOs_4Sb_{12}$ are reviewed in refs. [7, 45]. In the following, we give a brief update of ongoing experiments that were undertaken to probe the unconventional superconductivity of $PrOs_4Sb_{12}$: measurements of the temperature dependence of the superconducting penetration depth of $PrOs_4Sb_{12}$ and the response of the superconducting properties of $PrOs_4Sb_{12}$ to the substitution of Ru for Os and Nd for Pr.

In order to understand better the ground state of $PrOs_4Sb_{12}$, transverse-field muon-spin rotation (TF-μSR) measurements were recently performed. In general, TF-μSR has proven to be an effective probe of the internal magnetic field distribution in the vortex state of conventional and unconventional type-II superconductors [46, 47]. TF-μSR measurements on the vortex-lattice of $PrOs_4Sb_{12}$ yield a temperature independent magnetic penetration depth at low temperatures (see Fig. 6a), consistent with a nonzero gap for quasiparticle excitations. The curve in the inset suggests the gap is isotropic [48, 49]. However, radio frequency (rf) inductive measurements of the surface penetration depth in the Meissner state [50] suggest point nodes in the energy gap. The results of μSR and rf measurements are

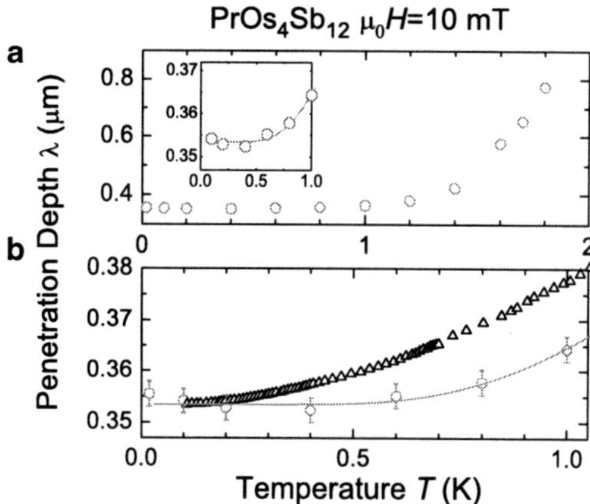

Fig. 6 (**a**) Temperature dependence of the vortex state penetration depth λ of $PrOs_4Sb_{12}$. Inset: $\lambda(T)$ at low temperature. The curve represents the function $\lambda(T) = \lambda(0)[1+\sqrt{\frac{\pi\Delta}{2T}}e^{-\Delta/T}]$, $\lambda(0) = 0.3534(4)(\mu m)$, $2\Delta/k_BT_c = 4.9(1)$. (**b**) $\lambda(T)$ at low temperatures for $PrOs_4Sb_{12}$. *Triangles*: rf measurements of the surface penetration depth in the Meissner state [50]. *Circles*: λ from μSR data.

compared in Fig. 6b. The discrepancy between the two measurements of the penetration depth at low temperatures has been explained in terms of a multiband superconductivity scenario [51,52].

Evidence for extreme multiband superconductivity in $PrOs_4Sb_{12}$ was extracted from heat transport measurements by Seyfarth et al. [53,54]. Their thermal conductivity and other data could be explained by small and large superconducting gaps on different sheets of the Fermi surface. A crossover field H_{c2}^{S}, which corresponds to the overlap of the vortex core electronic structure due to the band with the small gap, is about 10 mT [54]. This is of the order of the lower critical field H_{c1}. TF-μSR measurements were performed for an applied field $\mu_0H = 10$ mT $\approx H_{c2}^{S}$. At this field, the small-gap states and their contributions to screening supercurrents are nearly uniform; the vortex-state field inhomogeneity is mainly due to large-gap supercurrents, and λ exhibits an activated temperature dependence if the large gap is nodeless. In contrast, the rf surface measurement was performed in the Meissner state. Both large- and small-gap Cooper pairs contribute to the superfluid density. The temperature dependence of the superfluid density is controlled by both small- and large-gap contributions; the small-gap contribution dominates the temperature dependence at low temperatures. Thus, the temperature dependence of the penetration depth from μSR is weaker than that in the Meissner state from rf measurements.

In order to gain insight into the unconventional superconductivity, as well as to study the influence of the heavy fermion state, crystalline electric field, and the ferromagnetism on the unconventional superconductivity of $PrOs_4Sb_{12}$, substitutional studies in the pseudoternary systems $Pr(Os_{1-x}Ru_x)_4Sb_{12}$ and $Pr_{1-x}Nd_xOs_4Sb_{12}$ were undertaken [45, 55–61]. The non-magnetic analog compound $PrRu_4Sb_{12}$ has been identified as a conventional metal with conventional BCS-type superconductivity, and no high field ordered phase (HFOP) has been observed in this compound [62,63], while $NdOs_4Sb_{12}$ is a mean-field type ferromagnet with an enhanced electronic specific heat coefficient γ of $\sim 520\,\text{mJ/mol K}^2$ in the paramagnetic state due to some type of low-lying excitations, reminiscent of heavy fermion behavior [64].

3.1 $Pr(Os_{1-x}Ru_x)_4Sb_{12}$

Measurements of $\rho(T)$, $\chi_{dc}(T)$, $\chi_{ac}(T)$, $C(T)$, and zero-field muon-spin-relaxation (ZF-μSR) have been performed to characterize the physical properties of the $Pr(Os_{1-x}Ru_x)_4Sb_{12}$ system as a function of x [45, 55–61]. Figure 7a displays the x dependence of the zero field superconducting transition temperature T_c. T_c is suppressed nearly linearly from each end ($x = 0$, 1) and meets at a minimum at $x = 0.6$,

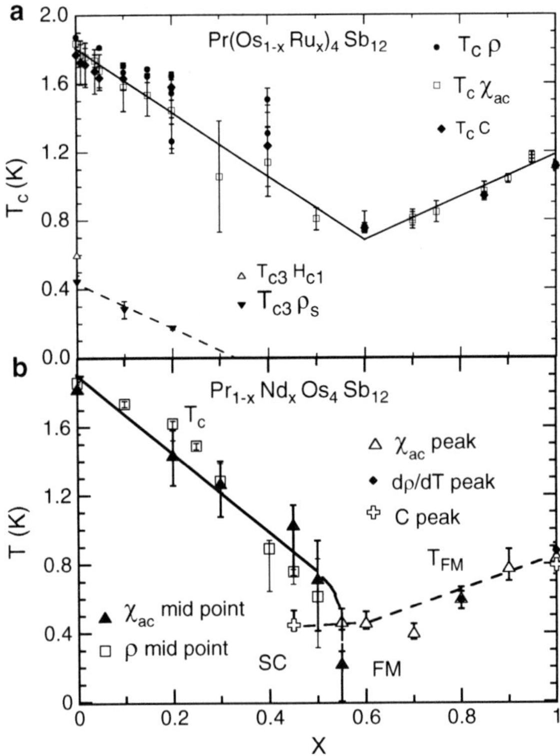

Fig. 7 (**a**) Superconducting transition temperature T_c versus Ru concentration x in the $Pr(Os_{1-x}Ru_x)_4Sb_{12}$ system. The two solid lines drawn from $x = 0$ and $x = 1$ toward $x = 0.6$ are guides to the eye. The possible low temperature phase boundary or crossover temperature identified as T_{c3}, which was determined from $H_{c1}(T)$ and magnetic penetration depth data [65, 66], is also plotted. (**b**) T_c and the ferromagnetic transition (Curie) temperature T_{FM} versus Nd concentration x in the $Pr_{1-x}Nd_xOs_4Sb_{12}$ system [58]. The vertical bars indicate the width of the superconducting and ferromagnetic transitions ($10-90\%$ of a ρ or χ_{ac} drop for superconductivity, and $10-90\%$ of a χ_{ac} peak for ferromagnetism). Data are from Refs. [45, 55–61].

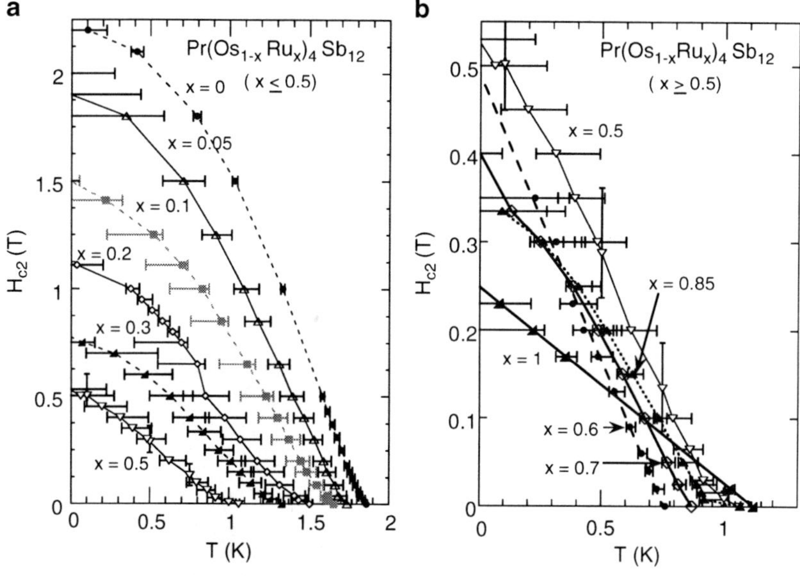

Fig. 8 (**a**) Upper critical field H_{c2} of $Pr(Os_{1-x}Ru_x)_4Sb_{12}$ versus temperature T for ruthenium concentrations x between 0 and 0.5. The *horizontal* or *vertical bar* associated with each data point represents the transition width, with endpoints corresponding to 10% and 90% of the drop in resistivity due to the superconductivity. (**b**) H_{c2} versus T for $0.5 \leq x \leq 1$. Data are from Refs. [45,56,57,60,61].

which suggests a competition between two types of superconductivity exhibited by the end-member compounds. Measurements of the lower critical field [65] $H_{c1}(T)$ and magnetic penetration depth [66] also reveal a possible low-temperature phase in the superconducting state below a characteristic transition or crossover temperature T_{c3}. Shown in Figs. 8a and b are upper critical field H_{c2} vs temperature T curves determined from $\rho(T,H)$ measurements at various values of x in the $Pr(Os_{1-x}Ru_x)_4Sb_{12}$ system [57]. The negative curvature of H_{c2} with respect to T gradually decreases as x increases, and when $x \geq 0.4$, H_{c2} becomes approximately linear in T. This is quite peculiar because $H_{c2}(T)$ of a BCS superconductor usually has a curved $H_{c2}(T)$ with negative curvature. In the $Pr(Os_{1-x}Ru_x)_4Sb_{12}$ system, the situation is reversed: an $H_{c2}(T)$ with negative curvature is found in the vicinity of the unconventional superconductivity and a linear $H_{c2}(T)$ is observed near the conventional superconductivity. It is unclear at this moment whether this linear behavior of $H_{c2}(T)$ results from crystalline electric field effects like those found in $(La_{1-x}Pr_x)_3In$ [67] and $La_{1-x}Tb_xAl_2$ [68], or from the two band superconductivity, since such linearity is also observed in $LaOs_4Sb_{12}$ [69], which is a BCS superconductor without the complications of the crystalline electric field effect (since there are no f-electrons in La).

From $\rho(T,H)$ isotherms and data at constant magnetic fields, the $H-x$ diagram for $Pr(Os_{1-x}Ru_x)_4Sb_{12}$ extrapolated to $T = 0\,\mathrm{K}$ is plotted in Fig. 9a: the superconducting phase exists below 2.2 T and T_c decreases monotonically as x increases;

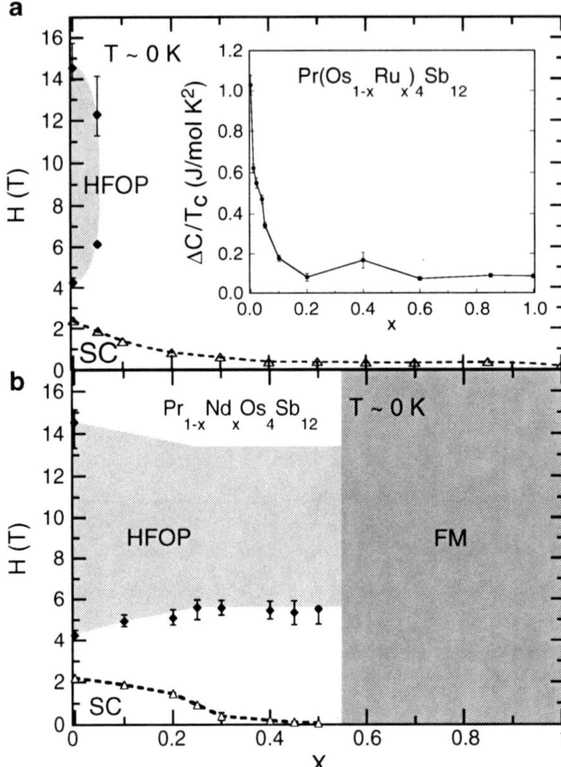

Fig. 9 0 K $H-x$ phase diagram for (**a**) $Pr(Os_{1-x}Ru_x)_4Sb_{12}$ and (**b**) $Pr_{1-x}Nd_xOs_4Sb_{12}$ [56–58]. The superconducting phase occurs below ∼2 T (*open triangles*). The high field ordered phase (HFOP) (*solid diamonds*) is located between 4.5 and 15 T and disappears quickly above a ruthenium concentration $x = 0.05$, but persists in the neodymium substitution study up to $x \sim 0.5$. The dashed lines and the gray areas are guides to the eye. Inset in (**a**): the ruthenium concentration x dependence of the specific heat jump at the superconducting transition divided by the transition temperature $\Delta C/T$, which is a measure of the strength of the superconducting coupling times γ. Data are from Refs. [56–58].

the HFOP lies between 4.5 and 15 T and disappears quickly above $x = 0.05$. The inset of Fig. 9a shows the specific heat jump ΔC at the superconducting transition divided by T_c as a function of x, which is a measure of the strength of the superconducting coupling times the electronic specific heat coefficient γ. The ratio $\Delta C/T$ is extremely large for $x = 0$ and then decreases very rapidly to a plateau at $x = 0.2$ (the slight jump at $x = 0.4$ is probably due to sample inhomogeneity at this concentration). An interesting correlation is also observed in μSR studies: the spontaneous magnetic moment in the superconducting state vanishes by $x \sim 0.2$ [59]. The connection between the HFOP, strong coupling superconductivity, and spontaneous moments in the very low x regime suggests that heavy fermion behavior and unconventional superconductivity in $PrOs_4Sb_{12}$ are related to the presence of strong quadrupolar interactions.

3.2 $Pr_{1-x}Nd_xOs_4Sb_{12}$

In order to determine the zero field $T-x$ phase diagram for the $Pr_{1-x}Nd_xOs_4Sb_{12}$ system, measurements of χ_{ac}, $\rho(T)$, and $C(T)$ were performed for various concentrations [58]. Figure 7b shows the T_c and ferromagnetic transition (Curie) temperature T_{FM} as a function of Nd concentration x in $Pr_{1-x}Nd_xOs_4Sb_{12}$. Although there is a slight break in the slope below and above the Nd concentration $x \sim 0.3$ in $T_c(x)$ and $x \sim 0.6$ in $T_{FM}(x)$, both T_c and T_{FM} decrease almost linearly toward $x = 0.55$. From the $C(T)$ data at $x = 0.45$ [58], the ferromagnetism seems to extend into the superconducting region. Note that the rate of decrease of T_c with x in $Pr_{1-x}Nd_xOs_4Sb_{12}$ from $x = 0$ is approximately the same as that for $Pr(Os_{1-x}Ru_x)_4Sb_{12}$. Therefore, substitution of Nd, an ion with a localized magnetic moment, at the Pr sites has a similar effect as the substitution of Ru, a nonmagnetic ion, at the Os sites. In addition, superconductivity disappears quite abruptly at $x \sim 0.55$ suggesting that this concentration may be near a quantum phase transition.

The temperature dependence of the upper critical field $H_{c2}(T)$ in $Pr_{1-x}Nd_xOs_4Sb_{12}$ has also been measured for various Nd concentration up to $x = 0.5$ and the curves are displayed in Fig. 10. Interestingly, in comparison to the $H_{c2}(T)$ data in the Ru substitution system (Fig. 8a), the stronger suppression of the $H_{c2}(T)$ curve as a result of Nd substitution only occurs above $x = 0.2$ in the $Pr_{1-x}Nd_xOs_4Sb_{12}$ system, in contrast to BCS superconductors, in which $H_{c2}(T)$ is immediately suppressed by the introduction of magnetic impurities.

Figure 9b summarizes the $H-x$ phase diagram including the superconducting phase, the ferromagnetic regime, and the lower phase boundary of the HFOP for $Pr_{1-x}Nd_xOs_4Sb_{12}$ at $T \sim 0$ K, extracted from $\rho(T,H)$ measurements [56, 58]. The

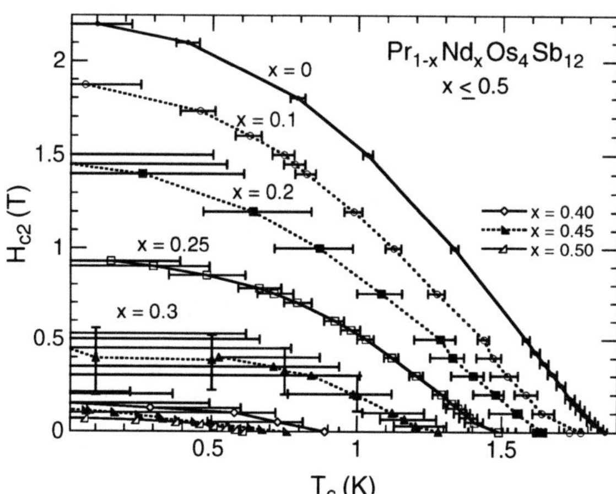

Fig. 10 Upper critical field H_{c2} of $Pr_{1-x}Nd_xOs_4Sb_{12}$ versus temperature T for neodymium concentrations x between 0 and 0.5. The horizontal or vertical bar associated with each data point represents the transition width, with endpoints corresponding to 10% and 90% of the drop in resistivity due to the superconductivity. Data are from Ref. [58].

HFOP in $Pr_{1-x}Nd_xOs_4Sb_{12}$ persists up to $x \sim 0.5$, in comparison with the rapid disappearance of the HFOP in $Pr(Os_{1-x}Ru_x)_4Sb_{12}$ above $x \sim 0.1$. This striking difference in the response of the HFOP to Ru and Nd substitutions may reflect the increase in the crystalline electric field energy level splitting between the Pr^{3+} Γ_1 ground state and $\Gamma_4^{(2)}$ first excited state in $Pr(Os_{1-x}Ru_x)_4Sb_{12}$, in contrast to Nd substitution, which dilutes the Pr concentration but preserves the configuration of ions around the Pr site and thereby has less effect on the crystalline electric field splitting between the Γ_1 and $\Gamma_4^{(2)}$ states. Further specific heat measurements on $Pr_{1-x}Nd_xOs_4Sb_{12}$ will help to elucidate this correspondence and reveal more information about the competition between superconductivity and ferromagnetism in this system.

3.3 ZF-μSR for $Pr(Os_{1-x}Ru_x)_4Sb_{12}$ and $Pr_{1-y}La_yOs_4Sb_{12}$

ZF-μSR measurements have also been carried out on various doping series of $PrOs_4Sb_{12}$, including $Pr(Os_{1-x}Ru_x)_4Sb_{12}$ and $Pr_{1-y}La_yOs_4Sb_{12}$. An important question addressed by these studies is whether time-reversal symmetry (TRS) is broken. ZF-μSR is a very sensitive method for detecting weak magnetism, which arises due to ordered magnetic moments, or random fields that are static or fluctuating in time [46]. Figure 11 shows the temperature dependence of the muon relaxation rate $\Delta(T)$ in $Pr(Os_{1-x}Ru_x)_4Sb_{12}$, $x = 0.1$, 0.2, and 1.0, and $Pr_{1-y}La_yOs_4Sb_{12}$, $y = 0$, 0.4, and 1.0. For $PrOs_4Sb_{12}$, Δ shows a significant increase with an onset temperature of T_c, indicating the appearance of a spontaneous internal field. This provides evidence that TRS is broken in the superconducting state of $PrOs_4Sb_{12}$. For Ru doping, $\Delta(T)$ is temperature-independent to within experimental uncertainty for

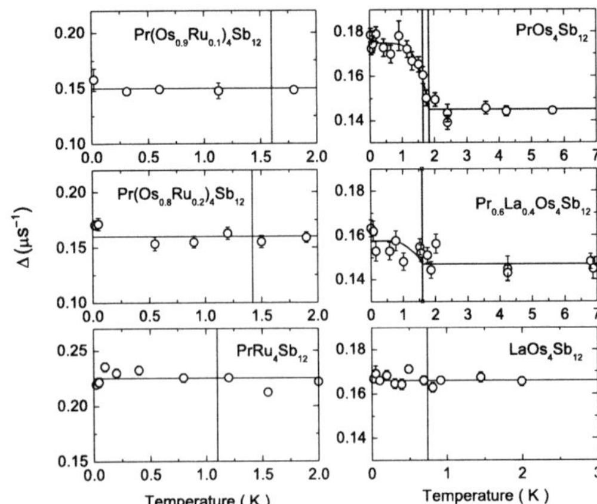

Fig. 11 Temperature dependence of ZF Kubo-Toyabe relaxation rate Δ in $Pr(Os_{1-x}Ru_x)_4Sb_{12}$, $x = 0.1$, 0.2, and 1.0, and $Pr_{1-y}La_yOs_4Sb_{12}$, $y = 0$. (Data are from Ref. [71]), 0.4, and 1.0. The *vertical lines* indicate the position of T_c.

$x \geq 0.1$. No extra spontaneous field is observed to set in below T_c. In contrast, for La doping the increased relaxation below T_c seen for $y = 0$ is also observed with reduced amplitude for $y = 0.4$, indicating the onset of a spontaneous field is correlated with the superconductivity. No spontaneous field is observed in the end compounds $PrRu_4Sb_{12}$ and $LaOs_4Sb_{12}$. Therefore, ZF-μSR experiments show that Ru doping is considerably more efficient than La doping in suppressing the TRS breaking found in $PrOs_4Sb_{12}$ [70]. Ru doping is more effective than La doping in (a) distorting the Sb cage that surrounds each Pr ion, which modifies the Pr^{3+} crystalline electric field splitting and restricts the "rattling" motion of the Pr ions, and (b) in changing the character and density of states of band electrons on the Fermi surface. One or more of these effects may be involved in quenching the TRS-breaking.

4 Summary

A major theme in modern condensed matter physics has been the study of correlated electron phenomena in solids, for which the filled skutterudites have provided a rich reservoir of new materials. As illustrated above, the delicate interplay between several types of competing interactions and the relatively straightforward synthesis requirements in the filled skutterudites have provided an opportunity to tune various interactions using "knobs" such as pressure and chemical substitution, yielding insight into the underlying mechanisms responsible for the correlated electron behavior. The richness of this family of materials is exemplified by the compound $PrOs_4Sb_{12}$, which exhibits heavy fermion behavior and some type of unconventional strong coupling superconductivity that may be mediated by a new type of pairing mechanism: quadrupolar fluctuations. The variety of unusual and remarkable types of behavior that have been observed in the known filled skutterudites, combined with the enormous number of uninvestigated skutterudite-like compounds, provides strong motivation for further investigations of this intriguing class of materials.

Acknowledgements At UCSD, crystal growth work was supported by the U. S. Department of Energy (DOE) under research Grant DE FG02-04ER46105 and low temperature measurements were funded by the National Science Foundation (NSF) under Grant 0802478. Work at the Institute of Low Temperature and Structure Research, Polish Academy of Sciences was supported by the Polish Ministry of Science and Higher Education grant NN2024129.

References

1. I. Oftedal. *Z. Kristallogr.*, A66:517, 1928.
2. W. Jeitschko and D. J. Braun. *Acta Cryst.*, B33:3401, 1977.
3. D. J. Braun and W. Jeitschko. *J. Less Com. Met.*, 72:147, 1980.
4. D. J. Braun and W. Jeitschko. *J. Solid State Chem.*, 32:357, 1980.
5. B. C. Sales. Handbook on the Physics and Chemistry of the Rare Earths, Vol 33. Elsevier Science Amsterdam, 2002, p.211.

6. M. B. Maple, E. D. Bauer, N. A. Frederick, P.-C. Ho, W. M. Yuhasz, and V. S. Zapf. *Physica B*, 328:29, 2003.
7. Y. Aoki, H. Sugawara, H. Hisatomo, and H. Sato. *J. Phys. Soc. Jpn.*, 74:209, 2005.
8. M. B. Maple, Z. Henkie, R. E. Baumbach, T. A. Sayles, N. P. Butch, P.-C. Ho, T. Yanagisawa, W. M. Yuhasz, R. Wawryk, T. Cichorek, and A. Pietraszko. *J. Phys. Soc. Jpn.*, 77:7, 2008.
9. H. Sato, D. Kikuchi, K. Tanaka, M. Ueda, H. Aoki, T. Ikeno, S. Tatsuoka, K. Kuwahara, Y. Aoki, M. Koghi, H. Sugawara, K. Iwasa, and H Harima. *J. Phys. Soc. Jpn.*, 77:1, 2008.
10. B. C. Sales, D. Mandrus, and R. K. Williams. *Science*, 272:1325, 1996.
11. B. C. Sales. *Science*, 295:1248, 2002.
12. G. S. Nolas, G. Yoon, H. Sellingschegg, A. Smalley, and D. C. Johnson. *Appl. Phys. Lett.*, 86:042111, 2005.
13. W. Schnelle, A. Leithe-Jasper, H. Rosner, R. Cardoso-Gil, R. Gumeniuk, D. Trots, J. A. Mydosh, and Y. Grin. *Phys. Rev. B*, 77:094421, 2008.
14. K. Kihou, I Shirotani, Y. Shimaya, C. Sekine, and T. Yagi. *Mat. Res. Bul.*, 39:317, 2004.
15. I. Shirotani, N. Araseki, Y. Shimaya, R. Nakata, K. Kihou, C Sekine, and T. Yagi. *J. Phys.:Condens. Matter*, 17:4383, 2005.
16. C. Sekine, H. Ando, Y. Sugiuchi, I. Shirotani, K. Matsuhira, and M. Wakeshima. *J. Phys. Soc. Jpn.*, 77 Suppl. A:135, 2008.
17. I. Shirotani, S. Sato, C. Sekine, T. Keiki, I. Inagawa, and T. Yagi. *J. Phys.: Condens. Matter*, 17:7353, 2005.
18. R. Gumeniuk, W. Schnelle, H. Rosner, M. Nicklas, A. Leithe-Jasper, and Y. Grin. *Phys. Rev. Lett.*, 100:017002, 2008.
19. A. Grytsiv, X.-Q. Chen, N. Melnychenko-Koblyuk, P. Rogl, E. Bauer, G. Hilscher, H. Kaldarar, H. Michor, E. Royanian, R. Podloucky, M. Rotter, and G. Geister. *J. Phys. Soc. Jpn.*, 77 Suppl. A:121, 2008.
20. A. N. Vasil'ev and O. S. Volkova. *Low Temp. Phys.*, 33:895, 2007.
21. H. Sugawara, S. Osaki, T. Namiki, S. R. Saha, Y. Aoki, and H. Sato. *Phys. Rev. B*, 71:125127, 2005.
22. E. D. Bauer, A. Slebarski, R. P. Dickey, E. J. Freeman, C. Sirvent, V. S. Zapf, N. R. Dilley, and M. B. Maple. *J. Phys.: Condens. Matter*, 13:5183, 2001.
23. R. E. Baumbach, P.-C. Ho, T. A. Sayles, M. B. Maple, R. Wawryk, T. Cichorek, A. Pietraszko, and Z. Henkie. *Proc. Natl. Acad. Sci.*, 105:17307, 2008.
24. R. E. Baumbach, P.-C. Ho, T. A. Sayles, M. B. Maple, R. Wawryk, T. Cichorek, A. Pietraszko, and Z. Henkie. *J. Phys.: Condens. Matter*, 20:075110, 2008.
25. I. Shirotani, T. Uchiumi, C. Sekine, M. Hori, and S. Kimura. *J. Solid State Chem.*, 142:146, 1999.
26. G. P. Meisner, M. S. Torikachvili, K. N. Yang, M. B. Maple, and R. P. Guertin. *J. Appl. Phys.*, 57:3073, 1985.
27. D. Berardan, E. Alleno, and C. Godart. *Physica B*, 359:865, 2005.
28. E. D. Bauer, A. Slebarski, E. J. Freeman, N. A. Frederick, B. J. Taylor, C. Sirvent, and M. B. Maple. *Physica B*, 312:230, 2002.
29. M. B. Maple, C. L. Seaman, D. A. Gajewski, Y. Dalichaouch, V. B. Barbetta, M. C. de Andrade, H. A. Mook, H. G. Lukefahr, O. O. Bernal, and D. E. MacLaughlin. *J. Low Temp. Phys.*, 95:225, 1994.
30. M. B. Maple, M. C. de Andrade, J. Hermann, Y. Dalichaouch, D. A. Gajewski, C. L. Seaman, R. Chau, R. Movshovich, M. C. Aronson, and R. Osborn. *J. Low Temp. Phys.*, 99:223, 1995.
31. G. R. Stewart. *Rev. Mod. Phys.*, 73:797, 2001.
32. G. R. Stewart. *Rev. Mod. Phys.*, 78:743, 2006.
33. M. B. Maple, L. E. DeLong, and B. C. Sales. page 797. Handbook on the Physics and Chemistry of the Rare Earths, Vol 1. North-Holland, Amsterdam, 1978.
34. A Jayaraman. page 575. Handbook on the Physics and Chemistry of the Rare Earths, Vol 2. North-Holland, Amsterdam, 1979.
35. E. D. Bauer, R. Chau, N. R. Dilley, M. B. Maple, D. Mandrus, and B. C. Sales. *J. Phys.: Condens. Matter*, 12:1261, 2000.

36. A. Slebarski and J. Spalek. *Phys. Rev. Lett.*, 95:046402, 2005.
37. T. A. Sayles, W. M. Yuhasz, J. Paglione, T. Yanagisawa, J. R. Jeffries, M. B. Maple, Z. Henkie, A. Pietraszko, T. Cichorek, R. Wawryk, Y. Nemoto, and T. Goto. *Phys. Rev. B*, 77:144432, 2008.
38. M. B. Maple, N. P. Butch, N. A. Frederick, P.-C. Ho, J. R. Jeffries, T. A. Sayles, T. Yanagisawa, W. M. Yuhasz, S. Chi, H. J. Kang, J. W. Lynn, P. Dai, S. K. McCall, M. W. McElfresh, M. J. Fluss, Z. Henkie, and A. Pietraszko. *Proc. Natl. Acad. Sci.*, 103:6783, 2006.
39. N. P. Butch, W. M. Yuhasz, P.-C. Ho, J. R. Jeffries, N. A. Frederick, T. A. Sayles, M. B. Maple, J. B. Betts, A. H. Lacerda, F. M. Woodward, J. W. Lynn, P. Rogl, and G. Giester. *Phys. Rev. B*, 71:214417, 2005.
40. M.S. Torikachvili, J.W. Chen, Y. Dalichaouch, R.P. Guertin, M.W. McElfresh, C. Rossel, M.B. Maple, and G.P. Meisne. *Phys. Rev. B*, 36:8660, 1987.
41. Y. Aoki, T. Namiki, T. D. Matsuda, K. Abe, H. Sugawara, and H. Sato. *Phys. Rev. B*, 65:064446, 2002.
42. H. Sato, H. Sugawara, T. Namiki, S. R. Saha, S. Osaki, T. D. Matsuda, Y. Aoki, Y. Inada, H. Shishido, R. Settai, and Y. Onuki. *J. Phys.: Condens. Matter*, 15:S2063, 2003.
43. C. Sekine, T. Uchiumi, and I. Shirotani. *Phys. Rev. Lett.*, 79:3218, 1997.
44. W. M. Yuhasz, P.-C. Ho, T. A. Sayles, T. Yanagisawa, N. A. Frederick, M. B. Maple, P. Rogl, and G. Giester. *J. Phys.: Condens. Matter*, 19:076212, 2007.
45. M. B. Maple, N. A. Frederick, P.-C. Ho, W. M. Yuhasz, and T. Yanagisawa. *J. Superconductivity Novel Magnetism*, 19:299, 2006.
46. A. Amato. *Rev. Mod. Phys.*, 69:1119, 1997.
47. J. E. Sonier, J. H. Brewer, and R. F. Kiefl. *Rev. Mod. Phys.*, 72:769, 2000.
48. D. E. MacLaughlin, J. E. Sonier, R. H. Heffner, O. O. Bernal, B.-L. Young, M. S. Rose, G. D. Morris, E. D. Bauer, T. D. Do, and M. B. Maple. *Phys. Rev. Lett.*, 89:157001, 2002.
49. L. Shu, D. E. MacLaughlin, R. H. Heffner, F. D. Callaghan, J. E. Sonier, G. D. Morris, O. O. Bernal, A. Bosse, J. E. Anderson, W. M. Yuhasz, N. A. Frederick, and M. B. Maple. *Physica B*, 374–375:247, 2006.
50. E. E. M. Chia, M. B. Salamon, H. Sugawara, and H. Sato. *Phys. Rev. Lett.*, 91:247003, 2003.
51. D. E. MacLaughlin, L. Shu, R. H. Heffner, J. E. Sonier, F. D. Callaghan, G. D. Morris, O. O. Bernal, W. M. Yuhasz, N. A. Frederick, and M. B. Maple. *Physica B*, 403:1132, 2008.
52. L. Shu, D. E. MacLaughlin, W. P. Beyermann, R. H. Heffner, G. D. Morris, O. O. Bernal, F. D. Callaghan, J. E. Sonier, W. M. Yuhasz, N. A. Frederick, and M. B. Maple. *Phys. Rev. B*, 79:174511, 2009.
53. G. Seyfarth, J. P. Brison, M.-A. Méasson, J. Flouquet, K. Izawa, Y. Matsuda, H. Sugawara, and H. Sato. *Phys. Rev. Lett.*, 95:107004, 2005.
54. G. Seyfarth, J. P. Brison, M.-A. Méasson, D. Braithwaite, G. Lapertot, and J. Flouquet. *Phys. Rev. Lett.*, 97:236403, 2006.
55. N. A. Frederick, T. D. Do, P.-C. Ho, N. P. Butch, V. S. Zapf, and M. B. Maple. *Phys. Rev. B*, 69:024523, 2004.
56. P.-C. Ho, T. Yanagisawa, N. P. Butch, W. M. Yuhasz, C. C. Robinson, A. A. Dooraghi, and M. B. Maple. *Physica B*, 403:1038, 2008.
57. P.-C. Ho, N. P. Butch, V. S. Zaph, T. Yanagisawa, N. A. Frederick, S. K. Kim, W. M. Yuhasz, M. B. Maple, J. B. Betts, and A. H. Lacerda. *J. Phys.: Condens. Matter*, 20:215226, 2008.
58. P.-C. Ho, M. B. Maple, T. Yanagisawa, W. M. Yuhasz, N. P. Butch, A. A. Dooraghi, and C. C. Robinson. *manuscript in preparation.*
59. L. Shu, D. E. MacLaughlin, Y. Aoki, Y. Tunashima, Y. Yonezawa, S. Sanada, D. Kikuchi, H. Sato, R. H. Heffner, W. Higemoto, K. Ohishi, T. U. Ito, O. O. Bernal, A. D. Hillier, R. Kadono, A. Koda, K. Ishida, H. Sugawara, N. A. Frederick, W. M. Yuhasz, T. A. Sayles, T. Yanagisawa, and M. B. Maple. *Phys. Rev. B*, 76:014527, 2007.
60. M. B. Maple, Z. Henkie, W. M. Yuhasz, P.-C. Ho, T. Yanagisawa, T. A. Sayles, N. P. Butch, J. R. Jeffries, and A. Pietraszko. *J. Magn. Magn. Mater.*, 310:182–187, 2007.
61. M. B. Maple, Z. Henkie, R. E. Baumcach, T. A. Sayles, N. P. Butch, P.-C. Ho, T. Yanagisawa, W. M. Yuhasz, R. Wawryk, T. Cichorek, and A. Pietraszko. *J. Phys. Soc. Jpn.*, 77 Suppl. A:7, 2008.

62. N. Takeda and M. Ishikawa. *J. Phys. Soc. Jpn.*, 69:868, 2000.
63. M. Yogi, H. Kotegawa, Y. Imamura, G. q. Zheng, Y. Kitaoka, H. Sugawara, and H. Sato. *Phys. Rev. B*, 67:180501, 2003.
64. P.-C. Ho, W. M. Yuhasz, N. P. Butch, N. A. Frederick, T. A. Sayles, J. R. Jeffries, M. B. Maple, J. B. Betts, A. H. Lacerda, P. Rogl, and G. Giester. *Phys. Rev. B*, 72:094410, 2005.
65. T. Cichorek, A. C. Mota, F. Steglich, N. A. Frederick, W. M. Yuhasz, and M. B. Maple. *Phys. Rev. Lett.*, 94:107002, 2005.
66. E. E. M. Chia, D. Vandervelde, M. B. Salamon, D. Kikuchi, H. Sugawara, and H. Sato. *J. Phys.: Condens. Matter*, 17:L303, 2005.
67. F. Heiniger, E. Bucher, J. P. Maita, and L. D. Longinotti. *Phys. Rev. B*, 12:1778, 1975.
68. G. Pepperl, E. Umlauf, A. Meyer, and J. Keller. *Solid State Commun.*, 14:161, 1974.
69. M. B. Maple, P.-C. Ho, N. A. Frederick, V. S. Zapf, W. M. Yuhasz, E. D. Bauer, A. D. Christianson, and A. H. Lacerda. *J. Phys.: Condens. Matter*, 15:S2071, 2003.
70. L. Shu, W. Higemoto, Y. Aoki, D. E. MacLaughlin, R. H. Heffner, K. Ohishi, K. Ishida, R. Kadono, A. Koda, D. Kikuchi, N. A. Frederick, M. B. Maple, H. Sato, H. Sugawara, S. Sanada, Y. Tunashima, and Y. Yonezawa. *J. Magn. Mag. Mat.*, 310:551, 2007.
71. Y. Aoki, A. Tsuchiya, T. Kanayama, S. R. Saha, H. Sugawara, H. Sato, W. Higemoto, A. Koda, K. Ohishi, K. Nishiyama, and R. Kadono. *Phys. Rev. Lett.*, 91:067003, 2003.

Role of Structures on Thermal Conductivity in Thermoelectric Materials

C. Godart, A. P. Gonçalves, E. B. Lopes, and B. Villeroy

Abstract The figure of merit $ZT = \sigma\alpha^2 T/\lambda$ (α the Seebeck coefficient, σ and λ the electrical and thermal conductivity, respectively) is an essential element of the efficiency of a thermoelectric material for applications, which convert heat to electricity or, conversely, electric current to cooling. From the expression of the power factor, $\sigma\alpha^2$, it was deduced that a highly degenerated semiconductor is necessary. In order to reduce the lattice part of the thermal conductivity, various mechanisms, mainly related to the structure of the materials, were tested in new thermoelectric materials and had been the topics of different reviews. These include cage-like materials, effects of vacancies, solid solutions, complex structures (cluster, tunnel,...), nano-structured systems. We plan to review structural aspects in the modern thermoelectric materials and to include results of the very recent years. Moreover, as micro- and nano-composites seem to be promising to increase ZT in large size samples, we will also briefly discuss the interest of spark plasma sintering technique to preserve the micro- or nano-structure in highly densified samples.

Symbols

α	Seebeck coefficient
λ	Thermal conductivity
λ_e	Electronic part of the thermal conductivity
λ_L	Phonon (lattice) part of the thermal conductivity
μ	Mobility

C. Godart and B. Villeroy
CNRS, ICMPE, CMTR, 2/8 rue Henri Dunant, 94320 Thiais, France
e-mail: godart@icmpe.cnrs.fr, villeroy@icmpe.cnrs.fr

A. P. Gonçalves and E. B. Lopes
Dep. Química, Instituto Tecnológico e Nuclear/CFMC-UL, P-2686-953 Sacavém, Portugal
e-mail: apg@itn.pt, eblopes@itn.pt

V. Zlatić and A. C. Hewson (eds.), *Properties and Applications of Thermoelectric Materials*,
NATO Science for Peace and Security Series B: Physics and Biophysics,

ρ	Electrical resistivity
σ	Electrical conductivity
τ	Thomson coefficient
Π	Peltier coefficient
ADP	Atomic Displacement Parameter
I	Intensity
PGEC	Phonon Glass Electron Crystal
q	Charge of a carrier
Q	Heat quantity
S	Entropy
TE	Thermoelectric
V	Potential
VEC	Valence Electron Concentration
ZT	Figure of merit

1 Introduction

Thermoelectric (TE) effects include the transformation of caloric energy to electric energy or its reverse, and their applications consequently include the two aspects: (micro)cooling or electricity generation from heat sources. Search for new friendly environmental energy sources recently became of major importance in our modern societies, specially after the signing of the Kyoto protocol. Consequently, the electricity generation from waste heat via thermoelectric modules (see Seebeck effect) can be seen as a new "green" energy source. Moreover, thermoelectric materials can be used to extract heat (see Peltier effect) from micro-electronic components. In the first case, the low efficiency of the thermoelectric systems has prevented their use for many years, but the situation is changing with increase of the needs (and cost) of energy and improved materials. In the latter case, classical methods (air/water) are no more appropriate due to miniaturization and high local power to dissipate. In both cases, new concepts lead, from ~1995, to rather remarkable progress, as well as the appearance of the topics in many conferences and national research programs in various countries.

The understanding of physical phenomena involved in thermoelectricity and the development of conventional TE materials occur during two main active periods.

From 1821 to 1851, the three TE effects (Seebeck, Peltier and Thomson) were discovered and understood from a macroscopic point of view. Their potential for applications to temperature measurements, cooling and electricity production have also been established. Later on, from 1930 to the beginning of the 1960s of the previous century, there has been important progress both in the understanding of phenomena at a microscopic scale and in the discovery and optimization of presently used TE materials. However, the efficiency was not sufficient to compete with the cooling by compression/relaxation cycles or for economically profitable electricity production.

More recently, from the beginning of 1990s, a renewal of interest in TE appears due to environmental concern with refrigerant gas and greenhouse effects, as well as the need to develop alternative energy sources.

The two main research directions happen to be – the development of new materials with complex or open structures, and – the development of already known materials with new low dimensional forms (quantum wells, nano-wires, nanograins, thin films, nano-composites, etc.). Among new materials having interesting TE characteristics for electricity generation from heat sources in the range 200–800°C we will emphasize the role of the structure and the potential of micro- and nanostructures to exceed the 14% efficiency reported by Caillat [16].

2 The Thermoelectric Effects

The first thermoelectric effect has been discovered by Seebeck [79] and understood later by Oersted: a potential appears at the junction of two materials a, b submitted to a gradient of temperature. The well known application of the Seebeck effect is the measurement of temperature by thermocouple. The Seebeck coefficient, $\alpha_{\mathbf{ab}} = \mathbf{\Delta V/\Delta T}$, is the entropy S per charge carrier q divided by the charge: $\alpha = S/q$. Later on, Peltier [69] has discovered the second effect: a temperature difference appears when a current I pass though the junctions of two different materials. There is absorption of heat Q at one junction and liberation of Q at the other junction, the Peltier coefficient is defined by $\mathbf{\Pi_{ab} = Q/I}$. The third effect [90] occurs when both a gradient of temperature and an electric current exists simultaneously. When a current passes through a material submitted to a gradient of temperature, then the material exchanges heat with the outside medium. Conversely, a current is produced when a heat flux passes through a material submitted to a temperature gradient. The thermal flux, $\mathbf{dQ/dx}$, is given by $\tau\mathbf{.I.dT/dx}$ with τ the Thomson coefficient. The Peltier and Seebeck coefficients are related by: $\mathbf{\Pi} = \alpha\mathbf{T} = \mathbf{Q/I}$.

3 Efficiencies, Figure of Merit

For both cooling or generation, a TE module is made of couples n,p electrically connected. Each couple includes a p-type material ($\alpha > 0$) and a n-type material ($\alpha < 0$) in which charge carriers are holes and electrons, respectively. These two materials are joined by a conducting material assumed to be with $\alpha = 0$.

The two branches of the couple and all other couples of the module are connected electrically in series and thermally in parallel. The scheme of principle of a couple for generation is shown in Fig. 1. In the case of cooling, the load is replaced by a dc-current source, which causes the charge carriers to move from the cold zone to the hot zone in both branches. This implies a thermal flux opposite to the normal heat conduction.

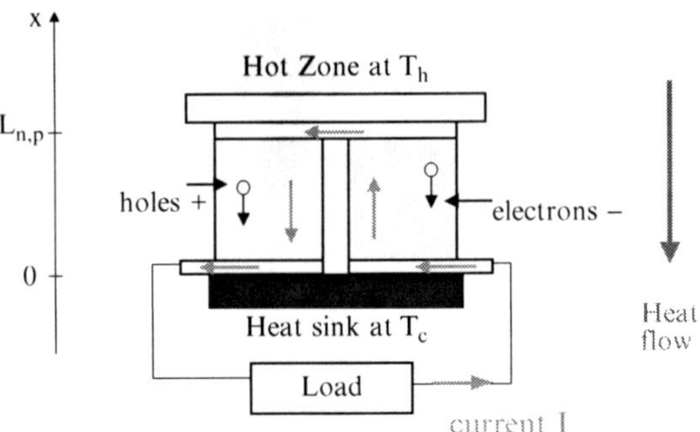

Fig. 1 Scheme of a TE couple for generation.

The optimization of the materials is needed to enhance their transport properties and maximize the figure of merit:

$$ZT = \sigma\alpha^2 T/\lambda \tag{1}$$

with $\sigma = 1/\rho$ and λ the electrical and thermal conductivities, respectively.

For cooling applications, the efficiency of the system is given by the coefficient of performance COP which, in an optimized system, is:

$$COP = \frac{\gamma T_c - T_h}{(T_c - T_h)(1+\gamma)} \tag{2}$$

with $\gamma = \sqrt{1+ZT}$, T_h and T_c the hot and cold temperatures respectively. For electricity generation, the maximum conversion efficiency is given by:

$$\eta_{\max} = \frac{(T_h - T_c)(\gamma - 1)}{T_c + \gamma T_h}. \tag{3}$$

From this expression, it is possible to estimate the maximum efficiency which can be expected for the couple as a function of the hot temperature (the cold one is taken as 300 K) with different values of ZT (Fig. 2).

The double arrow shows that for a ZT = 1 and a temperature gradient of 100°, the efficiency is no more than 5%. To increase it, it is necessary to increase ZT or to increase the temperature gradient: in both cases, this leads to a problem of materials: (1) to increase ZT; (2) to find materials stable at higher temperature.

In fact, the previous simple calculations assume that ZT is constant versus temperature, which is not correct specially if the temperature gradient is large (see below the curves of ZT for n- and p-type materials).

The expression of the figure of merit clearly sums up the difficulty in optimizing the materials. Intuitively, it seems difficult for a material to possess simultaneously a good electrical conductivity, characteristic of metals, and a bad thermal conductivity, characteristic of insulator. Closer examination of the power factor $\sigma\alpha^2$

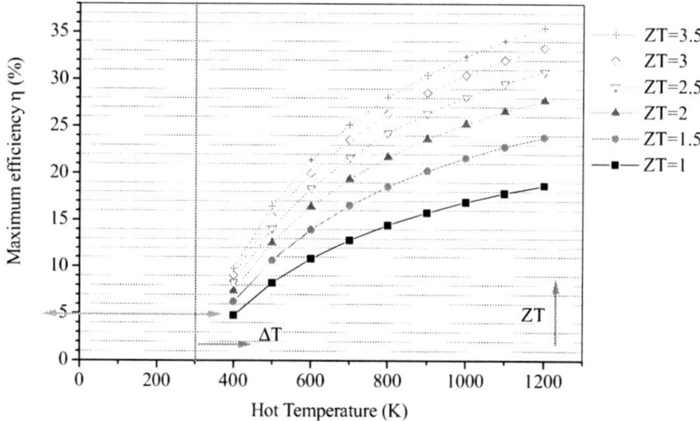

Fig. 2 Maximum efficiency for generation.

(Gonçalves et al., in this volume, pages 51–67) leads to consider semiconductors with a small gap and a carrier concentration in the range $\left[10^{18} - 10^{21}\,\text{cm}^{-3}\right]$. The second parameter is the thermal conductivity which, in such semiconductors, must be reduced.

4 Thermal Conductivity, Phonons

Intuitively, a good thermal conductivity would prevent a large temperature gradient. However, the thermal conductivity includes mainly two components: an electronic contribution λ_e, due to the movement of carriers, and a lattice contribution via the phonons λ_L: $\lambda = \lambda_e + \lambda_L$. The electronic part of the thermal conductivity is related to the electronic conductivity via the Wiedemann–Franz law:

$$\lambda_e = L_o T\sigma, \tag{4}$$

with L_o the Lorentz factor. In metals, it is equal to the Lorentz number: $L_0 = \frac{\pi^2}{3}\left(\frac{k}{e}\right)^2 = 2,45 \cdot 10^{-8}\ V^2.K^{-2}$, value generally used as a first approximation for the TE semiconductors. Replacing λ by its two components and with the Wiedemann–Franz, leads to:

$$ZT = \frac{\alpha^2 T\sigma}{LT\sigma + \lambda_L} \quad \text{or} \quad ZT = \frac{\alpha^2}{L}\frac{\lambda_e}{\lambda_e + \lambda_L}. \tag{5}$$

This last expression shows that the optimization of ZT implies the minimization of the phonons contribution to λ. Assuming a negligible lattice contribution, from $ZT = \alpha^2/L$, a ZT of 1 or 3[1] would be obtained from a Seebeck coefficient

[1] ZT = 3 correspond to a COP of refrigeration of 1.1, typical of a compression fridge.

of 156 μV/K or 270μV/K, respectively. However, the minimization should not modify the electrical conductivity (as ZT increases with λ_e/λ_L). This leads to the proposition of Slack [84] to check materials which conduct electricity as a crystal and heat as a glass, so-called "Phonon Glass Electron Crystal" (PGEC). This implies finding a selective diffusion process which affects the phonons more than the carriers.

Various physical process for inducing significant phonon diffusion have been tested on TE materials, let us quote those that are in some way related to the structure:

- A complex crystalline structure will increase the number of optical phonon modes, whereas the heat is mainly conducted by acoustic phonons.
- The insertion of heavy atoms in empty cages of crystalline structure may let them rattle independently of the lattice and so, create new phonon modes (weakly dispersive optical modes).
- Solid solutions between different materials with the same structure increase the disorder and create an important phonon diffusion by mass fluctuations on the sites [1]; such mass fluctuations can also be obtained by vacancy creation in the material.
- Impurities and point defects will diffuse phonons [17, 48, 95]: this leads to study micro- or nano-composites materials (mixing of a good TE material with another neutral for TE), or use "exotic" synthesis or shaping techniques inducing large amount of such defects.
- Grain boundaries will affect phonons (and also electrical conductivity), this leads to the study of nano-crystalline materials to increase the number of such boundaries or even to reduce the mean free path of phonons (when the sizes of the nano-grains are comparable with that of the mean free path of phonons).

That different processes are not exclusives and some of them can be created simultaneously in the same TE material; some examples are given by Gonçalves et al., in this volume, pages 51–67.

5 Conventional TE Materials

From 1960, all actually used TE materials (see Table 1) were known and their performance, bound to a stagnant ZT to ~1, have not much changed up to 1990.

Table 1 ZT of conventional materials at their optimal temperature of use T_u.

	Bi–Sb	Bi_2Te_3–Sb_2Te_3	$(Bi,Sb)_2(Te,Sb)_3$	-	PbTe	Te–Ag–Ge–Sb	Si–Ge
Type	n	n,p	n,p	-	n	p	n,p
T_u (K)	200	< 300	~300 – 400	-	700	750	1,000
ZT at T_u	1.1 (H)	0.8	0.9	-	0.8	1.1	0.6

H means a value obtained under a magnetic field.

The main market applications, were based on Bi_2Te_3-type of material mainly used for cooling. Let us remark that no material was efficient in the temperature range 400–700 K. At higher temperature, PbTe and TAGS (Te-Ag-Ge-Sb) have ZT $\sim$1 in n- and p-type respectively.

6 Recent Bulk TE Materials

6.1 Atomic Displacement Parameter

In solids, the atoms do not move only at zero temperature. At finite temperature, the atoms vibrate around their equilibrium position on their crystallographic site, the mean square displacement amplitude of such atoms is called the Atomic Displacement Parameter (ADP) which is useful in estimating the minimum value of the lattice contribution to the thermal conductivity and is thus of interest in TE materials [76].

The vibrations of the atoms are generally non-isotropic and this anisotropy is taken into account by a tensor of the ADP, noted U_{ij}. These U_{ij} are described by the ellipsoids of vibrations centered on each atom with an extension in one direction corresponding to the amplitude of the vibration in that direction. An isotropic ADP U_{iso} can be defined for each site, which is the mean value of the U_{ij} in all the directions of the space. Its amplitude allows one to compare the overall movements of the different atoms of the structure. Generally the U_{iso} tends toward zero at zero temperature, and if not, this may indicate some static disorder on the site.

The minimal value of the lattice part of the thermal conductivity is given by $\lambda_L = (1/3)C_v v_S d$, with v_S the speed of sound, d the mean free path of phonons and C_v the specific heat estimated from the Dulong and Petit law

$$C \underset{T \to \infty}{\to} 3R = 3N_A k_B \tag{6}$$

(with R the constant of gas, N_A the Avogadro number). In the case of a cage like material, like LaB_6, (Fig. 3), the specific heat has been calculated from ADP of La and B atoms.

The La atom is considered as an Einstein oscillator in a solid of Debye of B-atoms. $C_V = f.C_D + (1 - f)C_E$ with f the fraction of atoms contributing to the solid of Debye and $(1 - f)$ that of Einstein oscillators; θ_D and θ_E, the Debye and Einstein temperatures respectively:

$$C_D = 9N_A k_B \left(\frac{T}{\theta_D}\right)^3 \int_0^{\theta_D/T} \frac{x^4 e^x}{(e^x - 1)^2} dx \tag{7}$$

$$\text{and } C_E = 3N_A k_B \left(\frac{\theta_E}{T}\right)^2 \frac{e^{\theta_E/T}}{(1 - e^{\theta_E/T})^2} \tag{8}$$

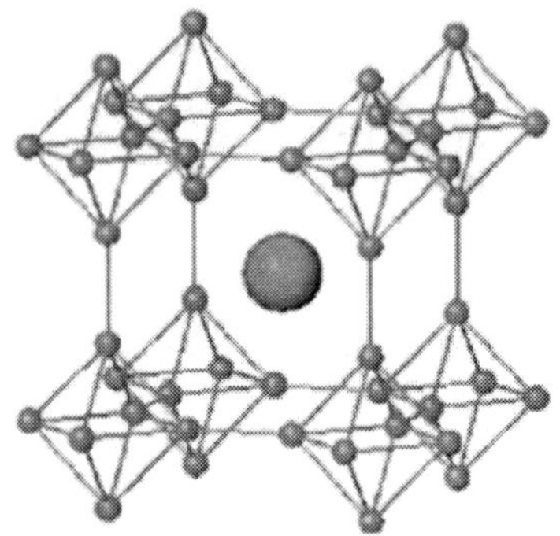

Fig. 3 Cage like structure of LaB_6, the La atom • is surrounded by B • groups.

6.2 *Cage Like TE Materials*

Cage-like TE materials have a rigid sub-lattice responsible for the electrical conductivity and large empty cages. When the cages are filled with heavy atoms, these atoms, weakly bound to the cage, can vibrate inside with a strong amplitude of vibration. Initially it has been proposed that these vibrations are incoherent ("rattling") and act as traps for the acoustic phonons and consequently decrease the thermal conductivity. According to more recent work, the vibrations of the inserted atom are optical phonons (coherent vibrations) mainly without dispersion (localized character) and with a weak energy which strongly interfere with acoustic phonons to decrease the thermal conductivity.

The two most studied families of cage like TE materials are the filled skutterudites $A_yM_4X_{12}$ (A: an electropositive element, M a metal and X = P, As, Sb) and the clathrates of the type $A_8Y_{16}X_{30}$ (X and Y being actually mainly Ga and Ge), both of them leading to increased values of ZT.

6.2.1 Skutterudites

In the skutterudite, the atom A fills the cubic structure (space group Im$\bar{3}$) of the binary CoX_3, which leads to the ternary filled structure $A_yM_4X_{12}$ [42] (Fig. 4).

Atoms of various oxidation numbers (single valence K, Na; divalent Ca, Sr, Ba, trivalent La, Ce, Pr, Nd, Th, U; mixed valent Yb [6]) can fill the cage with an occupation rate which depends on the nature of A ($\sim$1 with K and Na, [54] but smaller with rare earths) and strongly on the nature of M ($\sim$0.2 with Co and $\sim$0.9 with Fe) in rare earth based series (Table 2).

The cage in $CeFe_4Sb_{12}$ is clearly evident when looking on the distances between atoms (Fig. 5). The distance d (Ce–Sb) = 3.39 Å, and the covalent radius R of Sb (1/2 d(Sb–Sb) = 1.46 Å) gives $d - R = 1.93$ Å $\gg$ ionic radius of Ce = 1.14 Å.

Consequently, two effects can affect λ_L: a strong vibration of A, seen by neutron diffraction [19, 73] (Fig. 6), and a variable vacancy concentration rate in $A_y\,(Fe,Co)_4\,Sb_{12}$.

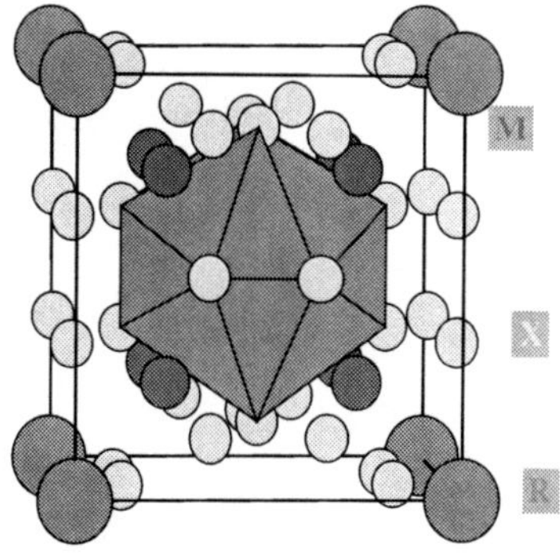

Fig. 4 Structure of filled skutterudite $A_yM_4X_{12}$.

Table 2 Occupation limits of A in $A_yCo_4Sb_{12}$.

A = Ba	La	Ce	Eu	Yb	Tl
0.44	0.23	0.1	0.54	0.25	0.22
[22]	[63]	[21]	[9]	[2]	[75]

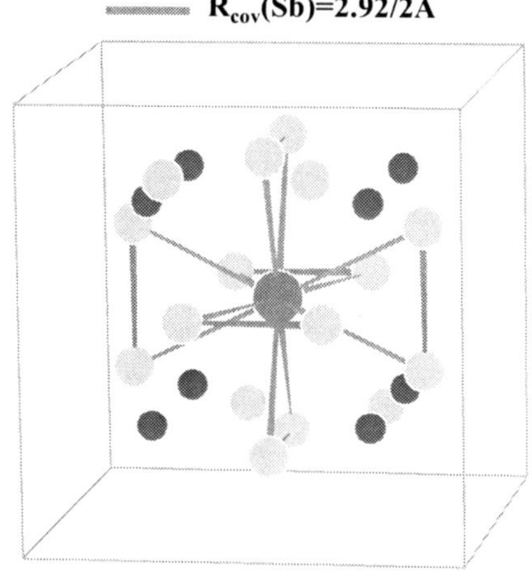

Fig. 5 Distances and cage in $CeFe_4Sb_{12}$.

As can be seen in Fig. 6, the ADP of La in $La_y(Fe,Co)_4Sb_{12}$ is much higher than that of M and Sb atoms. The thermal conductivity is strongly reduced in the ternaries as compared to binaries. The smallest λ_L appears in antimonides, i.e., when the cage is larger (due to increase lattice parameter) and consequently when the amplitude of vibration is stronger (Fig. 7) [31].

In the phenomenological model of (Sales [74,75] for skutterudites, the filler atom is treated as an Einstein oscillator and the (M,X) lattice as a solid of Debye.

By appropriate substitutions, keeping the semiconducting (semi-metallic) state, the values of ZT have been strongly improved from a maximum value of ~0.8 (Table 3) in $CoSb_3$ to higher than 1 in various series, with both n- and p-type (Table 3).

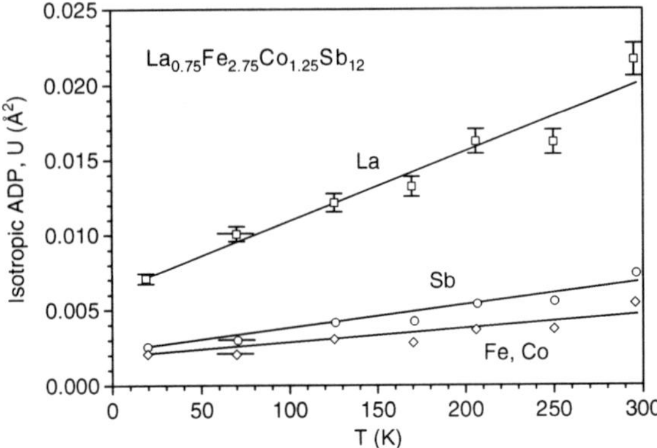

Fig. 6 Isotropic ADP in $La_y(Fe,Co)_4Sb_{12}$.

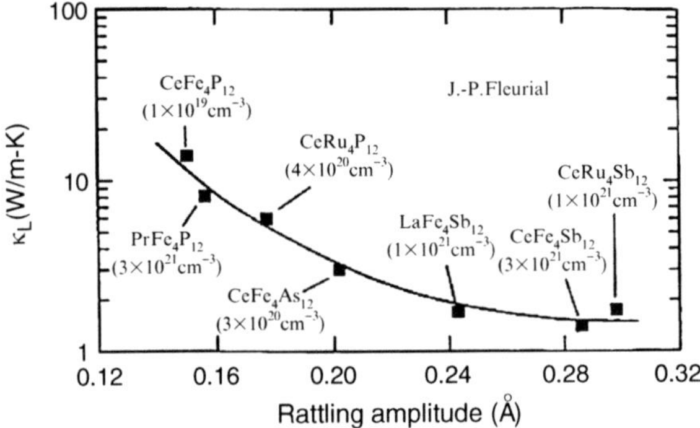

Fig. 7 Thermal conductivity versus the amplitude of rattling.

Let us note that the possibility of obtaining both types, with high ZT, in the same temperature range, with chemically rather similar materials is a positive point of these series, as similar behaviors are expected when used in applications.

6.2.2 Clathrates $A_8Ga_{16}Ge_{30}$ (A = Ba, Eu, Sr)

Clathrates are compounds made by inclusion of molecules of one species in a cavity of a crystalline lattice of another species: they consequently are cage-like compounds. In this article we are only concerned with semiconducting clathrates. The family of clathrates [44], derived from the binary A_8X_{46} (A = alkaline or alkaline earth, X = Si, Ge), comprise numerous structures built from complex polyhedron of X elements (X_{20}, X_{24}, ...) filled by the electropositive element A (Fig. 8).

The type I structure, made from two polyhedron X_{20} and $6X_{24}$, leads to a metallic character; a semiconducting state is obtained by substituting X by an element which accepts the electrons from A-metal. If A is divalent this means to replace 16 Ge (Si)

Table 3 Highest ZT in binary (B) and ternary (T) skutterudites.

	Samples	Type	Highest ZT	T_{ZT}(K)	Reference
B	$(CoSb_3)_{0.75}$+ $(FeSb_2)_{0.25}$	n	0.37	773	[46]
B	$Co_{0.94}Ni_{0.04}$ Sb_3	n	0.5	750	[47]
B	$CoSb_3$	p	0.21	600	[45]
B	$CoSb_3$ + 0.7+5%Te	n	0.5	600	[93]
B	$CoSb_3$ + 4%Te	n	0.8	750	[61]
B	$IrSb_3$	p	0.15	800	[83]
T	$Ba_{0.24}Co_4Sb_{11.87}$	n	1.1	850	[22]
T	$Ba_{0.3}Co_{3.95}Ni_{0.05}Sb_{12}$	n	1.2	800	[28]
T	$Ca_yCo_{4-x}Ni_xSb_{12}$	p	1	800	[71]
T	$Ce_{0.28}Co_{2.5}Fe_{1.5}$ Sb_{12}	p	1.1	800	[88]
T	$Ce_xFe_{3.5}Co_{0.5}Sb_{12}$	p	1.4 (1.2)[a]	870	[30]
T	$Eu_{0.42}Co_4Sb_{11.37}Ge_{0.50}$	n	1.1	700	[53]
T	$In_{0.25}Co_4Sb_{12}$	n	1.2	570	[36]
T	$In_{0.2}Ce_{0.2}Co_4Sb_{12}$	n	1.7	570	[35]
T	$La_xFe_{4-y}Co_ySb_{12}$	p	1	800	[72]
T	$Nd_xCo_4Sb_{12}$	n	0.45	700	[52]
T	$Tl_xCo_4Sb_{12}$	n	0.2	300	[75]
T	$Yb_{0.8}Fe_{3.4}Ni_{0.6}Sb_{12}$	p	1	800	[3]
T	$Yb_{0.19}Co_4Sb_{12}$	n	>1	600	[65]
T	$(Ce,Yb)_{0.4}Fe_3CoSb_{12}$	p	1	800	[7]

[a]Calculated values of 1.2 from experimental data.

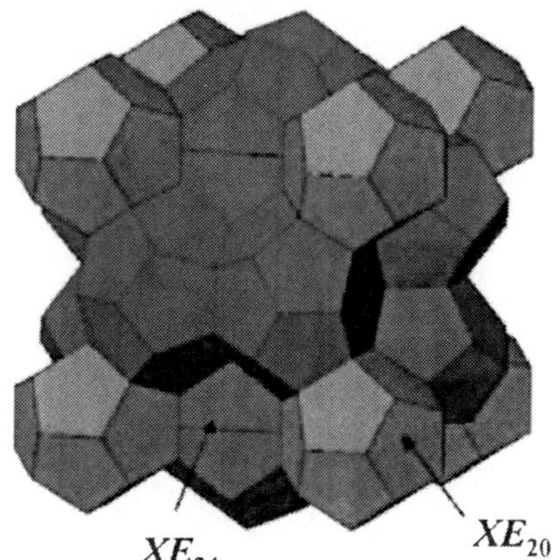

Fig. 8 The two types of polyhedron X_{20} and X_{24} around A atom in the type I clathrate $A_8Ga_{16}Ge_{30}$.

by 16 Ga which leads to $A_8^{2+}Ga_{16}^{-}Ge_{30}^{o}$ (cubic, space group Pm3n). In the cage X_{24}, the inserted atom has strong vibration amplitudes (Fig. 9) [20].

Consequently, the ADP values on the site Sr2 are higher than that on site Sr1 (Fig. 10), the strong residual component at low temperature being due to static disorder on this site. The filling of the cages induces strong decreases of the thermal conductivity which becomes as low as in glass or quartz [64].

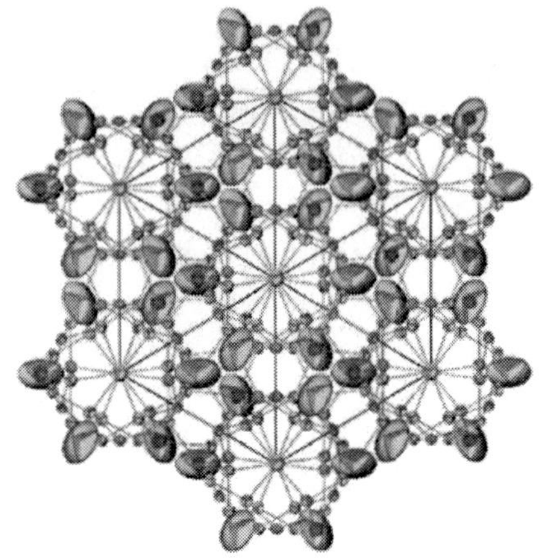

Fig. 9 View, along 111 projection, of the ellipsoids of movement of Sr in $Sr_8Ga_{16}Ge_{30}$.

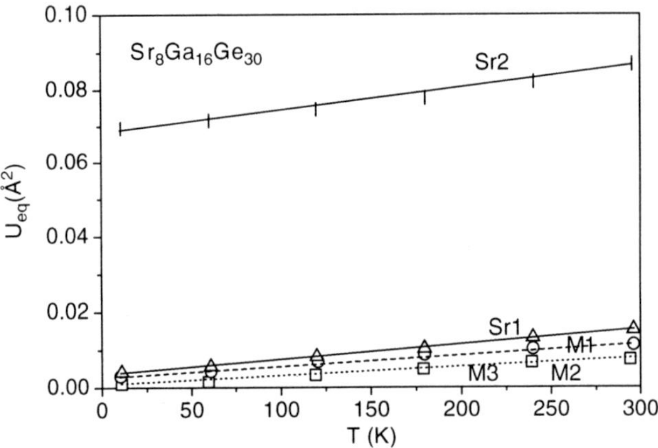

Fig. 10 Values of ADP for the different sites in $Sr_8Ga_{16}Ge_{30}$.

Presently, the type I has lead to ZT values highest than 1 [4, 5].

Other iso-electronic substitutions (Al, In) [11, 60] lead to semiconducting materials. Another possibility is to play on stoichiometry, for instance: $Eu_8Ga_{16\pm x}Ge_{30\pm x}$ [34] or to compensate non iso-electronic substitution by vacancies in $Ba_8Zn_xGe_{46-x-y}[]_y$, with [] a vacancy [26]. Presently, only Eu- and Ba-based clathrates have lead to good ZT values in these materials, in which the control of the stoichiometry is critical, not only for TE properties but also for the appearance of parasitic phases.

The chemistry of these phases is very rich (Fig. 11, simplified version of [60], the studies of their physical properties should lead to new interesting compounds for TE properties.

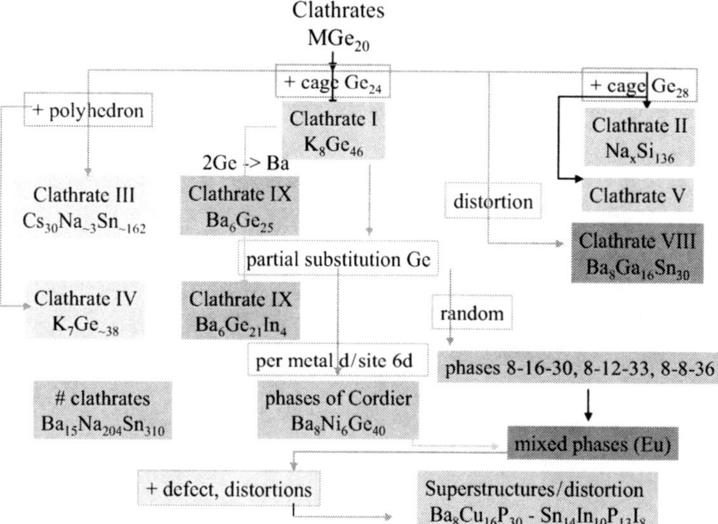

Fig. 11 Clathrates filiations.

6.2.3 Penta-Hexa-Tellurides

Three series of penta-tellurides and one hexa-telluride have been examined for TE properties.

In Tl_2MTe_5 (M = Ge, Sn), one Tl site (Tl_1) does not have close nearest neighbors whereas the (Tl_2) site has (Fig. 12).

Consequently, the ADP of Tl1 are larger than that of Tl in site 2. ZT values of ~0.6 have been obtained [80]. No solid solutions have been checked until now.

The second system Re_2Te_5 (Re_6Te_{15}) has a complex structure (space group Pbca), with 84 atoms/unit cell and has some similitude with Chevrel's phases (octahedral clusters Re_6 and large empty cages). n- and p-type are possible by appropriate doping. The last penta-telluride Ag_9TlTe_5 has a complex structure. The thermal conductivity of this material is smaller than that of other good TE materials like Bi_2Te_3 or TAGS [50]. Ag_9TlTe_5 has a ZT of 1.2 at 700 K. The hexa-telluride Tl_9BiTe_6 has also a very weak thermal conductivity and a ZT of 1.2 at 500 K, a temperature where Bi_2Te_3-based materials do not work [94]. The potential of these materials is interesting from academic point of view but environmental constraints will prevent their use in devices.

6.2.4 Phases with Mo Clusters

The structure of Chevrel's phases is built from a three-dimensional network of pseudo-cubic clusters (Fig. 13) Mo_6X_8 (X = S, Se, Te) and its versatility authorizes the formation of various representatives.

The clusters form cavities or channels which can be intercalated by various elements in such a way that it is possible to modify them from metallic in Mo_6X_8

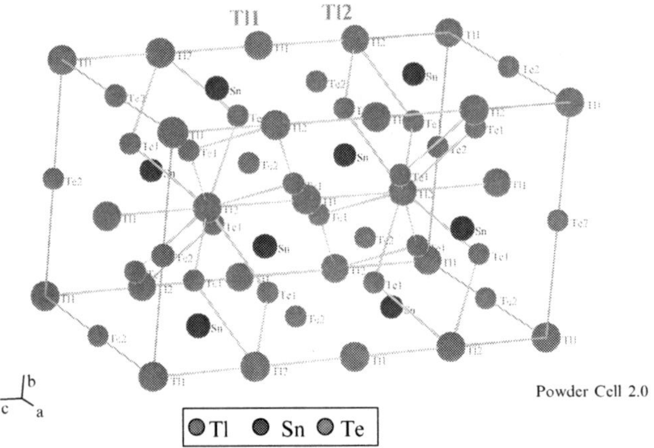

Fig. 12 Structure of Tl_2SnTe_5 and neighboring of Tl in site 2 and 1.

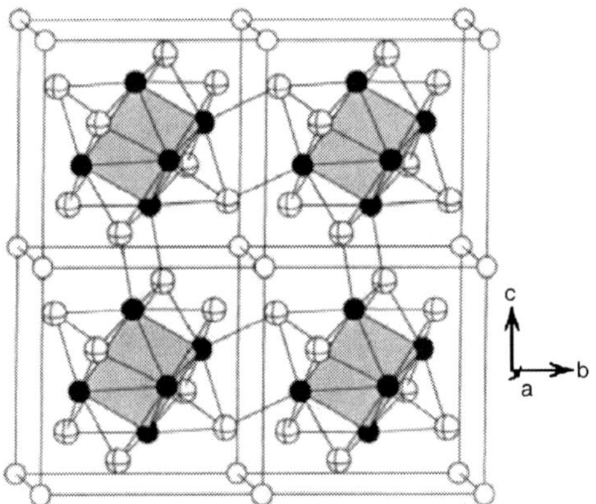

Fig. 13 Structure of Mo_6X_8 (Mo: •, X: ⊘, empty sites ○ can be filled by Pb, Cu.

to semiconductor for a Valence Electron Concentration (VEC) of 24. Under these conditions, interesting TE properties can be expected and a ZT of 0.6 (at the high temperature of 1,150 K) has been found in $Cu_{3.1}Mo_6Se_8$ by studying the series $M_xMo_6Se_8$ (M = Cu, Cu/Fe, Ti) [14]. Other compounds of these series have high power factors, like $Ti_{0.3}Mo_5RuSe_8$ [58] or $(Cu_yMo_6Se_8)_{1-x}(Mo_4Ru_2Se_8)_x$ [77]. This shows the potential of these series, specially with the possibility to replace Mo by Ru, Rh, Re to find good TE materials, of n- and p-type, stable at high temperatures (~1,200 K).

In the phase with Mo_9Se_{11} clusters, the structure possesses voids, and the insertion of atoms (Ag, Cs, Cl) in cages [33] (Fig. 14) leads to semiconducting or semi-metallic compounds, depending on the number of electrons per clusters. These characteristics should be favorable for TE properties, which have not yet been studied in detail.

However, a Seebeck coefficient of 72 μV/K and a weak resistivity have been reported in $Ag_{3.6}Mo_9Se_{11}$ [70].

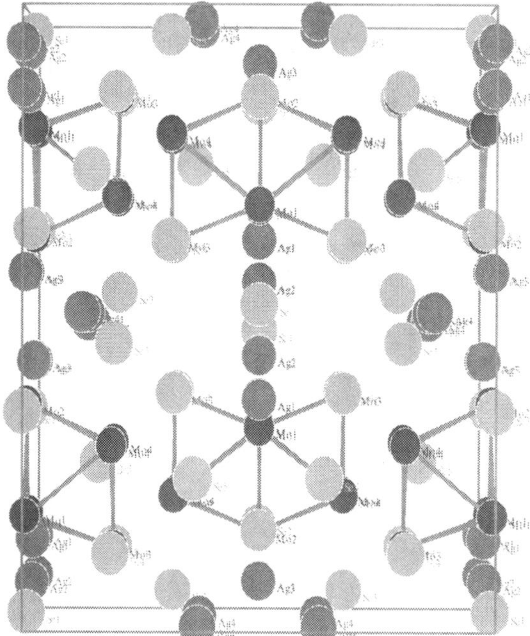

Fig. 14 Structure of $Ag_{3.6}Mo_9Se_{11}$ (• Mo, • Se, • Ag), –Mo clusters.

Many other structural families may have empty cages in which the insertion of atoms with various oxidation degrees may lead to a semiconducting state, a prerequisite for good TE properties.

6.3 Phases with Vacancies

6.3.1 Half Heusler

The cubic structure of the Heusler phases X_2YZ (space group Fm*3*m) consists in four interpenetrated face centered cubic lattices in A(0,0,0); B = (1/4, 1/4, 1/4), C = (1/2, 1/2, 1/2), et D = (3/4, 3/4, 3/4), with one occupancy of the sites A = Y; B & D = X and C = Z. If one of the equivalent sites (1/4, 1/4, 1/4) or (3/4, 3/4, 3/4) is empty the Half Heusler phase (F-*4*3m) forms (Fig. 15).

Among these compounds, those with 18 valence electrons have a band structure of semiconductor and often high Seebeck coefficients.

To decrease the thermal conductivity, complex substitutions with increased phonons diffusion by mass fluctuations, have been used with success (Fig. 16).

For instance, from TiNiSn (ZT = 0.4 at 750 K) and ZrNiSn, the n-type compound $Hf_{0,5}Zr_{0,5}Ni_{0,8}Pd_{0,2}Sn_{0,99}Sb_{0,01}$ has maximum of ZT of 0.7 at 800 K and the n-type compound $Ti_{0,5}(Zr_{0,5}Hf_{0,5})NiSn_{0,98}Sb_{0,02}$ has a ZT > 1.4 at 700 K [82].

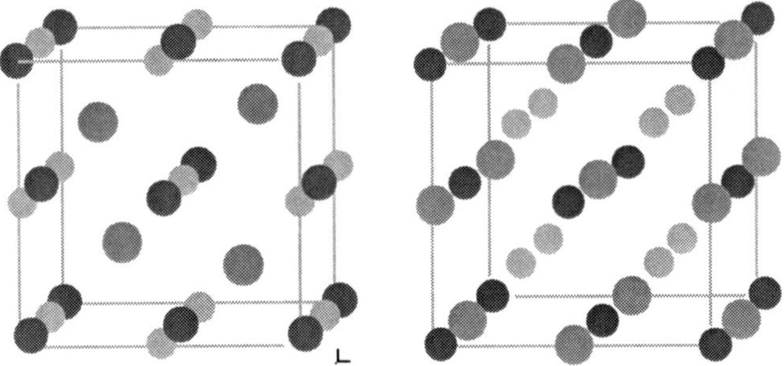

Fig. 15 Structures of the Heusler phases of Zr, Ni, Sn: half Heusler on left, Heusler on right (• Sn, • Zr, • Ni).

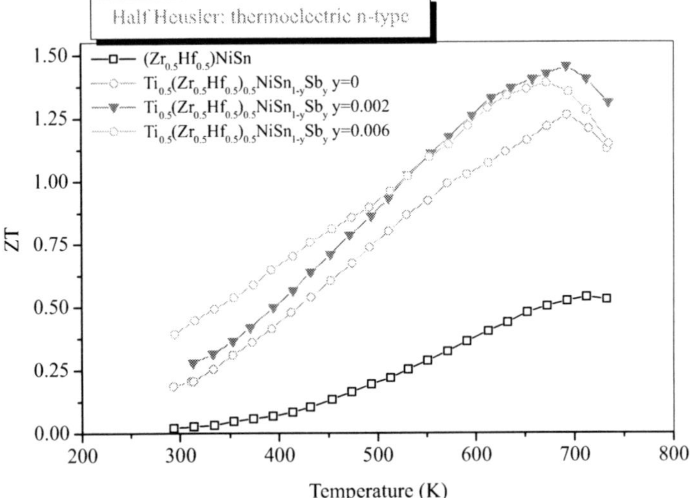

Fig. 16 Effect of substitutions in the half Heusler ZrNiSn.

Other systems like VFeSb (ZT = 0.11 at 900 K) have been intensively studied [43], but other substitutions should be examined, as simple substitutions like $V_{0.95}Ti_{0.05}$ FeSb lead to a ZT (p-type) of 0.4 at 600 K.

In these series, LnPdSb-type compounds (Ln = Sr, Y, rare earths, specially Er) have high Seebeck coefficients below 300 K [57,66], which could lead to interesting substituted materials for cooling.

6.3.2 Zn_4Sb_3

Zn_4Sb_3 exists under three crystallographic varieties: α-phase stable below −10°C, β-phase from −10°C to 492°C and γ-phase from 492°C to the melting temperature of 566°C. The structural and TE properties are summarized in Fig. 17.

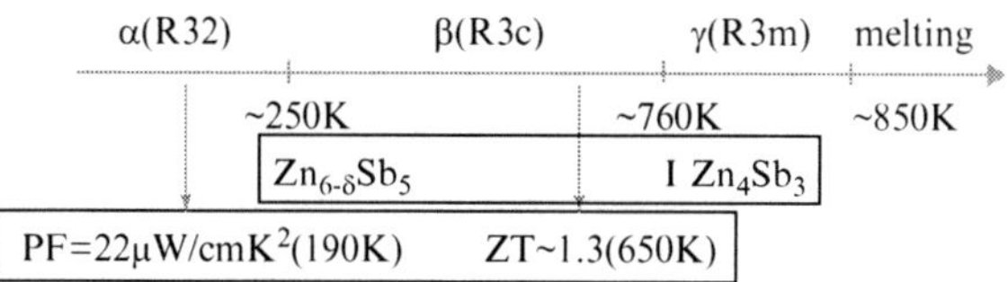

Fig. 17 Scheme of structural and Te properties of Zn_4Sb_3.

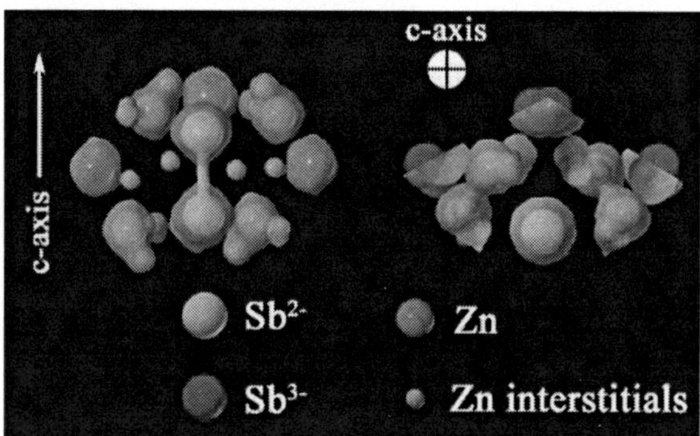

Fig. 18 Disorder in the structure of Zn_4Sb_3.

The β-phase leads to TE materials (p-type), with a high ZT of 1.3 at ~650 K [13], further increased to 1.4 at 525 K by substitution in $Zn_{3.2}Cd_{0.8}Sb_3$ [15].

In Zn_4Sb_3 the observation of vacancies, Zn-interstitials and of 2 types of Sb atoms (spherical ion Sb^{3-} and dimers Sb^{4-}) (Fig. 18) induce a strong disorder [85] (in fact the composition is $\sim Zn_{6-w}Sb_5$) which contribute to the decrease of the thermal conductivity.

A phonon mode of low energy has been associated to Sb dimers [78]. It does not seem that doping leads to one achieving the n-type material.

6.4 *Complex Solid Solutions Derived from Conventional Materials*

6.4.1 Bi_2Te_3-Derivatives

As Bi_2Te_3 is the archetype of TE materials used for cooling applications, few derivatives systems have been studied.

At low temperature, TE properties of $CsBi_4Te_6$ doped by 0.05% SnI_3 are equivalent to old $Bi_{2-x}Sb_xTe_{3-y}Se_y$ compositions with a ZT ~ 0.8 at 225 K, of possible use for cooling [24].

Strongly doped single crystals lead to higher values than ZT ~ 0.8–0.9 at ~300 K. The best values concern n-type $(Bi_{0.25}Sb_{0.75})_2Te_3$ doped with 0.07%

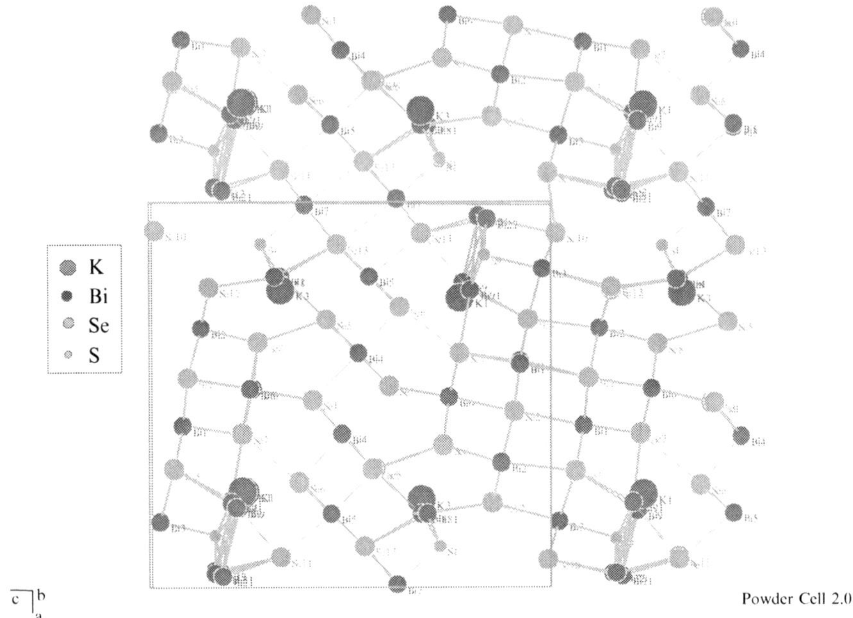

Fig. 19 Structure of $K_2Bi_8Se_9S_4$.

I (in mass), 0.02% Te, et 0.03% CuBr [97] leading to a ZT > 1.1 at ∼300 K. For the p-type, doping by 8% in mass of Te leads to ZT = 1.4 at the same temperatures [96]. The effects of Te doping as well as those of texturation induced by extrusion were previously studied in $Bi_{0.5}Sb_{1.5}Te_3$ [40]. On the contrary, the addition of few percents of PbTe in $Bi_{0.4}Sb_{1.6}Te_3$ do not lead to an improvement of ZT.

$K_2Bi_8Se_{13}$ forms two phases: α-$K_2Bi_8Se_{13}$ (triclinic, space group *P*-1d) and β-$K_2Bi_8Se_{13}$ (monoclinic, space group *P* 21/m). The β-phase (Fig. 19) has an architecture built from fragments structure type of Bi_2Te_3, CdI_2 and NaCl, and a mixed occupancy of the sites K/Bi [10].

In $K_2Bi_8Se_9S_4$, all the S sites are with a mixed occupancy and the representation (Fig. 19) is made according to the nature of the main occupant of the site. The thermal conductivity is very weak (complex structure, atoms vibrating in tunnels, disorder in the site occupancy) and the doping increases the Seebeck coefficient, however the ZT values are presently too small for cooling applications. Other substitutions could possibly lead to p-type materials.

6.4.2 Derivatives from PbTe and TAGS

The series of cubic materials $AgPb_mSbTe_{2+m}$ (space group Fm**3**m) (Fig. 20) has led to type n semiconductor with high ZT values when m = 10 (ZT = 1 at 700 K) or m = 18 (ZT = 2.2 at 800 K) [41].

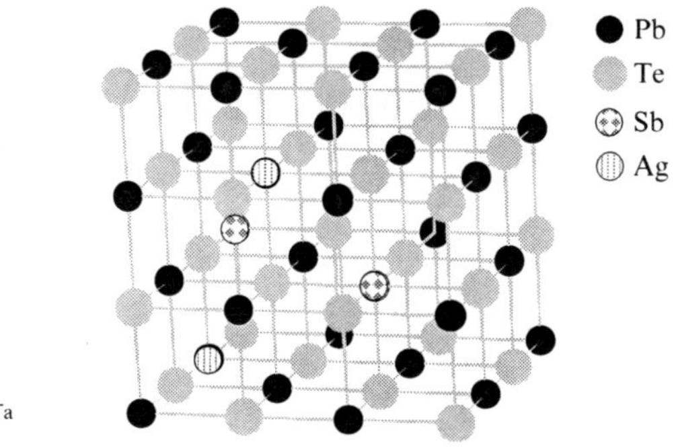

Fig. 20 Cubic structure of $AgPb_mSbTe_{2+m}$.

The smallest values of ZT have been observed in annealed Ag-deficient samples (ZT = 1.1 at 670 K with x = 0.4 and ZT = 0.3 with x = 0.3). However these materials are not conventional bulk but composites (see below In_2O_3 + Ge).

High resolution transmission microscopy have shown that Ag-Sb rich phase nano-crystals exist in a Pb–Te rich matrix, that has also been described as nano-clusters of $AgPb_3SbTe_5$ in PbTe matrix [55]. Such nano-phases could be responsible of high ZT values, theoretically predicted, and observed in quantum wells super-lattice, nano-dots etc. (see below nano-materials). The difficult control of the nature, size and dispersion of these nano-phases could be responsible of the mismatch of literature results. The effect of annealing and of eventual temperature cycling which could induce a growth of these nano-objects have not yet been checked.

6.5 *Other Intermetallic Compounds*

Recently, few new Sb-based compounds have been identified with good TE characteristics, a property which can be due to the property of Sb to diffuse phonons more efficiently than lighter ions with the same electronic configuration (we already mention that for skutterudites and for Zn_4Sb_3). Moreover, materials made from metalloid and slightly more electropositive elements (Zintl phase) seem to be able to have high ZT.

6.5.1 $Yb_{14}MnSb_{11}$

The tetragonal structure of $Yb_{14}MnSb_{11}$ (space group $I4_1/acd$) is complex (Fig. 21) and built from various structural units. The flexibility to accommodate various elements (possibility of future improvements?) have led to a ZT = 1 at the high temperature of 1,220 K in this p-type material [12].

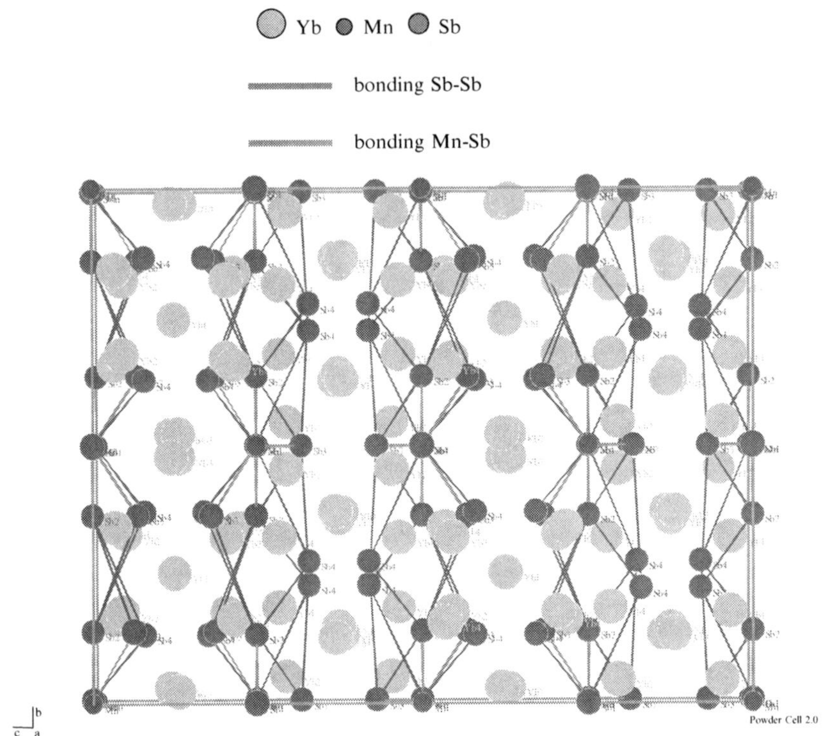

Fig. 21 (100) Projection of the structure of $Yb_{14}MnSb_{11}$.

6.5.2 $Mo_3Sb_{7-x}Te_x$ and Derivatives

The structure Ir_3Ge_7 (space group Im3m) of $Nb_3Te_{7-x}Sb_x$ [92] and of $Mo_3Sb_{7-x}Te_x$ ($x = 1,5$ et $1,6$) [32] compounds forms infinite chains in the three directions of space. $Mo_3Sb_{7-x}Te_x$ only exists for a VEC between 51 and 56, a value of 55 being a prerequisite to obtain a semiconductor.

With a ZT of 0.8 in $Mo_3Sb_{5,4}Te_{1,6}$ at 1,050 K [32] and in $Mo_{3-y-z}Ru_yFe_zSb_7$ at 1,000 K [18] for $(y,z) = (0.16, 0.5)$ and $(0.1, 0.7)$ these materials are better than the best p-type Si-Ge.

6.5.3 Other Sb-Based Materials

A ZT of 0.55 at 600 K was initially attributed to the phase Mg_3Sb_2 (space group P3m1) which was invalidated with a maximum of ZT of 0.2 at 875 K [25]. Moreover, the phase loses Mg and oxidizes above 900 K.

The structure of $CaZn_2Sb_2$ is the same as that of Mg_3Sb_2 and the material easily oxidizes. Due to its simple crystalline structure, its TE properties are poor with a ZT of 0.5 at 770 K [86].

6.5.4 $Mg_2Si_{1-x}Sn_x$

During the years 1960, the Ioffé Institute showed that the compounds Mg_2X ($X = Si$, Ge or Sn) are semiconductors with a band structure which should favor TE properties. Various substitutions have been tested and the best results have been obtained when the mass difference between elements is the highest, i.e., with Si and Sn [29]. A ZT of 1.1 has then been found at 800 K, a better value than with n-type Si-Ge [99].

6.6 Oxides

The main expected advantage of oxides is their chemical stability in oxidizing atmosphere, even though the stability and transport properties of some oxides depend on the partial pressure of oxygen. If many oxides have high Seebeck coefficients ($\alpha > 100\mu V/K$) their performance are presently limited by electrical resistivities much higher than in previously described materials.

6.6.1 Cobaltites

Among the most promising p-type oxides, cobaltites with conductive layers of CoO_2 (CdI_2 type) have been the starting point of interest for TE oxides with Na_xCoO_2: a metallic oxide with a high Seebeck coefficient [89].

6.6.2 Misfit Oxides

Among cobaltites, the layered compound $Ca_3Co_4O_9$ is in fact a lamellar oxide with a misfit structure [56] and a general formulae: $[A(O)_n]RS[CoO_2]_{b1/b2}$ made of CoO_2 planes (as in Na_xCoO_2) split by AO layers of NaCl type (RS: rock salt) with possible n values of 2, 3 and 4. Considering the sub-lattice CoO_2 as a mono-layer NaCl (quoted S_2) which matches with the sub-lattice AO (S_1) along the a-axis but mismatches along the b-axis. The ratio of the b-axis of S_1 and S_2, i.e., b_1/b_2 measures the incommensurability of the two lattices (Fig. 22). In the case of $Ca_3Co_4O_9$, this ratio is 1.625 (13/8) which leads to a formulae: $[Ca_2CoO_3]RS[CoO_2]_{1.62}$.

The family of lamellar oxides with misfit structure is very large (choice on n, and A, for instance Ca, Bi, Sr,... and replacement of Co by Rh,...) and many new materials are conceivable, however the actual ZT values are low due to high electrical resistivity [49].

The highest ZT value in an oxide has been obtained in a small size single crystal of Na_xCoO_2 (estimated value of 1.2 at 800 K). The impossibility of growing large

Fig. 22 Structure of the misfit compound $Ca_3Co_4O_9$ (b ∼ 8b1 ∼ 13b2).

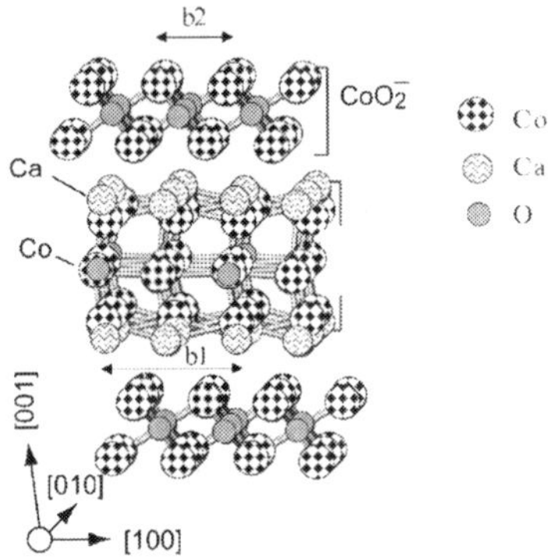

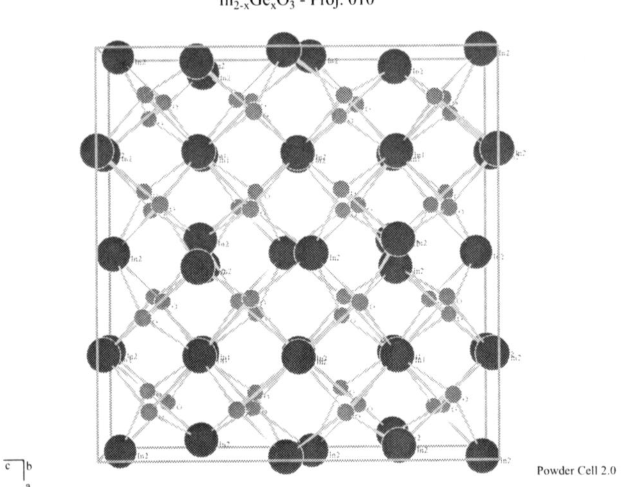

Fig. 23 (010) Projection of In_2O_3 structure.

enough single crystal for applications has led to many studies concerned by texturation processes, with the drawback that the electrical conductivity in the CoO_2 planes decrease as compared to crystals.

6.6.3 Other Oxides

Other oxides families have also been studied like perovskites $ACoO_3$-type, manganites $AMnO_3$ [51], delafossites $CuMO_2$ [67], YBCO ($YBa_2Cu_3O_{6+y}$), ZnO [68, 87], ruthenates $ARuO_3$ [37] and A_2RuO_4, In_2O_3 etc. The best ZT values reach 0.4 for the p-type and 0.3 for the n-type in conventional oxides and of 0.55 in misfits oxides.

The structure of In_2O_3 (space group Ia-3) shows alternatively ordered and disordered In-planes (see Fig. 23).

The solubility limit of Ge is no more than 0.5 atom %, for higher rate of Ge the material forms a composite In_2O_3+ inclusions of micronic size (~100 μm) of $In_2Ge_2O_7$. In that composite, the thermal conductivity is notably reduced and a ZT of 0.45 at 1,243 K has been observed which is actually the best value for a n-type oxide.

This shows the potential of micro-composites, as well as nano-composites (see $Ag_{1-x}Pb_{18}SbTe_{20}$), to improve ZT values as compared to "parent" materials.

6.7 *ZT of New Materials*

The use of the various processes, previously described, to minimize the thermal conductivity has lead to a noticeable increase of ZT from ~1995. Let us remark that the new families of materials have led to the improvement of ZT, moreover, they have increased the temperature ranges where the TE can be used. The value of ZT in few families is higher than 1.3 (30% increase as compared to conventional materials). The best value in a bulk material is actually of 1.7 in skutterudites. The curves of the best values of ZT versus temperature in bulk materials are shown in Figs. 24 and 25 for the p- and n-type materials, respectively.

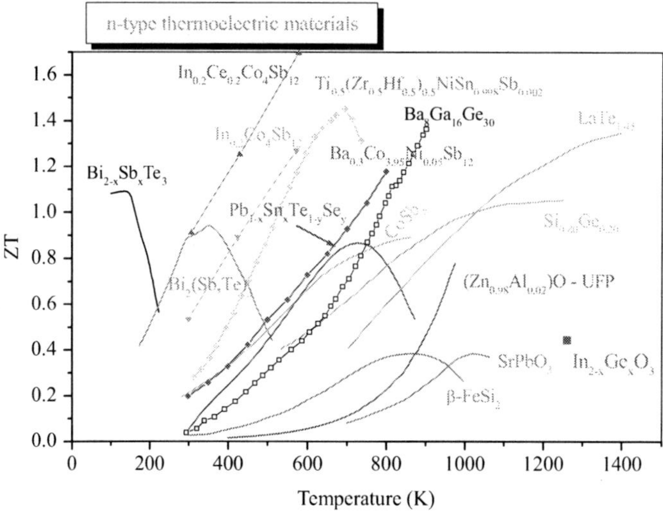

Fig. 24 Best ZT values for representatives of n-type materials.

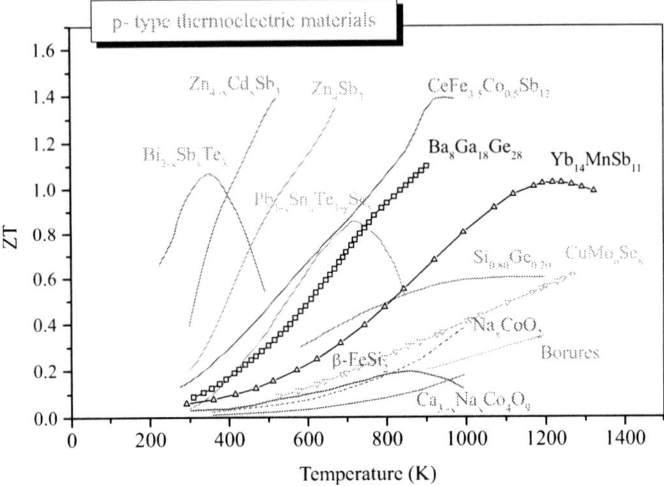

Fig. 25 Best ZT values for representatives of p-type materials.

6.8 *Semiconducting Glasses*

A certain number of characteristics leading to a weak thermal conductivity and to good TE properties, such as complex structures, inclusions, impurities, mass fluctuations, disorder,... can be found in glasses. However, a glass is generally an insulator, except the metallic glasses which have low Seebeck coefficients (as metals). Numerous semiconducting glasses, with a small gap to favor high Seebeck coefficients, exist specially in pnictides and chalcogenides. Their preliminary studies have shown encouraging results, even though the electrical resistivity is generally too high (Gonçalves et al., in this volume, pages 51–67).

7 Nano TE Materials

Transport properties in micro-, and nano-structures differ from that in bulk 3D (three-dimensional) materials. The thermal conductivity of nano-structures like super-lattices (material with periodic change of nano-metric layers of various elements or substances) is weaker than in bulk materials. This is a positive aspect for TE properties. However, these nano-structures are generally obtained by thin films techniques (Molecular Beam Epitaxy, Knudsen cells, Pulsed laser, CVD,...) too costly to be used for the production of materials for large scale applications. Moreover, their weak thickness are not very compatible with the formation of large temperature gradients. Conversely, they are well adapted to micro-cooling in electronics.

Electronic transport properties are also strongly modified by dimensionality effects on the band structure (Fig. 26) [27]).

To summarize, in a 3D material, α, σ and λ are constrained and it is difficult to optimize both of them, in the case of lower dimension new possibilities exist to independently adjust them.

In addition, new interfaces are created which can increase the phonon diffusion, more than that of carriers, leading to an increase value of ZT (Fig. 27).

It is remarkable that the improvement of ZT has been theoretically predicted [38] (Fig. 28) before its experimental observation.

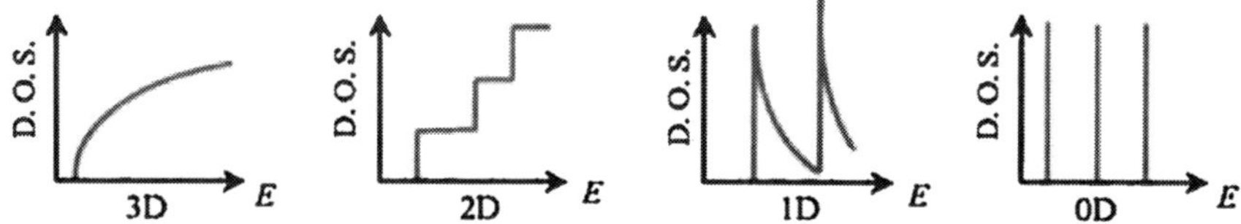

Fig. 26 DOS for a semiconductor 3D, a quantum well 2D, a nano-wire or nano-tube 1D and a quantum dot 0D.

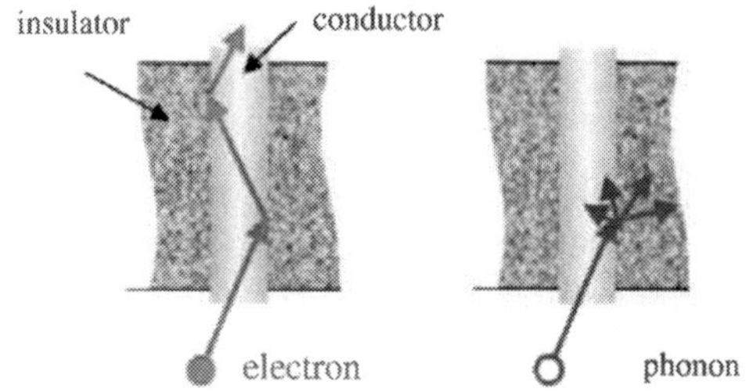

Fig. 27 Proposed model for the diffusion of phonons and carriers in a 1D structure.

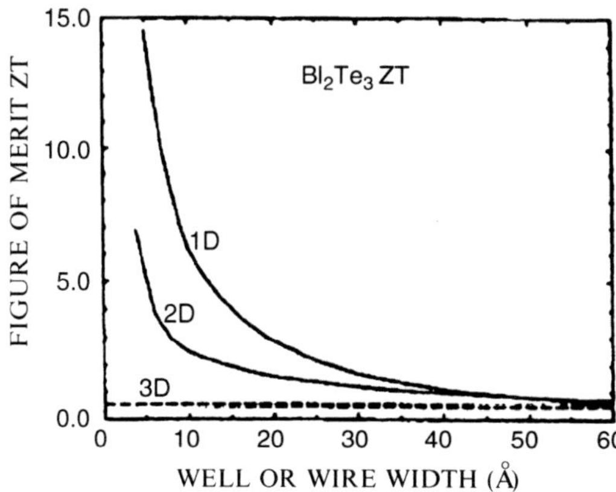

Fig. 28 Evolution of ZT with the dimensionality.

The experimental proofs, in various forms of nano-structures, concern, for instance:

- Super lattices including quantum wells PbTe-PbSeTe [39], in which ZT value of 1.6 has been calculated (in front of 0.4 for the bulk).
- Super lattices of Bi_2Te_3/Sb_2Te_3, with a ZT of 2.4 (in front of 1.0 for the bulk) [91].
- A decrease of the thermal conductivity of a Ge matrix by inclusions of Si-nano-wires which depends on the diameter of the nano-wires [98].
- An increase of the ZT value in various compositions of Bi_2Te_3 by addition of Bi_2Te_3-nano-powders [62].

These two last examples, as well as the case of $AgPb_mSbTe_{2+m}$ (see above) show that the study of nano-composites may allow one to benefit from TE performances related to the nano-sizes with samples of large sizes (greater than centimeter) requested for applications. The next year should be crucial for the understanding of mechanisms in nano-composites and consequently to be able to increase beneficial effects of such systems as compared to detrimental effects, for instance, on resistivity.

8 Shaping of Nano-Composites TE Materials

The shaping of such samples can be realized by hot pressing techniques. In the recent studies, this shaping is generally obtained by Spark Plasma Sintering (SPS) technique, a technique of fast sintering under pressure, which mainly avoids grain growth observed with a conventional hot press. A small size SPS machine, as we acquired in CNRS Thiais (Syntex 515S), is able to perform sintering under conditions up to 50 MPa and 2,000°C, with non costly graphite crucibles. Sample diameters of up to 20–25 mm are obtained for sintering temperatures up to $\sim 1,500°C$ for ceramics and $\sim 1,000°C$ for intermetallics; for higher temperatures smaller diameter can be produced. These temperatures are generally sufficient as the pressure helps to sinter. Pressures of up to 500 MPa have been realized with other crucibles (W-C) at lower temperatures and diameter.

Such techniques have been used with various TE conventional powders (grains smaller than a few 10 μm) of Bi_2Te_3, skutterudites [8], clathrates, half Heusler, Zn_4Sb_3, $Mg_2(Si, Sn)$, PbTe, Zno derivatives, etc.. Electric contact or protection against corrosion can be made, in few cases, by direct sintering of a thin metal foil with the powder.

More recently, the technique has been successfully applied in various composites systems, the starting nano-particles being produced by sol-gel method, electrochemistry, ... but more often from ball milling:

- Various compositions of Bi_2Te_3 with addition of Bi_2Te_3-nano-powders [62]
- Synthesis of $Yb_yCo_4Sb_{12}/Yb_2O_3$ composites [100]
- Shaping of composite made of nano-particles of SiC dispersed in Bi_2Te_3 [101]

- Shaping of binary skutterudites with C_{60} inclusions [81]
- Mixing of micro-nano-skutterudites (Zhang 2006)
- Bi addition in Heusler V_2FeAl [59]
- Nano-inclusions of ZrO_2 in the half Heusler $Zr_{0.5}Hf_{0.5}Ni_{0.8}Pd_{0.2}Sn_{0.99}Sb_{0.01}$ [23]

It is remarkable that such sintering process takes place within a short period of time (typically few minutes) and lead to density higher than 95% of the theoretical x-ray density, a very crucial parameter for TE transport properties. This explains the increasing use of this technique for the synthesis and/or shaping of TE materials and micro-, nano-composites from 2004.

As a conclusion, within the last 15 years, bulk materials with an increased complexity (structure, solid solutions) have led to an improvement of the TE figure of merit of $\sim$30%, and sometime more, in various series. With the complexity of the micro-, nano-structure it seems promising that these values of ZT can be exceeded. Studies of nano-composites coupled by fast shaping process by SPS, should lead to a real potential of using thermoelectricity for large scale applications.

Acknowledgements This work was supported by the bi-lateral French–Portuguese GRICES 2007–2008 program and European COST P16 program. One of us (CG) acknowledges NATO support to his participation to ARW Workshop "Properties and Applications of Thermoelectric Materials" 20–26 September 2008, Hvar, Croatia.

References

1. Abeles B.: Lattice thermal conductivity of disordered semiconductor alloys at high temperature, Phys. Rev. **131**, 1906 (1963)
2. Anno H., Nagamoto Y., Ashida K., et al.: Effect of Yb filling on thermoelectric properties of Co1-xMxSb3 (M = Pd,Pt) skutterudites, 19th International Conference on Thermoelectrics, Cardiff, Wales, UK, 20–24 Aug., p. 90 (2000)
3. Anno, H., Nagao, J. and Matsubara, K.: Electronic and thermoelectric properties of YbyFe4-xNixSb12, 21st International Conference on Thermoelectrics, Long Beach, CA, 25–29 Aug., p. 56 (2002)
4. Anno H., Hokazono M., Kawamura M., et al.: Thermoelectric properties of Ba8GaxGe46-x clathrate compounds, 21st International Conference on Thermoelectrics, Long Beach, CA, 25–29 Aug., p. 77 (2002)
5. Bentien A., Paschen S., Pachero V., et al.: High temperature thermoelectric properties of alpha and beta-Eu8Ga16Ge30, 22nd International Conference on Thermoelectrics ICT2003, La Grand Motte, France, Aug., p. 131 (2003)
6. Bérardan D., Godart C., Alleno E., et al.: Chemical properties and thermopower of the new series of skutterudite $Ce_{1-p}Yb_pFe_4Sb_{12}$, J. Alloys Compd. **351**, 18–23 (2003)
7. Bérardan D., Alleno E., Godart C., et al.: Improved thermoelectric properties in the double filled $(Ce,Yb)_y(Fe,Co,Ni)_4Sb_{12}$ skutterudites, J. Appl. Phys. **98**, 033710 (2005)
8. Bérardan D., Alleno E., Godart C., et al.: Synthesis of nanocrystalline filled skutterudites by mechanical alloying, 24th International Conference on Thermoelectrics ICT2006,Vienna, Austria, 6–10 Aug., p. 151 (2006)
9. Berger S., Paul C., Bauer E., et al.: Physical properties of novel thermoelectric skutterudites EuyFe4-xCoxSb12 and EuyFe4-xNixSb12, 20th International Conference on Thermoelectrics, Beijing, China, 8–11 June, p. 77 (2001)

10. Bilc D. I., Mahanti S. D., Kyratsi T., Chung D. Y., Kanatzidis M. G. and Larson P.: Electronic structure of $K_2Bi_8Se_{13}$, Phys. Rev. B **71**, 085116 (2005)
11. Blake N. P., Bryan J. D., Latturner S., et al.: Structure and stability of the clathrates Ba8Ga16Ge30, Sr8Ga16Ge30, Ba8Ga16Si30, and Ba8In16Sn30, J. Chem. Phys. **114**, 10063 (2001)
12. Brown S. R., Kauzlarich S. M., Gascoin F., et al.: Yb14MnSb11: New high efficiency thermoelectric material for power generation, Chem. Mater. **18**, 1873 (2006)
13. Caillat, T., Fleurial, J. P. and Borshchevsky, A.: Preparation and thermoelectric properties of semiconducting Zn4Sb3, J. Phys. Chem. Solids **58**, 1119 (1997)
14. Caillat, T., Fleurial, J. P. and Snyder, G. J.: Potential of Chevrel phases for thermoelectric applications, Solid State Sci. **1**, 535–544 (1999)
15. Caillat, T., Borshchevsky, A. and Fleurial, J. P.: High performance thermoelectric materials based on beta Zn4Sb3, Report NPO19851, JPL, Pasadena, CA (1999)
16. Caillat T., Kisor A., Lara L., et al.: Progress Status of Skutterudite-Based Segmented Thermoelectric Technology Development, 23rd International Conference on Thermoelectrics, Adelaide, Australia, 25–29 July, (2004)
17. Callaway, J. and von Baeyer, H. C.: Effect of point imperfections on lattice thermal conductivity, Phys. Rev. **120**, 1149 (1960)
18. Candolfi, C.: Synthèse, caractérisation physico-chimique et propriétés de transport de composés de type Mo3Sb7, Thesis University of Nancy, France, Oct. 6 (2008)
19. Chakoumakos B. C., Sales B. C., Mandrus D., et al.: Disparate atomic displacements in skutterudite type LaFe3CoSb12, a model for thermoelectric behaviour, Acta Crystallogr. **55**, 341 (1999)
20. Chakoumakos, B. C., Mandrus, D. and Sales, B. C.: Structural disorder and magnetism of the semiconducting clathrate Eu8Ga16Ge30, J. Alloys Compd. **322**, 127 (2000)
21. Chen B., Xu J. H., Uher C., et al.: Low temperature transport properties of the filled skutterudites CeFe4-xCoxSb12, Phys. Rev. B **55**, 1476 (1997)
22. Chen L. D., Kawahara T., Tang X. F., et al.: Anomalous barium filling fraction and n-type thermoelectric performance of BayCo4Sb12, J. Appl. Phys. **90**, 1864 (2001)
23. Chen L. D., Huang X. Y., Zhou M., et al.: The high temperature thermoelectric performances of Zr0.5Hf0.5Ni0.8Pd0.2Sn0.99Sb0.01 alloy with nanophase inclusions, J. Appl. Phys. **99**, 064305 (2006)
24. Chung D. Y., Hogan T. P., Brazis P., et al.: CsBi4Te6: a high performance thermoelectric material for low temperature applications, Science **287**, 1024 (2000)
25. Condron C. L., Kauzlarich S. M., Gascoin F., et al.: Thermoelectric properties and microstructure of Mg3Sb2, J. Solid State Chem. **179**, 2252 (2006)
26. Deng S., Tang X., Zhang Q.: Synthesis and thermoelectric properties of p-type Ba8Ga16ZnxGe30-x type-I clathrates, J. Appl. Phys. **102**, 043702 (2007)
27. Dresselhaus M. S., Dresselhaus G., Sun X., et al.: Low-dimensional thermoelectric materials, Phys. Solid State **41**, 679 (1999)
28. Dyck J. S., Chen W., Uher C., et al.: Thermoelectric properties of the n-type filled skutterudite BaO.3Co4Sb12, J. Appl. Phys. **91**, 3698 (2002)
29. Fedorov M. I., Pshenaj-Severin D. A., Zaitsev V. K., et al.: Features of conduction mechanism in n-type Mg2Si1-xSnx solid solutions, 22nd International Conference on Thermoelectrics ICT2003, La Grand Motte, France, Aug., p. 142 (2003)
30. Fleurial J. P., Borshchevsky A., Caillat T., et al.: High figure of merit in Ce filled skutterudites, 15th International Conference on Thermoelectrics, Pasadena, CA, 26–29 March, pp. 91–95 (1996)
31. Fleurial, J. P., Caillat, T. and Borshchevsky, A.: Skutterudites: an update, 16th International Conference on Thermoelectrics, Dresden, Germany, 26–29 Aug., pp. 1–11 (1997)
32. Gascoin, F., Rasmussen, J. and Snyder, G. J.: High temperature thermoelectric properties of Mo3Sb7-xTex x = 1.6 and 1.5, J. Alloys Compd. **427**, 324 (2007)
33. Gougeon, P., Potel, M. and Gautier, R.: Syntheses and structural, physical, and theoretical studies of the novel isostructural Mo9 cluster compounds Ag2.6CsMo9Se11, Ag4.1ClMo9Se11, and h-Mo9Se11 with tunnel structures, Inorg. Chem. **43**, 1257 (2004)

34. Grin, Y.N.: Chemistry and physics of intermetallic clathrates, 14th International Conference on Solids Compounds of Transition Elements, Linz, Austria, 6–11 July (2003)
35. He, T., Krajewski, J. J. and Subramanian, M. A.: High performance thermoelectric materials and their method of preparation, US Patent, US 2005/0229963 A1 (2005)
36. He T., Chen J., Rosenfeld H. D., et al.: Thermoelectric properties of indium filled skutterudites, Chem. Mater. **18**, 759 (2006)
37. Hébert S., Flahaut D., Pelloquin D., et al.: Cobalt oxides as potential thermoelectric elements: the influence of the dimensionality, 22nd International Conference on Thermoelectrics ICT2003, La Grand Motte, France, Aug., p. 161 (2003)
38. Hicks L. D. and Dresselhaus M. S.: Thermoelectric figure of merit of a one-dimensional conductor, Phys. Rev. B **47**, 16631 (1993)
39. Hicks L. D., Harman T. C., Sun X., et al.: Experimental study of the effect of quantum-well structures on the thermoelectric figure of merit, Phys. Rev. B **53**, R10493 (1996)
40. Hong, S. J. and Chun, B. S.: Microstructure and thermoelectric properties of extruded n-type (Bi2Te3-Bi2Se3) alloy along bar length, Mater. Sci. Eng., A **356**, 345 (2003)
41. Hsu K. F., Loo S. L., Guo F., W., et al.: Cubic AgPbmSbTe2 + m: bulk thermoelectric materials with high figure of merit, Science **303**, 818 (2004)
42. Jeitschko, W. and Braun, D. J.: LaFe4P12 with filled CoAs type structure and isotypic LnxMyPz., Acta Crystallogr. **33**, 3401 (1977)
43. Jodin L., Tobola J., Pecheur P., et al.: Effect of substitutions and defects in half-Heusler FeVSb studied by electron transport measurements and KKR-CPA electronic structure calculations, Phys. Rev. B **70**, 184207 (2004)
44. Kaspar, J. S., et al.: Intermetallic clathrates, Science **150**, 1713 (1965)
45. Katsuyama S., Schichijo Y., Ito M., et al.: Thermoelectric properties of the skutterudite Co1-xFexSb3 system, J. Appl. Phys. **84**, 6708 (1998)
46. Katsuyama S., Kanayama T., Ito M., et al.: Thermoelectric properties of CoSb3 with dispersed FeSb2 particles, J. Appl. Phys. **88**, 3484 (2000)
47. Katsuyama S., Watanabe M., Kuroki M., et al.: Effect of NiSb on the thermoelectric properties of skutterudite CoSb3, J. Appl. Phys. **93**, 2758 (2003)
48. Klemens, P. G.: Thermal resistance due to point defects at high temperature, Phys. Rev. **119**, 507 (1960)
49. Koumoto, K., Terasaki, I. and Funahashi, R.: Complex oxide materials for potential thermoelectric applications, MRS Bull. **31**, 206 (2006)
50. Kurosaki K., Kosuga A., Muta H., et al.: Ag9TlTe5: a high performance thermoelectric bulk material with extremely low thermal conductivity, Appl. Phys. Lett. **87**, 061919 (2005)
51. Kuwahara H., Hirobe Y., Nagayama J., et al.: Thermoelectric properties of filling controlled manganites, J. Magn. Magn. Mater. **272–276**, e1393 (2004)
52. Kuznetsov V. L., Kuznetsova L. A. and Rowe D. M.: Effect of partial void filling on the transport properties of NdxCo4Sb12 skutterudites, J. Phys.: Condens. Matter **15**, 5035 (2003)
53. Lamberton G. A., Bhattacharya S., Littleton R. T., et al.: High figure of merit in Eu filled CoSb3 based skutterudites, Appl. Phys. Lett. **80**, 598 (2002)
54. Leithe-Jasper A., Schnelle W., Rosner H., et al.: Ferromagnetic ordering in alkali-metal iron antimonides: NaFe4Sb12 and KFe4Sb12, Phys. Rev. Lett. **91**, 037208 (2003)
55. Lin H., Bozin E. S., Billinge S. J. L., et al.: Nanoscale clusters in the high performance thermoelectric AgPbmSbTem + 2, Phys. Rev. B **72**, 174113 (2005)
56. Masset A. C., Michel C., Maignan A., et al.: Misfit-layered cobaltite with an anisotropic giant magnetoresistance: Ca3Co4O9, Phys. Rev. B **62**, 166 (2000)
57. Mastronardi K., Young D., Wang C. C., et al.: Antimonides with the half-Heusler structure: new thermoelectric materials, Appl. Phys. Lett. **74**, 1415 (1999)
58. McGuire M. A., Schmidt A. M., Gascoin F., et al.: Thermoelectric and structural properties of a new Chevrel phase: Ti0.3Mo5RuSe8, J. Solid State Chem. (2006)
59. Mikami M. and Kobayashi K.: Effect of Bi addition on microstructure and thermoelectric properties of Heusler Fe2VAl sintered alloy, J. Alloys Compd. **466**, 530–534 (2008)
60. Mudryk Y., Rogl P., Paul C., et al.: Thermoelectricity of clathrate I type Si- and Ge-phases, J. Phys., Condens. Matter **14**, 7991–8004 (2002)

61. Nagamoto, Y., Tanaka, K. and Koyanagi, T.: Transport properties of heavily doped n-type CoSb3, 17th International Conference on Thermoelectrics, Nagoya, Japan, 24–28 May, p. 302 (1998)
62. Ni H. L., Zhao X. B., Zhu T. J., et al.: Synthesis and thermoelectric properties of Bi2Te3 based nanocomposites, J. Alloys Compd. **397**, 317 (2005)
63. Nolas, G. S., Cohn, J. L. and Slack, G. A.: Effect of partial voigt filling on the lattice thermal conductivity of skutterudites, Phys. Rev. B **58**, 164 (1998)
64. Nolas G. S., Cohn J. L., Slack G. A., et al.: Semiconducting Ge clathrates: promising for thermoelectric applications, Appl. Phys. Lett. **73**, 178 (1998)
65. Nolas G. S., Kaeser M., Littleton R. T., et al.: Partially filled skutterudites – optimizing the thermoelectric properties, MRS Spring Meeting, Boston, MA, April **626**, p. Z10.1.1 (2000)
66. Oestreich J., Probst U., Richardt F., et al.: Thermoelectric properties of the compounds ScMSb and YMSb (M = Ni, Pd, Pt), J. Phys. Condens. Matter **15**, 635 (2003)
67. Ono Y., Satoh K., Nozaki T., et al.: Structural, magnetic and thermoelectric properties of Delafossite type oxide CuCr1-xMgxO2, 24th International Conference on Thermoelectrics ICT2006, Vienna, Austria, 6–10 Aug., (2006)
68. Orikasa Y., Hayashi N. and Muranaka S.: Effects of oxygen gas pressure on structural, electrical, and thermoelectric properties of (ZnO)3In2O3 thin films deposited by rf magnetron sputtering, J. Appl. Phys. **103**, 113703 (2008)
69. Peltier, J. C.: Mémoire sur la formation des tables des rapports qu'il y a entre la force d'un courant électrique et la déviation des aiguilles des multiplicateurs; suivi de recherches sur les causes de pertubation des couples thermo-électriques et sur les moyens de s'en garantir dans leur emploi a la mesure des températures moyennes, Ann. Chim. (Paris) **1**, 371 (1834)
70. Potel M., Gougeon P., Merdrignac-Conanec O., et al.:, Meeting of the French GDR "Thermoélectricité", Juillet, Paris (2008)
71. Puyet M., Dauscher A., Lenoir B., et al.: Beneficial effect of Ni substitution on the thermoelectric properties in partially filled CayCo4–xNixSb12 skutterudites, J. Appl. Phys. **97**, 083712 (2005)
72. Sales, B. C., Mandrus, D. and Williams, R. K.: Filled skutterudite antimonides: a new class of thermoelectric materials, Science **272**, 1325 (1996)
73. Sales B. C., Mandrus D., Chakoumakos B. C., et al.: Filled skutterudite antimonides: electron crystals and phonon glasses, Phys. Rev. B **56**, 15081 (1997)
74. Sales B. C., Chakoumakos B. C., Mandrus D., et al.: Atomic displacement parameters and the lattice thermal conductivity of clathrate like thermoelectric compounds, J. Solid State Chem. **146**, 528 (1999)
75. Sales, B. C., Chakoumakos, B. C. and Mandrus, D.: Thermoelectric properties of thallium filled skutterudites, Phys. Rev. B **61**, 2475 (2000)
76. Sales, B. C., Chakoumakos, B. C. and Mandrus, D.: Connections between crystallographic data and new thermoelectric compounds, MRS Spring Meeting, Boston, MA, April **626**, p. Z7.1.1 (2000)
77. Schmidt A. M., McGuire M. A., Gascoin F., et al.: Synthesis and thermoelectric properties of $(Cu_yMo_6Se_8)_{1-x}(Mo_4Ru_2Se_8)_x(Cu_yMo_6Se_8)_{1-x}(Mo_4Ru_2Se_8)_x$ alloys, J. Alloys Compd. **431**, 262 (2007)
78. Schweika W., Hermann R. P., Prager M., et al.: Dumbbell rattling in thermoelectric zinc antimony, Phys. Rev. Lett. **99**, 125501 (2007)
79. Seebeck, T. J.: Magnetisch polarisation der metalle und erze durch temperatur differenz, Abh. Akad. Wiss. Berlin, 1820–1821, pp. 289–346 (1822)
80. Sharp J. W., Sales B. C., Mandrus D., et al.: Thermoelectric properties of Tl2SnTe5 and Tl2GeTe5, Appl. Phys. Lett. **74**, 3794 (1999)
81. Shi X., Chen L., Yang J., et al.: Enhanced thermoelectric figure of merit of CoSb3 via large defect scattering, Appl. Phys. Lett. **84**, 2301 (2004)
82. Shutoh, N. and Sakurada, S.: Thermoelectric properties of Tix(Zr,Hf)1-xNiSn half Heusler compounds, 22nd International Conference on Thermoelectrics ICT2003, La Grand Motte, France, Aug., p. 312 (2003)

83. Slack, G. A. and Tsoukala, V. G.: Some properties of semiconducting IrSb3, J. Appl. Phys. **76**, 1665 (1994)
84. Slack G.A., Thermoelectric Handbook, Ed. Rowe D.M., Chemical Rubber, Boca Raton, FL, p. 407 (1995)
85. Snyder G. J., Christensen M., Nishibori E., et al.: Disordered zinc in Zn4Sb3 with phonon-glass and electron-crystal thermoelectric properties, Nature Mater. **3**, 458 (2004)
86. Stark, D. and Snyder, G. J.: The synthesis of CaZn2Sb2 and its thermoelectric properties, 21st International Conference on Thermoelectrics, Long Beach, CA, 25–29 Aug., p. 181 (2002)
87. Tanaka Y., Ifuku T., Tsuchida K., et al.: Thermoelectric properties of ZnO based materials, J. Mater. Sci. Lett. **16**, 155 (1997)
88. Tang X., Chen L., Goto T., Hirai T.: Effects of Ce filling fraction and Fe content on the thermoelectric properties of Co rich $Ce_yFe_xCo_{4-x}Sb_{12}$, Journal of Materials Research, **16**, 3, 837 (2001)
89. Terasaki, I., Sasago, Y. and Uchinokura, K.: Large thermoelectric power in NaCo2O4 single crystals, Phys. Rev. B **56**, 12685 (1997)
90. Thompson, W.: On the dynamical theory of heat. Part VI. Thermoelectric Currents., Proceedings of the Royal Society of Edinburgh **91**, (1851)
91. Venkatasubramanian R., Sivola E., Colpitts T., et al.: Thin film thermoelectric devices with high room temperature figures of merit, Nature **413**, 597 (2001)
92. Wang, S., Snyder, G. J. and Caillat, T.: Thermoelectric properties of Nb3SbxTe7-x, 21st International Conference on Thermoelectrics, Long Beach, CA, 25–29 Aug., p. 170 (2002)
93. Wojciechowski K. T., Malecki A., Leszczynski J., et al.: Physical properties of Te doped CoSb3, 6th Europ. Workshop on Thermoelectrics, Freiberg im Breisgau, Austria, 20–21 Sept. (2001)
94. Wolfing B., Kloc C., Teubner J., et al.: High performance thermoelectric Tl9BiTe6 with an extremely low thermal conductivity, Phys. Rev. Lett. **86**, 4350 (2001)
95. Worlock J. M.: Thermal conductivity in sodium chloride containing silver colloids, Phys. Rev. **147**, 636 (1966)
96. Yamashita, O., Tomiyoshi, S. and Makita, K.: Bismuth telluride compounds with high thermoelectric figures of merit, J. Appl. Phys. **93**, 368 (2003)
97. Yamashita, O. and Tomiyoshi, S.: High performance n-type bismuth telluride with highly stable thermoelectric figure of merit, J. Appl. Phys. **95**, 6277 (2004)
98. Yang, R., Chen, G. and Dresselhaus, M. S.:Thermal conductivity of simple and tubular nanowire composites in the longitudinal direction, Phys. Rev. B **72**, 125418 (2005)
99. Zaitsev V.K., Fedorov M. I., Gurieva E. A., et al.: Highly effective Mg2Si1-xSnx thermoelectrics, Phys. Rev. B **74**, 045207 (2006)
100. Zhao X.Y., Shi X., Chen L. D., et al.: Synthesis of YbyCo4Sb12/Yb2O3 composites and their thermoelectric properties, Appl. Phys. Lett. **89**, 092121 (2006)
101. Zhao L.D., Zhang B.P., Li J.F., et al.: Thermoelectric and mechanical properties of nano-SiC-dispersed Bi2Te3 fabricated by mechanical alloying and spark plasma sintering, J. Alloys Compd. **455**, 259 (2008)

New Approaches to Thermoelectric Materials

A. P. Gonçalves, E. B. Lopes, E. Alves, N. P. Barradas, N. Franco, O. Rouleau, and C. Godart

Abstract A deeper understanding of the parameters that affect the dimensionless figure of merit, the development of new concepts and the use of innovative synthesis techniques has recently led to systems with better thermoelectric performances. Here we present part of the work that has been recently performed in our groups in order to get new and improved thermoelectric systems. Two new systems, electrical conducting glasses and doped tellurium films, are proposed as new families of thermoelectric materials.

Abbreviations

S	Seebeck coefficient
V	Electrical potential
T	Absolute temperature
I	Electrical current
Q	Heat quantity
Π	Peltier coefficient
τ	Thomson coefficient
ZT	Figure of merit
σ	Electrical conductivity
λ	Thermal conductivity

A. P. Gonçalves and E. B. Lopes
Dep. Química, Instituto Tecnológico e Nuclear/CFMC-UL, P-2686-953 Sacavém, Portugal
e-mail: apg@itn.pt, eblopes@itn.pt

E. Alves, N. P. Barradas, and N. Franco
Dep. Física, Instituto Tecnológico e Nuclear/CFN-UL, P-2686-953 Sacavém, Portugal
e-mail: ealves@itn.pt, nunoni@itn.pt, nfranco@itn.pt

O. Rouleau and C. Godart
CNRS, ICMPE, CMTR, 2/8 rue Henri Dunant, 94320 Thiais, France
e-mail: rouleau@icmpe.cnrs.fr, godart@icmpe.cnrs.fr

V. Zlatić and A. C. Hewson (eds.), *Properties and Applications of Thermoelectric Materials*,
NATO Science for Peace and Security Series B: Physics and Biophysics,

$S^2\sigma$	Power factor
λ_e	Electronic contribution to the thermal conductivity
λ_L	Lattice (phonon) contribution to the thermal conductivity
PGEC	Phonon Glass Electron Crystal
T_g	Glass temperature
T_C	Crystallization temperature
ρ	Electrical resistivity
GIXRD	Grazing incidence x-ray diffraction
SEM	Scanning electron microscopy
EDS	Energy dispersive x-ray spectroscopy
RBS	Rutherford backscattering spectroscopy
ρ_{Bi}	Electrical resistivity of bismuth doped tellurium regions

1 Introduction

The climate changes that are being observed on a planetary scale, together with the recent oil crisis, has woken up public opinion to the serious consequences of an exaggerated consumption of primary resources and increase of pollution. A better use of energy became top priority for most of the developed countries, in particular of the European Union.

Following its general environmental politics, which has as primary objective to keep the maximum increase of the average world temperature below 2°C, the European Union has recently established as main targets a 20% decrease of greenhouse effect gases emission (when compared with the 1990 levels), a 20% increase of the energy efficiency and a rise of the renewable energy sources proportion to 20% by 2020.

The research of new environment friendly energy sources and the energy consumption optimization are therefore priority targets. A huge and almost unexplored reservoir of "green" energy is the electricity generation from temperature gradients. Thermoelectric materials are able to convert directly electrical energy into thermal energy and, reversibly, thermal energy into electrical. In many applications, thermoelectric energy generation uses zero-cost input energy (waste, heat of exhaust pipe of cars, etc.). Moreover, the complete absence of moving parts and the absence of substances as fluorinated cooling agents makes thermoelectric devices highly attractive.

The actual thermoelectrical devices use materials developed until the early 1960s, mainly based on the Bi_2Te_3 and Si–Ge phases. However, their efficiency is small ($<10\%$). It is therefore fundamental to identify new systems which could lead to higher device efficiencies.

A deeper understanding of the parameters that affect the thermoelectric performance, the development of new concepts and the use of innovative synthesis techniques have recently led to the identification of materials, as skutterudites, clathrates, half-Heusler and low dimensional systems, with improved thermoelectric characteristics (see review by Godart et al. in this volume [1]). However, the efficiency increase is limited in bulk materials and the cost of low dimensional systems is very high and therefore new systems need to be identified. In the next section the fundamentals of the thermoelectric effects and figure of merit will be presented, together with the new concepts on the parameters that affect the thermoelectric performance. Section 3 is devoted to the presentation of two new systems with promising thermoelectric performances. Finally, this chapter will finish with a conclusions section.

2 The Thermoelectric Effects and Figure of Merit

The three thermoelectric effects, Seebeck, Peltier and Thomson, were discovered during the first half of the nineteenth century. The first one was discovered by Seebeck, in 1821, and consists in the formation of an electrical potential difference, ΔV, when a circuit made of two different electrical conducting materials is submitted to a temperature gradient, ΔT. The Seebeck coefficient is defined as $\mathbf{S} = \Delta\mathbf{V}/\Delta\mathbf{T}$.

Peltier effect was discovered in 1834: when an electrical current, I, is applied to the circuit there is a heat absorption, Q, in one of the junctions and its liberation on the other. The Peltier coefficient is given by $\Pi = \mathbf{Q}/\mathbf{I}$.

The third thermoelectric effect, the Thomson effect, is observed when there are simultaneously present in an electrical circuit, a temperature gradient and an electrical current. There is absorption or liberation of heat in each individual segment of the circuit, being the thermal flux given by $\mathbf{dQ/dx} = \tau\mathbf{I}\ \mathbf{dT/dx}$, where $\mathbf{x}$ represents the spatial coordinate and τ is the Thomson coefficient of the material. Thomson also show that the Seebeck and Peltier effects are related by $\Pi = \mathbf{S.T} = \mathbf{Q}/\mathbf{I}$.

The optimization of a compound or material for thermoelectrical applications mainly implies the maximization of a dimensionless number, the figure of merit ZT, that depends only on the material and is given by

$$ZT = \mathbf{S}^2\mathbf{T}\sigma/\lambda,$$

where $\mathbf{T}$ represents the absolute temperature, $\mathbf{S}$ is the Seebeck coefficient, and σ and λ represent the electrical and thermal conductivities, respectively (for a deeper examination of the efficiencies of a thermoelectric material see review by Godart et al. in this volume [1]). ZT maximization can be done via both the maximization of the numerator, $\mathbf{S}^2\sigma$ (also called power factor), and the minimization of the denominator (thermal conductivity).

Figure 1 shows the Seebeck coefficient, electrical conductivity and power factor as a function of the charge carriers concentration on a logarithmic scale. The Seebeck coefficient decreases with the increasing concentration, whereas the electrical conductivity increases. Therefore, a maximum on the power factor is observed for concentrations of $\sim 10^{18}$–10^{21} carriers/cm^3, which correspond to low gap semiconductors or semimetals. The identification of Bi_2Te_3-based materials, which

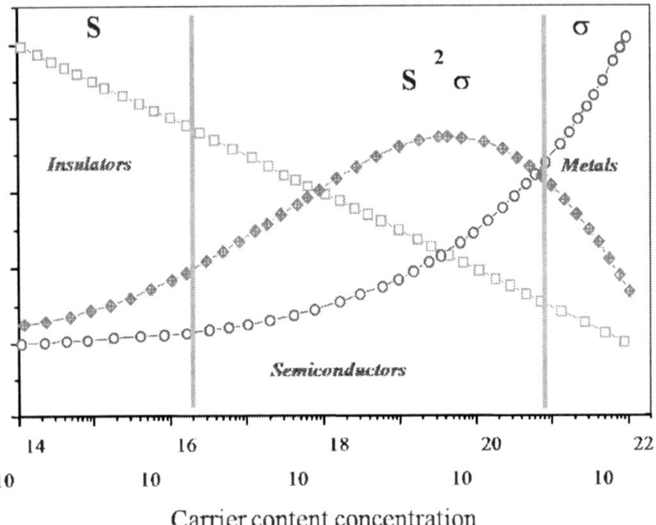

Fig. 1 Maximization of the power factor, $\mathbf{S^2\sigma}$, through carrier concentration tuning.

have $ZT \sim 1$ and are one of the most used materials for commercial thermoelectrical devices, followed this criterion.

The thermal conductivity is a second factor that must be optimized in order to maximize ZT: a material will increase its thermoelectric performance with the decrease of its thermal conductivity.

The thermal conductivity can be seen as a sum of two main different contributions, $\lambda = \lambda_e + \lambda_L$, where λ_e represents the electronic contribution and λ_L is the contribution from the lattice vibrations (phonons). The electronic contribution to thermal conductivity is related with the electrical conductivity via the Wiedemann–Franz law, $\lambda_e = \mathbf{LT}\sigma$, where $\mathbf{L}$ represents the Lorentz factor. Therefore, ZT can be rewritten as

$$ZT = \frac{S^2}{L}\frac{\lambda_e}{\lambda_e + \lambda_L}$$

being evident that its maximization implies the minimization of the phonons contribution.

At the beginning of the 1990s Slack presented the new concept of "Phonon Glass Electron Crystal" (PGEC) [2], which proposes the research of compounds that conduct the electricity as a crystalline material and the heat as a glass. Based on this concept new improved thermoelectric materials, as skutterudites and clathrates, have been identified. The research of new thermoelectric systems based on the PGEC concept has lead to an intense effort to understand better the mechanisms that affect the phonons propagation without changing significantly the electrical charge propagation. Some general rules have been identified from these works, the most important ones being presented in Table 1. It must be stressed that these rules are not exclusive and several can exist simultaneously in the same material.

Table 1 Main general approaches to develop improved thermoelectrical systems by using the "Phonon Glass Electron Crystal" concept.

Approach	Effects on phonons	Materials (examples)
Heavy atoms weakly bounded to the structures	Phonon-scattering centers	Skutterudites, clathrates
Complex structures	Increase the optical phonon modes	Clathrates, $Yb_{14}MnSb_{11}$
Inclusions, impurities	Increase diffusion (affects more phonons than carriers)	Composites
Solid solutions	Increase mass fluctuations (higher phonon scattering)	Half-Heusler systems
Grain boundaries	Reduce the phonons mean free path	Low dimensional systems

3 New Systems

3.1 Conducting Glasses

A careful analysis of the main general approaches recently developed to obtain improved thermoelectric systems shows that the new materials must have complex structures, including the presence of inclusions and impurities, and should have mass fluctuations and disorder. A type of materials that follows all these principles are the glasses. Indeed, they have extremely complex structures, with a certain degree of order only at small distances, and present mass fluctuations, easily allowing high concentrations of inclusions and impurities.

Metallic glasses have been widely studied in the 1970s–1980s. The reported electrical conductivity and Seebeck coefficient values indicate that these types of materials are not suitable for thermoelectrical applications: in fact, they show Seebeck coefficients values typical of metals, ranging between $\pm 5\ \mu V/K$ [3–7]. Moreover, their electrical conductivity is smaller than the crystalline counterparts due to the intrinsic disorder. However, it is known that the maximum of the power factor is observed for low gap semiconductors and semimetals, and not for metals. Therefore, to identify glasses with improved thermoelectric performances it is necessary to center the investigation on the low gap semiconductor and semimetal glasses.

Numerous semiconducting glasses have been reported in the literature (see, for instance, [8, 9]), most of them based on pnictides and chalcogenides. One in particular, $Ge_{20}Te_{80}$, is based only on two elements, is formed mainly by heavy atoms, and has been described as easy to prepare [10]. Moreover, its electrical transport properties have already been studied, being reported as having high Seebeck coefficients, albeit presenting small electrical conductivity values [11]. Previous works have also shown that doping it with copper or silver can increase dramatically the

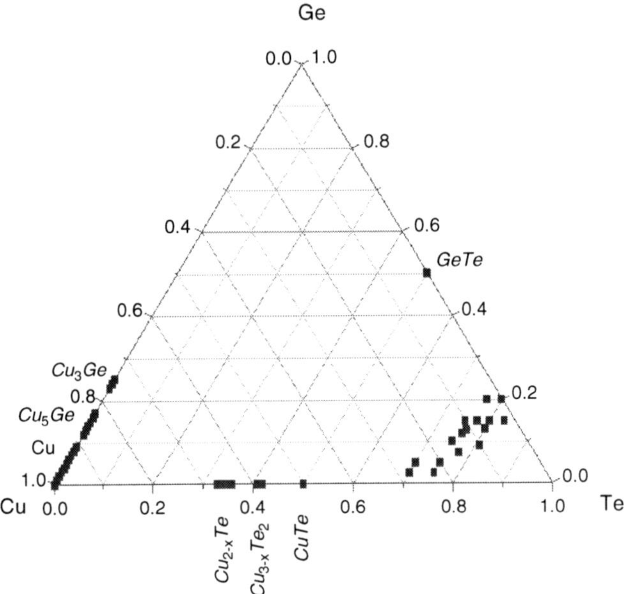

Fig. 2 $Cu_xGe_yTe_z$ general composition glasses (*black squares*) studied in the work presented in [14].

electrical conductivity [12,13]. For all these reasons we decided to use the $Ge_{20}Te_{80}$ glass as a starting test material, in order to check the possibility of optimizing the thermoelectrical performance of a glass by changing its composition [14].

Several samples with $Cu_xGe_yTe_z$ general compositions close to $Ge_{20-x}Te_{80-y}Cu_{x+y}$ (Fig. 2), together with the $Cu_{25}T_5Te_{70}$ (T = Si, Ga) ones, have been prepared by melt spinning, as described elsewhere [14]. The quality of the samples was checked by powder x-ray diffraction and differential thermal analysis measurements, together with optical microscope observations.

The materials have in general a glassy aspect (Fig. 3a), the disordered state being confirmed by the x-ray diffraction results for most of the samples (Fig. 4) [14]. However, it has been observed that some materials with higher copper concentrations frequently present regions where crystallization already starts (Fig. 3b), most probably being a sign of inhomogeneities on the cooling. X-ray diffractograms, made on powders representative of the totality of the sample, also reflect this reality, many of the higher copper concentration samples showing small crystallization peaks. The extreme composition $Cu_{30}Te_{70}$, albeit still presenting some disorder, is already crystallized and its electrical transport properties were not studied. The higher facility for samples with larger copper concentration to crystallize can be also deduced from the differential thermal analysis measurements: a decrease of the glass and crystallization temperatures are observed for the doped materials, when compared with the $Ge_{20}Te_{80}$ (Fig. 5). Therefore, it can be concluded that there is a decrease of the glass stability with the increase of the copper concentration, which implies higher

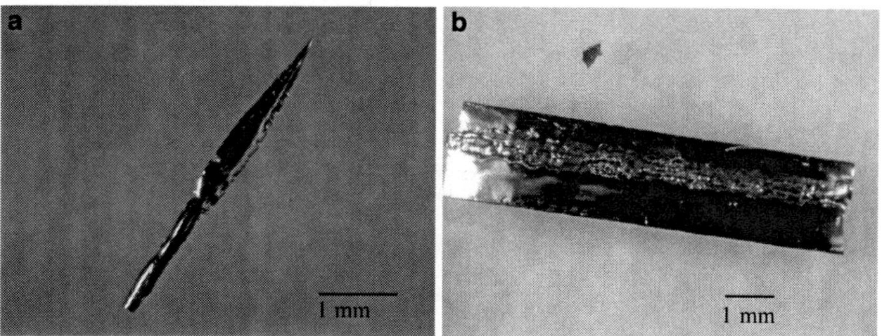

Fig. 3 Optical micrographs of the $Cu_{15}Ge_{7.5}Te_{77.5}$ (**a**) and $Cu_{22.5}Ge_{2.5}Te_{75}$ (**b**) melt-spinning samples.

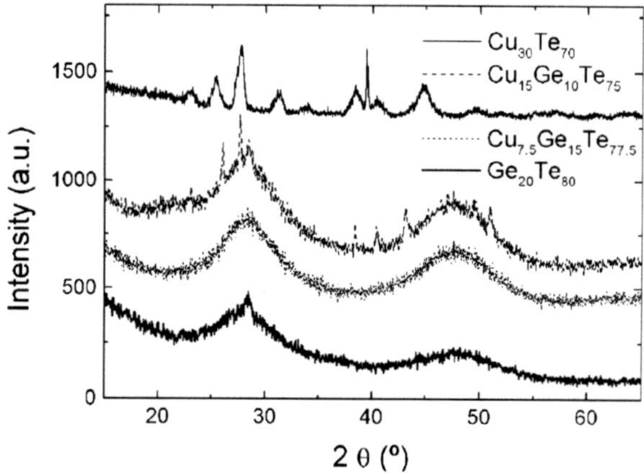

Fig. 4 Powder x-ray diffractograms of the $Ge_{20}Te_{80}$, $Cu_{7.5}Ge_{15}Te_{77.5}$, $Cu_{15}Ge_{10}Te_{75}$ and $Cu_{30}Te_{70}$samples.

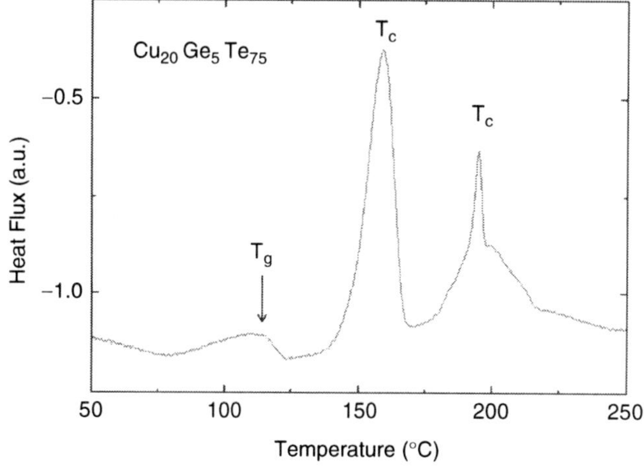

Fig. 5 Differential scanning calorimetric thermogram of the $Cu_{20}Ge_5Te_{74}$ glass sample. T_g and T_C represent, respectively, the glass and crystallization temperatures.

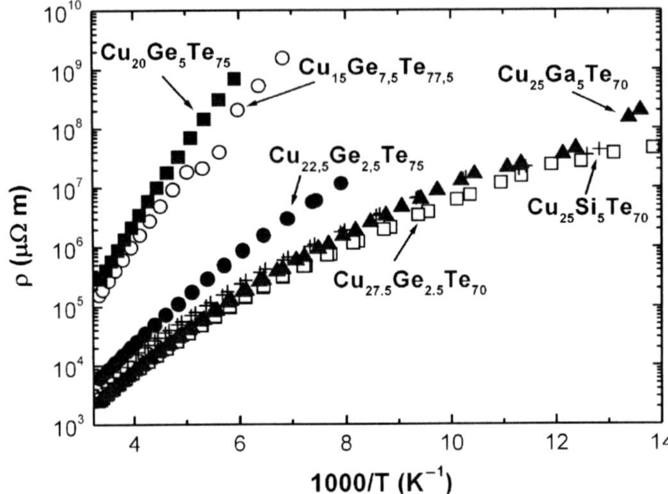

Fig. 6 Temperature dependence of the electrical resistivity for the $Ge_{20}Te_{80}$-based glasses.

cooling rates to obtain the disordered state. Moreover, the low glass transition temperature limits their use to temperatures close to the ambient. For all these reasons, the study of the effect of glass stabilizers addition on $Ge_{20}Te_{80}$-based glasses needs to be performed.

The thermoelectric performance of $Ge_{20}Te_{80}$-based conducting glasses was tested by measuring their Seebeck coefficients and electrical resistivities as a function of the temperature [14]. Special care was taken in order to make the measurements on purely glassy pieces, preferably with a needle-like shape.

All samples show a semiconductor behavior, with the electrical resistivity increasing with the decreasing temperature (Fig. 6). A strong dependence of the resistivity with the composition is observed. The analysis of the influence of the different components concentration in the electrical resistivity (Table 2) indicates that copper concentration is the main factor: a decrease on the room temperature electrical resistivity of five orders of magnitude can be observed when the composition changes from $Ge_{20}Te_{80}$ ($\sim 3 \times 10^8$ μΩm [11]) to $Cu_{27.5}Ge_{2.5}Te_{70}$ ($\sim 3 \times 10^3$ μΩm) [14]. Moreover, a decrease of the energy gap from ~0.47 to ~0.13 eV is also observed when changing from $Ge_{20}Te_{80}$ to $Cu_{27.5}Ge_{2.5}Te_{70}$ [11, 14].

Given the strong dependence of the electrical conductivity with composition one could also expect a corresponding huge change of the Seebeck coefficient. However, only a decrease from 960 μV/K to 394 mV/K has been observed for the extreme compositions ($Ge_{20}Te_{80}$ and $Cu_{27.5}Ge_{2.5}Te_{70}$). Furthermore, the measurements of the Seebeck coefficient as a function of temperature (Fig. 7) show an almost constant variation, particularly for the higher copper concentration samples.

The large decrease of the electrical resistivity together with low drop of the Seebeck coefficient, observed during the $Ge_{20}Te_{80}$-based conducting glasses composition optimization, lead to a huge increase of the power factor: the $Ge_{20}Te_{80}$

Table 2 Electrical transport properties of the $Ge_{20}Te_{80}$-based glasses.

Glass composition	$\rho_{300\,K}$ ($\mu\Omega$m)	$E_{a(High\ T)}$ (meV)	$S_{300\,K}$ (μV/K)	S^2/ρ (μW/K^{2}m)	Reference
$Ge_{20}Te_{80}$	2.77×10^8	470	960	3.3×10^{-3}	[11]
$Cu_{15}Ge_{7.5}Te_{77.5}$	1.6×10^5	244	540	2	[14]
$Cu_{20}Ge_5Te_{75}$	2.9×10^6	263	453	1×10^{-1}	[14]
$Cu_{22.5}Ge_{2.5}Te_{75}$	6×10^3	164	415	29	[14]
$Cu_{27.5}Ge_{2.5}Te_{70}$	2.5×10^3	126	394	62	[14]
$Cu_{25}Si_5Te_{70}$	5.2×10^3	125	357	25	[15]
$Cu_{25}Ga_5Te_{70}$	2.5×10^3	134	344	47	[15]

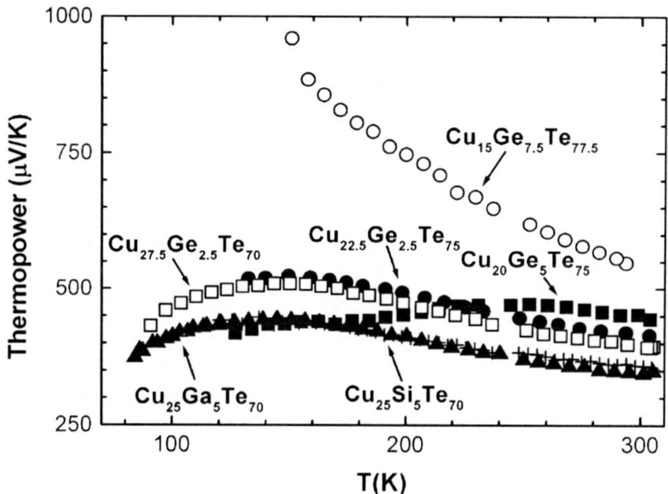

Fig. 7 Variation of the Seebeck coefficient as a function of the temperature for the $Ge_{20}Te_{80}$-based glasses.

presents a value of 3×10^{-3} μW/K^{2}m, five orders of magnitude lower than the $Cu_{27.5}Ge_{2.5}Te_{70}$, which has a value higher than 6×10^1 μW/K^{2}m. The very low values of thermal conductivity observed in this type of glasses (<0.2 W/Km for $Ge_{20}Te_{80}$ [14]) lead to ZT > 0.1 at 300 K for the $Cu_{27.5}Ge_{2.5}Te_{70}$ sample, making them interesting materials for thermoelectric applications.

The work described was only a first study of the possibility of using conducting glasses for thermoelectrical applications. Much work on the exploration of other systems and optimization of compositions is still to do, in order to get the proper materials. However, this study clearly indicates that conducting glasses are very promising materials and a strong efforts should be made to investigate their thermoelectric properties.

3.2 Bismuth Doped Tellurium Films

Low dimensional systems are actually considered one of the most important types of thermoelectric materials. Indeed, *ZT* values as high as 2.4 have already been reported in the literature for Bi_2Te_3/Sb_2Te_3 superlattice devices [16]. However, they are difficult to prepare, frequently not reproducible, and their preparation costs are very high. Therefore, the identification of new, simple and easy-to-prepare systems is also needed for these kinds of materials.

Many of the most promising recently discovered thermoelectric materials have tellurium as one of the main constituents [1]. This element has been widely studied, mainly in Japan in the late 1940s and early 1950s, for semiconductor applications, its thermoelectric properties being also explored [17–23].

Tellurium is an element that has a small value of thermal conductivity at room temperature and ambient pressure ($\sim$2 W m^{-1} K^{-1}) [24], and presents a semiconductor behavior, with an energy gap of $\sim$340meV) [21, 25]. Moreover, its Seebeck coefficient was reported to be high, albeit very sensitive to impurity concentrations [22]. The electrical conductivity of pure tellurium crystals is low, but can be increased significantly with doping [22], making it a good candidate for thermoelectrical applications if higher power factor values are obtained.

Studies on tellurium low dimensional systems, in particular on films, have been performed from the beginning of the last century [17, 26–28]. However, non-reproductive behavior was observed on tellurium thin films, which was later identified as being due to the absorption of atoms and molecules at the surface [29–31]. Only for thicknesses greater than 180 nm does the effect of gas absorption become negligible [29]. Doping effect studies on tellurium films are scarce [32, 33] and, to the authors' best knowledge, no work has been made with bismuth in low dimensional systems. Moreover, only one work was recently reported on bulk materials, with the carrier concentration significantly increasing with the bismuth doping content [34]. On the other hand, several studies have been made on tellurium-doped bismuth materials [35–38]. Most probably, this discrepancy is due to the small solubility limit of bismuth on tellurium ($\sim$0.1 atom %) [34].

One way of overcoming the solubility limits imposed by thermodynamic rules can be achieved using ion implantation by introducing bismuth ions into the tellurium. Ion implantation is a well established technique in the semiconductor field and allows the incorporation of any element in a matrix to concentrations well above the equilibrium values [39]. Therefore, we decided to use the ion implantation technique to prepare films with high bismuth doping concentrations and study their effect on the electrical transport properties. Additionally, being a ballistic process, ion implantation also increases the structural disorder, with a consequent decrease of the thermal conductivity, as observed in Bi_xTe_3/Sb_2Te_3 multilayers implanted with silicon [40, 41].

The films were prepared by evaporation of high purity tellurium (99.999%, metal basis) on glass substrates and under a pressure of $<10^{-6}$ mbar, as describe elsewhere [42]. Their thicknesses were controlled during deposition in order to be insensitive to the atmospheric gases and Rutherford Backscattering Spectrometry

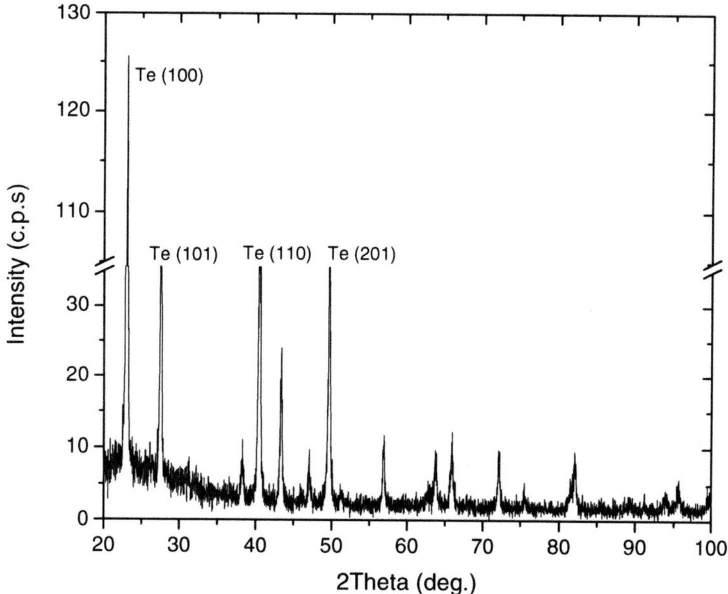

Fig. 8 Grazing incidence x-ray diffractogram of a deposited tellurium film with the major peaks labeled.

(RBS) analysis of the evaporated films confirms an average thickness of 500 nm. The quality of the films was checked by grazing incidence x-ray diffraction (GIXRD) and by scanning electron microscopy (SEM) observation, and their composition homogeneity was studied by energy dispersive x-ray spectroscopy (EDS).

Figure 8 shows a typical x-ray diffractogram of the deposited tellurium films. All peaks can be indexed to the ambient pressure tellurium trigonal crystal structure [43], albeit their intensities indicate some texture with a preferential orientation along the *a* axis.

The tellurium films were implanted with bismuth with fluences ranging from 5×10^{15} to 2×10^{16} Bi^{+}/cm^{2}. The energy was 170 keV corresponding to an implantation range of 37 nm, according the theoretical predictions of the simulation code SRIM2006 [44].

After implantation GIXRD measurements of films show no structural changes, with the diffractograms identical to the undoped ones and with all peaks belonging to the tellurium trigonal crystal structure. SEM/EDS observations of the implanted films point to a homogeneous distribution of the bismuth atoms, with no signs of visible precipitates. In any case we cannot rule out the presence of precipitates with dimensions below the detection limits of the technique. In fact the presence of pure Bi or Bi_2Te_3 precipitates in the nanometer range is a possibility to consider since the maximum concentration of Bi in the Te films highly exceeds the solubility limit.

The thickness of the implanted Te layer after the implantation and the concentration profile of the Bi were measured by RBS. The spectrum obtained for the film

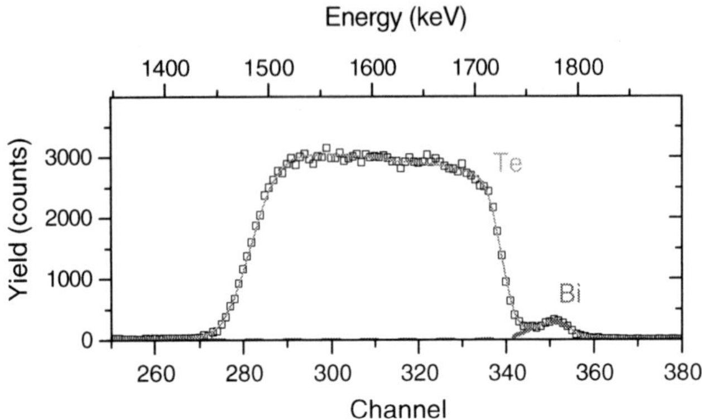

Fig. 9 Experimental RBS spectrum of a tellurium film implanted with nominally 1×10^{16} Bi^{+}/cm^{2} implantation fluence (*open squares*) and calculated curve by assuming a Gaussian implantation profile. The measured fluence is 8.6×10^{15} Bi^{+}/cm^{2}.

implanted with $1 \times 10^{16}\,cm^{-2}$ is shown in Fig. 9. The results reveal a reduction of about 50% of the thickness of the films implanted with the highest fluences indicating that a significant sputtering effect occurred during the implantation. Also the measured fluences show a decrease of 5–10% relative to the nominal values compatible with a self-sputtering process. The Bi profile, due to the sputtering process, extends from the surface up to ∼70 nm with a maximum around 20 nm (Fig. 10). The maximum concentration values measured for the implanted fluences were in the range of 3–8 atom % in the implanted regions (Fig. 10). In order to have a better control of the implantation profile further studies are necessary to establish the implantation conditions that minimize the sputtering yield. This a relevant parameter considering the low surface binding energy (2.02 eV) of Te responsible for the high sputtering yields determined.

Even though the sputtering effects limit the maximum concentration attainable of Bi in Te by ion implantation, the results clearly show that values much higher than the ∼0.1 atom % reported for an equilibrium situation [34] are possible, confirming that this technique is suitable for doping tellurium with large bismuth concentrations.

The electrical transport properties of the films were studied by measuring the electrical resistivity and Seebeck coefficient variation as a function of the temperature.

A change from a semiconductor to a semimetallic behavior can be seen with the bismuth doping (Fig. 11). Annealing the implanted films at 200°C for 10 min recovers the semiconductor behavior, which is most probably due to the formation of Bi_2Te_3 precipitates. However, this point still needs further clarification. An energy gap of 56 meV (Table 3) is obtained for the unimplanted tellurium film, much lower than what would be expected for this element (∼340 meV). The annealed films also show energy gaps of the same order of magnitude as the undoped sample. These

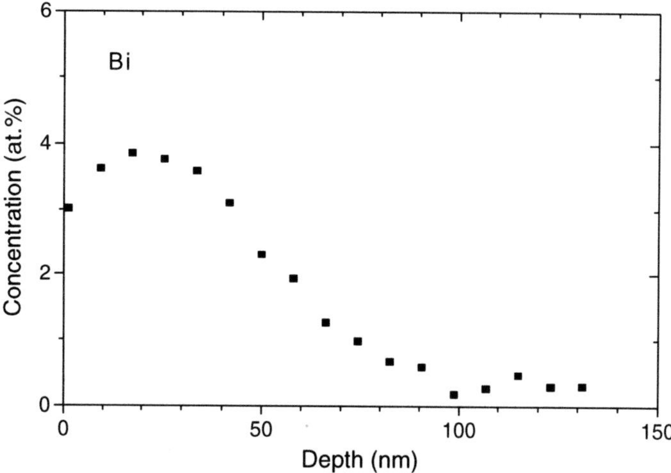

Fig. 10 Concentration profile of bismuth in a Te film implanted with a fluence of $1 \times 10^{16}\,\mathrm{cm}^{-2}$.

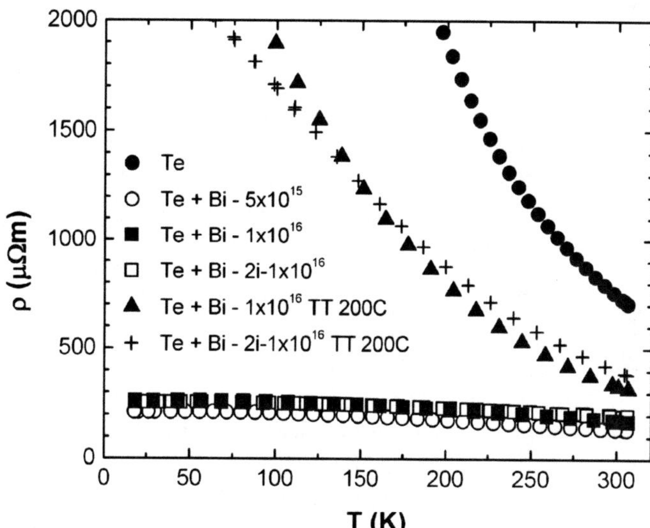

Fig. 11 Electrical resistivity of tellurium films as a function of temperature (TT 200° C represent implanted films after heat treatments at 200° C for 10 min).

lower calculated values possibly indicate that the tellurium used in the films preparation had already some impurities.

It is important to emphasize that bismuth exists only in a limited volume of the tellurium film, close to the surface and, consequently, the samples can be considered as a double layer. Therefore, a more careful analysis of the resistivity data implies the consideration of the samples as two resistances in parallel, which correspond to pure and bismuth doped tellurium regions (ρ_{Bi}). By using this approximation

Table 3 Electrical transport properties of the bismuth doped tellurium films.

Film type	ρ_{Bi300K} ($\mu\Omega$m)	E_a (meV)	S_{300K} (μV/K)	S^2/ρ (μW K^{-2} m^{-1})
Pure Te	738	56	187	47
Bi impl. 5×10^{15} cm^{-2}	28	–	188	1262
Bi impl. 1×10^{16} cm^{-2}	26	–	145	809
Bi impl 2×10^{16} cm^{-2}	55	–	29	15
Bi impl. 1×10^{16} cm^{-2} HT 200°C	66	53	147	327
Bi impl 2×10^{16} cm^{-2} HT 200°C	152	50	172	194

it is possible to calculate more accurate resistivity values for the bismuth doped tellurium regions (Table 3) than the average values obtained if the total thicknesses are considered. A significant decrease of the resistivity at room temperature (of more than one order of magnitude) can be seen in the bismuth doped tellurium regions, when compared with pure tellurium. The annealing increases the resistivity values, albeit not recovering the pure tellurium ones.

Measurements of the Seebeck coefficient as a function of the temperature indicate that the conduction is mainly hole-type for all films (Fig. 12). Bismuth doping decreases the room temperature Seebeck coefficient by ~25%, from ~200 μV/K, for the unimplanted and 5×10^{15} cm^{-2} implanted film, to ~150μ V/K for $1-2 \times 10^{16}$ cm^{-2} implanted films. These results are in agreement with the presence of small amounts of impurities already in the undoped films, as a Seebeck value of −200 μ V/K has been reported for high-purity bulk tellurium, together with an extreme sensitivity of thermopower to doping [22].

In all films is observed a linear decrease of the Seebeck coefficient with the decreasing temperature. Albeit they present similar values, there is a change from a convex to a concave curve with implantation, the annealing partially recovering the convex curvature, in agreement with the complex band structure of tellurium [25].

Thermoelectric power is mainly sensitive to the less electrical resistive part of the samples and, therefore, the measured Seebeck coefficients are mostly coming from those regions. In the case of tellurium films, and considering the electrical resistivity differences in the bismuth doped and undoped regions, this means that the measured Seebeck coefficients mainly come from the contribution of the bismuth doped tellurium regions, and consequently, as a first approximation, can be considered as only coming from those regions. As a result, a power factor of those regions can be roughly estimated if the calculated electrical resistivity of the bismuth doped tellurium regions and the measured Seebeck coefficients are considered.

An increase of almost two orders of magnitude in the power factor of the best doped film (bismuth implanted with 5×10^{15} cm^{-2} fluence) is observed, when compared with the pure tellurium film (Table 3). The best doped film has a power factor of ~1,260 μ W K^{-2} m^{-1}, which is already close to those obtained on the Bi_2Te_3 based phases (2,000 – 4,000 μ W K^{-2} m^{-1}) and points to a good possibility of obtaining high performance thermoelectric materials by implantation (if the thermal conductivity of bulk tellurium is considered, a ZT value of ~0.2 is already obtained

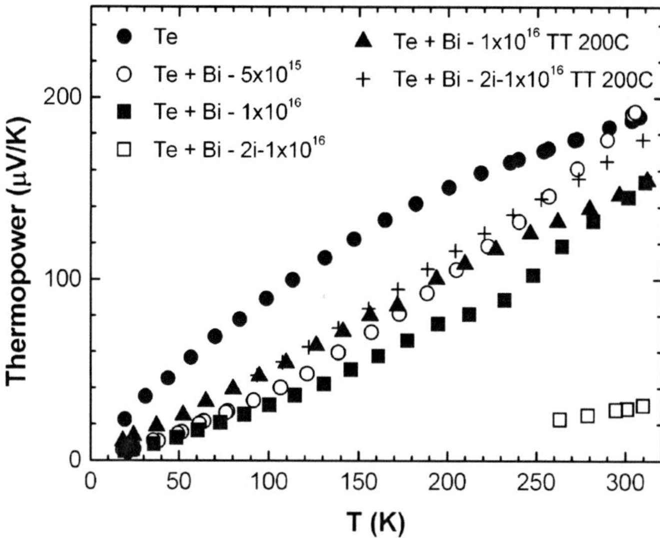

Fig. 12 Seebeck coefficients of the tellurium films, as a function of temperature (TT 200° C represent implanted films after heat treatments at 200° C for 10 min).

at 300 K for the best doped film). However, a better understanding and optimization of the implantation process is still need in order to get bismuth doped tellurium films with thermoelectric performances suitable to practical applications.

4 Conclusions

The deeper understanding of the parameters that affect the thermoelectric performance, together with the development of new concepts and the use of novel synthesis techniques, have recently led to the identification of materials with improved thermoelectric characteristics. Nevertheless, there are still strong limitations on those materials and the discovery and investigation of new systems are highly desirable.

The identification and use of general rules based on the "Phonon Glass Electron Crystal" concept, which affect the phonons propagation without changing significantly the electrical charge propagation, can lead to the development of new improved thermoelectrical systems. Here were presented two new systems, electrical conducting glasses and doped tellurium films, which we think that should be studied more deeply. Albeit not yet optimized, both show interesting thermoelectric performances, with room temperature ZT values higher than 0.1. However, much work is still to do on these two new systems, from the development of *n*- and *p*-type conducting materials and increasing of the thermoelectric performances, to the enhancement of the glass stabilization and the optimization of the implantation conditions, before they could be considered as high-efficiency thermoelectric materials with practical use. The rising of the global need for sustainable energy should encourage not only the optimization of the already identified promising systems, but also the development of new families with better thermoelectric properties.

Acknowledgments This work was partially supported by the bilateral French–Portuguese GRICES/CNRS 2007–2008 program, European COST P16 program and FCT, Portugal, under contract nr. PTDC/QUI/65369/2006. APG and CG acknowledge NATO ARW support for their participation to ARW Workshop "Properties and Applications of Thermoelectric Materials" September 20–26, 2008, Hvar, Croatia.

References

1. Godart, C., Gonçalves, A.P., Lopes, E.B. et al., this volume, 19–49.
2. Slack, G.: New materials and performance limits for thermoelectric cooling. In: Rowe D.M. (ed.) CRC Handbook of Thermoelectrics, pp. 407–440. CRC, Boca Raton, FL (1995)
3. Nagel, S.R.: Thermoelectric-power and resistivity in a metallic glass. Phys. Rev. Lett. **41**, 990–993 (1978)
4. Teoh, N., Teoh, W., Arajs, S. et al.: Absolute thermoelectric-power of amorphous metallic glass $Fe_{80}B_{20}$ between 300-K and 1000-K. Phys. Rev. B **18**, 2666–2667 (1978)
5. Carini, J.P., Basak, S., Nagel, S.R. et al.: The thermoelectric-power of the metallic glass $Ca_{0.8}Al_{0.2}$. Phys. Lett. A **81**, 525–526 (1981)
6. Pekala, K., Pekala, M., Trykozko, R.: Magnetic thermoelectric-power of $Fe_{20}Ni_{60}B_{10}Si_{10}$ metallic-glass. Solid State Commun. **46**, 413–415 (1983)
7. Bhatnagar, A.K., Prasad, B.B., Rathnam, N.R.M.: Magnetic, electric and thermoelectric studies on metallic-glass $Fe_{39}Ni_{39}Mo_4Si_6B_{12}$. J. Non-Cryst. Solids **61–2**, 1201–1206 (1984)
8. Petersen, K.E., Birkholz, U., Adler, D.: Properties of crystalline and amorphous silicon telluride. Phys. Rev. B **8**, 1453–1461 (1973)
9. Sidorov, V.A., Brazhkin, V.V., Khvostantsev, L.G. et al.: Nature of semiconductor-to-metal transition and volume properties of bulk tetrahedral amorphous GaSb and GaSb-Ge semiconductors under high pressure. Phys. Rev. Lett. **73**, 3262–3265 (1994)
10. El-Oyoun, M.A.: A study of the crystallization kinetics of $Ge_{20}Te_{80}$ chalcogenide glass. J. Phys. D: Appl. Phys. **33**, 2211–2217 (2000)
11. Parthasarathy, G., Bandyopadhyay, A.K., Asokan, S. et al.: Effect of pressure on the electrical-resistivity of bulk $Ge_{20}Te_{80}$ glass. Solid State Commun. **51**, 195–197 (1984)
12. Ferhat, A., Ollitrault-Fichet, R., Mastelaro, V. et al.: Etude des verres du système Ag-Ge-Te. J. de Physique IV **2**, C2–201-C2–206 (1992)
13. Ramesh, K., Asokan, S., Sangunni, K.S. et al.: Compositional dependence of high pressure resistivity behaviour of Cu-Ge-Te glasses. Phys. Chem. Glasses **37**, 217–220 (1996)
14. Gonçalves, A.P., Lopes, E.B., Rouleau, O. et al.: Conducting glasses as new potential thermoelectric materials: the Cu-Ge-Te case. Submitted to J. Mater. Chem.
15. Gonçalves, A.P., Lopes, E.B., Rouleau, O. et al.: to be published.
16. Venkatasubramanian, R., Siivola, E., Colpitts, T. et al.: Thin-film thermoelectric devices with high room-temperature figures of merit. Nature **413**, 597–602 (2001)
17. Fukuroi, T., Tanuma, S., Tobisawa, S.: Electric resistance, Hall effect, magneto-resistance and Seebeck effect in pure tellurium film. Sci. Rep. Res. **A1**, 365–372 (1949)
18. Fukuroi, T., Tanuma, S., Tobisawa, S.: On the electro-magnetic properties of single crystals of tellurium. I electrical resistance, Hall effect, magneto-resistance and thermo-electric power. Sci. Rep. Res. **A1**, 373–386 (1949)

19. Fukuroi, T., Tanuma, S., Tobisawa, S.: On the electro-magnetic properties of single crystals of tellurium. II Ettingshausen–Nernst effect. Sci. Rep. Res. **A2**, 233–238 (1950)
20. Fukuroi, T., Tanuma, S., Tobisawa, S.: On the electro-magnetic properties of single crystals of tellurium. III Adiabatic and isothermal Hall effect, and Ettingshausen effect. Sci. Rep. Res. **A2**, 239–248 (1950)
21. Fukuroi, T.: On the width of forbidden energy zone of tellurium. Sci. Rep. Res. **A3**, 175–181 (1951)
22. Fukuroi, T., Tanuma, S., Tobisawa, S.: Electrical properties of antimony-doped tellurium crystals. Sci. Rep. Res. **A4**, 283–297 (1952)
23. Fukuroi, T., Tanuma, S., Yoshio, M.: Electrical properties of tellurium crystals at very low temperatures. Sci. Rep. Res. **A5**, 18–29 (1953)
24. Fischer, G., White, G.K., Woods, S.B.: Thermal and electrical resistivity of tellurium at low temperatures. Phys. Rev. **106**, 480–483 (1957)
25. Deng, S., Köhler, J., Simon, A.: The flat/steep band condition created in Te-II. Physica C **460–462**, 1020–1021 (2007)
26. Warburton, F.W.: The Hall effect and resistance in sputtered tellurium films. Phys. Rev. **30**, 673–680 (1927)
27. Sakurai, T., Munesue, S.: Resistivity of evaporated tellurium films, Phys. Rev. **85**, 921–921 (1952)
28. Ghosh, S.K.: Variation of field effect mobility and Hall effect mobility with the thickness of the deposited films of tellurium. J. Phys. Chem. Solids **19**, 61–65 (1961)
29. Silbermann, R., Landwehr, G., Köhler, H.: Field effect in tellurium. Solid State Commun. **9**, 949–951 (1971)
30. Tsiulyanu, D., Tsiulyanu, A., Liess, H.-D. et al.: Characterization of tellurium-based films for NO_2 detection. Thin Solid Films **485**, 252–256 (2005)
31. Tsiulyanu, D., Stratan, I., Tsiulyanu, A. et al.: Sensing properties of tellurium based thin films to oxygen, nitrogen and water vapour. 2006 International Semiconductor Conference, Vol. **2**, pp. 287–290, Sinaia, (2006)
32. Mathur, P.C., Dawar, A.L., Taneja, O.P.: Electrical transport properties of copper-doped tellurium films. Thin Solid Films **66**, 281–285 (1980)
33. Rao, L.V., Naidu, B.S., Reddy, P.J.: Electrical-conductivity and thermopower measurements on silver doped tellurium-films. Phys. Status Solidi A **65**, K135–K138 (1981)
34. Orlov, A.M., Gonchar, L.I., Salanov, A.A.: Solubility limit and thermoelectric properties of bismuth-supersaturated tellurium. Tech. Phys. Lett. **32**, 38–41 (2006)
35. Morimoto, T.: Heavy electrons in tellurium-doped bismuth. J. Phys. Soc. Japan **21**, 1008 (1966)
36. Vangoor, J.M.N.: Hall coefficients of tellurium doped bismuth. Phys. Lett. A **25**, 442 (1967)
37. Heremans, J., Morelli, D.T., Partin, D.L. et al.: Properties of tellurium-doped epitaxial bismuth-films. Phys. Rev. B **38**, 10280–10284 (1988)
38. Red'ko, N.A., Kagan, V.D., Rodionov, N.A.: Low-temperature thermal conductivity of tellurium-doped bismuth. Phys. Solid State **47**, 416–423 (2005)
39. Ryssel, H., Ruge, I.: Ion Implantation. Wiley, New York (1986)
40. Xiao, Z., Zimmerman, R.L., Holland, L.R., et al.: MeV Si ion bombardments of thermoelectric Bi_xTe_3/Sb_2Te_3 multilayer thin films for reducing thermal conductivity. Nucl. Instr. Meth. Phys. Res. B **241**, 568–572 (2005)
41. Xiao, Z., Zimmerman, R.L., Holland, L.R., et al.: Nanoscale Bi_xTe_3/Sb_2Te_3 multilayer thin film materials for reduced thermal conductivity. Nucl. Instr. Meth. Phys. Res. B 242, 201–204 (2006)
42. Lopes, E.B., Gonçalves, A.P., Alves, E. et al.: to be published
43. Adenis, C., Langer, V., Lindqvist, O.: Reinvestigation of the structure of Tellurium. Acta Cryst. C 45, 941–942 (1989)
44. Ziegler, J.F., Biersack, J.P., Littmark, U: The Stopping and Range of Ions in Matter, Vol. 1. IBM Research, New York (2006)

Thermoelectric Effect in Transition Metal Oxides

W. Koshibae and S. Maekawa

1 Introduction

Thermoelectric materials generate electric energy from heat without pollution. The potential applications include the utilization of waste heat and power sources for deep-space probes etc.

Terasaki et al. [1] have shown the unique thermoelectric response in the layered-hexagonal oxide $NaCo_2O_4$. The in-plane resistivity of the oxide shows metallic behavior; the electrical resistivity (ρ) at 300 K is 200 $\mu\Omega$cm, i.e., that of a good metal. Although the thermopower of good metals is usually small, this oxide shows a large thermopower(Q), increasing with temperature to reach 100 μV/K at 300 K.

A real advantage of the cobalt oxide is the large thermopower at higher temperatures. Large high-temperature thermopower has also been observed in many related compounds [2–7]. The thermoelectric performance of a material is quantified by the figure of merit $ZT(= Q^2T/\rho\kappa$ where T and κ are temperature and thermal conductivity, respectively). The maximum ZT in sintered samples of $NaCo_2O_4$ is around $\sim$1,000 K [8], and the value $ZT \geq 1$ has been measured [9] at $T \geq 873$ K for layered-hexagonal oxide $Ca_2CO_2O_5$.

Traditionally semiconductors have been useful as thermoelectric materials. The electronic structure of semiconductors is well described by the band theory. The design of conventional thermoelectric materials is made by compromising between low electrical resistivity and large thermopower. Both depend on carrier density with the thermopower increasing logarithmically with decreasing carrier density. However, a low carrier density also implies a larger resistivity. The optimum carrier

W. Koshibae
Cross-Correlated Materials Research Group (CMRG), RIKEN, Saitama 351-0198, Japan
e-mail: wataru@riken.jp

S. Maekawa
Institute for Materials Research, Tohoku University, Sendai 980-8577, Japan
and
CREST, Japan Science and Technology Agency, Sanbancho, Tokyo 102-0075, Japan
e-mail: maekawa@imr.tohoku.ac.jp

V. Zlatić and A. C. Hewson (eds.), *Properties and Applications of Thermoelectric Materials*,
NATO Science for Peace and Security Series B: Physics and Biophysics,

density is estimated to be $10^{18}\sim10^{19}\,\mathrm{cm}^{-3}$. Thermoelectric materials for application at the present stage are synthesized with this restriction on the carrier density.

For conventional thermoelectric materials, attention is focused on the charge degree of freedom of electrons. However, as will be seen in the next section, thermopower does not always reflect the charge. It has to do with the entropy transported by an electric current. From this point of view, we may ask: what are the key ingredients to be used in the design of new thermoelectric materials?

Let us focus on the electronic structure of $NaCo_2O_4$. The lamellar-crystal structure of $NaCo_2O_4$ has hexagonal layer of CoO_2 sheets. Similarly $Ca_2Co_2O_5$, (Bi,Pb)-Sr-Co-O systems and other related cobalt oxides also contain CoO_2 layers. It has been observed [10, 11] that there exist Co^{3+} and Co^{4+} sites in $NaCo_2O_4$ with the configurations of $t^6{}_{2g}$ and $t^5{}_{2g}$, respectively. The CoO_6 octahedra in the hexagonal layer share edges with each other, so that the angle of the Co-O-Co bond is about 90°. In the 90° configuration, the hopping-matrix-elements of adjacent $3d$ orbitals due to σ bond are zero, and consequently, the $3d$ electrons have small hopping-matrix-elements due to a π bond with neighboring sites. Consequently, the bands derived from these t_{2g} orbitals are narrower than in other metallic transition metal oxides with perovskite structure whereas the effective on-site Coulomb interaction is comparable with them.

In contrast to semiconductors, strongly correlated t_{2g} electrons characterize the electronic properties of the cobalt oxides. Before discussing the thermo-electronic behavior of $NaCo_2O_4$ in more detail, let us review [12] the thermoelectric response of a solid.

2 Thermoelectric Effect

Thermodynamic systems are either in equilibrium or are not. In turn, non-equilibrium systems may be either in the linear or non-linear regime. The thermoelectric effects which are of interests here are based on the approximation where the deviation from the thermodynamic equilibrium values are small and are linear in the driving fields.

It is known that not only the electric field $\nabla\varphi$ but also the gradient of the temperature ∇T drive the electric current, where φ is the electrostatic potential. When an electric current flows, there is a concurrent heat current. This situation is reflected by the following linear response equations,

$$\begin{cases} j_e = l^{11}\nabla\varphi + l^{12}\nabla T \\ j_{heat} = l^{21}\nabla\varphi + l^{22}\nabla T \end{cases} \tag{1}$$

where j_e and j_{heat} are the electric and heat currents, respectively. The linear dynamical laws of thermoelectric response normally exhibit the Onsager reciprocity. However, Eqs. (1) do not express the Onsager reciprocity relation by $l^{12} = l^{21}$, since the generalized forces, which correspond to $\nabla\varphi$ and ∇T, and the currents, j_e and

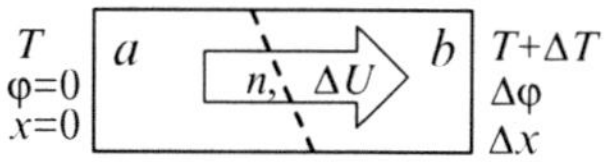

Fig. 1 Thermoelectric system.

j_{heat}, are not chosen appropriately. Let us derive expressions for the dynamic thermoelectric linear response which do not have this defect.

Shown in Fig. 1 is a finite but very small area in a thermoelectric system. For simplicity, it is assumed that all flows and forces are parallel to the x-direction. The low- and high-temperature sides are labeled a and b, respectively, and between these there exist differences in the electro-static potential $\Delta\varphi$ and temperature ΔT, where n and ΔU denote the amount of transferred particles and energy, respectively.

The system is irreversible, i.e., there is a net generation of entropy. The change in the local entropy density Δs_a on side a is expressed as

$$\Delta s_a = -\frac{\Delta U}{T} + \frac{\mu(T)}{T} n,$$

where $\mu(T)$ is the chemical potential at temperature T. On side b,

$$\Delta s_b = -\frac{\Delta U}{T+\Delta T} - \frac{\mu(T+\Delta T) - e\Delta\varphi}{T+\Delta T} n,$$

where $e(>0)$ is the absolute value of electric charge. The rate of change of the entropy is therefore,

$$\frac{ds}{dt} = \frac{d(s_a+s_b)}{dt} = \frac{d(\Delta U)}{dt}\left(-\frac{\Delta T}{T^2}\right) + \frac{dn}{dt}\left[\frac{e}{T}\Delta\varphi - \Delta T \frac{\partial}{\partial T}\left(\frac{\mu}{T}\right)\right]. \qquad (2)$$

Note that dn/dt is the rate at which particles move through the boundary of a and b, i.e., the particle current. In the same way, $d(\Delta U)/dt$ is the energy current. The generalized forces are identified from the coefficients which multiply the energy and number currents in Eq. (2). These are written as,

$$\frac{e}{T}\Delta\varphi - \Delta T \frac{\partial}{\partial T}\left(\frac{\mu}{T}\right) = \Delta x \left[\frac{e}{T}\frac{\partial\varphi}{\partial x} - \frac{\partial}{\partial x}\left(\frac{\mu}{T}\right)\right]$$

for the particle current and

$$-\frac{\Delta T}{T^2} = \Delta x \frac{\partial}{\partial x}\left(\frac{1}{T}\right)$$

for the energy current, respectively. Finally, the appropriate form for the linear dynamical laws is

$$\begin{cases} j_1 = M^{11}\left[\frac{e}{T}\frac{\partial \varphi}{\partial x} - \frac{\partial}{\partial x}\left(\frac{\mu}{T}\right)\right] + M^{12}\frac{\partial}{\partial x}\left(\frac{1}{T}\right), \\ j_2 = M^{21}\left[\frac{e}{T}\frac{\partial \varphi}{\partial x} - \frac{\partial}{\partial x}\left(\frac{\mu}{T}\right)\right] + M^{22}\frac{\partial}{\partial x}\left(\frac{1}{T}\right), \end{cases} \tag{3}$$

where the j_l are the particle, ($l = 1$), and energy, ($l = 2$), flux densities. Here, the quantities M^{lm} are called the kinetic coefficients and satisfy the Onsager reciprocity relation $M^{12} = M^{21}$.

The kinetic coefficients M^{lm} are determined from the experimental data through the relevant phenomenological laws, i.e., Ohm's law of electrical conduction,

$$j_e = -ej_1 = \frac{1}{\rho}\left(-\frac{\partial \varphi}{\partial x}\right)$$

(where j_e is the electric current and ρ is the electrical resistivity), Fick's law of diffusion,

$$j_1 = -D\frac{\partial n_p}{\partial x}$$

(where n_p is particle density and D is the diffusion coefficient), Fourier's law of heat conduction,

$$j_{heat} = j_2 - \mu j_1 = -\kappa\frac{\partial T}{\partial x}$$

(where j_{heat} is the heat-current density and κ is the thermal conductivity). The final law corresponds to the Seebeck effect which is to be discussed below.

When the temperature and concentration gradients are not allowed ($\partial T/\partial x = 0$ and $\partial \mu/\partial x = 0$), the first equation in Eq. (3) gives

$$j_1 = M^{11}\frac{e}{T}\frac{\partial \varphi}{\partial x}$$

and the relation

$$\frac{1}{\rho} = e^2\frac{M^{11}}{T}$$

is obtained. In the case that $\partial T/\partial x = 0$ and $\partial \varphi/\partial x = 0$, the first equation in Eq. (3) gives

$$j_1 = -M^{11}\frac{1}{T}\frac{\partial \mu}{\partial x} = -M^{11}\frac{1}{T}\frac{\partial \mu}{\partial n_p}\frac{\partial n_p}{\partial x}$$

and we find

$$D = M^{11}\frac{1}{T}\frac{\partial \mu}{\partial n_p}.$$

The thermal conductivity κ is expressed as

$$\kappa = \frac{1}{T^2}\left[M^{22} - \frac{\left(M^{12}\right)^2}{M^{11}}\right]$$

with the condition that $j_1 = 0$ and $\partial \varphi/\partial x = 0$.

An expression for the thermopower (Q) is obtained from the first Eq. (3) for an open circuit ($j_1 = 0$), i.e.,

$$-\left(\frac{\partial\varphi}{\partial x} - \frac{1}{e}\frac{\partial\mu}{\partial x}\right) = \left(-\frac{1}{eT}\frac{M^{12}}{M^{11}} + \frac{\mu}{eT}\right)\frac{\partial T}{\partial x} \equiv Q\frac{\partial T}{\partial x}$$

This defines the *absolute* Seebeck coefficient, i.e., the thermopower is defined to be the ratio of the gradients of electro-chemical potential $(\varphi - \mu/e)$ and the temperature, i.e.,

$$Q = -\frac{1}{eT}\frac{M^{12}}{M^{11}} + \frac{\mu}{eT}. \tag{4}$$

When we apply a temperature gradient to a solid, the hot electrons will spread into the cold end of the sample due to the diffusive force originating from the temperature gradient. Consequently, electrons tend to pile up at the cold end. The resulting charge imbalance causes an electric field. The force due to the electric field will compete with the diffusive force. The difference between the forces appears as a net electro-motive force between the ends of the sample. This is the physics implied by Eq. (4).

The steady state ($j_1 = 0$) thermoelectric response follows from the linear dynamical law, Eq. (3), i.e.,

$$0 = M^{11}\left[\frac{e}{T}\frac{\partial\varphi}{\partial x} - \frac{\partial}{\partial x}\left(\frac{\mu}{T}\right)\right] + M^{12}\frac{\partial}{\partial x}\left(\frac{1}{T}\right). \tag{5}$$

The steady state reflects the balance among the various terms. There are not only electric and diffusive forces but an interference between the electric and heat currents (the last term of the left hand side of Eq. (5)). This is the nature of the thermoelectric effect.

We are now in the position to expose an interesting insight into the physical meaning of the thermopower. Given that the heat Δq is related to the entropy change by $T\Delta s$, then the relation between heat, J_{heat}, and entropy, J_s, currents is

$$J_{heat} = TJ_s.$$

On the other hand, the heat current-density is defined by

$$j_{heat} = j_2 - \mu j_1. \tag{6}$$

Inserting Eq. (3) into Eq. (6), the entropy current-density can be expressed as

$$\begin{aligned} J_s &= \frac{1}{T}\left(-\frac{M^{12}}{M^{11}} + \mu\right) j_1 + \frac{1}{T}\left[M^{22} - \frac{\left(M^{12}\right)^2}{M^{11}}\right]\frac{\partial}{\partial x}\left(\frac{1}{T}\right) \\ &= Qj_e + \frac{1}{T}\left(-\kappa\frac{\partial T}{\partial x}\right). \end{aligned} \tag{7}$$

This expression shows that there are two contributions to the flow of entropy: one is the entropy carried by the Fourier heat and the other is carried by electric current. An electric current is a flow of charge carriers. Equation (7) demonstrates why the thermopower is often said to be the entropy of electric current and to measure the entropy per carrier.

It is often said that the thermopower reflects the sign and the density of charge carriers in metals and semiconductors. However, Eq. (7) should convince the reader that the thermopower is not *solely* determined by carrier charge degrees of freedom. The thermopower is, in fact, a direct probe of the charge carrier entropy. An electron has not only charge but also spin and orbital. Although the spin and orbital degrees of freedom may exist in metals and degenerate semiconductors, they do not always contribute to their entropy, since the electronic state is well described by the degenerate electron gas.

To understand this physics in more detail, let us discuss a microscopic formulation for the thermopower in the following sections.

3 Spin and Orbital Degrees of Freedom and Thermopower

A real macroscopic body can be divided into a set of sub-systems which are relatively small but yet macroscopic. These sub-systems interact with each other, i.e., the extensive parameters (such as energy) flow to and from the neighboring sub-systems and the interesting extensive parameters undergo spontaneous fluctuation *even* in the thermal equilibrium state. Evidently, the smaller the sub-system is, the larger the spontaneous fluctuation is. The fluctuation will then dissipate through the spontaneous flow of the relevant extensive parameters to or from the surrounding sub-systems. The Onsager reciprocity theorem follows if these spontaneous fluctuations obey the same dynamical laws as do the irreversible process described in the previous sections. With this assumption, the linear response theory based on statistical mechanics has been constructed, based on the phenomenological law Eq. (3).

The formula for the linear response to an external perturbation is known as the Kubo formula. In this formula, the kinetic coefficients are expressed by correlation functions of the currents j_1 and j_2. It can be shown [13–15] that M^{12}/M^{11} in the expression for the thermopower equation (Eq. (4)) becomes temperature independent at high-temperatures, and the thermopower at high temperatures is given by the simple expression

$$Q = \frac{\mu}{eT}. \tag{8}$$

Thermodynamics tells us that μ/T is given by the partial derivative with respect to the number of particles N of the entropy s, so from Eq. (8)

$$Q = -\frac{1}{e}\left(\frac{\partial s}{\partial N}\right)_{E,V} \tag{9}$$

where E and V are the internal energy and the volume of the electron system, respectively. In other words, Eq. (9) states that the thermopower is the entropy per carrier.

In the high temperature limit, Eq. (9) is valid. In the strongly correlated electron systems, there exist characteristic energies, i.e., inter-site interaction (J), kinetic energy (t), and Coulomb interaction (U). The electron systems like transition metal oxides are characterized by $J,t \ll U$. Let us consider the electronic state, in particular, the state of d shell and the entropy or μ/T at high temperatures ($J,t \ll k_BT \ll U$).

The expectation value of the number of electrons (n_e) is given by

$$n_e = \frac{\mathrm{Tr}\left[Ne^{-(H-\mu N)/k_BT}\right]}{\mathrm{Tr}\left[e^{-(H-\mu N)/k_BT}\right]}, \tag{10}$$

where H and N are Hamiltonian and the number operator, respectively. The chemical potential μ is a function of n_e. Because of the large Coulomb interaction U, the chemical potential is restricted into a Hubbard band for given n_e.

Figure 2 shows a schematic representation of the relation between the chemical potential and electron and/or hole concentration: In the case that all of the sites take d^m states, the Hubbard band are completely filled, i.e., the system is a Mott insulator. By introducing holes with amount n_h, the chemical potential is in the Hubbard band and n_e is given by $m - n_h$. This situation is expressed by the Boltzmann factor $\exp(-U/k_BT)$. Because of the Boltzmann factor, only the d^m and d^{m-1} electronic states are available and other electronic states are excluded in the calculation of Tr[$\sim$] in Eq. (10). At high temperatures ($J,t \ll k_BT \ll U$), kinetic energy and inter-site interactions are neglected. As a result, Eq. (10) turns into the following expression:

$$n_e = \frac{(m-1)g_h + mg_e e^{\frac{\mu}{k_BT}}}{g_h + g_e e^{\frac{\mu}{k_BT}}}. \tag{11}$$

Here, g_e and g_h are the local degeneracies of d^m and d^{m-1} electronic states, respectively. Finally, we derive the formula of thermopower at high temperatures [15, 16]:

$$Q = \frac{\mu}{eT} = -\frac{k_B}{e}\ln g_e + \frac{k_B}{e}\ln g_h - \frac{k_B}{e}\ln\frac{n_h}{1-n_h}, \tag{12}$$

where n_h denotes the hole concentration.

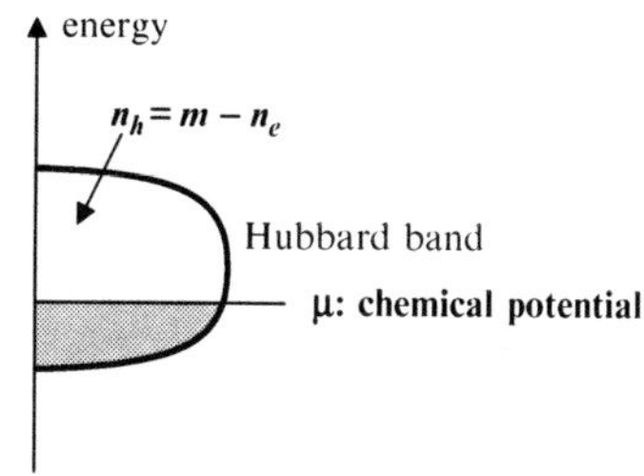

Fig. 2 Chemical potential and number of electrons in 3d shell. Here, n_e and n_h denote numbers of electrons and holes, respectively.

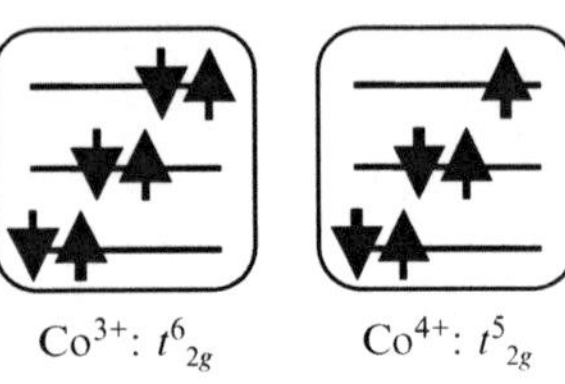

Fig. 3 Electronic configurations of cobalt ions.

In the oxide $NaCo_2O_4$, for example, the electronic states of cobalt ions are $t^6{}_{2g}$ and $t^5{}_{2g}$, i.e., $m = 6$ for available electronic states in the d shell. Figure 3 shows the electronic configurations of cobalt ions.

The electronic state $t^5{}_{2g}$ is able to be understood as the state obtained by removing an electron from the state $t^6{}_{2g}$. There exist six ways to remove an electron from the state $t^6{}_{2g}$, so that the state $t^5{}_{2g}$ has sixfold degeneracy. On the other hand, the state $t^6{}_{2g}$ has no degeneracy. Therefore, in the cobalt oxide, g_e and g_h are 1 and 6, respectively. Since the average valence of a cobalt ion in the stoichiometric compound is +3.5, the ratio of Co^{3+} to Co^{4+} ions is unity, i.e., $n_h = 0.5$. Thus, the theory (Eq. (12)) gives 154 μV/K for the high-temperature value of the thermopower and well explains the experiment. This result clearly shows why $NaCo_2O_4$ has the large thermopower with positive sign regardless of the carrier-density. The last term in Eq. (12) vanishes for $n_h = 0.5$, and it is solely the spin and orbital degrees of freedom reflecting the strong Coulomb interaction ($U \gg t$) [15–17] which gives rise to the large thermopower.

The underlying physics is easily understood. Suppose that the d^m and d^{m-1} electronic states are next to each other. When an electron moves between them, the d^m (d^{m-1}) state changes into the d^{m-1} (d^m) state. In this process, the charge $-e$ moves and the degeneracy is exchanged. This displacement of the degeneracy corresponds to the motion of entropy. This is reflected as a large thermopower since the thermopower is a direct probe of the entropy per carrier. This is in contrast to the resistivity.

We can extend the theory to more complicated oxides. The double Perovskite systems, for examples, involve two kinds of transition metal ions. In this case, possible configurations of local electronic state may realize interesting entropy flow in the strongly correlated systems. Let us use A and B for the two different transition metal ions of the double Perovskite system.

Figure 4 shows the relation between number of electrons and chemical potential. The chemical potential on A and B should be the same in an electron system. Suppose the ion A (B) takes electronic states with d^{m_A} and d^{m_A-1} (d^{m_B} and d^{m_B-1}). The numbers of electrons on A and B are given by $n_A = m_A - n_h$ and $n_B = m_B - n_h$, respectively, where n_h is the hole concentration. Therefore, the expectation value of electrons is expressed as

$$n_A + n_B = \frac{(m_A - 1)A_h + m_A A_e e^{\frac{\mu}{k_B T}}}{A_h + A_e e^{\frac{\mu}{k_B T}}} + \frac{(m_B - 1)B_h + m_B B_e e^{\frac{\mu}{k_B T}}}{B_h + B_e e^{\frac{\mu}{k_B T}}}$$

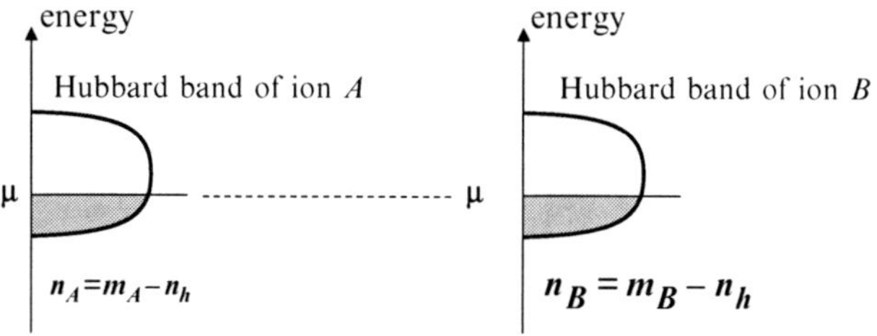

Fig. 4 Chemical potential and number of electrons in 3*d* shell in the double Perovskite systems. Here, n_A and n_B denote numbers of electrons of ion *A* and *B*, respectively.

where A_h and A_e (B_h and B_e) are the local degeneracies by spin and orbital degree of freedom on ion *A* (*B*), respectively. Finally, the formula of thermopower at high temperatures in the double Perovskite systems is expressed as follows:

$$\begin{aligned} Q &= \frac{\mu}{eT} \\ &= \frac{k_B}{e} \ln \frac{(1-2n_h)(A_e B_h + A_h B_e) + \sqrt{(1-2n_h)^2 (A_e B_h + A_h B_e)^2 + 16 n_h (1-n_h) A_e A_h B_e B_h}}{4 n_h A_e B_e} \end{aligned} \tag{13}$$

The expression of Eq. (13) is complicated. However, the underlying physics can be understood in the same way as the previous formula (Eq. (12)). In the case that $n_h = 0.5$, the contribution from the charge degree to the thermopower disappears because of the particle-hole symmetry. As a result, Eq. (13) in this case gives

$$Q = \frac{1}{2}\left[-\frac{k_B}{e}(\ln A_e + \ln B_e) + \frac{k_B}{e}(\ln A_h + \ln B_h) \right]. \tag{14}$$

It is noted that Eq. (14) is the average of the results given in the previous formula (Eq. (12)) with $n_h = 0.5$. Even in the complex oxides, we can see that the entropy flow by spin and orbital degrees of freedom results in the thermopower and its essential physics in strongly correlated electron systems.

4 Summary

We have seen that the strong Coulomb interaction induces the spin and orbital degrees of freedom which enhance the thermopower. This realization indicates a new path by which novel high performance thermoelectric materials will be developed. The thermopower can be enhanced with little effect on resistivity when, due to strong Coulomb interactions, there is a large degeneracy in spin and orbital degrees of freedom. This is completely different from the physics of conventional thermoelectric materials based on semiconductors.

It should, however, be noted that the electron-phonon and superexchange interactions can lead to the long-range order which suppresses the spin and orbital degeneracies at temperatures lower than that at which short-range correlations set in, and that this might be considerably higher than the critical temperature for long-range order. Such correlations are detrimental to a substantial spin and orbital thermopower especially in the bipartite lattices. However, for a triangular lattice there is a frustration of the spin and orbital degrees of freedom and this helps prevent the electron–phonon and superexchange interactions from lifting the degeneracies. Thus, frustrated lattices favor a large thermopower.

The carrier doping is usually needed in order to have a good conductivity. Although the effects of the spin and orbital degrees of freedom in the thermopower have been also observed for some other doped transition metal oxides, the effects are lost with increasing carrier concentration since Coulomb correlations are less important as the doping is increased. Upon doping, sufficient strong Coulomb interaction (U) is required to retain its internal degrees of freedom which are reflected in the thermopower. The layered-hexagonal structure of $NaCo_2O_4$ with the Co–O–Co bond angle of about 90° provides the narrow electron band width (W) [18], i.e., U/W is large and the effects of Coulomb interaction are important.

In this sense, small angle *TM*–O–*TM* bonds (*TM* denotes a transition metal ion) leading to a narrow band, strong correlation of the electrons and frustration are considered to be the key ingredients for potential transition metal oxide thermoelectric materials.

The thermopower derived from the first and second terms in Eq. (12) for several transition metal oxides is listed in Table 1.

Equation (14) tells us that in the strongly correlated complex oxides, the possible thermopower is given by the average of the list in Table 1. Therefore, there is no further enhancement in comparison with the previous formula (Eq. (12)). However, there exist other useful functions in the complex oxides and they can be effective in assisting an enhancement of the thermopower. The following is one possibility: Although the formula (Eq. (12)) gives large negative thermopower for the manganese oxides, the oxides do not always show such behavior even at high temperatures. This is also well understood by the formula (Eq. (12)). Some manganese oxides show strong cooperative Jahn–Teller effects. In those cases, the orbital degree couples with lattice distortion strongly, so that the thermopower due to the degree is strongly suppressed. Therefore, the formula (Eq. (12)) gives small

Table 1 Possible thermopower induced by spin and orbital in transition metal oxides.

	g_e/g_h	$-(k_B/e)\ln(g_e/g_h)$, μV/K
$Ti^{3+}(3d^1)$, $Ti^{4+}(3d^0)$	6/1	−154
$V^{3+}(3d^2)$, $V^{4+}(3d^1)$	9/6	−35
$Cr^{3+}(3d^3)$, $Cr^{4+}(3d^2)$	4/9	70
$Mn^{3+}(3d^4)$, $Mn^{4+}(3d^3)$	10/4	−79
$Co^{3+}(3d^6)$, $Co^{4+}(3d^5)$	1/6	154
$Rh^{3+}(4d^6)$, $Rh^{4+}(4d^5)$	1/6	154

The first column shows the combination of ions. The second column shows the ratio of the local degeneracies g_e and g_h.

thermopower in the system without orbital degree of freedom. By composing complex oxides using manganese and un-cooperative transition metal ions, it is expected that the cooperative effects is weak. For this strategy to enhance the thermopower, the formula (Eq. (12)) and Table 1 summarize the important physics of strongly correlated systems.

References

1. I. Terasaki, Y. Sasago, and K. Uchinokura, Phys. Rev. B **56**, R12685 (1997).
2. T. Itoh and I. Terasaki, Jpn. J. Appl. Phys. **39**, 6658 (2000).
3. A. C. Masset, C. Michel, A. Maignan, M. Hervieu, O. Toulemonde, F. Studer, B. Raveau, and J. Hejtmanek, Phys. Rev. B **62**, 166 (2000).
4. T. Yamamoto, I. Tsukada, K. Uchinokura, M. Takagi, T. Tsunobe, M. Ichihara, and K. Kobayashi, Jpn. J. Appl. Phys. **39**, L747 (2000).
5. Y. Ono, K. Satoh, T. Nozaki and T. Kajitani, Jpn. J. Appl. Phys. **46**,1071 (2007).
6. T. Okuda, N. Jufuku, S. Hidaka, and N. Terada: Phys. Rev. B **72**, 144403 (2005).
7. Y. Okamoto, M. Nohara, F. Sakai, and H. Takagi, J. Phys. Soc. Jpn. **75**, 023704 (2006).
8. M. Ohtaki and E. Maeda, J. Jpn. Soc. Powder Powder Metallurgy **47**, 1159 (2000).
9. R. Funahashi, I. Matsubara, H. Ikuta, T. Takeuchi, U. Mizutani, and S. Sodeoka, Jpn. J. Appl. Phys. **39**, L1127 (2000).
10. T. Tanaka, S. Nakamura, and S. Iida, Jpn. J. Appl. Phys. **33**, L581 (1994).
11. R. Ray, A. Ghoshray, K. Ghoshray, and S. Nakamura, Phys. Rev. B **59**, 9454 (1999).
12. As the typical text books, one can find: H. B. Callen, Thermodynamics and an Introduction to Thermostatistics, 2nd Edition (Wiley, New York, 1985). D. K. Kondepudi and I. Prigogine, Modern Thermodynamics: From Heat Engines to Dissipative Structures (Wiley, New York, 1998). G. D. Mahan, Many-Particle Physics, 3rd Edition (Plenum Press, New York, 2000). C. Kittel, Elementary Statistical Physics (Wiley, New York, 1958). S. Maekawa et al., Physics of Transition Metal Oxides, (Springer Series in Solid-State Sciences, Vol. 144 (2004)).
13. P. M. Chaikin, and G. Beni, Phys. Rev. B **13**, 647 (1976).
14. J.-P. Doumerc, J. Sold State Chem. **109**, 419 (1994).
15. W. Koshibae, K. Tsutsui, and S. Maekawa, Phys. Rev. B **62**, 6869 (2000).
16. W. Koshibae and S. Maekawa, Phys. Rev. Lett. **87**, 236603 (2001).
17. W. Koshibae and S. Maekawa, Phys. Rev. Lett. **91**, 257003 (2003).
18. D. J. Singh, Phys. Rev. B **61**, 13397 (2000).

Thermoelectric Power of Correlated Compounds

N. Oeschler, S. Hartmann, U. Köhler, M. Deppe, P. Sun, and F. Steglich

Abstract In the search for effective thermoelectric materials the class of strongly correlated electron systems has become one the main research topics for low-temperature Peltier cooling applications.

In this paper we present results for a number of different compounds which exhibit strong correlations. The described heavy fermion metals have a particularly large thermopower $S(T)$ below 30 K. The enhancement of the thermoelectric power is related to the large degeneracy of the ground-state f multiplet. The maximum in $S(T)$ can be shifted by volume changes which cause a change of the hybridization. Some correlated semiconductors which have a huge thermoelectric power at low temperatures are also described. As an example, we consider $FeSb_2$ which has a thermoelectric power exceeding $-30,000\mu V/K$ at 10 K. Due to its semiconducting behavior, this material exhibits a record power factor $PF = S^2\sigma$, where σ is the electrical conductivity.

1 Introduction

The development and characterization of effective thermoelectric materials is a major topic of applied research in condensed matter physics. Whereas most groups have focused on the search for high-temperature applications above room

N. Oeschler, S. Hartmann, U. Köhler, M. Deppe, and P. Sun
Max Planck Institut für Chemische Physik fester Stoffe, Nöthnitzer Str. 40, 01187 Dresden, Germany
e-mail: oeschler@cpfs.mpg.de, Hartmann@cpfs.mpg.de, u.koehler@ifw-dresden.de, deppe@cpfs.mpg.de, sun@cpfs.mpg.de

U. Köhler
Present address: Leibniz-Institut für Festkörper- und Werkstoffforschung Dresden, Helmholtzstr. 20, 01069, Dresden, Germany

F. Steglich
Max Planck Institut für Chemische Physik fester Stoffe, Nöthnitzer Str. 40, 01187 Dresden, Germany
e-mail: Steglich@cpfs.mpg.de

V. Zlatić and A. C. Hewson (eds.), *Properties and Applications of Thermoelectric Materials*, 81
NATO Science for Peace and Security Series B: Physics and Biophysics,

temperature, we consider here the class of correlated electron systems, which might prove to be useful as Peltier coolers at cryogenic temperatures below the boiling point of nitrogen.

Important members of this class are the heavy-fermion compounds, which contain elements, such as cerium or ytterbium with partially filled *f* shells and associated magnetic moments. Below a characteristic temperature T_K that ranges typically between 10 and 100 K, a sharp peak of the density of states develops at the Fermi energy E_F, associated with the screening of the f spins by the conduction electrons [1]. The interaction between the local moment of the f electrons and the conduction electrons can be described by the Kondo exchange interaction. Similar physical phenomena are observed in d-band materials, in particular, Fe-based compounds.

Due to the large density of states at E_F an enhanced Sommerfeld coefficient, $\gamma = C_{el}/T(T \rightarrow 0)$, is found in these materials with values that can exceed those for ordinary metals by two to three orders of magnitude [1]. The thermoelectric power, S, is found to be similarly enhanced to values above 30 μV/K [2], while ordinary metals exhibit thermoelectric powers in the order of 1 μV/K at low T.

The electron-diffusion part of the thermoelectric power S_d is described by the Mott formula based on the linearized Boltzmann equation:

$$S_d = \frac{\pi^2}{3} \frac{k_B^2 T}{e} \left(\frac{\partial \ln \sigma(\varepsilon)}{\partial \varepsilon} \right)_{E_F} \tag{1}$$

where σ is the electrical conductivity. σ can be approximated by the density of states at the Fermi energy assuming a temperature-independent relaxation time.

$$S_d \approx \left(\frac{\partial \ln N(\varepsilon)}{\partial \varepsilon} \right)_{E_F} \tag{2}$$

Thus, the diffusion thermoelectric power is given by the slope of the density of states at the Fermi energy. For strongly correlated electron systems, S_d may be significantly enhanced due to the narrow peak in the density of states at the Fermi level. By contrast, the phonon-drag contributions are usually of minor importance compared to the diffusion thermoelectric power. In clean samples with long phonon mean free paths phonons with higher energy "drag" charge carriers from the hot side to the cold side of the sample. The effect is most pronounced at low temperatures at a fraction of the Debye temperature (typically $0.2\Theta_D$).

The thermal conductivity consists of two relevant parts. One is determined by the heat transport by phonons, the other by charge carriers. The electronic part of thermal conductivity, κ_{el} is related to the electrical conductivity via the Wiedemann–Franz law ($L_0 = 2.45 \cdot 10^{-8}$ WΩ/K^2 being the Lorenz number):

$$\kappa_{el}/T = L_0 \sigma \tag{3}$$

As a measure of the thermoelectric performance the dimensionless figure of merit has been introduced:

$$ZT = (S^2\sigma/\kappa)T \tag{4}$$

A ZT value that exceeds unity is essential for practical use as thermoelectric generator or Peltier cooler. For metallic systems, on the assumption that the phononic thermal conductivity $\kappa_{\text{ph}} = 0$, the figure of merit is given by $ZT = S^2Ł_0$. Then, enhancing S to above $150\,\mu\text{V/K}$ yields a ZT of more than 1. However, particular effort is required to ensure that the contribution from κ_{ph} can be neglected.

A number of compounds have reached values above $ZT = 1$ at high T [3]. For low temperatures such a system has yet to be discovered. Whereas for the temperature range above room temperature, heavily doped semiconductors are the best candidates for high ZT values, strongly correlated electron system with metallic conductivity seem to have considerable potential at low temperatures. However, an extremely large thermoelectric power has also been observed for correlated semiconductors with d elements.

In the following, examples for heavy-fermion metals are presented and discussed in Section 2. The two systems $Yb_2Pt_6Al_{15}$ and CeTiGe have a degenerate ground-state f multiplet which seems advantageous for enhanced thermoelectric power values below 30 K. Chemical substitution affects the hybridization strength, and thus leads to a shift of the maximum in $S(T)$. This raises the possibility of achieving a good thermoelectric performance over a specific T range, as will be shown for $Lu_{1-x}Yb_xRh_2Si_2$. In Section 3, the semiconductor $FeSb_2$ with correlated narrow gaps is introduced. We present evidence that this compound has a very high thermoelectric potential. The thermoelectric power and the power factor exhibit record values, which are most likely enhanced due to strong correlations. Our conclusions are given in Section 4.

2 Strongly Correlated *f*-electron Systems

Large thermoelectric power values are commonly observed for heavy fermion systems at low temperatures [4]. Two well separated maxima (minima) are found for a number of Ce (Yb)-based heavy fermion compounds such as $CeCu_2Si_2$ [5] and $CeCu_6$ [6]. The low-T extremum is attributed to the Kondo scattering on the ground state doublet around its characteristic temperature [7]:

$$T_{\text{low}} \approx T_{\text{K}}. \tag{5}$$

The peak at higher temperatures results from Kondo scattering on higher multiplets which are split by crystal electric field (CEF) effects. Its relation to the CEF levels is given by [8, 9]:

$$T_{\text{high}} = (0.3\ldots0.6)\Delta_{\text{CEF}}/k_{\text{B}}. \tag{6}$$

The thermoelectric power is known to be rather insensitive to the number of magnetic moments. Even dilute systems show high thermoelectric power values. Therefore, substitution with isoelectronic elements of different atomic radii does not reduce the absolute values of S significantly. Instead, it allows for shifting the extrema of S to the desired temperature ranges due to the continuous change in the hybridization strength upon doping. As an example the thermoelectric power of the substitution series $Lu_{1-x}Yb_xRh_2Si_2$ with $x = 0.23$ and $x = 1$ is shown in Fig. 1. $YbRh_2Si_2$ is a well established heavy fermion system which attracted considerable interest due to its unusual low-temperature properties [10, 11].

The lattice Kondo scale of 17 K [12] is deduced from entropy considerations according to Rajan's calculation [13]. The ground-state doublet is well separated from excited crystal electric field split multiplets by $\Delta_{CEF}/k_B \approx 200$ K [14]. However, the thermoelectric power of $YbRh_2Si_2$ exhibits only one peak close to 80 K with large absolute values of close to 60 μV/K [15], cf. Figure 1. The minimum is caused by Kondo scattering on the excited CEF levels. According to Eq. (6) the position of the peak in S(T) is in good agreement with the observed CEF splitting.

By reducing the hybridization strength the signatures of the Kondo scale can be enhanced. In $Lu_{1-x}Yb_xRh_2Si_2$ this is achieved by substitution of the non-magnetic, slightly larger Lu on the Yb site [16]. As an example, we show in Fig. 1 the thermoelectric power of $Lu_{1-x}Yb_xRh_2Si_2$ with $x = 0.23$. Indeed, the second minimum becomes apparent in the thermoelectric power data at lower temperatures. The peak at 80 K does not shift significantly with respect to the $x = 1$ data, but decreases in size. This observation agrees with the expectation that the CEF levels remain unaffected by the substitution. The Kondo scattering on the ground-state doublet gives rise to a minimum around 10 K. The Kondo scale can be estimated from the min-

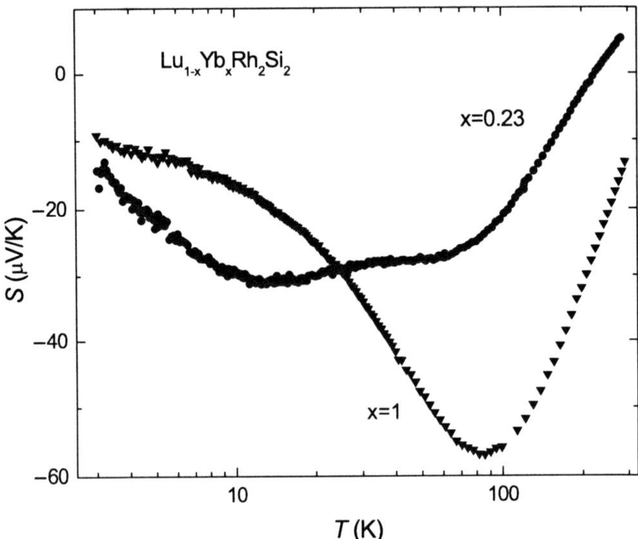

Fig. 1 Thermopower S of $Lu_{1-x}Yb_xRh_2Si_2$ with $x = 0.23$ and 1 vs. T.

imum temperature of the thermoelectric power, cf. Eq (5), whereas the observed lowering of T_K with reduction of Yb reflects the weakened hybridization.

The absolute values of the thermoelectric power, although reduced compared to those of $YbRh_2Si_2$, are still enhanced. While the resistivity changes largely upon substitution, the thermoelectric power decreases only slightly. A fraction of magnetic ions (Yb) is sufficient to generate large thermopower values.

Although $Lu_{1-x}Yb_xRh_2Si_2$ does not exhibit thermoelectric power values sufficiently large for application, this study shows that the temperature of the extrema of the thermoelectric power can be shifted by modification of the hybridization strength. Here, such shifts in the Kondo scales are realized by a volume change through substitution. Such an approach can commonly be applied to strongly correlated electron system.

The highest values of the thermoelectric power among f electron systems have been found for compounds with intermediate valent behavior [17]. The Kondo scale of these systems rises above the CEF levels. Consequently, the f electrons become partly itinerant with Ce valences between +3 and +4 or Yb valences between +2 and +3. A large thermopower exceeding 80 μV/K has been observed for instance in $CePd_3$ [17]. For Yb systems even larger negative values are reported for $YbAl_3$ [17]. Due to the high T_K, all these compounds exhibit large thermoelectric power above 100 K.

However, for applications at cryogenic temperatures, large thermopower values are required at lower T. As will be shown here, such behavior can be observed also for heavy fermion compounds with highly degenerate ground states. $Yb_2Pt_6Al_{15}$ crystallizes in a hexagonal structure with a large c/a ratio of around 4. The magnetic susceptibility χ and the magnetic part of the specific heat C_{4f} are described in terms of the Coqblin–Schrieffer model with full Yb momentum $J = 7/2$ and enhanced Sommerfeld coefficient of $330\,\mathrm{mJ/mol\,K^2}$ [18]. From the characteristic maximum in χ and C_{4f}/T around 20–30 K a Kondo temperature of 64 K is deduced. In Fig. 2 the thermoelectric power of single crystalline $Yb_2Pt_6Al_{15}$ is plotted for the heat flow along the a-axis. The minimum in the thermoelectric power of approximately 70 μV/K lies in the same temperature range as for χ. Such large values at low temperatures are rarely observed and presumably characteristics for correlated compounds with multiplet ground states.

A similar behavior has been detected for the tetragonal compound CeTiGe. Again, the magnetic contribution to the specific heat is well described by the Coqblin-Schrieffer model. Above a constant specific heat divided by T with a Sommerfeld coefficient of $270\,\mathrm{mJ/mol\,K^2}$ a maximum occurs below 20 K [19]. This feature indicates the involvement of the full Ce momentum $J = 5/2$. The thermoelectric power of a polycrystalline sample shown in Fig. 3 is also dominated by a single maximum at 17 K. Again the maximum temperature in $S(T)$ corresponds to the one in C_{4f}/T. The absolute values of $S(T)$ are comparable to those of $Yb_2Pt_6Al_{15}$ which has a similar Sommerfeld coefficient as CeTiGe. The high slope of S/T for $T \to 0$ supports the enhanced effective masses of the quasiparticles [19].

Interestingly, the compounds presented have a full multiplet as ground state. Due to the CEF splitting one expects the multiplet of Ce (Yb) ions for systems with non-

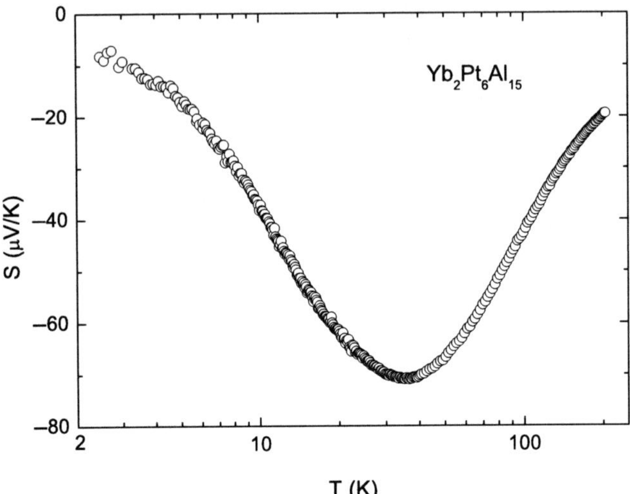

Fig. 2 Thermopower S of $Yb_2Pt_6Al_{15}$ vs. T for $Q||a$.

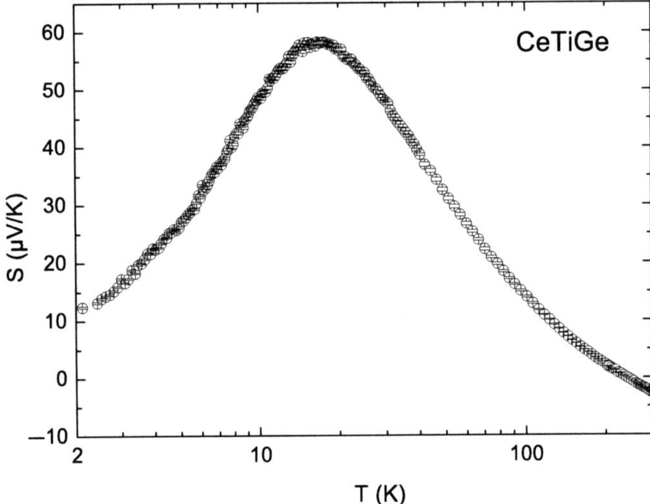

Fig. 3 Thermopower S of CeTiGe vs. T.

cubic symmetry to split into three (four) doublets. However, the CEF effects seem to be of minor importance for $Yb_2Pt_6Al_{15}$ and CeTiGe. Thus, the ground state involves all states of the full moment. This appears to be essential for a large thermoelectric power at low temperatures. No heavy-fermion metal with a ground-state doublet reaches thermopower values of such magnitude at low T. However, the thermoelectric power of CeB_6 forming a quartet ground state exceeds values of 60 $\mu V/K$ around 10 K [20]. Theoretical calculations for a sixfold degenerate ground state re-

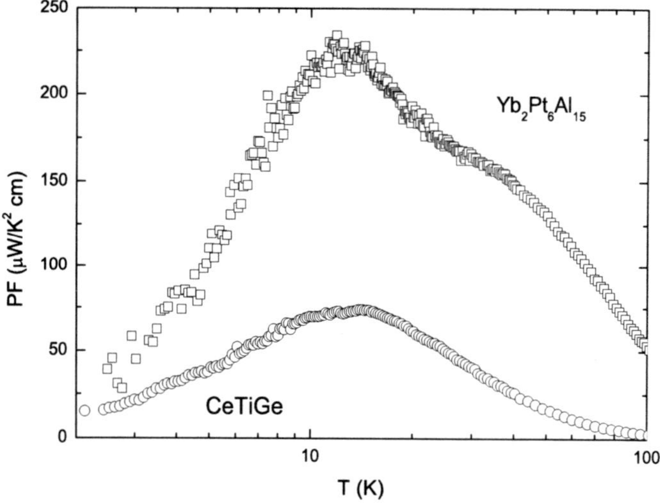

Fig. 4 Power factor of CeTiGe and $Yb_2Pt_6Al_{15}$ vs. T.

veal values of up to 100 $\mu V/K$ around the Kondo temperature [7]. A theoretical comparison of different ground-state degeneracies are not available.

The power factors $PF = S^2/\rho$ of these compounds are comparably high since, besides the enhanced thermoelectric power, the electrical resistivity ρ is rather low (cf. Fig. 4). Peak values around 50 (CeTiGe) to 250 $\mu W/K^2cm$ ($Yb_2Pt_6Al_{15}$) are obtained for the compounds presented. They are in the same order of magnitude as those of commercial thermoelectric materials [3]. However, in order to achieve high figures of merit the thermal conductivity has to be reduced significantly. As the electronic part to the thermal conductivity, κ_{el} and the electrical resistivity are connected (Eq. (3)) the phononic contribution to κ has to be tremendously diminished.

Attempts to use heavy fermion systems for Peltier cooling have been already undertaken. Modules based on CeB_6 achieve a cooling power of 0.2 K at 4 K [21].

3 Correlated Semiconductors

The most promising thermoelectric materials among strongly correlated electron systems are semiconducting Kondo compounds. In some cases the hybridization of the *d* or *f* states and the conduction bands leads to an opening of a gap at the Fermi energy [22]. This phenomenon is observed in a few cubic rare-earth compounds as YbB_{12} and $Ce_3Bi_4Pt_3$ [23]. At room temperature they behave like ordinary metals with local moments. At intermediate temperatures typical Kondo signatures are observed. Upon lowering the temperature their resistivity increases even further showing semiconducting behavior. The gap size roughly corresponds to the Kondo temperature which is the cause for the gap opening. Also 3*d* systems might develop a narrow gap as was identified in FeSi [24]. At low T, the extended d states seem to hybridize with the conduction electrons to form a hybridization gap.

A new member of this class is $FeSb_2$. It is an orthorhombic system with marcasite structure. The Fe ions are surrounded by distorted octahedra of Sb atoms. The susceptibility is temperature independent up to 50 K indicative of a non-magnetic ground state of the Fe and becomes paramagnetic at higher temperatures [25]. The resistivity of $FeSb_2$ follows a semiconducting behavior below room temperature with two regions of thermal activation separated by a plateau around 20 K. The gap sizes are estimated as 320 K for the high-T part and 54 K for the low-T side. Hall effect measurements indicate the existence of multiple charge carriers with electron-like and hole-like character. Calculations within a two-band model by Hu et al. derived a huge carrier mobility [27]. The thermoelectric power of $FeSb_2$ is displayed in Fig. 5. The thermopower exhibits negative values below 30 K at which the smaller gap opens. At 10 K extremely large values in the order of $-45\,\mathrm{mV/K}$ at 10 K are observed [26]. The maximum values of $|S|$ are largely enhanced compared to the thermopower of related systems like $RuSb_2$ [28]. This indicates that the phonon-drag effect can only be of minor importance for the unique thermoelectric properties of $FeSb_2$. In fact, the experimental $S(T)$ curve can be qualitatively described within a full decade in temperature (7–70K) by a 'classical' description of electron diffusion in a non-degenerate system. However, to reproduce the observed $S(T)$ behavior semiquantitatively, an enhancement factor of 30 is required [29]. It is assumed that strong electron–electron correlations have to be taken into account to explain this enhancement factor. The origin of the strong correlations is not yet fully understood.

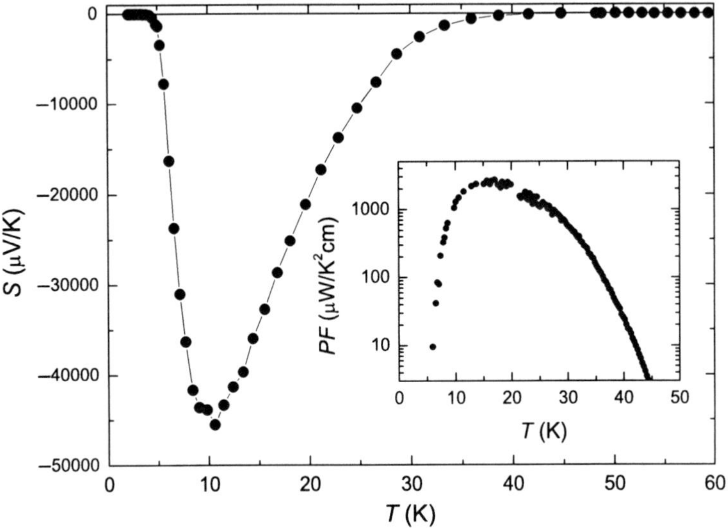

Fig. 5 Thermopower S of $FeSb_2$ vs. T. Inset: Power factor PF of $FeSb_2$.

$FeSb_2$ meets the requirements for high thermoelectric performance. It is a semiconductor with a very low charge carrier concentration and high mobility. The thermoelectric power at low temperature is extremely large. For the power factor which is shown in the inset of Fig. 5, a broad maximum with high values below 30 K is deduced. The values at the maximum of 2300 $\mu W/cm\,K^2$ depict record values among all thermoelectric materials [26]. However, in order to obtain large ZT values the thermal conductivity of the single crystals exhibit too high values due to the large phononic contribution. The aim of achieving high ZT values are pursued by minimization of the phononic mean free path. The first studies on doped $FeSb_2$ revealed a reduced thermal conductivity. However, concomitantly the electronic properties changed drastically [30].

4 Conclusion

The thermoelectric power of strongly correlated electron systems has been shown to reach very high values at low temperatures. Different routes are presented for possible high-performance materials. It has been demonstrated, how chemical substitution of heavy-fermion systems can be applied to tune the position of the maximum in S(T) to a specific temperature region to meet defined requirements. However, even more promising for low-temperature applications are Ce or Yb compounds with a highly degenerate ground state. As an example the thermoelectric properties of $Yb_2Pt_6Al_{15}$ and CeTiGe have been introduced. Both compounds exhibit a fully degenerate ground state without CEF splitting. The thermoelectric powers exceed values of 60 $\mu V/K$ below 30 K. Although those values are not sufficient according to the simple estimation presented in the introduction, they suggest a possible route for improvement. Rare cases of high S at such low T are known in the literature. However, the phononic thermal conductivity of the crystal has to be lowered in order to improve the thermoelectric performance.

Another highly promising class of materials are $3d$ semiconductors, in particular, $FeSb_2$. The record power factor generated by an extremely large thermoelectric power of $-45\,mV/K$ fulfills the requirements for a good thermoelectric. Further effort has to made to reduce the thermal conductivity. If κ is reduced by two or three orders of magnitude, which is realistic considering the high values on single crystals, ZT would rise to values around 1.

Acknowledgements Useful discussions with C. Geibel and B.B. Iversen are acknowledged. The work was partially supported by COST action P16.

References

1. G.R. Stewart, Rev. Mod. Phys. **56**, 755–787 (1984).
2. K. Behnia, D. Jaccard, J. Flouquet, J. Phys.: Condens. Matter **16**, 5187–5198 (2004).

3. G.J. Snyder and E.S. Toberer, Nature Materials **7**, 105–114 (2008).
4. S. Paschen: Thermoelectric Handbook, chapter 15, ed. D.M. Rowe (CRC Press, Taylor & Francis, Boca Raton, FL 2006).
5. G. Sparn, W. Lieke, U. Gottwick, F. Steglich, N. Grewe, J. Magn. Magn. Mater. **47, 48**, 521–523 (1985).
6. Amato, D. Jaccard, J. Sierro, F. Lapierre, P. Haen, P. Lejay, J. Flouquet, J. Magn. Magn. Mater. **76, 77**, 263–264 (1988).
7. S. Maekawa, S. Kashiba, M. Tachiki, S. Takahashi, J. Phys. Soc. Japan **55**, 3194–3198 (1986).
8. A.K. Bhattarjee and B. Coqblin, Phys. Rev. B **13**, 3441–3451 (1976).
9. V. Zlatić and R. Monnier, Phys. Rev. B **71**, 165109 (2005).
10. O. Trovarelli, C. Geibel, S. Mederle, C. Langhammer, F.M. Grosche, P. Gegenwart, M. Lang, G. Sparn, F. Steglich, Phys. Rev. Lett. **85**, 626–629 (2001).
11. P. Gegenwart, Q. Si, F. Steglich, Nat. Phys. **4**, 186–197 (2008).
12. J. Ferstl, PhD thesis, TU Dresden (2007).
13. V. T. Rajan, Phys. Rev. Lett. **51**, 308–311 (1983).
14. O. Stockert, M.M. Koza, J. Ferstl, A.P. Murani, C. Geibel, F. Steglich, Physica B **378–380**, 157–158 (2006).
15. S. Hartmann, U. Köhler, N. Oeschler, S. Paschen, C. Krellner, C. Geibel, F. Steglich, Physica B **378–380**, 70–71 (2006).
16. U. Köhler, N. Oeschler, F. Steglich, S. Maquilon, Z. Fisk, Phys. Rev. B **77**, 104412 (2008).
17. D. Jaccard, J. Sierro: Valence Instabilities, eds. P. Wachter and H. Boppart (North Holland, Amsterdam, 1982).
18. M. Deppe, S. Hartmann, M.E. Macovei, N. Oeschler, M. Nicklas, C. Geibel, New J. Phys. **10**, 093017 (2008).
19. M. Deppe, N. Caroca-Canales, S. Hartmann, N. Oeschler, C. Geibel, J. Phys.: Condens. Matter **21**, 206001 (2009).
20. Y. Peysson, C. Ayache, B. Salce, S. Kunii, T. Kasuya, J. Magn. Magn. Mater. **59**, 33–40 (1986).
21. S.R. Harutyunyan, V.H. Vardanyan, A.S. Kuzanyan, V.R. Nikoghosan, S. Kunii, K.S. Wood, A.M. Gulian, Appl. Phys. Lett. **83**, 2142–2144 (2003).
22. P.S. Riseborough, Adv. Phys. **49**, 257–320 (2000).
23. G. Aeppli and Z. Fisk, Comments Cond. Mat. Phys. **16**, 155–165 (1992).
24. D. Mandrus, J.L. Sarrao, A. Migliori, J.D. Thompson, Z. Fisk, Phys. Rev. B **51**, 4763–4767 (1995).
25. C. Petrovic, J.W. Kim, S.L. Bud'ko, A.I. Goldman, P.C. Canfield, W. Choe, G.J. Miller, Phys. Rev. B **67**, 155205 (2003).
26. A. Bentien, S. Johnsen, G.K.H. Madsen, B.B. Iversen, F. Steglich, Europhys. Lett. **80**, 17008 (2007).
27. R. Hu, V.F. Mitrovic, C. Petrovic, Appl. Phys. Lett. **92**, 182108 (2008).
28. P. Sun, N. Oeschler, S. Johnsen, B.B. Iversen, F. Steglich, J. Phys.: Conf. Series **150**, 012049 (2009).
29. P. Sun, N. Oeschler, S. Johnsen, B.B. Iversen, F. Steglich, Phys. Rev. B **79**, 153308 (2009).
30. A. Bentien, G.K.H. Madsen, S. Johnsen, B.B. Iversen, Phys. Rev. B **74**, 205105 (2006).

Thermoelectric Power and Thermal Transport of Anomalous Rare-Earth Kondo Compounds

B. Coqblin, B. Chevalier, and V. Zlatić

Abstract This paper reviews the thermoelectric power (TEP) of cerium and ytterbium Kondo systems. We present first the experimental situation regarding the Kondo compounds and describe briefly the intermediate valence (IV) systems. We then define the Kondo temperature T_K and summarize the theoretical approaches that explain the behavior of Kondo systems much above, and much below, the Kondo temperature. The results obtained for $T \gg T_K$, explain the high-temperature TEP data of cerium (ytterbium) compounds which exhibit a broad positive (negative) maximum at $T \simeq \Delta/a$, where Δ is given by the crystalline field (CF) splitting and $a \geq 1$ is a numerical constant. For $T \ll T_K$, the theory yields a small positive (negative) TEP peak at about $T_K/2$ which explains the low-temperature TEP data on cerium (ytterbium) compounds. Combining these theoretical results, we can account for the overall temperature dependence of the TEP in most Kondo and IV compounds. We also explain the anomalous pressure dependence of the TEP in some cerium and ytterbium compounds and show that the charge transfer is of crucial importance for pressure experiments. We discuss in some detail the hydrogenation effects for cerium compounds and show that one should distinguish between the "negative pressure" effects and the chemical effects which are due to the cerium-hydrogen bonding. Finally, we present briefly the experiments of the thermal conductivity in anomalous rare-earth Kondo compounds and show that the data can be explained by the Kondo scattering in the presence of CF splitting.

B. Coqblin
L.P.S., CNRS UMR 8502, Université Paris-Sud, 91405-Orsay, France
e-mail: coqblin@lps.u-psud.fr

B. Chevalier
I.C.M.C.B., CNRS, Université Bordeaux I, 33608-Pessac, France
e-mail: chevalie@icmcb-bordeaux.cnrs.fr

V. Zlatić
Institute of Physics, Bijenicka cesta 46, P.O. Box 304, HR-10001 Zagreb, Croatia
e-mail: zlatic@ifs.hr

V. Zlatić and A. C. Hewson (eds.), *Properties and Applications of Thermoelectric Materials*, 91
NATO Science for Peace and Security Series B: Physics and Biophysics,

1 Introduction

The thermoelectric properties of intermetallic compounds with cerium, ytterbium and other anomalous rare-earth or actinide ions are attracting considerable attention and are widely studied by experimental and theoretical methods. The reason is that an understanding of thermal transport is important for possible applications and transport coefficients are directly related to the electronic structure of a given system. An understanding of heat and charge transport in Kondo systems is of a fundamental importance.

The thermoelectric power (TEP), or the Seebeck coefficient, has been measured for many magnetic systems [1–5], as the search for systems with a large thermopower is obviously of great practical interest [6, 7]. The efficiency of a thermoelectric material is given by the dimensionless quantity called "figure of merit" ZT:

$$ZT = \frac{S^2 T}{\rho \kappa}, \tag{1}$$

where S, ρ and κ are the thermoelectric power, the electrical resistivity and the thermal conductivity, respectively. The figure of merit is directly affected by all three transport coefficients [2], so that one can try maximizing ZT following many different routes. Efficient thermoelectric materials should have $ZT \geq 1$ and the largest ZT has recently been obtained in semiconducting materials in which the thermal conductivity was greatly reduced [8]. There are several recent reviews which discuss the thermoelectric power and the figure of merit of various materials, and describe different contributions due to electrons and phonons [9–11].

The thermoelectric powers of cerium compounds are large and show some typical features. In general, the TEP has two peaks: the high-temperature one due to the crystalline field (CF) splitting and the low-temperature one due to the Kondo scattering on the lowest CF state. In this paper we first review the experimental situation and, then, present various theoretical approaches which explain the experimental data. We also discuss the competition between the Kondo effect and magnetic ordering coming from the intersite magnetic interaction, and explain the effects of pressure in the case of cerium and ytterbium compounds. Finally, we describe the effects of hydrogenation and its influence on cerium compounds, where the hydrogenation can lead to different and even opposite effects.

2 The Thermoelectric Power of Cerium and Ytterbium Systems

2.1 *The Experimental Situation*

The TEP has been measured for many cerium and ytterbium compounds or other anomalous rare-earth systems. A review of the different behaviors can be found in Refs. [12, 13]. The data on cerium and ytterbium intermetallics exhibit some characteristic features which can be used to classify these systems into various groups, as explained in Ref. [12].

In many cerium compounds, a large positive TEP peak is observed at about 100 K. For example, the TEP of $CeAl_3$ and $Ce_{1-x}La_xAl_3$ shows large positive peaks of $40\,\mu V/K$ and $55\,\mu V/K$ at 50 and 40 K, respectively [14, 15]. A very large and negative peak of $-90\,\mu V/K$ has been observed in $YbAl_3$ at about 200 K [16]. Such large positive (negative) peaks have been observed in many cerium (ytterbium) compounds at sufficiently high temperatures and, as shown in the theoretical part of this review, can been explained in terms of the CF effects in Kondo systems [17]. This strong connection between experiment and theory was a good motivation for the study of the TEP in Kondo systems.

Another interesting example is provided by the pressure-dependence of the TEP in $CeCu_2Si_2$, which is shown in Fig. 1 as a function of temperature for pressures up to 6 GPa [18, 19]. A positive peak of the order $20-50\,\mu V/K$ is observed at about 150 K. The compound $CeCu_2Si_2$ is a superconductor at very low temperatures [20] and, as we can see in the inset of Fig. 1, the TEP vanishes in the superconducting state. In the presence of a large magnetic field which destroys the superconductivity, the TEP exhibits a low temperatures a small positive peak [19]. Such a positive peak has also been observed at very low temperatures in $CeAl_3$ and in $Ce_{1-x}La_xAl_3$ alloys [19, 21, 22]. The emergence of a low-temperature positive peak in cerium Kondo compounds is a theoretical challenge, which is addressed in the theoretical part of this review [13].

In CeM_2Si_2 compounds [23], the $CeAu_2Si_2$ behaves as a normal metal with an integer valence and has a very small thermoelectric power. The thermoelectric power of the Kondo compound $CePd_2Si_2$ has at normal pressure a positive peak at 95 K, a negative minimum at 23 K, and then rises to a positive peak at 4 K, which is below the Néel temperature $T_N = 10$ K. The compound $CeRh_2Si_2$ orders magnetically at ambient pressure at $T_N = 39$ K; the TEP has a positive peak at 165 K and another positive peak around 20 K. Between these two peaks the TEP is negative, reaching

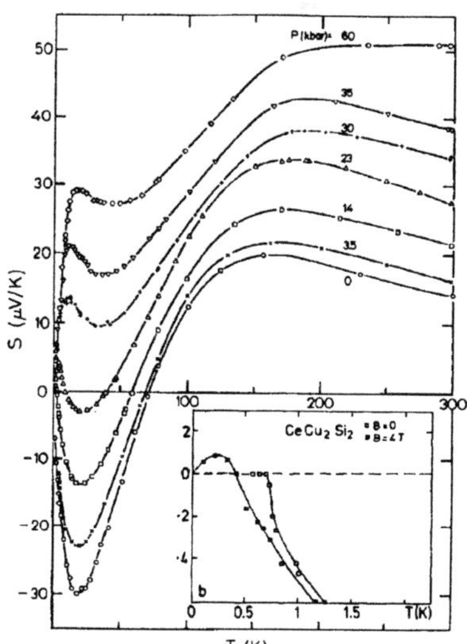

Fig. 1 The experimental curves of the thermoelectric power of $CeCu_2Si_2$ versus temperature at different pressures. (Reproduced from Ref. [13].)

a minimum at the Néel temperature. The TEP of $CeRu_2Si_2$ exhibits also a two-peak structure but is always positive; the high-temperature maximum is at 220 K and the low-temperature one at 22 K [23].

The TEP of many cerium compounds has been measured under very high pressure by Jaccard and coworkers [12]. In $CeCu_2Ge_2$ and $CePd_2Si_2$, the TEP increases with increasing pressure to yield a large positive peak which remains roughly at the same temperature at all pressures. In $CeCu_2Ge_2$, this peak reaches 100 $\mu V/K$ at 150 K, while in $CePd_2Ge_2$, it reaches 60 $\mu V/K$ at 100 K. In both compounds, the negative minimum becomes less pronounced with pressure and disappears at very high pressure [12] (the minimum disappears when the low-temperature peak merges with the high-temperature one).

Another example is provided by the TEP of $CeRu_2Ge_2$, plotted in Fig. 2 as a function of temperature for pressures up to 15 GPa. The change of shape of the TEP follows the same pattern as in other Ce compounds. At ambient pressure, there is a large positive peak at 200 K and a small positive peak at low temperature, separated by a negative minimum. As pressures increases, this minimum is less pronounced and finally disappears completely. At very high pressures, above 10 GPa, the TEP exhibits only a single peak with the maximum of $70-80$ $\mu V/K$ at 200 K [24]. The compound $CeRu_2Ge_2$ shows an antiferromagnetic order at $T_N = 8.55\ K$ and a ferromagnetic order below $T_c = 7.40\ K$ [25–27]. The onset of the long-range magnetic order is pushed with pressure to lower temperatures and disappears at 7.8 GPa [24]. In $CeRu_2(Si_{1-x}Ge_x)_2$, the Néel temperature decreases with decreasing germanium concentration, which can be understood as a chemical pressure effect; the compound $CeRu_2Si_2$ is non-magnetic [26,27].

Although doping produces in cerium alloys similar features to pressure, the connection is not completely straightforward, as can be seen from the magnetic resistivity and the thermopower of $Ce_xLa_{1-x}Cu_{2.05}Si_2$ [28,29] and $Ce_xY_{1-x}Cu_{2.05}Si_2$

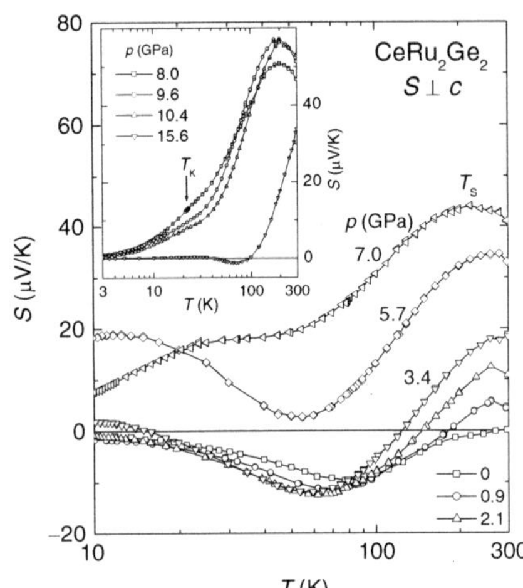

Fig. 2 Experimental dependence of the thermoelectric power of the compound $CeRu_2Ge_2$ with pressure [24].

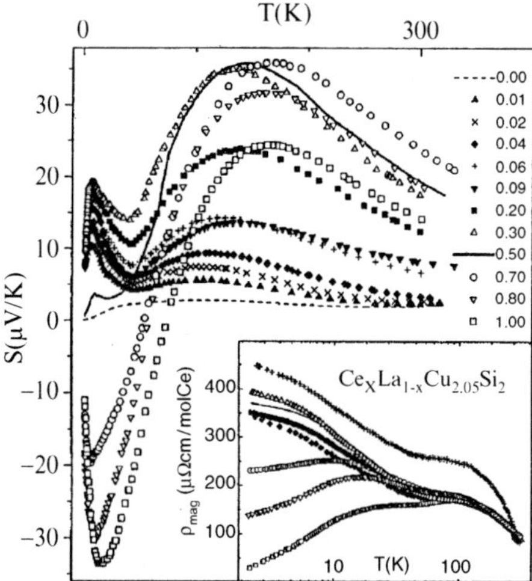

Fig. 3 Thermopower versus temperature for different concentrations x of cerium in $Ce_xLa_{1-x}Cu_{2.05}Si_2$ alloys. Inset : The magnetic resistivity ρ_{mag} versus temperature for different x values [28].

alloys [30]. Figure 3 shows the thermopower of $Ce_xLa_{1-x}Cu_{2.05}Si_2$ plotted versus temperature for different concentrations x of cerium. The data show that for $x = 1$, the thermopower $S(T)$ has a negative minimum at about 20 K, equals to zero at 70 K, and passes through a large positive maximum at about 170 K, where $S = 25\ \mu V/K$. The corresponding magnetic resistivity ρ_{mag} is typical of a stoichiometric compound; it passes through a maximum at approximately 100 K and decreases at higher temperatures as $\log T$. When x decreases to $x = 0.7$, the system is still coherent down to lowest temperatures. For $x \leq 0.5$, however, the resistivity increases with decreasing temperature, which is typical of an alloy-like incoherent behavior. When x decreases below $x = 0.5$, the TEP is always positive but has two peaks. As shown in Fig. 3, the positions of these peaks are almost x-independent but the values at the maxima first increase with decreasing x and then decrease for very small x; the maximum of the TEP is $35\,\mu V/K$ which is found at 150 K for $x = 0.5$ [28, 29], Similar behavior is observed in the case of $Ce_xY_{1-x}Cu_{2.05}Si_2$ alloys [30].

It is now well established that among cerium Kondo compounds we have to distinguish between Kondo and intermediate valence (IV) systems. For example, very large positive peaks of the TEP are observed in the intermediate valence compound $CePd_3$ and in some concentrated $Ce_{1-x}Y_xPd_3$ alloys, where the TEP reaches $90\ \mu V/K$ [31]. The temperature dependence of the TEP of pure cerium metal and $Ce_{0.9}Th_{0.1}$ alloy has also been measured under pressure. The data show a TEP peak at the temperature where the $\gamma - \alpha$ transition takes place, i.e., the magnitude and the position of the TEP maximum varies with pressure or thorium concentration [32].

The distinction between Kondo and IV behavior is complicated by the fact that a smooth transition between the two behaviors can be induced by changing the experimental parameters. The resistivity and thermopower have been measured in intermediate valence compounds $CeNi_x$. In CeNi, one finds a very large positive TEP peak of 55 μV/K at 110 K. In $CeNi_2$ [33] there is an even broader TEP peak of 20 μV/K at 200 K The thermopower of $Ce_xLa_{1-x}Ni$, $Ce_xLa_{1-x}Ni_2$ and $Ce_xLa_{1-x}In_{3+y}$ shows a positive peak at high temperatures, followed by a smaller peak or a shoulder at lower temperatures. Such a behavior can be interpreted in terms of the Kondo scattering [34]. The thermoelectric power of $Ce_xLa_{1-x}Ni_{0.8}Pt_{0.2}$ and $Ce_xY_{1-x}Ni_{0.8}Pt_{0.2}$ alloys shows positive peaks at temperatures of the order 100 K, which can also be related to the Kondo effect [35].

The TEP of $CePd_{1-x}Ni_xAl$ single crystals has also been investigated [36]. The compound CePdAl is a heavy fermion antiferromagnet with Néel temperature $T_N = 2.7$ K and an electronic specific heat constant $\gamma = 270$ mJ/mol K^2. The compound CeNiAl is a non-magnetic valence fluctuator. The TEP of CePdAl shows along the a- and c-axis a positive peak at 180 K and a negative minimum at 30 K, while the TEP along the a-axis shows a second positive peak at 5 K. The TEP along the c-axis shows only an upward curvature at the same temperature. On the other hand, the resistivity of CePdAl along the a- and c-axis shows an increase up to about T_N, then a $\log T$ decrease, and finally a continuous increase up to room temperature; the high temperature maximum of the resistivity lies clearly above the positive peak of the TEP. In $CePd_{1-x}Ni_xAl$ alloys, the magnetic ordering disappears for $x \geq 0.1$ and the electronic specific heat constant of $CePd_{0.9}Ni_{0.1}Al$ approaches $\gamma = 1100\,\mathrm{mJ/mol\,K^2}$. The positive TEP peak continues to be present for $x \geq 0.1$, indicating a non-magnetic heavy fermion system.

The TEP of CeTGe (T = Ni, Pd, Pt) shows large positive values at sufficiently large temperatures [37]. In CePtGe and CePdGe, which are Kondo systems with an antiferromagnetic order, the TEP has a large positive peak at roughly 100 K and another positive peak at 1–2 K. On the other hand, CeNiGe has also a very large TEP which is almost constant between 150 and 300 K; this compound appears to be an intermediate valence compound [37, 38]. Finally, the TEP of the Kondo compound CePdSn, which has a Néel temperature $T_N = 7$ K, exhibits a relatively small positive peak around 100 K, a negative minimum around 20 K, and finally another positive peak around 2 K [38, 39].

The thermopower of several intermediate valence Ce-based compounds, namely Ce_2Ni_2In, Ce_2Rh_2In, Ce_2Ni_2Ga and $CeNi_5Sn$, remains positive between 6 and 300 K. A large positive peak is often observed at about 100–200 K [40]. In Kondo compound $CePt_4In$ [41], a small positive peak occurs at 70 K. This value should be compared with 210 K which is the theoretical value deduced for the CF splitting [17] and is close to the CF splitting derived from the analysis of the magnetic susceptibility [41]. All these compounds are heavy fermions, with very large electronic specific heat constants, reaching $\gamma = 1.75\,J/\mathrm{mol\,K^2}$ in $CePt_4In$ [41]. None of these compounds order magnetically at low temperatures. $CePt_4In$ is apparently a Kondo compound, like $CeAl_3$, while the four other compounds seem to be valence fluctuators with valence close to 3. In cerium compounds, typical Kondo behavior is indicated by a large positive thermopower peak at relatively high temperatures, a small positive peak at low temperatures, and a well defined negative minimum in-between.

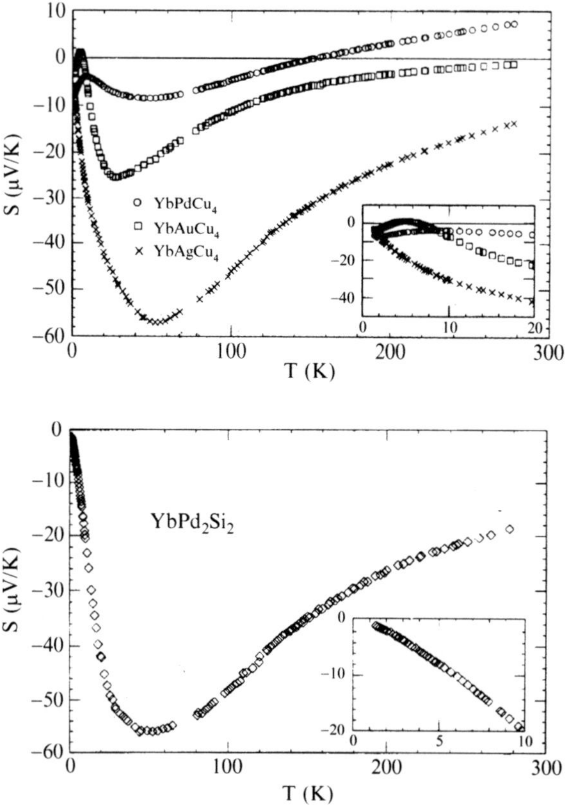

Fig. 4 The thermoelectric powers of the compounds $YbMCu_4$ (with M = Pd, Au, Ag) and $YbPd_2Si_2$ [42].

We now examine the experimental situation regarding the ytterbium Kondo compounds; a review on these compounds can be found in Ref. [13]. We have already mentioned a large and negative TEP peak of $-90\ \mu V/K$ found in $YbAl_3$ at 200 K [16]. A large negative peak, approaching $-60\ \mu V/K$, is shown in Fig. 4, where we plotted the temperature dependence of the TEP of $YbPd_2Si_2$ or $YbMCu_4$, with M = Pd, Au, Ag [42]. The TEP of these Yb compounds approaches zero at very low temperatures and only $YbAuCu_4$ shows a small positive maximum at 6 K. The $YbPd_2Si_2$ and $YbAgCu_4$ appear to be intermediate valence compounds, while $YbAuCu_4$ and $YbPdCu_4$ are Kondo systems which order magnetically below 1 K [42].

A negative TEP peak has been observed in several other ytterbium compounds. The thermopower of $Yb_2Ir_3Al_9$ has a negative peak of $-25\ \mu V/K$ at 40 K and a positive one at 5 K. The variation of the TEP of $Yb_2Rh_3Al_9$ is similar but smaller [43]. A smaller TEP is also found in $YbAu_2$ and $YbAu_3$, where a negative peak is seen around 60–80 K [44]. The thermoelectric power of YbPtIn reaches a minimum of

$-9\,\mu V/K$ at 150 K, another minimum of $-6\,\mu V/K$ at 4 K, and a positive maximum of $+6\,\mu V/K$ at 30 K. The TEP of YbPtIn with two negative minima separated by a positive maximum mirrors the behavior of cerium compounds, although the TEP is here much smaller [45].

The thermoelectric powers of $YbRhIn_5$ and $YbIrIn_5$ are relatively small and change to a small negative value around 100 K. However, these two compounds are divalent non-magnetic ones, and are not really relevant of the present description [46].

The resistivity and the thermoelectric power of $YbInAu_2$, $YbCu_2Si_2$, YbCuAl and YbSi have also been measured under very high pressure [47]. For ytterbium ions, pressure has the opposite effect to that for cerium ions. On the Doniach diagram, pressure tends to decrease the exchange integral J_K and, consequently, pressure changes a non-magnetic heavy fermion into a magnetically ordered one, as observed in YbCuAl [48] or in $YbCu_2Si_2$ above the critical pressure of 8 GPa [47]. The ordering temperature increases with pressure in YbSi [47]. At normal pressure, the thermopowers of $YbCu_2Si_2$ and YbSi are very large and negative around 100 K. In $YbCu_2Si_2$, the position and absolute value of the TEP at the minimum decrease with pressure but there is no positive value at very low temperature even at 9.6 GPa. On the other hand, the thermopower of YbSi has, at very low temperatures, a positive peak which increases with pressure [47].

The thermoelectric power has also been studied in Kondo Uranium compounds such as UBe_{13} [49] and UPd_2Sb [50]. However, they are not relevant of our present study which does not discuss the case of Kondo compounds with more than one f electron [51].

In summary, we have presented in this section the TEP of many cerium and ytterbium systems. The analysis of the data shows clearly the occurrence of a large peak, positive for cerium and negative for ytterbium, at relatively large temperature, typically of about 100 K. In cerium compounds, the low-temperature data often show a second positive peak, like in $CeAl_3$, and a change of sign at intermediate temperatures. In ytterbium compounds, only a few of them show a second negative peak at low temperatures. In the next sections we present the theoretical interpretations of these experimental data.

2.2 Calculation of the Thermopower in the "High Temperature" Regime

The most prominent feature of Kondo systems is the crossover from the high-temperature regime, where the conduction electrons are weakly scattered on the local moments of the $4f$ ions, into the low-temperature Fermi liquid regime, where the conduction and $4f$ electrons are tightly-bound into a non-magnetic singlet. Since the crossover takes place in a rather narrow temperature range, it can be used to define the Kondo temperature T_K of a given system. The properties of the system above and below the Kondo temperature are fundamentally different and can be conveniently described by different effective models. A unified description of the high- and low-temperature regimes has been a major theoretical problem for a long time. The exact solution of the single-impurity Kondo effect valid at all temperatures has, first, been obtained by the renormalization group technique [55] and, then, by the Bethe Ansatz method [56, 57]. The evidence for the "heavy fermion" behaviour has also been provided by the experiments on cerium compounds, like $CeAl_3$ [58], and there are now many reviews describing the experimental and theoretical aspects of the Kondo problem and physics of strongly correlated electrons [59–62].

This section presents the calculation of the thermopower valid in the high-temperature regime. Before providing the details, we remark that the Kondo effect originates from the theoretical attempts to explain the resistivity minimum in dilute alloys with magnetic impurities, like Mn or Fe in copper. A logarithmic increase of the resistivity has been explained by the perturbative treatment of the exchange interaction between the conduction and f electrons [52]. The thermopower has been computed by the same perturbation calculation [53], after previous calculation of the thermopower for normal magnetic systems [54].

In cerium alloys and compounds with $4f$ electrons, the Kondo effect takes place together with the CF effects which split the $(2j+1) = 6$ degenerate ground state either in three doublets, or in a doublet and a quartet, depending on the symmetry of the lattice. The splitting Δ between different CF levels is often of the order of 100 K, which makes the CF effects so important for Ce and Yb systems. The CF effects are particularly important when one goes from a sixfold degenerate level at high temperatures ($T \geq \Delta$) to a two- or fourfold degenerate level at low temperatures ($T \leq \Delta$).

The starting point of the high-temperature ($T > T_K$) calculation is the Anderson Hamiltonian [63], which describes the hybridization between the conduction and correlated 4f electrons; we consider the CF effects in the f-energy term by taking the sum $\sum_M E_M n_M$, where the new energies E_M are the energies of the configuration $4f^1$ split by the crystalline field effect. By making the Schrieffer–Wolff transformation [64], this Hamiltonian is reduced to an effective exchange Hamiltonian which describes both the Kondo and the CF effects. It is defined by Hamiltonian [65, 66]:

$$
\begin{aligned}
H = & \sum_{kM} \varepsilon_k n^c_{kM} \\
& + \sum_{kk'MM'} J_{MM'} c^+_{kM} c_{k'M'} (f^+_{M'} f_M - \delta_{MM'} n_M) \\
& + \sum_{kk'M} \mathscr{V}_{MM} c^+_{kM} c_{k'M} ,
\end{aligned}
$$

where

$$\mathscr{V}_{MM} = V_{MM} - J_{MM} < n_M > . \tag{2}$$

The V_{MM} are classical potential integrals and the $J_{MM'}$ are the exchange integrals given by:

$$J_{MM'} = \frac{| V_{kf} |^2}{2} \left[\frac{1}{| E_M |} + \frac{1}{| E_{M'} |} \right] . \tag{3}$$

Since the CF splitting is much smaller than the average distance E_0 from the 4f level to the Fermi level, different exchange integrals can be approximated by,

$$J_K = \frac{|V_{kf}|^2}{|E_0|} \tag{4}$$

The high temperature electrical resistivity [66], thermoelectric power [17] and thermal conductivity [67] of cerium or ytterbium Kondo compounds have been successfully explained using the above described "Coqblin–Schrieffer" Hamiltonian (2). We do not give the details but present here only the main lines of the calculation for the thermoelectric power [17]. The transport coefficients are calculated using the Boltzmann equation and the perturbation expansion of the Hamiltonian (2) up to the third order in exchange integrals $J_{MM'}$.

The resistivity ρ and the electrical conductivity σ due to the Kondo effect are given by [66]:

$$\frac{1}{\rho} = \sigma = \frac{e^2}{3\pi^2 m}\int k^3\left(-\frac{\partial f_k}{\partial \varepsilon_k}\right)\tau_k d\varepsilon_k \tag{5}$$

and the thermopower is [17]:

$$S = \frac{1}{eT}\frac{\int \varepsilon_k\left(-\frac{\partial f_k}{\partial \varepsilon_k}\right)\tau_k d\varepsilon_k}{\int\left(-\frac{\partial f_k}{\partial \varepsilon_k}\right)\tau_k d\varepsilon_k} \tag{6}$$

where f_k is the Fermi function and $e < 0$. The result for the relaxation time τ_k reads,

$$\frac{1}{\tau_k} = \frac{mkv_0c}{\pi\hbar^3(2j+1)}(R_k + S_k) \tag{7}$$

where the second order term R_k is given by:

$$R_k = \sum_M(|V_{MM}|^2 - 2V_{MM}J_{MM} < n_M >) + \sum_{MM'}\frac{|J_{MM'}|^2 < n_{M'} >}{1 - f_k(1 - \exp[(E_M - E_{M'})/T])} \tag{8}$$

and the third order term S_k is

$$\begin{aligned} S_k = &\sum_{MM'}\sum_m J_{MM'}J_{mM}J_{mM'} < n_{M'} > (1 - \delta_{mM}\delta_{mM'})\frac{g(\varepsilon_k + E_m - E_M)}{1 - f_k(1 - \exp[(E_M - E_{M'})/T])} \\ &-2\sum_M\sum_m(1 - \delta_{mM})V_{MM}|J_{MM}|^2(< n_M > - < n_m >)g(\varepsilon_k + E_m - E_M) \end{aligned} \tag{9}$$

The thermopower calculated in such a way has a pronounced peak which depends on the different parameters Δ, $J_{MM'}$, and also on direct scattering potential $\mathcal{V}$. The peak is located at temperature corresponding to a fraction of the CF splitting, typically $\Delta/3$ [17].

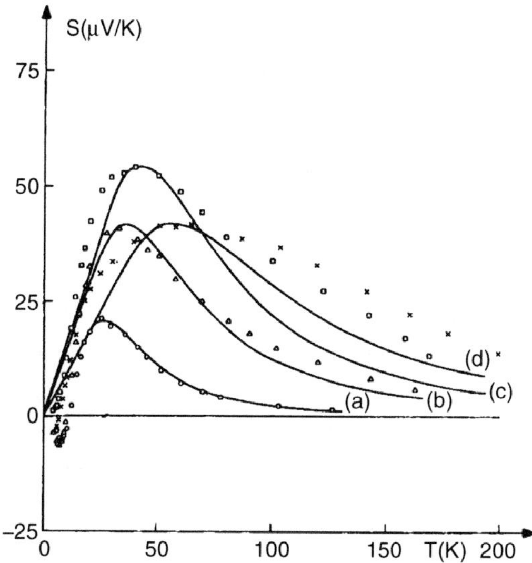

Fig. 5 Comparison between experiment and theory for the thermoelectric power S of dilute $Ce_{1-x}La_xAl_3$ [17]. The experimental points correspond to the differences between the values of S of $Ce_{1-x}La_xAl_3$ and those of $LaAl_3$ for $x = 0.99(O)$, $0.9(\triangle)$, $0.5(\square)$, $0(X)$. The theoretical curves correspond to the two-level case with a doublet ground state for $V_{kf} = 0.07$ eV, $n(E_F) = 2.2$ states/eV at., D = 850 K and the following parameters. Curve (a): $\Delta = 130$ K, $\mathcal{V} = -0.27$ eV, $J_{11} = -0.145$ eV and $E_1 = -395$ K; curve (b): $\Delta = 166$ K, $\mathcal{V} = -0.7$ eV, $J_{11} = -0.2$ eV and $E_1 = -285$ K; curve (c): $\Delta = 203$ K, $\mathcal{V} = -0.85$ eV, $J_{11} = -0.195$ eV or $E_1 = -290$ K; curve (d): $\Delta = 255$ K, $\mathcal{V} = -0.93$ eV, $J_{11} = -0.16$ eV and $E_1 = -355$ K.

Taking a ground state doublet and assuming that the two excited doublets at a distance Δ above the ground state are nearly degenerate, we can fit the TEP of $CeAl_3$ and $Ce_{1-x}La_xAl_3$ [14] with reasonable parameters [17]. Figure 5 shows the TEP data of $Ce_{1-x}La_xAl_3$ for different values of $x = 0$, 0.5, 0.9 and 0.99, together with four theoretical curves obtained for the different parameters. The temperature of the maximum provides an estimate of Δ and the form of the curves allows us to determine $\mathcal{V}$ and J_{11}. The theoretical curve (d) of Fig. 5 is close to the experimental curve of $CeAl_3$. The values of Δ deduced by fitting of the thermopower [17] are in rough agreement with the values required to fit the electrical resistivity [66]. We can conclude that the agreement between the theory and the experiment is satisfying and that the high-temperature TEP peak is due to the Kondo scattering on the $4f$ state split by the CF effect.

However, a closer inspection of the theoretical curves shows that below the CF maximum the TEP monotonically decreases as temperature is reduced, while the experimental curves show a non-monotonic behavior. The TEP of $CeAl_3$ and $Ce_{1-x}La_xAl_3$ [14] goes below the high-temperature maximum through a minimum and, then rises again to a low-temperature maximum. A unified description of Ce systems with the CF splitting valid at all temperatures is difficult but the description of the low-temperature thermopower maximum is relatively straightforward and is presented in the following subsection. By interpolating between the low- and high-temperature calculations, an overall description of the thermopower can be obtained.

2.3 Theoretical Approach to the "Low Temperature" Regime

At low temperatures, the Kondo systems approach the Fermi liquid ground state [55]. In the case of cerium Kondo systems, the Kondo temperature T_K is always much smaller than the CF splitting Δ, so that for $T \ll T_K$ the excited CF states can be neglected [13]. The Coqblin–Schrieffer (CS) model can, therefore, be approximated by the spin$-1/2$ Kondo model or, alternatively, by the single-impurity Anderson model (SIAM) with the number of $4f$ electrons (or holes) close to 1. The TEP of such a model is calculated by the modified perturbation theory [68] which interpolates between the weak coupling limit and the exact atomic solution of the SIAM.

The results are displayed in Fig. 6, where the TEP is plotted as a function of T/T_K for different values of the number of $4f$ electrons n_{4f}. The TEP obtained for $n_{4f} \simeq 1$, has a positive peak at a temperature of order $T_K/2$ and changes sign at about T_K. These curves describe the TEP of a cerium compound with a ground state doublet. A similar, but negative, TEP peak is obtained for $n_{4f}^{holes} \simeq 1$, which corresponds to ytterbium compounds. We point out that the magnitude of the low-temperature TEP peak increases as n_{4f} (or n_{4f}^{holes}) decreases below 1, and that the sign change

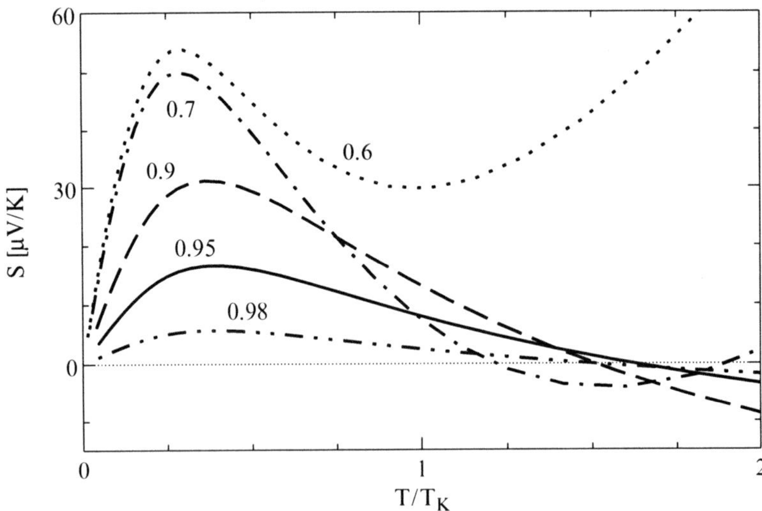

Fig. 6 The thermoelectric power S versus T/T_K for different values of the number n_f of 4f electrons (as indicated on the figure). The corresponding values of T_K are respectively $k_B T_K/\mathrm{W} = 0.011$ (for $n_f = 0.98$), 0.011, 0.012, 0.025 and 0.035 (for $n_f = 0.6$) [13].

above the maximum occurs only for $n_{4f} \simeq 1$. Such a behavior is easy to understand by the following argument. The system with $n_{4f} \simeq 1$ has a low Kondo temperature but is relatively close to the electron–hole symmetry, where the thermopower vanishes exactly. For $n_{4f} \ll 1$, the Kondo temperature is large but the system is further away from the electron–hole symmetry and the TEP increases. This behavior is most clearly seen in pressure experiments which reduce the number of f electrons in Ce compounds. The pressure data show clearly that (i) an increase of pressure enhances the low-temperature Kondo peak of the TEP and moves it to higher temperatures, and that (ii) the sign change occurs only for small pressure and vanishes once n_{4f} becomes small. (A more detailed discussion of pressure effects is given in Refs. [13,69].)

Combining the low- and high-temperature calculations we can explain the experimental results which show the thermopower of Kondo systems with two asymmetric humps, the small one at about $T_k/2$ and the large one at $\Delta/3$. The high-temperature positive (negative) peak is often seen in cerium (ytterbium) systems. The low-temperature peak is also found in some cerium compounds, such as $CeAl_3$ or $CeCu_2Si_2$. However, it is only rarely observed in ytterbium compounds [13].

3 The Kondo Lattice and the Effect of Pressure

3.1 The Doniach Diagram

The single-impurity Kondo effect has been extensively studied after the perturbation calculation of Kondo [52]. An exact solution of the Kondo Hamiltonian has been obtained by the renormalization group method [55] and by the Bethe Ansatz [56, 57]. The single impurity multichannel Kondo model [70, 71], and more realistic impurity models like the Anderson and the Coqblin–Schrieffer model, have also been solved exactly. A Fermi liquid or heavy fermion behaviour has been observed experimentally in cerium compounds like $CeAl_3$ [58].

In the case of a lattice, the problem is much more difficult, as there exists a competition between the Kondo effect on each site which tends to decrease the magnetic moment below T_K and the indirect Rudermann–Kittel–Kasuya–Yosida (RKKY) type interaction which favors magnetic order below the Néel T_N or Curie T_c temperature. This situation is illustrated by the Doniach diagram [72] shown in Fig. 7, where several characteristic temperatures of the system are plotted versus $J_K\rho$. Here, J_K is the exchange intra-site integral and ρ the density of states of the conduction band at the Fermi energy. On Fig. 7, T_{K0} is the Kondo temperature for a single impurity and T_{N0} the Néel (or Curie) temperature without the Kondo effect, which is proportional to J_K^2. For small values of $J_K\rho$, T_{N0} is larger than T_{K0} and the system continues to be ordered magnetically often with a reduction of the magnetic moment due to the Kondo effect. In the opposite case, for large values of $J_K\rho$, T_{N0} is smaller than T_{K0} and the system tends to become non-magnetic with

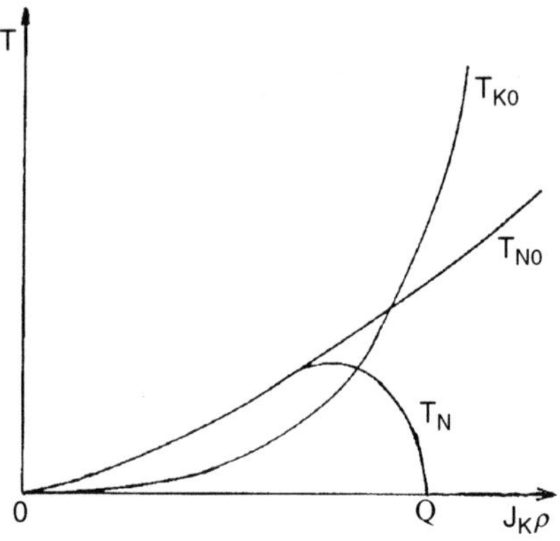

Fig. 7 The Doniach diagram, as explained in the text.

an heavy fermion behaviour. The Doniach diagram provides a simple explanation of the transition between the magnetically ordered state and a non-magnetic heavy fermion state, which takes place at the "Quantum Critical Point" (QCP).

Pressure tends to decrease the distance $| E_0 |$ between the 4f level and the Fermi level and to increase the hybridization term V_{kf} in the case of cerium compounds. It results from the expression (4) that the effect of pressure is to increase J_K in cerium systems and has an inverse effect in ytterbium systems. A transition from a magnetic order to an heavy fermion state has been observed in many cerium compounds [62], like $CeAl_2$, $CeRh_2Si_2$ [73], $CePd_2Si_2$ or the ferromagnetic CeAg [74]. The inverse transition from an heavy fermion state to a magnetic order has been observed for example in $YbCu_2Si_2$ [47] or in YbCuAl [48].

We have to remark finally that the effect of pressure is complex. The Doniach diagram describes the case of Kondo materials where the number of 4f electrons does not change too much under pressure. But, it is well known that the pressure can also induce a large change of the number of 4f electrons and yield, therefore, a change from a magnetic state with n_f close to 1 to an intermediate valence behavior. The difference between the two approaches has been studied very often and, for example, the first explanations of the $\gamma - \alpha$ transition with pressure in cerium metal have invoked either a large decrease of n_f [75] or a large increase of T_K [76,77].

Detailed results can be found in many reviews, cited in Ref. [62]. Then, in the next two sections, we will present two cases showing the effect of pressure on the thermopower of the cerium compounds, such as $CeRu_2Ge_2$ [24], and of the ytterbium compounds.

3.2 The Effect of Pressure in Cerium Compounds

The compound $CeRu_2Ge_2$ is a Kondo system, which orders magnetically at 10 K at normal pressure and the magnetic order disappears at roughly 7.8 GPa. But, we are studying here the pressure dependence of the thermoelectric power S of $CeRu_2Ge_2$ up to the very high pressure of 15.6 GPa [24]. As shown on Fig. 2, at normal pressure, one observes a positive peak at roughly 300 K and another small positive peak at roughly 10 K, separated by an important and negative minimum. S increases with pressure and there are always two positive peaks still separated by an important minimum which becomes positive above 5 GPa, as shown on Fig. 2, At very high pressure, one observes only one very important positive peak. The temperature of order 200 K of the maximum of S does not vary practically with pressure, which corresponds to a roughly constant crystal field splitting versus pressure.

We have presented in Ref. [24] a reasonable theoretical explanation of these experimental results based on the two models in Ref. [13, 17]. In our model to account for the pressure dependence of the thermopower S with pressure, there are in fact two parameters, the crystalline field splitting Δ which is taken constant with pressure and equal to $\Delta = 0.07$eV and $\Gamma = |V_{kf}|^2/D$, where D is the half-width of the conduction band. In our model, Γ varies from 0.06 eV at normal pressure to 0.2 eV at the maximum pressure of 15 GPa and this increase of Γ with pressure corresponds in fact to an increase of the hybridization parameter $|V_{kf}|$.

The spectral density of the single impurity Anderson model (SIAM) with the CF split f states has for small Γ two low-energy peaks which correspond to the CF excitations and a Kondo peak which is located right at the chemical potential. For larger values of Γ the CF resonances merge with the Kondo peak, i.e., the effective degeneracy of the f level increases. Figures 8 and 9 give the theoretical results for the temperature dependence of the TEP obtained for various values of Γ. We note a

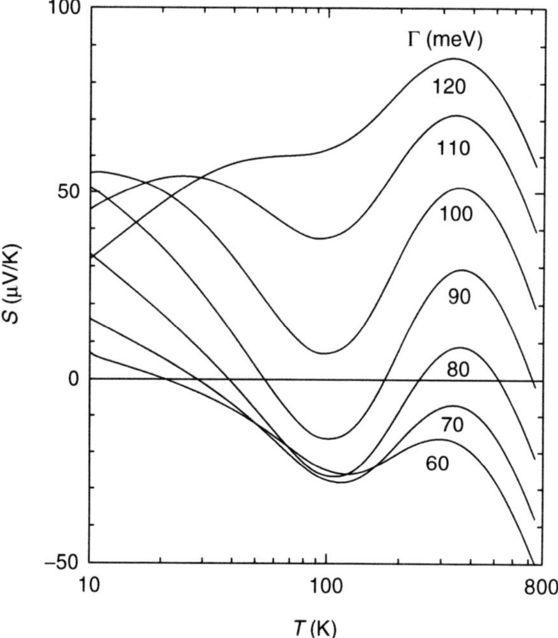

Fig. 8 Theoretical calculations of the pressure dependence of the thermopower S of the compound $CeRu_2Ge_2$ for small Γ values corresponding to low pressures [69].

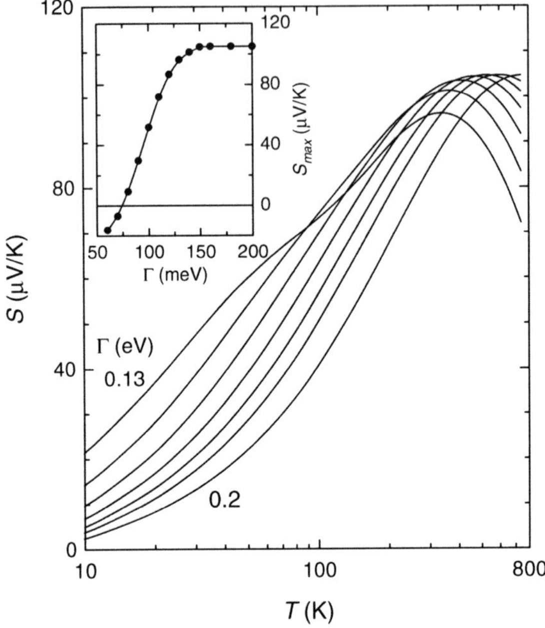

Fig. 9 Theoretical calculations of the pressure dependence of the thermopower S of the compound $CeRu_2Ge_2$ for large Γ values corresponding to high pressures [69].

rather good agreement between the theoretical curves shown in Fig. 8 and 9 and the experimental curves in Fig. 2 which show the temperature dependence of the TEP at various pressures.

Our model can, therefore, account for the increase of the TEP with pressure. It explains the evolution from a relatively small TEP with two positive peaks separated by a negative minimum to a large positive TEP with a single positive peak centered at about $\Delta/3$ [24]. The $CeRu_2Ge_2$ provides a good example for the effects of pressure on the thermopower but similar behavior is also seen in other cerium compounds, like $CeCu_2Si_2$ which is shown on Fig. 1.

3.3 *The Effect of Pressure in Ytterbium Compounds*

The thermopower of intermetallic compounds with ytterbium and cerium ions is quite different, as can seen by comparing the data in Fig. 4 and the first three figures in Section 2. We already noticed that the TEP of Yb systems has a negative minimum at relatively high temperature, which is given by some fraction of the CF splitting. The pressure effects for Yb and Ce systems are also different. In Yb compounds, pressure stabilizes the $4f^{13}$ configuration with one magnetic f hole, while in Ce it stabilizes the non-magnetic $4f^0$ configuration. Thus, the tendency to magnetism is suppressed by pressure in Ce and reinforced in Yb compounds. In the case of ytterbium compounds with n_f^h holes in the $4f$ shell, the $4f^{13}$ configuration corresponds to $n_f^h = 1$ and Kondo effect is found for $n_f^h \simeq 1$. The $4f^{13}$ configuration with two f holes is forbidden due to large Coulomb repulsion.

The thermopower of Yb compounds has recently been computed using the single impurity Anderson model which assumes that the conduction electrons scatter incoherently on magnetic ions [13,69]. The $4f$ hole is placed at negative energy E_{f0} (below the Fermi energy) and it is assumed that $|E_{f0}|$ increases with pressure. Thus, pressure increases the number of $4f$ holes and makes the system more magnetic. It is also assumed that the $4f^{13}$ magnetic configuration is split by the CF into four doublets at energies E_{f0} and $E_f^i = E_{f0} + \Delta_i$ (with $i = 1, \ldots, 3$). The hybridization of the conduction electron and 4f hole is described by the function Γ, introduced in the previous section.

The transport properties of such a model have been calculated by the non-crossing approximation (NCA). The Kondo temperature T_K of the system is estimated from the position of the Kondo peak in the spectral function with respect to the chemical potential. The details regarding the NCA solution of the Anderson model with infinite f-f Coulomb integral U_{ff} can be found in Refs. [69, 70].

For small T_K, i.e., for $n_f^h \simeq 1$, the spectral function exhibits a narrow Kondo peak at the chemical potential μ and several low-energy resonances at energies $\mu + \Delta_i$. The zero-temperature width of these low-energy resonances is much smaller than Γ. The CF splitting is important, because the effective low-temperature degeneracy is determined by the degeneracy of the CF ground state. For large T_K, which is obtained for $n_f^h < 1$, the spectral function does not show the CF excitations and the effective degeneracy of the f-state becomes large.

Figure 10 shows the thermopower due to the scattering of conduction electrons on four CF doublets. We are showing the TEP curves obtained for various values of

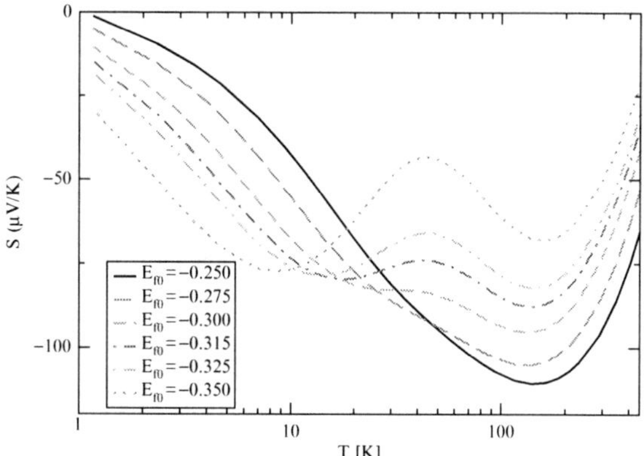

Fig. 10 Thermopower of an ytterbium 4f-ion with the ground state doublet and three excited doublets separated from the ground state by 200, 300 and 500 K, respectively. The thermopower is plotted as a function of $\log T$ for several values of the f-level position E_{f0}, as indicated in the figure. We have T_K = 24, 27, 30, 40, 60, and 90 K, as E_{f0} changes from −0.35 to −0.25 eV. The corresponding room temperature values of n_f^h are 0.95, 0.94, 0.935, 0.93, 0.91, and 0.89, respectively.

E_{f0} and for three excited doublets separated from the ground state by 200 , 300 and 500 K, respectively. The hybridization Γ and the total number of electrons (the sum of conduction electrons and localized f holes) is kept constant. Different values of E_{f0} produce different values of T_K, and n_f^h; the actual values of E_{f0}, T_K and n_f^h are shown in the caption of Fig. 10.

All the TEP curves in Fig. 10 have a broad negative minimum at about 150 K, which corresponds to about one-third of the overall CF splitting Δ. The TEP curves obtained for large $|E_{f0}|$ ($n_f^h \simeq 1$ and relatively small T_K) have an additional negative minimum at temperature given by some fraction of T_K. A similar figure, showing the TEP of Yb systems with larger values of $|E_{f0}|$ (smaller T_K), can be found in Ref. [69]. The overall agreement with experimental data on Yb systems is rather good, despite the quantitative differences regarding the value of the TEP and the magnitude of the high-temperature minimum.

As noticed already, the effect of pressure is complex, as it affects both the hybridization function Γ and the position of the f level $|E_{f0}|$. In the case of $CeRu_2Ge_2$ Kondo compound, we assume that pressure increases the overlap between the $4f$ states and the ligands, i.e., we assume that the main effect of pressure is to increase the hybridization function Γ. This agrees with the experimental data which show that pressure enhances T_K and makes Ce ions less magnetic. In the case of Yb ions the situation is more complicated. Since Yb is a very small ion, we assume that pressure does not affect Γ but increases $|E_{f0}|$, which reduces T_K and makes Yb ions more magnetic. As usual, we assume that pressure does not appreciably change the CF splitting. With these assumptions, we obtain the curves in Fig. 10 which describe the pressure effects in intermetallic compounds with Yb ions.

The TEP curves shown in Fig. 10 can also be used to describe the chemical pressure effects due to the substitution on the ligand sites in ternary Yb systems. For small $|E_{f0}|$, which gives $n_f^h < 1$ and large T_K, the theoretical TEP curves can be compared with the experimental data on $YbAgCu_4$ and $YbPd_2Si_2$ which show a single large negative minimum at high temperatures. The experimental thermopower of $YbAuCu_4$, which shows a small positive maximum at very low temperatures, and a large negative minimum at high temperatures, can be explained by performing the calculation with larger $|E_{f0}|$ or smaller Γ, as discussed in Ref. [69].

4 The Hydrogenated Cerium Compounds

The effect of the insertion of hydrogen in cerium compounds has been extensively studied in the recent years and here we describe the main results regarding the thermopower.

We notice, first, that CeNiIn is an intermediate valence system with a spin fluctuation temperature estimated to about 100 K, while according to magnetization measurements [78, 79] the hydride $CeNiInH_{1.8}$ is a ferromagnetic below the Curie temperature $T_c = 6.8$ K. The insertion of hydrogen in CeNiIn expands the unit-cell volume by 9.6% [78]. The hydrogenation induces a transition from an intermediate valence to a ferromagnetic order and has clearly a role opposite to that of pressure, according to the Doniach diagram.

The compound CeNiGa exists in two crystallographic forms (LTF and HTF, at low and high temperatures, respectively). The hydrogenation which leads to $CeNiGaH_{1.1}$ is accompanied by an increase of the unit cell volume by 19.3% and 15.6% for the LTF and HTF, respectively [80, 81]. The CeNiGa is an intermediate valence compound, while the hydride becomes a trivalent Kondo compound. The electrical resistivity shows a clear Kondo behavior with a maximum at 100 K and two well pronounced $\log T$ behaviors below 20 K and between 100 and 200 K. A value of 100 K has been estimated to be the crystal field splitting Δ from the different experimental data. At very low temperatures below 4 K, there is a possibility to obtain a non-magnetic atom-disorder spin glass behavior, very close to the occurrence of a magnetic ordering [81]. The thermoelectric power S of $CeNiGaH_{1.1}$ is relatively small and has a maximum of $6\,\mu V/K$ around 40 K [81], which corresponds to an important fraction of Δ [17]. However, it is difficult to interpret the change of S under hydrogenation, because the thermopower is very different in the two LTF and HTF phases, but any way there is a global decrease of S under hydrogenation.

Both CeNiGe and its hydride $CeNiGeH_{1.6}$ are intermediate valence compounds, but their magnetic susceptibilities are quite different: the behavior of the compound is characteristic of a nearly Pauli paramagnet, while the susceptibility of the hydride is larger and shows a broad maximum centered around 130 K. The hydrogenation of CeNiGe has been interpreted by considering an important decrease of the spin fluctuation temperature [82] and the thermopower S of the two compounds is relatively large, as shown on Fig. 11. The thermopower of the hydride presents a positive peak of about 38 $\mu V/K$ near 125 K and a weak negative peak of $-5\,\mu V/K$ around 15 K, which can be considered as characteristic of a Kondo-type behaviour with a large Kondo temperature [82]. A similar change under hydrogenation from an intermediate valence behaviour to a Kondo-type behavior with an almost Ce^{3+} state has been also observed in CeIrAl [83, 84] and in CeNiAl [85].

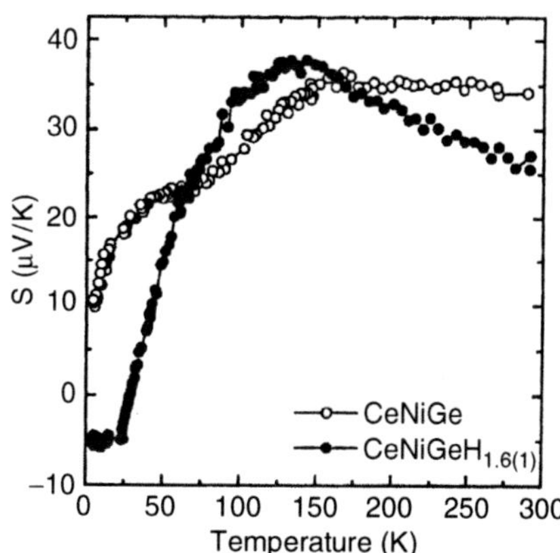

Fig. 11 The thermoelectric powers of the compounds $CeNiGeH_{1.6}$ and CeNiGe versus temperature [82].

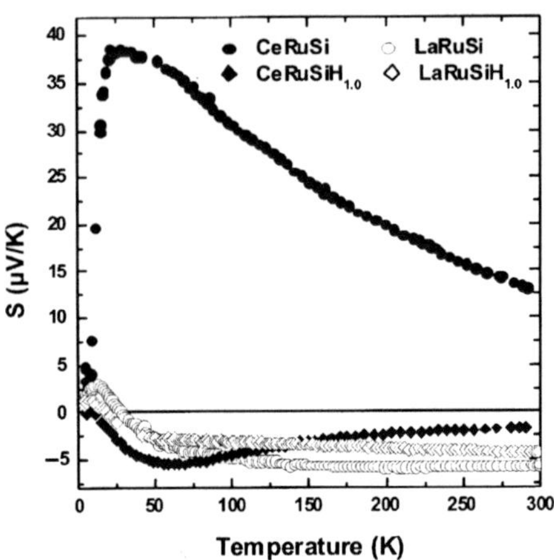

Fig. 12 The thermoelectric powers of the compounds CeRuSiH, CeRuSi, LaRuSiH and LaRuSi versus temperature [86].

CeRuSi is a compound which has a moderate heavy fermion behaviour with an electronic specific heat coefficient $\gamma = 220\ \mathrm{mJ/mol\ K^2}$ and which does not order magnetically at low temperature. On the contrary, the hydrogenated compound CeRuSiH has a smaller $\gamma = 26\ \mathrm{mJ/mol\ K^2}$ value, but orders magnetically with two antiferromagnetic transitions at $T_{N1} = 7.5$ K and $T_{N2} = 3.1$ K, according to magnetization and specific heat measurements [86] or neutron diffraction [87]. Thus, the hydrogenation has here a clear effect opposite to that of pressure, which corresponds to a decrease of J_K in the Doniach diagram. Figure 12 gives the thermoelectric powers of CeRuSi, CeRuSiH and the two equivalent compounds with lanthanum [86]. The thermoelectric power S of CeRuSi has a maximum at a temperature roughly equal to half the value of 45 K of the crystalline field splitting deduced from the resistivity measurements [86]. The curve of S shown on Fig. 12 has a similar shape, in the regime $T > T_K$, to that of $CeAl_3$ which is obviously a stronger heavy fermion compound and the thermopowers of CeRuSiH, LaRuSi and LaRuSiH are negative and very small.

In conclusion of this part, we can say that the previously described compounds, CeNiIn, CeNiGa, CeNiGe, CeIrAl and CeRuSi undergo the type of change under hydrogenation, which corresponds to an effect opposite to that of pressure. More precisely, the behaviour of the two compounds CeNiIn and CeRuSi is the most clear, with a change from a non-magnetic state to a magnetically ordered state and with a decrease of the heavy fermion character, which corresponds to a decrease of J_K in the Doniach diagram. The change of the three compounds CeNiGa, CeNiGe and CeIrAl is less obvious, because they remain non-magnetic but with a clear change from an intermediate valence to an almost trivalent one, which goes in the same direction as a "negative" pressure. In these compounds, the hydrogenation gives a dilatation of the atomic volume cell, opposite to the pressure effect.

But, the hydrogenation can yield a different effect on cerium compounds and we will describe it now. The two ternary compounds CeCoSi and CeCoGe exhibit an antiferromagnetic ordering at 8.8 and 5 K respectively, while the two hydrides

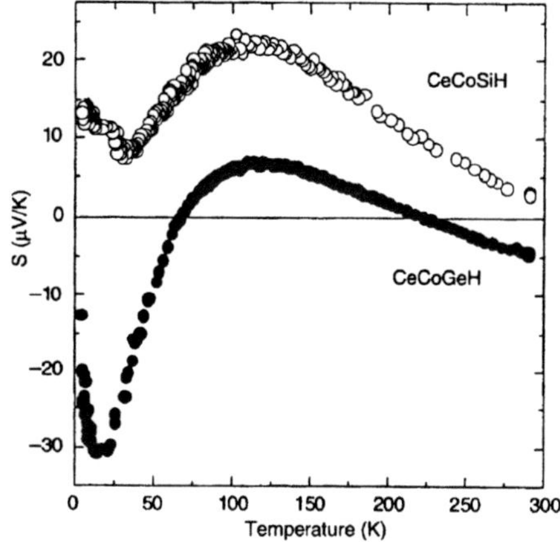

Fig. 13 The thermoelectric powers of the compounds CeCoSiH and CeCoGeH versus temperature [89].

CeCoSiH and CeCoGeH present a non-magnetic spin fluctuation behavior [88, 89]. A spin fluctuation temperature has been estimated from magnetic susceptibility and specific heat measurements to be 130 K in CeCoSiH and 15 K in CeCoGeH. Figure 13 shows the temperature dependence of the thermoelectric power of the two hydrides CeCoSiH and CeCoGeH [89]. The thermopower of CeCoSiH is always positive, with a minimum at low temperatures ($S = 8$ μV/K near 30–35 K) and a broad peak at high temperatures ($S = 22$ μV/K around 110 K), while S of CeCoGeH has a negative minimum of -30 μV/K around 16 K and a small positive peak of 7 μV/K near 110 K. The positive peak observed at 110 K can be interpreted as corresponding to a fraction of the CF splitting Δ, while the low temperature behavior can be related to the Kondo effect, but the determination of a precise Kondo-type temperature is not obvious, especially in this spin fluctuation regime. However, the thermopower of CeCoGeH seems to be shifted from that of CeCoSiH and the difference in behavior between the two hydrides looks like the pressure effect observed in Fig. 2 on $CeRu_2Ge_2$.

The compound CeCuGe is magnetically ordered with a Curie temperature $T_c =$ 10.5 K, while the hydride CeCuGeH does not present any magnetic ordering [82]. The difference in behaviors between the resistivities is surprising, since the resistivity of CeCuGe has a very weak minimum around 20 K above T_c and is then continuously increasing up to 270 K, while the resistivity of CeCuGeH is continuously decreasing with a $\log T$ decrease from 70 to 270 K [82].

Thus, the three compounds CeCoSi, CeCoGe and CeCuGe undergo under hydrogenation a transition from a magnetically ordered behavior to a non-magnetic spin fluctuation one and this change is going in the same sense as the pressure and corresponds to an increase of J_K in the Doniach diagram.

In conclusion, the effect of hydrogenation can have two opposite effects, either a dilatation of the unit cell volume, or a Ce–H bonding which has an opposite effect. The first effect is relatively simple to understand when the hydrogen atoms are sufficiently far from the cerium atoms and have a simple minded effect of dilatation or "negative pressure". The second effect originates from a strong Ce–H interaction, which is bonding throughout the conduction band and tends to destabilize the $4f^1$ configuration and, therefore, to make a possible change from a magnetic order compound to a non-magnetic one in the same direction as pressure in the Doniach diagram. A critical distance Ce–H between Ce and H atoms has been invoked to separate the effect of negative pressure above it from the Ce–H bonding below it. This critical distance has been estimated to be roughly 2.42 Å.

The electronic and magnetic structures of CeCoSi and CeCoSiH have been self-consistently calculated within the local spin density functional theory [90]. These calculations have shown that, in the case of CeCoSi, the chemical effect of hydrogen prevails over the cell expansion effect. On the other hand, the electronic properties of the valence intermediate compound CeNiIn and its ferromagnetic hydride have been self-consistently calculated within a local spin density theory, in order to describe the opposite effect of a negative pressure [91]. Thus, the competition between the cell expansion effect and the Ce–H bonding effect in the insertion of hydrogen is not completely understood from a theoretical point of view and has to be studied more quantitatively in the different cerium compounds.

5 Thermal Conductivity of Anomalous Rare-earth Systems

Another important transport property is the thermal conductivity, which has been much less studied in anomalous rare-earth systems than the electrical resistivity and the TEP [92]. However, this property is important for the figure of merit given by Eq. (1). The separation of the thermal conductivity in the Kondo contribution and the other ones is difficult. In this case, we have to separate the different contributions which enter either the thermal conductivity K if they are coming from different carriers or the "thermal resistivity" W if they are coming from different scattering mechanisms for a given carrier. In fact, when the Kondo contribution is large, we can start directly from W and write it as a sum of a magnetic contribution W_{mag} and of a phonon contribution W_{ph}, taken, as usual, as the thermal resistivity of a similar non-magnetic compound (e.g. with lanthanum):

$$K^{-1} = W = W_{mag} + W_{ph} \tag{10}$$

The magnetic thermal resistivity W_{mag} has been computed for $T > T_K$ with the Hamiltonian (2) describing the Kondo effect and the influence of the crystalline field effect [67]. The most interesting result is that the product $W_{mag}.T$ has a $\log T$ behaviour, for temperatures larger than the crystalline field splitting Δ, in analogy to the well known result of the magnetic electrical resistivity. This contribution in $\log T$ is given by:

$$W_{mag}T = 2\frac{A}{L_0}n(E_F)J^3\frac{\alpha_i^2 - 1}{(2j+1)}\log\frac{k_B T}{D_i} \tag{11}$$

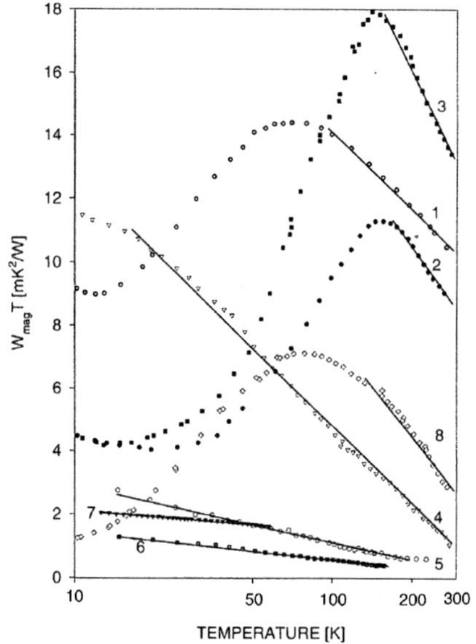

Fig. 14 $W_{mag}T$ (in units of mK^2/W) versus $\log T$ for different heavy fermion compounds. The straight lines give the temperature intervals in which the $\log T$ of $W_{mag}T$ is observed. The numbers correspond to the following compounds: 1 for $CeAl_2$, 2 for $CeCu_2$, 3 for $CeCu_5$, 4 for $CeCu_4Al$, 5 for $PrSn_3$, 6 for $TmGa_3$, 7 for $TmIn_3$ and 8 for $YbCu_4Ag$ [95].

where A is given by:

$$A = \frac{3\pi(m^*)^2 v_0 c}{e^2\hbar^3 k_F^2} = \frac{2\pi m^* \rho c}{ze^2\hbar} \tag{12}$$

and $L_0 = (\pi k_B)^2/3e^2$ is the Sommerfeld value of the Lorenz number.

Figure 14 gives the experimental results on the thermal conductivity of the Kondo compounds $CeAl_2$, $CeCu_2$, $CeCu_5$, $CeCu_4Al$, $YbCu_4Ag$, $PrSn_3$, $TmGa_3$ and $TmIn_3$ and the theoretical $\log T$ behaviour of $W_{mag}.T$ is also plotted for comparison [93–95]. A good agreement between experiment and theory for $T > \Delta$ is, therefore, obtained for these eight compounds of cerium, praseodymium, thulium and ytterbium [95]. A good agreement is also obtained between the J_K values deduced from the measurements of the thermal conductivity and of the electrical resistivity [95]. Finally, in some cerium compounds like $CeCu_2$, the measured ratio of the slopes in $\log T$ of $W_{mag}.T$ for $T > \Delta$ and $T < \Delta$ is clearly much larger than 1 and is in rough agreement with the theoretical value of $35/3$ deduced from Eq. (11) for α_i equal to 6 and 2.

6 Conclusions

We have reviewed the thermoelectric power and the thermal conductivity of anomalous rare-earth systems. The experimental data on cerium Kondo systems often show a thermopower which has a broad positive maximum at about 100 K, a small positive peak at very low temperatures, and a negative minimum between these two peaks.

The thermopower of some ytterbium Kondo compounds has a negative maximum at high temperatures, while a small negative peak is more rarely observed at very low temperatures. The experimental behavior of the thermopower is explained by the theoretical models which give a large peak at temperatures $T_K \ll T \leq \Delta$ and a smaller peak at about $T \simeq T_K/2$. The same models accounts for the $\log T$ behavior observed in the product $T \cdot W_{mag}$ for the magnetic contribution to "thermal resistivity" of many rare-earth Kondo compounds.

We have also described the effect of pressure, using as an example the thermopower of $CeRu_2Ge_2$ which shows at ambient pressure two positive peaks separated by a large negative minimum. Our theoretical analysis explains very well the pressure-induced increase of the TEP from relatively small values at ambient pressure to a single positive peak centered at about $T \simeq \Delta/3$. These results show clearly that the charge transfer is of crucial importance for explaining the pressure experiments.

We have also reviewed the TEP measurements on hydrogenated cerium compounds. The hydrogenation effects are complicated by the fact that the insertion of hydrogen leads in some cases to "negative pressure" and in other cases to the chemical bonding between Ce and H atoms. These two effects influence the thermopower in opposite ways. Different behaviors induced by the hydrogenation are related to a critical distance Ce–H, which has been estimated to be 2.42 Å. However, further studies are needed for a better understanding of these effects.

The search for cerium and ytterbium intermetallics with large thermoelectric powers is of considerable interest for application. In the case of heavy fermion compounds, the thermopower is often large, the thermal conductivity can be relatively small, but as long as the electrical resistivity remains large, the figure of merit cannot be very large. We hope that further studies of cerium and ytterbium alloys and compounds might improve the situation, so as to reduce the thermal conductivity and increase the thermoelectric power, to finally increase the figure of merit.

Acknowledgements We acknowledge financial support from the European COST-ECOM (P16) Action and The International Institute for Complex and Adaptive Matter (I2CAM). The NATO support to the ARW Workshop *Properties and Applications of Thermoelectric Materials*, September 20–26, 2008 Hvar, Croatia, is acknowledged as well. One of us (VZ) acknowledges support from the Ministry of Science of Croatia under Grant No. 0035-0352843-2849 and the NSF under Grant No. DMR-0210717.

References

1. N. F. Mott and H. Jones, *The Theory of the Properties of Metals and Alloys*, Clarendon Press, Oxford (1936).
2. A. F. Ioffe, *Physics of Semiconductors*, Publishing House of the U.S.S.R, Moscow (1957).
3. R. D. Barnard, *Thermoelectricity in Metals and Alloys*, Taylor & Francis, London (1972).
4. J. M. Ziman, *Electrons and Phonons*, Clarendon Press, Oxford (1960).
5. G. D. Mahan, *Many-Particle Physics*, Plenum Press, New York (1981).
6. H. J. Born, S. Legvold and F. H. Spedding, J. Appl. Phys. **32**, 2543 (1961).

7. B. Coqblin, *The Electronic Structure of Rare-Earth Metals and Alloys : The Magnetic Heavy Rare-Earths* Academic Press, London (2007).
8. C. B. Vining, Nature, **451**, 132 (2008) and references therein.
9. G. Mahan, B. Sales and J. Sharp, Phys. Today, 50, 42 (1997).
10. A. Majumdar, Science, **303**, 777 (2004).
11. A. Dauscher, B. Lenoir, H. Scherrer and T. Caillat, Recent Res. Devel. Mat. Sci., **3**, 181 (2002).
12. P. Link, D. Jaccard and P. Lejay, Physica B, **225**, 207 (1996).
13. V. Zlatic, B. Horvatic, I. Milat, B. Coqblin, G. Czycholl and C. Grenzebach, Phys. Rev. B **68**, 104432 (2003).
14. K. H. J. Buschow and H. J. van Daal, Solid State Comm. **8**, 363 (1970).
15. P. B. van Aken, H. J. van Daal and K. H. J. Buschow, Phys. Lett., **49A**, 201 (1974).
16. H. J. van Daal, P. B. van Aken and K. H. J. Buschow, Phys. Lett., **49A**, 246 (1974).
17. A. K. Bhattacharjee and B. Coqblin, Phys. Rev. B **13**, 3441 (1976).
18. D. Jaccard, J. M. Mignot, B. Bellarbi, A. Benoit, H. F. Braun and J. Sierro, J. Mag. Mag. Mater. **47–48**, 23 (1985).
19. G. Sparn, W. Lieke, U. Gottwick, DF. Steglich and N. Grewe, J. Mag. Mag. Mater. **47–48**, 521 (1985).
20. F. Steglich, J. Aarts, C. D. Bredl, W. Lieke, D. Meschede, W. Franz and H. Schafer, Phys. Rev. Lett. **43**, 1892 (1979).
21. D. Jaccard and J. Flouquet, Helv. Phys. Acta, **60**, 108 (1987).
22. R. Cibin, D. Jaccard and J. Sierro, J. Mag. Mag. Mater. **108**, 107 (1992).
23. A. Amato and J. Sierro, J. Mag. Mag. Mater. **47–48**, 526 (1985).
24. H. Wilhelm, D. Jaccard, V. Zlatic, R. Monnier, B. Delley and B. Coqblin, J. Phys.: Condens. Matter, **17**, S823 (2005).
25. J. D. Thompson, Y. Uwatoko, T. Graf, M. F. Hundley, D. Mandrus, C. Godart, L. C. Gupta, P. C. Canfield, A. Migliori and H. A. Borges, Physica **B 199–200**, 589 (1994).
26. H. Wilhelm and D. Jaccard, Physica, **B 259–261**, 79 (1999).
27. P. Haen, H. Bioud and T. Fukuhara, Physica, **B 259–261**, 85 (1999).
28. M. Ocko, B. Buschinger, C. Geibel and F. Steglich, Physica, **B 259–261**, 87 (1999).
29. M. Ocko, D. Drobac, B. Buschinger, C. Geibel and F. Steglich, Phys. Rev. B, **64**, 195106 (2001).
30. M. Ocko, C. Geibel and F. Steglich, Phys. Rev. B, **64**, 195107 (2001).
31. D. Jaccard, M. J. Besnus and J. P. Kappler, J. Mag. Mag. Mater. **63–64**, 572 (1987).
32. C. Fierz, M. Decroux and J. Sierro, J. Mag. Mag. Mater. **47–48**, 517 (1985).
33. U. Gottwick, K. Gloos, S. Horn, F. Steglich and N. Grewe, J. Mag. Mag. Mater. **47–48**, 536 (1985).
34. J. Sakurai, T. Ohyama and Y. Komura, J. Mag. Mag. Mater. **63–64**, 578 (1987).
35. J. Sakurai, J. C. Gomez-Sal and J. Rodriguez Fernandez, J. Mag. Mag. Mater. **140–144**, 1223 (1995).
36. D. Huo, T. Kuwai, T. Mizushima, Y. Isikawa and J. Sakurai, Physica, **B 312–313**, 232 (2002).
37. J. Sakurai, D. Huo, D. Kato, T. Kuwai, Y. Isikawa and K. Mori, Physica, **B 281–282**, 98 (2000).
38. D. Huo, K. Mori, T. Kuwai, H. Kondo, Y. Isikawa and J. Sakurai, J. Phys. Soc. Japan, **68**, 3377 (1999).
39. D. Huo, K. Mori, T. Kuwai, S. Fukuda, Y. Isikawa and J. Sakurai, Physica, **B 281–282**, 101 (2000).
40. D. Kaczorowski and K. Gofryk, Solid State Comm. **138**, 337 (2006).
41. A. P. Pikul, D. Kaczorowski, Z. Bukowski, K. Gofryk, U. Burkhardt, Yu. Grin and F. Steglich, Phys. Rev. B **73**, 092406 (2006).
42. R. Casanova, D. Jaccard, C. Marcenat, N. Hamdaoui and M. J. Besnus, J. Mag. Mag. Mater. **90–91**, 587 (1990).
43. O. Trovarelli, C. Geibel, B. Buschinger, R. Borth, S. Mederle, M. Grosche, G. Sparn, F. Steglich, O. Brosch and L. Donnevert, Phys. Rev. B **60**, 1136 (1999).
44. O. Nakamoto, T. Nobata, S. Ueda, Y. Nakajima, T. Fujioka and M. Kurisu, Physica, **B 259–261**, 154 (1999).

45. O. Trovarelli, C. Geibel, R. Cardoso, S. Mederle, R. Borth, B. Buschinger, M. Grosche, Y. Grin, G. Sparn and F. Steglich, Phys. Rev. B **61**, 9467 (2000).
46. Z. Bukowski, K. Gofryk and D. Kaczorowski, Solid State Comm. **134**, 475 (2005).
47. K. Alami-Yadri, H. Wilhelm and D. Jaccard, Physica, **B 259–261**, 157 (1999).
48. J. M. Mignot and J. Wittig, *"Valence Instabilities"*, ed. P. Wachter and H. Boppart, North-Holland, Amsterdam p. 203 (1982).
49. U. Gottwick, R. Hel, G. Sparn, F. Steglich, K. Vey, W. Assmus, H. Rietschel, G. R. Stewart and A. L. Giorgi, J. Mag. Mag. Mater. **63–64**, 341 (1987).
50. K. Gofryk, D. Kaczorowski and A. Czopnik, Solid State Comm. **133**, 625 (2005).
51. N. B. Perkins, M. D. Nunez-Regueiro, J. R. Iglesias and B. Coqblin, Phys. Rev. B **76**, 125101 (2007).
52. J. Kondo, Progr. Theoret. Phys. (Kyoto) **32**, 37 (1964).
53. J. Kondo, Progr. Theoret. Phys. (Kyoto) **34**, 372 (1965).
54. T. Kasuya, Progr. Theoret. Phys. (Kyoto) **22**, 227 (1959).
55. K. G. Wilson, Rev. Mod. Phys., **47**, 773 (1975).
56. N. Andrei, Phys. Rev. Lett. **45**, 379 (1980).
57. P. B. Wiegmann, Sov. Phys. JETP Lett. **31**, 392 (1980) and Phys. Lett. **A 80**, 163 (1980).
58. K. Andres, J. E. Graebner and H. Ott, Phys. Rev. Lett. **35**, 1779 (1975).
59. A. C. Hewson, *"The Kondo Problem to Heavy Fermions"*, Cambridge University Press, Cambridge (1993).
60. J. M. Lawrence, P. S. Riseborough and R. D. Parks, Rep. Mod. Phys. **44**, 1 (1981).
61. P. Schlottmann, Phy. Rep. **181**, 1 (1989).
62. B. Coqblin, *Lectures on the Physics of Highly Correlated Electron Systems X*, Vietri sul Mare (2005), AIP Conference Proceedings, **846**, pp. 3–93 (2006).
63. P. W. Anderson, Phys. Rev. **124**, 41 (1961)
64. J. R. Schrieffer and P. A. Wolff, Phys. Rev. **149**, 491 (1966).
65. B. Coqblin and J. R. Schrieffer, Phys. Rev. **185**, 847 (1969).
66. B. Cornut and B. Coqblin, Phys. Rev. B **5**, 4541 (1972).
67. A. K. Bhattacharjee and B. Coqblin, Phys. Rev. B **38**, 338 (1988).
68. A. Martin Rodero, E. Louis, F. Flores and C. Tejedor, Phys. Rev. B **33**, 1814 (1986).
69. V. Zlatic and R. Monnier, Phys. Rev. B **71**, 165109 (2005).
70. N. E. Bickers, D. L. Cox, and J.W. Wilkins, Phys. Rev. B **36**, 2036 (1987); N. E. Bickers, Rev. Mod. Phys. **59**, 845 (1987).
71. P. Sacramento and P. Schlottmann, Phys. Rev **40**, 431 (1989) and Phys. Rev **43**, 13294 (1991).
72. S. Doniach, in *Valence Instabilities and Related Narrow-Band Phenomena*, ed. R. D. Parks, Plenum Press, New York (1976), p. 168.
73. T. Graf, J. D. Thompson, M.F. Hundley, R. Movshovich, Z. Fisk, D. Mandrus, R. A. Fischer and N. E. Phillips, Phys. Rev. Lett. **78**, 3769 (1997).
74. A. Eiling and J. S. Schilling, Phys. Rev. Lett. **46**, 364 (1981).
75. B. Coqblin and A. Blandin, Adv. Phys. **17**, 281 (1968).
76. M. Lavagna, C. Lacroix and M. Cyrot, Phys. Lett. **90A**, 210 (1982) and J. Phys. F **13**, 1007 (1983).
77. J. W. Allen and R. M. Martin, Phys. Rev. Lett. **49**, 1106 (1982).
78. B. Chevalier, M. L. Kahn, J. L. Bobet, M. Pasturel and J. Etourneau, J. Phys.: Condens. Matter **14**, L 365 (2002).
79. J. Sanchez Marcos, J. Rodriguez-Fernandez, B. Chevalier and J. L. Bobet, Physica B **378–380**, 799 (2006).
80. B. Chevalier, J. L. Bobet, E. Gaudin, M. Pasturel and J. Etourneau, J. Solid State Chem. **168**, 28 (2002).
81. B. Chevalier, J. Sanchez Marcos, J. Rodriguez-Fernandez, M. Pasturel and F. Weill, Phys. Rev. B **71**, 214437 (2005).
82. B. Chevalier, M. Pasturel, J. L. Bobet, F. Weill, R. Decourt and J. Etourneau, J. Solid State Chem. **177**, 752 (2004).
83. S. K. Malik, P. Raj, A. Sathyamoorthy, K. Shashikala, N. Harish Kumar and Latika Menon, Phys. Rev. B **63**, 172418 (2001).

84. P. Raj, A. Sathyamoorthy, K. Shashikala, C. R. Venkateswara and S. K. Malik, Solis State Comm. **120**, 375 (2001).
85. J. L. Bobet, B. Chevalier, B. Darriet, M. Nakhl, F. Weill and J. Etourneau, J. Alloys and Compounds, **317–318**, 67 (2001).
86. B. Chevalier, E. Gaudin, S. Tencé, B. Malaman, J. Rodriguez-Fernandez, G. André and B. Coqblin, Phys. Rev. B **77**, 014414 (2008).
87. S. Tencé, G. André, E. Gaudin and B. Chevalier, J. Phys.: Condens. Matter **20**, 255239 (2008).
88. B. Chevalier, S. F. Matar, J. Sanchez Marcos and J. Rodriguez-Fernandez, Physica B **378–380**, 795 (2006).
89. B. Chevalier, S. F. Matar, M. Menetrier, J. Sanchez Marcos and J. Rodriguez-Fernandez, J. Phys.: Condens. Matter **18**, 6045 (2006).
90. B. Chevalier and S. F. Matar, Phys. Rev. B **70**, 174408 (2004).
91. S. F. Matar, B. Chevalier, V. Eyert and J. Etourneau, Solid State Sci. **5**, 1385 (2003).
92. E. Bauer, Adv. Phys. **40**, 417 (1991).
93. E. Bauer, E. Gratz, G. Hutflesz, A. K. Bhattacharjee and B. Coqblin, J. Magn. Magn. Mater. **108**, 159 (1992).
94. E. Bauer, E. Gratz, G. Hutflesz, A. K. Bhattacharjee and B. Coqblin, Physica B **186–188**, 494 (1993).
95. Z. Kletowski and B. Coqblin, Solid State Comm., **135**, 711 (2005).

Thermoelectrics Near the Mott Localization–Delocalization Transition

K. Haule and G. Kotliar

1 Introduction

1.1 Weakly Correlated Systems

The dream of accelerating the discovery of materials with useful properties using computation and theory is quite old, but actual implementations of this idea is recent. Successes in material design using weakly correlated materials, are due, to a large degree, to a two important developments:

1. Approximate implementations of the first principles density functional theory, which are relatively accurate and computationally efficient
2. Robust implementation of algorithms which are highly reproducible and widely available

Density functional theory based approaches gives reliable estimates of the total energy, and are an excellent starting point for computing excited state properties of weakly correlated electron systems. These approaches allows the evaluation of transport coefficients using very limited, or no empirical information, and are beginning to be used in conjunction with data mining technique and combinatorial searches.

1.2 Strongly Correlated Electron Systems

Since a large number of interesting physical phenomena, such as high temperature superconductivity and large Seebeck coefficients, are realized in strongly correlated materials, there is a great interest in the possibility of carrying out rational material design with correlated materials.

K. Haule and G. Kotliar
Physics Department and Center for Materials Theory, Rutgers University,
136 Frelinghuysen Road, Piscataway, NJ
e-mail: haule@physics.rutgers.edu, kotliar@physics.rutgers.edu

V. Zlatić and A. C. Hewson (eds.), *Properties and Applications of Thermoelectric Materials*, 119
NATO Science for Peace and Security Series B: Physics and Biophysics,

The theoretical situation in this area, however, is a lot more uncertain. For example, the issue of whether the two dimensional one band Hubbard model supports superconductivity or not is still very open [10]. Given that this model is an extraordinary oversimplification of realistic materials, it is hard to contemplate explaining, let alone predicting experimental results in materials that require a much more elaborate models for their description. The prospect of predicting properties of materials which have not yet been synthesized is even more daunting. In this chapter we will argue that this assessment is overly pessimistic, and we will give some reasons why we expect a rapid progress in the coming years through the interplay of qualitative reasoning, new theoretical methods, and experiments. We will then describe some attempts in gaining experience in this field, and the lessons that we have learned in the process using thermoelectric performance as an example.

1.2.1 Dynamical Mean Field Theory

The advent of Dynamical Mean Field Theory (DMFT) removed many difficulties of the traditional electronic structure methods. DMFT describes Mott insulators, as well as correlated metals. It treats quasiparticle bands and Hubbard bands on the same footing, and, unlike simpler approaches such as LDA+U, is able to describe the multiplet structure of correlated solids. The latter is being inherited from open shell atoms and ions. DMFT has been successful in accounting for the behavior observed in correlated materials ranging from plutonium to vanadium oxides and has even made some predictions, which have been confirmed by experiment. This suggests that the approach is reasonably accurate, in the sense that it gives a zeroth orderpicture of correlated materials, not too close to criticality. Ten years ago, a combination of DMFT with electronic structure methods, LDA+DMFT, wasproposed [1, 8, 15] and accurate implementations are being actively developed across the world. Just like LDA, these tools connect the atomic positions with the physical observables using very little information from experiment, and therefore they have the potential to accelerate material discovery.

Predicting the phase diagram of strongly correlated materials is an extremely difficult problem. Correlated materials have many competing phases, which are very close in energy. This poses serious difficulties to traditional many body approaches. (i) Terms in the Hamiltonian, present in the actual material, but absent in the model Hamiltonian, can exchange the stability of two very different phases. (ii) Finite size effects or boundary conditions can artificially stabilize a phase, which is not stable in experiment.

DMFT divides the solution of the many-body problem of a solid state system into two separate distinct steps. Common to many mean field approaches, a given Hamiltonian can have many distinct DMFT solutions, describing various possible phases of a material. Which phase is realized for a given value of parameters (temperature, volume, stress, doping concentration of impurities, etc.) is determined by comparing the free energy of the different DMFT solutions. A lot of important information can be obtained from the first step alone, when combined withexperimental information.

If one knows that for some value of parameters certain phase is realized in material, one can use DMFT to explore the properties of that phase, and optimize desired physical property, side stepping the difficult issue of the comparing the free energies of competing phases. The free energy difference can be computed at a later stage.

2 The Process of Rational Material Design

Figure 1 describes schematically the rational material design process. It begins with a qualitative idea, which is then tested by a calculation. One of the major advances of realistic DMFT implementations such as LDA+DMFT or GW+DMFT is that now this calculation can be made material specific, resulting in a set of predictions that can be tested experimentally. The experimental results can either rule out the qualitative idea, in which case the process stops, or reinforce and refine the idea. Experiments also help to calibrate the computational methods, which in turn lead to an improved material specific prediction in the next iteration. Not only materials with improved properties $M_1, M_2, M_3, \cdot$ result from this approach, but in addition, this process tests theoretical ideas in an unbiased way, deepens our understanding of materials physics, and refines the accuracy of computational tools. Large databases of existing materials are created (e.g. http://icsd.ill.eu/icsd/index.html), which are starting to be used, in combination with the first principle methods, for data mining techniques. Using the crystal structure information from the database, the first principles methods can identify potentially promising materials, which can then be analyzed experimentally.

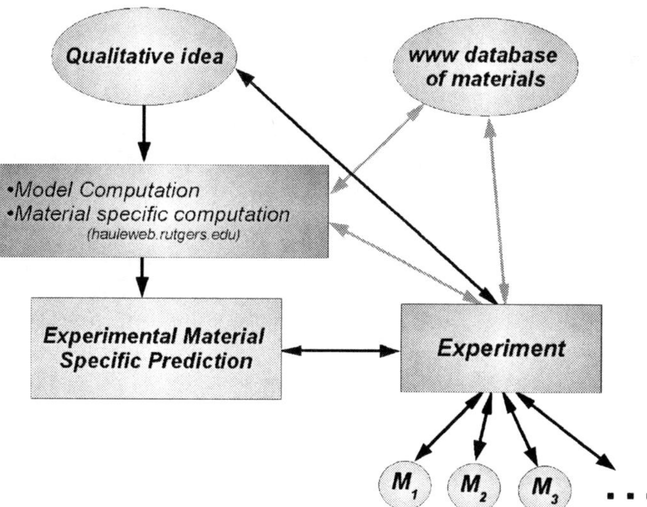

Fig. 1 A schematic drawing of the rational material design process. It relies on condensed matter theory, material databases and realistic DMFT implementations, and it involves a close and iterative interplay of theory and experiment.

3 Thermoelectricity of Correlated Materials

3.1 *Formalism*

The transport coefficients that govern the thermopower, electric and thermal conductivity can be expressed in terms of the matrix of kinetic coefficients A_m relating the electric and thermal currents J, J_Q to the applied external fields $\nabla\mu/T$, $\nabla T/T^2$. Transport quantities become $S = -(k_B/e)(A_1/A_0)$, $\sigma = (e^2/T)A_0$, $\kappa = k_B^2[A_2 - A_1{}^2/A_0]$. The thermoelectric response thus reduces to the evaluation of kinetic coefficients.

The thermoelectric figure of merit is defined by

$$ZT = \frac{S^2\sigma T}{\kappa + \kappa_{phonon}}, \tag{1}$$

where T is the absolute temperature, σ is the electrical conductivity, S is the Seebeck coefficient or thermopower, and κ (κ_{phonon}) is the electron (phonon) contribution to the thermal conductivity.

The Wiedemann–Franz law is an approximate relation that allows us to estimate the ratio of the electronic contribution to the thermal conductivity (κ) and electric conductivity (σ). It postulates that the Lorentz number, $L = \kappa/(\sigma T)$, is weakly material dependent.

Its value at low temperatures is given by $(\pi^2/3)(k_B/e)^2 = 2.44 \times 10^{-8} W\Omega/K^2$. We will return to the Lorentz number at higher temperatures later in this article. If we ignore the thermal conductivity of the lattice, the figure of merit can be written as $ZT = S^2/L$, hence to have a promising figure of merit (ZT close to or larger than one) it is necessary to have S bigger than the basic scale $k/e = 86 \times 10^6 \mathrm{V/K}$. The thermal current of an interacting electronic system was determined first by Mahan and Jonson [11]. Reference [11] discusses a model containing electrons interacting with phonons, and the review [16] discusses the general case of the electron–electron interactions (see also Ref. [22]).

DMFT expresses the one particle Greens function in terms of a local self energy of an impurity model, satisfying a self consistency condition. Practical evaluation of the transport coefficients becomes possible in the approximation of small vertex corrections. This was first done by Schweitzer and Czycholl [25] (see also Ref. [23]). For the Hubbard-like interactions, there are no contributions from the non-local Coulomb interactions, and the neglect of the vertex corrections can be justified rigorously in the limit of infinite dimensions [13]. The same is true, but far less obvious, for the thermal current, as it was shown in Ref. [22]. In the multi-orbital situation, the vertex corrections to the conductivity need to be examined on a case by case basis, and do not necessarily vanish, even in infinite dimensions. With this approximation, the LDA+DMFT transport coefficients reduce to

$$A_m^{\mu\nu} = \pi T \int d\omega \left(-\frac{df}{d\omega}\right)\left(\frac{\omega}{T}\right)^m \sum_{\mathbf{k}} \mathrm{Tr}[v_{\mathbf{k}}^{\mu}(\omega)\rho_{\mathbf{k}}(\omega)v_{\mathbf{k}}^{\nu}(\omega)\rho_{\mathbf{k}}(\omega)] \tag{2}$$

where $v_{\mathbf{k}}^{\mu} = -i \int d\mathbf{r} e^{-i\mathbf{kr}} \frac{\partial}{\partial x^{\mu}}$ are velocities of electrons and $\rho_{\mathbf{k}}$ is the electron spectral density

$$\rho_{\mathbf{k}}(\omega) = \frac{1}{2\pi i}[G_{\mathbf{k}}^{\dagger}(\omega) - G_{\mathbf{k}}(\omega)]. \tag{3}$$

The weakly interacting case appears as a limiting case where the spectral function becomes a delta function $\rho_{\mathbf{k}}(\varepsilon) = \sum_i \delta(\varepsilon - \varepsilon_{\mathbf{k}i})$. One can therefore formulate the problem of the optimization of the figure of merit as the problem of optimizing a functional of spectral functions, with self energies which are realizable from an Anderson impurity model, with a bath satisfying the DMFT self-consistency condition.

3.2 *Thermoelectricity near the Mott Transition: Qualitative Considerations*

Following the early developments of DMFT and its successful application to the theory of the Mott transition in three dimensional transition metal oxides [6], it was natural to use this approach to formulate and answer the question of whether we should look for good thermoelectrics near the Mott localization–delocalization transition. The theoretical answer to this central issue of this article is no, but perhaps yes.

There were several reasons to suspect that proximity to the localization–delocalization transition is good for thermoelectricity:

1. Sharp structures in the density of states lead to large S in simple theories [17]. The modern theory of the Mott transition predicts a quasiparticle peak, which narrows as the transition is approached. And this could result in a large thermoelectric response.
2. One can think on a qualitative level of the thermoelectric coefficient as the entropy per carrier. In the incoherent regime, one could imagine that each carrier can transport a large amount of entropy. The incoherent regime, above a characteristic coherence temperature T^*, is easy to access near a localization–delocalization transition, because the proximity to this boundary makes T^* low.
3. Orbital degeneracy increases the number of carriers and would be expected to increase the figure of merit. There are many orbitally degenerate three dimensional correlated transition metal oxides.

Reference [14] considered a model of the prototypical doped insulator $LaSrTiO_3$, which has been carefully investigated in a series of papers [26]. The thermoelectric properties of this system had not been investigated at that time. Early DMFT studies accounted for the divergence of the linear term of the specific heat, and the susceptibility, as well as the existence of a quasiparticle peak in the spectra [28].

The Hall coefficient, however, coincides with the band theory calculations, and is non-critical near the Mott transition [12]. It is possible to analyze the DMFT transport equations in two regimes: (i) $T \ll T^*$, where the electronic transport is controlled by band-like coherent quasiparticles, well described in momentum space, (ii) $T \gg T^*$ when the electron is better described as a particle in real space, and the transport is diffusive [14] (see below). The second regime is well described by the high temperature expansion, valid for $T > D$ (D is the bandwidth), and by comparison with approximate numerical solutions of the DMFT equations. In Ref. [14] it was noticed that the thermoelectric response of the high temperature regime matches smoothly with the low response at low temperatures, valid for $T \ll T^*$.

3.3 Application to $LaSrTiO_3$

An approximate numerical solution of the DMFT equations for the titanides was shown to interpolate smoothly between the high temperature and low temperature region. This is consistent with the idea that DMFT reconciles the band picture at low energies and low temperatures, with the particle picture at high energies and high temperatures. The temperature scale here is set by the coherence temperature T^*. Taking a tight binding parametrization suitable for the titanites, the figure of merit as a function of temperature and doping is reproduced in Fig. 2.

The behavior of the thermoelectric power near the Mott transition is shown in Fig. 3. Notice that at low doping, the contribution from the lower Hubbard

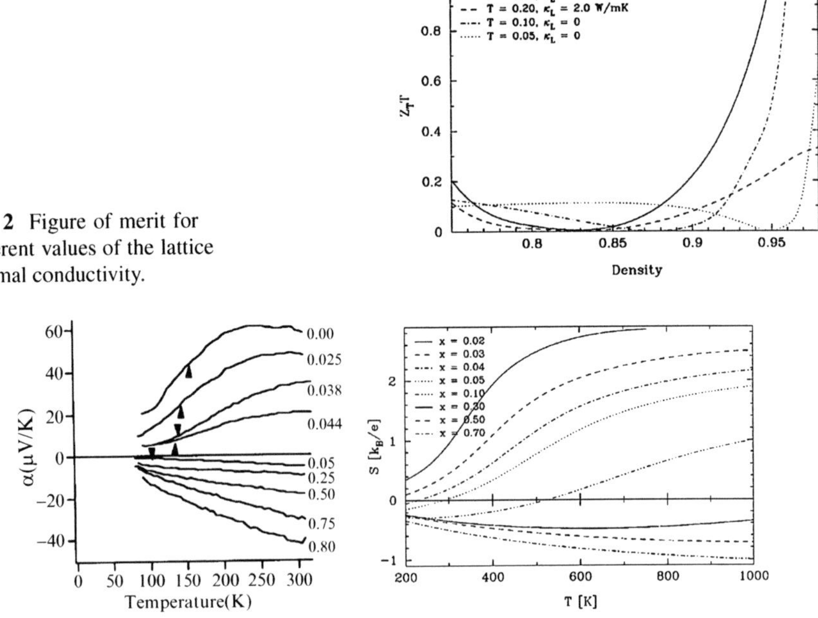

Fig. 2 Figure of merit for different values of the lattice thermal conductivity.

Fig. 3 Experimental (*left panel*) and theoretical computations of the thermoelectric power of the titanites from Refs. [7, 14].

band dominates and the thermoelectric power is positive while at high doping the quasiparticle contribution dominates and the thermoelectric power is electron-like. Measurements near the Mott transition were carried out a few years later [7], and they are qualitatively, but not quantitatively, similar to the theory. This is to be expected, given the various approximations that were made (the electronic structure, the lattice distortion, and crystal field effects ignored, the impurity solvers used were very approximate).

3.4 Low Temperature Regime

$LaSrTiO_3$ is described by a multi-band Hubbard model. At low temperatures, the Fermi liquid theory is valid. The slope of the real part of the self energy scales as $1 - 1/Z$, where Z is the quasiparticle residue. The quadratic part of the self energy is related to the quasiparticle lifetime, which is small in the Fermi liquid regime.

Under these assumptions, we can rewrite a simpler expressions for the transport coefficients A_n of a multiband Hubbard model at low temperatures

$$A_n^{\mu\nu} = \frac{Nk_BT}{8} \int_{-\infty}^{\infty} dx \frac{x^n}{\cosh^2(x/2)} \frac{\Phi_{\mu\nu}(xT + \mu - \Sigma'(xT))}{\Sigma''(xT)}, \tag{4}$$

where $\Phi_{\mu\nu}$ is the transport function defined by $\Phi_{\mu\nu} = \sum_{\mathbf{k}} v_{\mathbf{k}}^{\mu} v_{\mathbf{k}}^{\nu} \delta(\omega - \varepsilon_{\mathbf{k}})$ and $\Sigma''(\omega)$ is the imaginary part of the electron self-energy.

At low temperatures, A_0 and A_2 are simply estimated by replacing $\Sigma''(\omega)$ by its quadratic approximation, $\Sigma''(\omega) \sim \frac{\gamma_0}{Z^2}(\omega^2 + \pi^2 T^2) \equiv \Sigma^{(2)}(\omega)$. We then obtain

$$A_{2n} = \frac{Z^2}{T} \frac{Nk_B}{2\gamma_0 \pi^2} E_{2n}^1 \Phi_{\mu\nu}(\mu_0),$$

where $\mu_0 = \mu - \Sigma'(0)$ and

$$E_n^k = \int_{-\infty}^{\infty} \frac{x^n dx}{4\cosh^2(x/2)[1 + (x/\pi)^2]^k}$$

are numerical constants of the order unity.

On the other hand, this approximation neglects particle-hole asymmetry and gives zero thermoelectricity since $E_1^1 = 0$. There are two sources of particle-hole asymmetry. One is obtained by expanding the transport function in Eq. (4) to first order, which describes the particle-hole asymmetry in the electronic velocities, contained in the bare band structure of the problem. This term can be approximated by the LDA Seebeck coefficient divided by quasiparticle renormalization amplitude Z.

The second contribution is the result of the particle hole asymmetry of the scattering rate. It involves subleading *cubic* terms in the self energy, which scale near the Mott transition as

$$\Sigma''(\omega) = \Sigma^{(2)}(\omega) + \Sigma^{(3)}(\omega) + \cdots,$$
$$\Sigma^{3}(\omega) = \frac{(a_1\omega^3 + a_2\,\omega T^2)}{Z^3}, \tag{5}$$

and a_1, a_2 are constants of order unity (even terms in frequency are not important). This leads to the following expression for the thermoelectric coefficient:

$$A_1 = Z\frac{Nk_B}{2\gamma_0\pi^2}\left[\Phi'_{\mu\nu}(\mu_0)\,E_2^1 - \Phi_{\mu\nu}(\mu_0)\,(a_1E_4^2 + a_2E_2^2)/\gamma_0\right], \tag{6}$$

where $\Phi'(x) = d\Phi(x)/dx$.

Unfortunately it has proved to be very difficult to estimate the magnitude of the coefficients a_1 and a_2. It is important to develop intuition into when these terms are important and their sign. Since in many cases, LDA predicts the correct sign of the thermoelectric power at low temperatures, perhaps the scattering time particle-hole asymmetry Eq. (5) is not dominant in the $LaTiO_3$ system but should be investigated carefully in other materials.

At low temperature, the thermoelectric coefficients is

$$S = -\frac{k_B}{|e|}\frac{k_BT}{Z}\left[\frac{\Phi'(\mu_0)}{\Phi(\mu_0)}\frac{E_2^1}{E_0^1} - \frac{a_1E_4^2 + a_2E_2^2}{\gamma_0E_0^1}\right], \tag{7}$$

which clearly scales as T/Z with Z vanishing at the Mott transition. Since the linear term of the specific heat γ scales as $1/Z$ the ratio $S/(\gamma T)$ in a Hubbard-like model approaches a finite value as Z vanishes:

$$\frac{S}{\gamma T} = -\frac{3}{|e|}\frac{1}{D(\mu_0)}\left[\frac{\Phi'(\mu_0)}{\Phi(\mu_0)}\frac{E_2^1}{E_0^1} - \frac{a_1E_4^2 + a_2E_2^2}{\gamma_0E_0^1}\right]. \tag{8}$$

The first part of the ratio depends only on the bare band-structure quantities and is not effected by strong correlations. The second part, however, is due to the asymmetry of the quasiparticle lifetime, and might be less universal and more material and correlation specific. This question deserves further study.

For the $LaSrTiO_3$ system, we estimated its value numerically using LDA+DMFT [21] and we include its value in the plot of Behnia et al. [2] in Fig. 4. In Ref. [2] it was observed that the ratio $S/\gamma T$ is weakly material dependent in a large number of materials which they compiled. From the theoretical point of view, the weak dependence of the ratio of Behnia et al. on material can be view as a validation of the local approximation, since the most material dependence is embodied in the quantity Z, which cancels in the ratio $S/(\gamma T)$. This suggest that the DMFT approach holds great promise for the search of good thermoelectric materials. Deviations

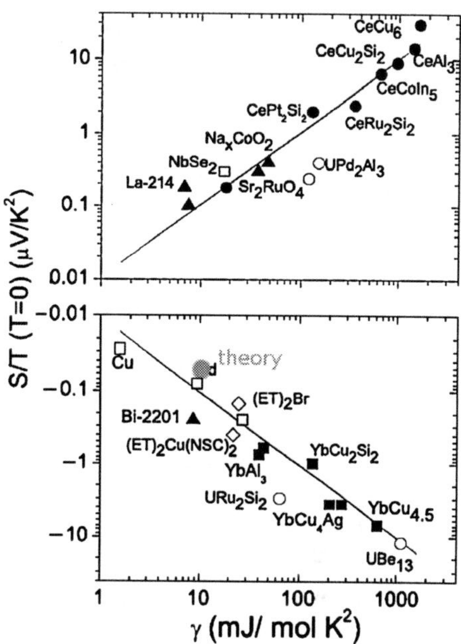

Fig. 4 Bhenia Jaccard Flouquet plot from Ref. [2]. The theoretical point obtained on the LaSrTiO3 system with 20% doping away from the Mott insulator is also shown in the same graph.

from universality arise from the variations of the bare density of states and from the effects of the cubic terms in the self energy that were not included in the analysis of Ref. [14]. It would be interesting to return to this problem using modern LDA+DMFT tools.

3.5 High Temperature Results

In the high temperature region, the expansion of the solution of the DMFT equations led to the celebrated Heikes formula for the Seebeck coefficient. In this limit, thermopower is given by $S = \mu/(eT)$, where μ is the chemical potential. The exact diagonalization of the atomic problem gives a set of atomic eigenvalues E_m and their degeneracies d_m. The chemical potential is then determined from the partition sum

$$n = \sum_m d_m e^{-\beta(E_m - \mu N)}/Z, \tag{9}$$

where n is the number of electrons in a correlated orbital. Hence, the valence of the solid, n, can be used to predict the high temperature value of thermopower.

For the case of $n \leq 1$, which is relevant for the titanides, the expressions for transport quantities take the explicit form:

$$\sigma = \frac{e^2}{a\hbar}\pi N(D\beta)\gamma_0 \frac{\frac{n}{N}(1-n)}{[\frac{n}{N} + (1-n^2)]^2}, \tag{10}$$

$$S = \frac{k_B}{e}\log\frac{n}{N(1-n)}, \tag{11}$$

$$\kappa = \frac{k_B D}{a\hbar} \pi N (D\beta)^2 \gamma_2 \frac{\frac{n}{N}(1-n)}{[\frac{n}{N} + (1-n)]^2}. \tag{12}$$

Here N is the spin and orbital degeneracy, and n is the electron density, D is half of the bare bandwidth and γ_0, γ_2 are numerical constants of order unity.

Notice that at high temperature the Lorentz number is given by $L = (k/e)^2 (D/kT)^2 \gamma_2/\gamma_0$. Hence the Lorentz number in a model with a fixed number of particles and finite bandwidth goes to zero at high temperatures. Thus eventually the electronic thermal conductivity becomes less than the lattice conductivity and the latter controls the figure or merit. This effect was modeled in the dashed curve of Fig. 2, where the effects of the lattice thermal conductivity was modeled by a constant 2.0 W/mK. The inclusion of the lattice thermal conductivity resulted in a dramatic reduction of the figure of merit. We can interpret the high temperature DMFT results for the thermal transport using a well known equation $\kappa = \frac{1}{3} v_F c_V l$, where v_F is the Fermi velocity, c_V the specific heat, and l the electron mean free path. Since the specific heat decreases as $(D/T)^2$, the mean free path has saturated to a lattice spacing, and the velocity of the electrons is of the order of v_F. This is consistent with the value of the conductivity if one uses the Einstein relation $\sigma = D_c dn/d\mu$ with $dn/d\mu \approx 1/T$ and the charge diffusion constant $D_c = v_F l$. Here the mean free path l is of the order of the lattice spacing, and the Fermi velocity v_F is approximately temperature independent.

4 Towards Material Design

4.1 *Rules for Good Correlated Thermoelectricity*

From the theoretical analysis it becomes clear why $LaSrTiO_3$ is not a good thermoelectric material. The contributions from the Hubbard bands and the quasiparticle peak have opposite signs, and they compete with each other in the interesting temperature regime, when T is comparable to T^*. This observation leads to empirical rules for the search for good correlated thermoelectric materials:

1. The optimal performance (when the thermal conductivity of the lattice is taken into account) occurs in the crossover region $T \approx T^*$. Hence one should tune T^* to the temperature region where the thermoelectric device operates. One should also reduce the electronic thermal conductivity (and therefore also the electric conductivity) until it becomes comparable to the lattice thermal conductivity, but not any further.
2. In the crossover regime, both the quasiparticle bands and the Hubbard bands contribute to the transport. Hence one should try to optimize *both* high temperature and low temperature expressions for the figure of merit. Therefore good candidates for thermoelectricity have quasiparticle carriers and Hubbard band carriers of the same sign.

We see that $LaSrTiO_3$ does *NOT* satisfy the second rule, and hence its figure of merit is not large. The quasiparticle contribution to the thermopower is electron-like while the lower Hubbard band contribution is hole-like.

In contrast, the cobaltates have one hole in the lower Hubbard band, and the quasiparticle contribution evaluated from the LDA [27] has a positive sign, hence it satisfies the second rule for good thermoelectricity (assuming that the contribution from the asymmetry in the scattering rate does not modify the sign of the Seebeck coefficient).

An investigation of the density driven Mott transition in the context of a two band Hubbard model, with one electron per site, was carried out in Ref. [20], and the qualitative analysis is very similar to the doping driven Mott transition.

4.2 Emergent Mottness

Interest in thermoelectricity near the doping driven Mott transition leads to theoretical and experimental investigations of $La_{1-x}Sr_xTiO_3$ and CoO_2Na_x for small values of the concentration parameter x. Both theory and experiment suggest that the thermoelectric figure of merit is not very large in this regime. On the other hand, the vicinity of the band insulator end, $La_{1-x}Sr_xTiO_3$ [19] and CoO_2Na_x (see Fig. 5

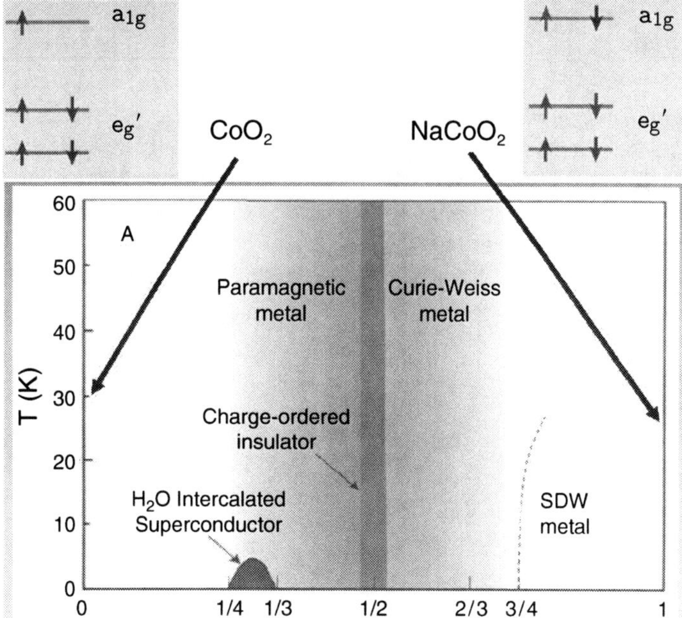

Fig. 5 Phase diagram of CoO_2Na_x compound from Foo et al. [5]. The Mott insulating side at $x = 0$ has low thermopower, while the thermopower is greatly enhanced in the vicinity of the band insulator at $x = 1$.

for the phase diagram) were shown to have promising thermoelectric performance. Should we conclude that Mottness is bad for thermoelectricity? Not necessarily, after all, clear signatures of correlation were found in more realistic modeling of doped band insulators, once the impurity potentials of the dopant atoms were taken into account [24]. The impurity potential was found to restrict the spatial regions available for the motion of the electricity and heat carriers. In this restricted configuration space, the occupancy of the electrons is close to integer and Mott physics is realized.

We have suggested that a similar situation occurs in the electron gas close to the metal insulator transition. Here, the long range Coulomb interaction generates short range charge crystalline lattice order. The occupancy of these lattice sites is close to integer filling, suggesting that the character of the metal to insulator transition is that of a Wigner–Mott transition [3]. The mechanism, spatial or orbital differentiation results in a restricted low energy configuration, making Mott physics relevant. This mechanism is quite general, and operates in other materials such as the ruthenates [18]. It could be called emergent Mottness or super-Mottness, and contains similar physics to the orbital selective Mott transition phenomena. Hence (super) Mottness might be relevant for high performance thermoelectricity after all!. It would be useful to reconsider the most recent advances in thermoelectric materials in this light, and investigate the local magnetic susceptibility at the impurity sites of the high performance thermoelectrics [4, 9].

5 Outlook

The outlook for material design in the field of thermoelectric is quite promising. DMFT seems to capture qualitative trends in oxides of practical interest, furthermore we have simple qualitative ideas, which can be refined and tested with tools of ever increasing precision. In this context, the new thermoelectric modules to be developed in conjunction with the new generation of LDA+DMFT codes, look very appealing. In conjunction with the renewed experimental efforts in this field, the future looks very promising.

Acknowledgements K. Hanle is supported by a grant of the ACS of the Petroleum Research Fund. G. Kotliar is supported by the NSF.

References

1. Anisimov, V.I., Poteryaev, A., Korontin, M., Anohkin A., and Kotliar, G., J. Phys. Condens. Matter **9**, 7359 (1997).
2. Behnia, K., Jaccard, D., and Flouquet, J., Phys.: Condens. Matter **16**, 5187 (2004).
3. Camjayi, A., Haule, K., Dobrosavljevic, V., Kotliar, G., Nature Phys. **4**, 932 (2008).

4. Chen, N., Gascoin, F., and Jeffrey Snyder, G., Mueller, E., Karpinski, G., and Stiewe, C., App. Phys. Lett. **87**, 171903 (2005).
5. Foo, M.L., Wang, Y., Watauchi, S., Zandbergen, H.W., He, T., Cava, R.J., and Ong, N.P., Phys. Rev. Lett. **92**, 247001 (2004).
6. Georges, A., Kotliar, G., Krauth, W., and Rozenberg, M., Rev. of Mod. Phys. **68**, 13–125 (1996).
7. Hays, C.C., Zhou, J.-S., Markert, J.T., and Goodenough, J.B., Phys. Rev. B **60**, 10367 (1999).
8. Held, K., Adv. Phys. **56**, 829 (2007).
9. Hsu, K.F., Loo, S., Guo, F., Chen, W., Dyck, J.S., Uher, C., Hogan, T., Polychroniadis, E.K., and Kanatzidis, M.G., Science **303**, 818 (2004).
10. Imai Y., Imada, M., J. Phys. Soc. Japan **75**, 094713 (2006).
11. Jonson, M., and Mahan, G. D., Phys. Rev. B, **21**, 4223 (1980).
12. Kajueter, H., Kotliar, G., and Moeller, G., Phys. Rev. B **53**, 16214 (1996).
13. Khurana, A., Phys. Rev. Lett. **64**, 1990 (1990).
14. Kotliar, G., and Palsson, G., Phys. Rev. Lett. **80**, 4775 (1998).
15. Kotliar, G., Savrasov, S.Y., Haule, K., Oudovenko, V.S., Parcollet, O., and Marianetti, C., Rev. Mod. Phys. **78**, 865 (2006).
16. Mahan, G.D., Solid State Phys. **51**, 81 (1998).
17. Mahan, G.D., and Sofo, J.O., Proc. Natl. Acad. Sci. U.S.A. **93**, 7436 (1996).
18. Neupane, M., Richard, P., Pan, Z.-H., Xu Y., Jin, R., Mandrus, D., Dai, X., Fang, Z., Wang, Z., Ding, H., arXiv:0808.0346.
19. Okuda, T., Nakanishi, K., Miyasaka, S., Tokura, Y., Phys. Rev. B **63**, 113104 (2001).
20. Oudovenko, V.S., and Kotliar, G., Phys. Rev. B **65**, 075102 (2002).
21. Oudovenko, V.S., Palsson, G., Haule, K., Kotliar, G., and Savrasov, S.Y., Phys. Rev. B **73**, 035120 (2006).
22. Paul, I., and Kotliar, G., Phys. Rev. B **67**, 115131 (2003).
23. Pruschke, T., Jarrell, M., and Freericks, J., Adv. Phys. **44**, 187 (1995).
24. Sarma, D.D., Barman, S.R., Kajueter, H., Kotliar, G., Europhys. Lett. **36**, 307 (1996).
25. Schweitzer, H., and Czycholl, G., Phys. Rev. Lett. **67**, 3724 (1991).
26. Tokura, Y., Taguchi, Y., Okada, Y., Fujishima, Y., Arima, T., Kumagai, K., Iye, Y., Phys. Rev. Lett. **70**, 2126 (1993).
27. Xiang, H.J., and Singh, D.J., Phys. Rev. B **76**, 195111 (2007).
28. Yoshida, T., Ino, A., Mizokawa, T., Fujimori, A., Taguchi, Y., Katsufuji, T., Tokura, Y., Europhys. Lett., **59**, 258 (2002).

Quasiparticles, Magnetization Dynamics, and Thermopower of Yb-Based Heavy-Fermion Compounds

G. Zwicknagl

Abstract We calculate the dispersion of the heavy quasiparticles which form in the heavy-fermion compound $YbRh_2Si_2$ at low temperatures due to the Kondo effect. We analyze the consequences of the tetragonal Crystalline Electric Field (CEF) which removes the degeneracy of the Yb 4f states. The calculated Fermi surface is compared to recent deHaas-vanAlphen (dHvA) data. The calculated quasiparticle bands may serve as starting point for the calculation of transport integrals and thus a more quantitative understanding of transport properties.

1 Introduction

Thermoelectricity of systems with correlated electrons has gained much interest during the past years [1]. The interest derives from the fact that thermoelectric power provides a sensitive tool to study low-energy excitations. A wide range of compounds including heavy-fermion systems, organic metals and various oxides display a striking correlation between the Seebeck coefficient and the electronic specific heat. Introducing a dimensionless ratio relating the two signatures of mass renormalization Behnia et al. [1] demonstrated that the absolute value of this ratio can characterize the ground state of correlated electron systems. A simple relation between α/T and γ is found for free electrons scattering off impurities. It was pointed out by Miyake and Kohno [2] that a similar behavior should be expected in heavy-fermion materials provided the low-energy excitations can be described in terms of quasiparticles. This result was confirmed by Zlatic et al. [3] starting from microscopic model Hamiltonians. Although the existence of a simple correlation between the Seebeck coefficient and the specific heat has been convincingly demonstrated for many systems the actual value of the above-mentioned ratio should depend upon

G. Zwicknagl
Institut fuer Mathematische Physik, Technische Universitaet Braunschweig, Mendelssohnstr. 3, 38116 Braunschweig, Germany
e-mail: g.zwicknagl@tu-bs.de

V. Zlatić and A. C. Hewson (eds.), *Properties and Applications of Thermoelectric Materials*, 133
NATO Science for Peace and Security Series B: Physics and Biophysics,

bandstructure effects in real materials. A quantitative analysis hence requires detailed information on the dispersion of the renormalized quasiparticles.

This paper presents calculations of the quasiparticle bands in the heavy-fermion compound $YbRh_2Si_2$. For this material, detailed studies of the low-temperature thermopower have been performed recently [4]. $YbRh_2Si_2$ belongs to the same family of ternary compounds as the archetype heavy fermion superconductor $CeCu_2Si_2$. The isostructural compounds of this family exhibit a great variety of ground states and were extensively studied to clarify the interplay between the formation of magnetic order and heavy fermion behavior [5]. $YbRh_2Si_2$ has gained much interest since its low-temperature behavior is highly anomalous. The compound undergoes an antiferromagnetic phase transition at a Néel temperature $T_N = 70\,mK$ [6,7]. By applying a tiny magnetic field of $B_c = 60\,mT$ the magnetic order is suppressed and the system is driven through a quantum critical point (QCP) into a strongly renormalized Landau Fermi Liquid (LFL) state. The changes of the ground state and the low-energy excitations across the QCP manifest themselves in numerous thermodynamic and transport properties. A prominent example is the Hall coefficient whose abrupt changes indicate Fermi surface reconstructions (see e. g. [6] and references therein). Pronounced anomalies are found in the thermoelectric power which changes sign when the system is tuned through the QCP [4].

The energy bands of $YbRh_2Si_2$ in the local-moment phase prevailing at high temperatures are well understood theoretically. Recent calculations adopting the LDA+U ansatz [8] yield an energy dispersion which is in reasonable agreement with the corresponding results from Angular-Resolved Photoemission Spectroscopy (ARPES). In this regime, the Fermi surface and the single-particle excitations are derived from the weakly correlated conduction states which can be described by standard bandstructure methods.

Our calculations focus on the Landau Fermi Liquid regime where the relevant excitations are strongly renormalized (heavy) quasiparticles. The dispersion of the quasiparticle bands is calculated by means of the Renormalized Band (RB) method [9] which has been shown to describe successfully quasiparticles and Fermi liquid instabilities in a great variety of heavy-fermion systems [9–12]. The renormalized bands provide a basis for evaluating transport integrals [13] which eventually may yield a more quantitative understanding of thermopower in heavy-fermion systems.

The present paper is organized as follows: In Section 2 we explain the technical aspects of our calculations. Section 3 briefly discusses the local 4f dynamics. The Fermi surface predicted by the RB method is analyzed in Section 4. A brief summary and an outlook will be given in Section 5.

2 Model and Calculational Scheme

The band-structure results reported in the present paper were obtained by the fully relativistic formulation of the linear muffin-tin orbitals (LMTO) method [14–16]. The spin-orbit interaction is fully taken into account by solving the Dirac equation.

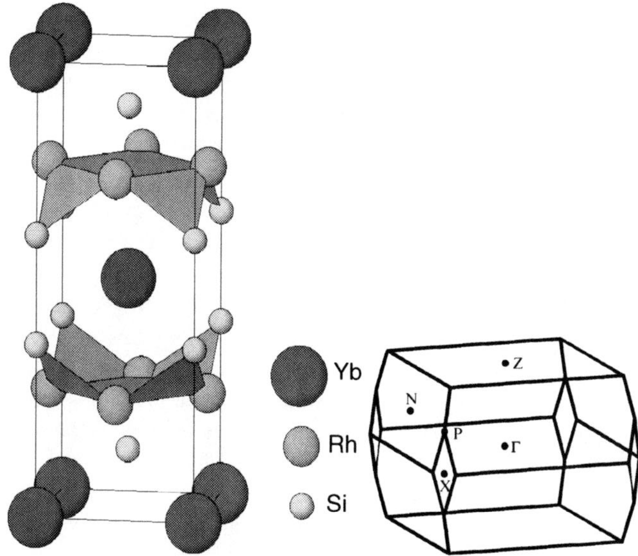

Fig. 1 Crystal structure and Brillouin zone of $YbRh_2Si_2$.

Because the heavy fermion compound $YbRh_2Si_2$ crystallizes in the tetragonal $ThCr_2Si_2$ structure, the crystal structure is relatively open. Therefore we took into account the combined correction term [14]. Exchange and correlation effects among the conduction electrons were introduced using the Barth-Hedin potential [17]. The band-structure was converged for 405 **k**-points in the irreducible wedge, whose volume equals $1/16$ of the Brillouin zone. The density of states (DOS) was evaluated by the tetrahedron method with linear interpolation for the energies. For the conduction band the DOS was calculated at 0.25 mRy (≈0.0034 eV) intervals. We used the experimentally determined lattice parameters a = b = 4.007Å, c = 9.858 Å [18]. The unit cell and the Brillouin zone are displayed in Fig. 1 where the notation for symmetry lines and points of the Brillouin zone according to Ref. [19] is explained.

The crystal potential seen by the conduction electrons is calculated self-consistently within the LDA. The calculations were done using two energy panels, i. e., two separate LMTO calculations were performed to determine self-consistently the Yb 5p states and the conduction bands, respectively. The charge contributions of the other core states were taken from atomic calculations and kept frozen during the iterative procedure. For the lower panel we included s-p-d angular momentum components in the basis at the Yb and Rh sites and s-p components at the Si sites. For the upper panel we included s-p-d-f components at the Yb site and s-p-d at the remaining sites.

The strongly renormalized "heavy" quasiparticle bands are determined by means of the renormalized band method. A detailed description is given in Ref. [9]. Within this approach the influence of the strong correlations is accounted for by introducing a small number of parameters. The key idea of the renormalized band method is to determine the quasiparticle states by computing the band structure for a given effective potential. Coherence effects which result from the periodicity of the lattice are then automatically accounted for. The quantities to be parameterized are the effective potentials which include the many-body effects. The parameterization of the quasiparticles is supplemented by information from conventional band structure calculations as they are performed for "ordinary" metals with weakly correlated electrons. Within the scattering formulation of the band structure problem the characteristic properties of a given material enter through the properly chosen set of phase shifts $\{\eta_\nu^i(E)\}$ specifying the change in phase of a wave incident on site i with energy E and symmetry ν with respect to the scattering center. Within the renormalized band method the effective potentials and hence the phase shifts of the weakly correlated conduction states are determined from first principles to the same level as in the case of "ordinary" metals. The f-phase shifts at the lanthanide and actinide sites, on the other hand, are described by a resonance type expression

$$\tilde{\eta}_f \simeq \arctan \frac{\tilde{\Delta}_f}{E - \tilde{\varepsilon}_f} \tag{1}$$

which renormalizes the effective quasiparticle mass. One of the two remaining free parameters, usually $\tilde{\varepsilon}_f$, is eliminated by imposing the condition that the charge distribution is not significantly altered as compared to the LDA calculation by introducing the renormalization. The renormalized band method devised to calculate the quasiparticles in heavy-fermion compounds is essentially a one-parameter theory. Spin-orbit and CEF splittings can be accounted for in a straightforward manner [9]. The method has been successfully used to study quasiparticle properties in a great variety of Ce-based compounds. In applying the method to Yb compounds we account for the fact that Yb can be considered as the hole analogue of Ce. Operationally this implies that we have to renormalize the 4f $j = 7/2$ channels at the Yb sites instead of the 4f $j = 5/2$ states in the Ce case. As the 4f hole count is slightly less than unity the center of gravity $\tilde{\varepsilon}_f$ will lie below the Fermi level. In addition, we have to reverse the hierarchy of the CEF scheme. We should like to mention that the effective band structure Hamiltonian constructed along these lines corresponds to a hybridization model which closely parallels the one obtained from the periodic Anderson model in mean-field approximation.

3 Local 4f Dynamics in $YbRh_2Si_2$

The central goal is to determine the CEF energies and states which enter the Renormalized Band calculation. In general, we use the Inelastic Neutron Spectra (INS) [20, 21] to determine the relative positions of the energy levels and the observed easy plane anisotropy in the magnetic susceptibility to select the CEF states. At low temperature, however, the properties of the heavy-fermion compound under consideration are determined by a strong interplay between the CEF effects and the formation of (local) Kondo singlets. The CEF scheme is therefore tested by calculating experimental quantities from a simplified Non-Crossing Approximation (NCA) [22, 23]. Proceeding along these lines yields the following values for the tetragonal CEF $B_{20} = 0.5246\,\text{meV}$, $B_{40} = 0.01195\,\text{meV}$, $B_{60} = 0.0004725\,\text{meV}$, $B_{44} = 0.03598\,\text{meV}$, $B_{64} = -0.01206\,\text{meV}$ [24]. The densities associated with these CEF states are displayed in Fig. 2. We use these parameters and a Kondo temperature of T_K 25 K to calculate the averaged imaginary part of the dynamical susceptibility $\chi''_{ave}(\omega) = 1/3\left(\chi''_{\|}(\omega) + 2\chi''_{\perp}(\omega)\right)$ where $\|$ and $\perp$ refer to the directions parallel and perpendicular to the tetragonal axis, respectively. The result agrees rather well with the INS data as can be seen from Fig. 3. The low energy peak in the calculated spectra is a consequence of the Kondo effect. The experiments did not measure in this energy regime for technical reasons. To further assess the validity of the suggested CEF scheme it were desirable to have data on the quadrupole moment Q(T) which we expect to vary non-monotonically with temperature.

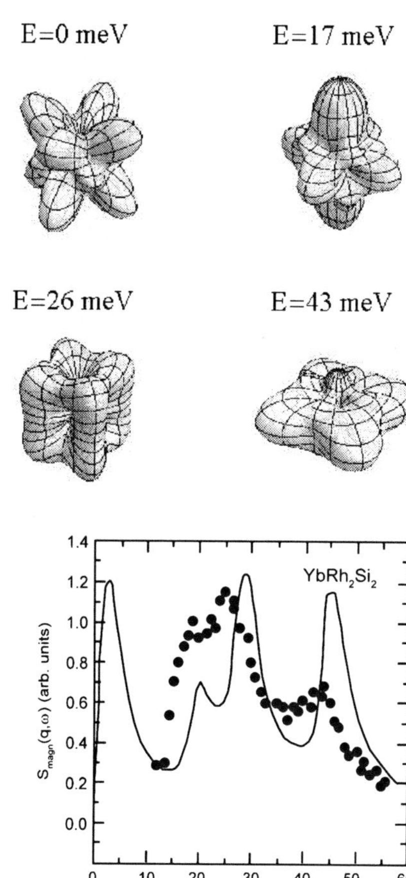

Fig. 2 $YbRh_2Si_2$: CEF states calculated from the CEF parameters given in the text.

Fig. 3 $YbRh_2Si_2$: Averaged imaginary part of dynamical susceptibility and vs INS [20, 21]. The theory curve (*solid line*) is calculated from the simplified NCA assuming a Kondo temperature of $TK = 20\,K$ and a 4f valence of $nf = 0.9$. The experimental data (*filled circles*) were taken at $T = 1.5\,K$.

4 Fermi Surface and Quasiparticles

The conduction states are determined from a self-consistent fully-relativistic calculation treating the Yb 4f electrons as localized core states. A common feature of the YbM_2X_2 (M = metal, X = Si, Ge) compounds are the broad s-p bands. In $YbRh_2Si_2$, the density of conduction states at the Fermi level of about 2.1 states/(eV cell). This value corresponds to a bare Sommerfeld coefficient $\gamma \simeq 5$ mJ/mol K^2 which is in reasonable agreement with the experimental γ-value found for the non-f sister compound $LuRh_2Si_2$.

Let us now turn to the Fermi liquid model as described by the renormalized band scheme. For the intrinsic band widths of the quasiparticle band, $\tilde{\Delta}_f$, a value of $\tilde{\Delta}_f \simeq 20$ K is chosen. The present calculations adopt a CEF scheme which is consistent with susceptibility data and the results from inelastic neutron scattering. The low-energy properties are mainly determined by the CEF ground state which is well separated from the excited states. The hybridization and hence the effective quasiparticle masses are rather anisotropic. The renormalized band calculation yields a DOS of 290 states/(eV unit cell) corresponding to a specific heat coefficient of 680 mJ/mol K^2.

We focus on the two most important FS sheets which are displayed in Fig. 4. There is a rather small pocket derived from a third band which we shall not discuss in detail here. The strongly anisotropic hybridization of the CEF split f states with the conduction bands affects the two sheets of the Fermi surface differently. The Fermi surface cross sections in selected directions are in good agreement with recent dHvA studies [25].

The characteristic feature $YbRh_2Si_2$ shares with other Yb-based heavy fermion compounds is that the effective masses are distributed over a remarkably broad range. This data suggests that two different types of long-lived quasiparticles with significantly different mass renormalizations coexist in these compounds. The anisotropy of the CEF ground state is reflected in strong anisotropies of the hybridization between f states and conduction bands. As a consequence, we anticipate a strong variation of the band curvature over the Fermi surface. Numerical calculations of the transport integrals relating to thermopower require high k- and energy resolution. This problem is currently under investigation.

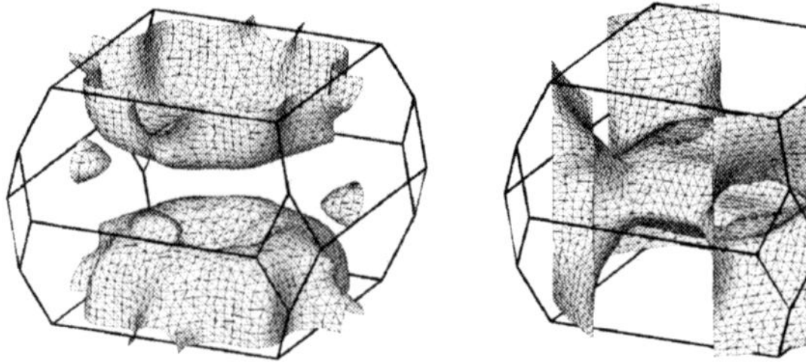

Fig. 4 $YbRh_2Si_2$: Fermi surface of the renormalized quasiparticles.

5 Summary and Discussion

In this contribution we presented renormalized band calculations for the heavy quasiparticles in the LFL state of $YbRh_2Si_2$. The results will be used as starting point for a quantitative study of transport properties.

Acknowledgements I gratefully acknowledge discussions with S. Friedemann, A. Hiess, F. Steglich, O. Stockert, T. Westerkamp, and V. Zevin.

References

1. K. Behnia, D. Jaccard, J. Flouquet, J. Phys.: Condens. Matter **16**, 5187 (2004)
2. K. Miyake, H. Kohno, J. Phys. Soc. Jpn. **74**, 254 (2005)
3. V. Zlatic, R. Monnier, K. Freericks, K.W. Becker, Phys. Rev. B **76**, 085122 (2008)
4. S. Hartmann, N. Oeschler, C. Krellner, C. Geibel, F. Steglich, arXiv:0807.3420 (2008)
5. P. Thalmeier, G. Zwicknagl, *Handbook on the Physics and Chemistry of Rare Earths*, (Elsevier, Berlin, 2005) vol. 34
6. P. Gegenwart, Q. Si, F. Steglich, nature physics **4**, 186 (2008)
7. J. Custers, P. Gegenwart, H. Wilhelm, K. Neumaier, Y. Tokiwa, O. Trovarelli, C. Geibel, F. Steglich, C. Pepin, P. Coleman, Nature **424**, 524 (2003)
8. G.A. Wigger, F. Baumberger, Z.X. Shen, Z.P. Yin, W.E. Pickett, S. Maquilon, Z. Fisk, Phys. Rev. B **76**, 035106 (2007)
9. G. Zwicknagl, Adv. Phys. **41**, 203 (1992)
10. G. Zwicknagl, Physica Scripta T **49**, 34 (1993)
11. O. Stockert, E. Faulhaber, G. Zwicknagl, N. Stüßer, H.S. Jeevan, M. Deppe, R. Borth, R. Küchler, M. Loewenhaupt, C. Geibel, F. Steglich, Phys. Rev. Lett **92**, 136401 (2004)
12. I. Eremin, G. Zwicknagl, P. Thalmeier, P. Fulde, Phys. Rev. Lett. **101**, 187001 (2008)
13. S. Friedemann, N. Oeschler, S. Wirth, C. Krellner, F. Steglich, S. MacQuilon, Z. Fisk, S. Paschen, G. Zwicknagl, arXiv:0803.4428 (2008)
14. O.K. Andersen, Phys. Rev. B **12**, 3060 (1975)
15. H. Skriver, *The LMTO Method, Springer Series in Solid State Sciences*, vol. 41 (Springer-Verlag, Berlin, 1984)
16. R. Albers, A.M. Boring, N.E. Christensen, Phys. Rev. B **33**, 8116 (1986)
17. U. von Barth, L. Hedin, J. Phys. C **5**, 1629 (1972)
18. O. Trovarelli, C. Geibel, S. Mederle, C. Langhammer, F.M. Grosche, P. Gegenwart, M. Lang, G. Sparn, F. Steglich, Phys. Rev. Lett. p. 626 (2000)
19. C. Bradley, A. Cracknell, *The Mathematical Theory of Symmetry in Solids* (Clarendon Press, Oxford, 1972)
20. O. Stockert, M. Koza, J. Ferstl, A. Murani, C. Geibel, F. Steglich, Physica B: Condensed Matter **378–380**, 157 (2006)
21. A. Hiess, O. Stockert, M. Koza, Z. Hossain, C. Geibel, Physica B: Condensed Matter **378–380**, 748 (2006)
22. G. Zwicknagl, V. Zevin, P. Fulde, Phys. Rev. Lett. **60**, 2331 (1988)
23. G. Zwicknagl, V. Zevin, P. Fulde, Z. Physik B **79**, 365 (1990)
24. V. Zevin, G. Zwicknagl, to be published (2009)
25. T. Westerkamp, private communication (2008)
26. P.M. Rourke, A. McCollam, G. Lapertot, G. Knebel, J. Flouquet, S.R. Julian, Phys. Rev. Lett. **101**, 237205 (2008)

The LDA+DMFT Route to Identify Good Thermoelectrics

K. Held, R. Arita, V. I. Anisimov, and K. Kuroki

Abstract For technical applications thermoelectric materials with a high figure of merit are desirable, and strongly correlated electron systems are very promising in this respect. Since effects of bandstructure and electronic correlations play an important role for getting large figure of merits, the combination of local density approximation and dynamical mean field theory is an ideal tool for the computational materials design of new thermoelectrics as well as to help us understand the mechanisms leading to large figures of merits in certain materials. This conference proceedings provides for a brief introduction to the method and reviews recent results for $LiRh_2O_4$.

1 Introduction

Against the background of climate change and the present energy crisis, the quest for alternative, green energy sources is more urgent than ever. In this regard, thermoelectric materials which transform waste heat (gradients) into electrical power through the Seebeck effect [1, 2] are particularly appealing. However, due to a low efficiency we have not yet witnessed a wider technological application almost 200 years after Seebeck's discovery. Instead, thermoelectrical applications are restricted

K. Held
Institute for Solid State Physics, Vienna University of Technology, A-1040 Vienna, Austria
http://www.ifp.tuwien.ac.at/cms

R. Arita
Department of Applied Physics, University of Tokyo, Tokyo 113-8656, Japan
e-mail: arita@ap.t.u-tokyo.ac.jp

V. I. Anisimov
Institute of Metal Physics, Russian Academy of Science-Ural division, 620219 Yekaterinburg, Russia
e-mail: via@imp.uran.ru

K. Kuroki
University of Electro-Communications 1-5-1 Chofugaoka, Chofu-shi Tokyo 182-8585, Japan
e-mail: kuroki@vivace.e-one.uec.ac.jp

V. Zlatić and A. C. Hewson (eds.), *Properties and Applications of Thermoelectric Materials*, 141
NATO Science for Peace and Security Series B: Physics and Biophysics,

to niche markets such as radioisotope power systems for satellites [3]. A possible first major application is the exhaust heat of cars and trucks, as automobile companies presently test thermoelectrical generators in prototypes [4]. Such efforts could be put on another level if novel materials with a higher figure of merit ZT, where Z is the power factor and T the temperature, and hence a higher efficiency, were available. Most present technical applications use semiconductors such as Bi_2Te_3 [2] where recently power factors Z considerably larger than 1 could be achieved through phonon [5] and bandstructure engineering [6].

Very promising are novel materials on the basis of strongly correlated electron systems (SCES) [7] which are at the core of the present conference proceedings. This class of materials is very diverse, ranging from metals to Kondo insulators and semiconductors, from d to f electron systems, from relative simple crystal structures such as $FeSb_2$ [8] to most complex metallic cage compounds.

Having such a wide field and the additional possibilities to nano- and heterostructure these systems, a better theoretical understanding and reliable tools to compute thermoelectric properties quantitatively are mandatory. Theoretical physicists from the SCES community have analyzed thermoelectric materials mainly on a model level, i.e., on the basis of the Falikov-Kimball, Hubbard and periodic Anderson model [9, 10], often employing dynamical mean field theory (DMFT) [11–13]. These calculations showed, among others, the importance of correlation-induced enhancements of the effective mass generating a high, but narrow density of states –or spectral function to be precise– close to –but not at– the Fermi level. As a consequence, the thermoelectric figure of merits can be strongly enhanced. On the other hand, theoreticians from the density functional theory (DFT) [14] community have been emphasizing the importance of a particularly high density of states (DOS) [15, 16] and of the large group velocities for certain shapes of the bandstructure [17].

Since both, correlations and bandstructure, can substantially contribute to enhanced thermoelectrical figures of merit, we need to deal with both of them on an equal footing. Only if both aspects are optimized we can expect to design materials or artificial heterostructures with really large figures of merit. Taking correlations and bandstructure into account is possible with the merger [18, 19] of DFT in its local density approximation (LDA) [20] and DMFT, for which the name LDA+DMFT was coined [21], see Refs. [22–24] for reviews. While LDA+DMFT has been applied already to many SCES materials, thermoelectrical properties have been calculated rarely in the past. Noteworthy exceptions are $LaTiO_3$ [25] and $LiRh_2O_4$ [26]. The main reasons for this is that a wider experimental interest in SCES thermoelectrics emerged rather recently and that the calculation of thermoelectric properties such as the Seebeck coefficient requires some additional postprocessing which is not yet standard in LDA+DMFT calculations.

1.1 Outline

In the following, we will give a brief, elementary introduction to the LDA+DMFT approach in Section 2. This section is divided into the three steps LDA (Section 2.1),

DMFT (Section 2.2), and the necessary postprocessing for calculating thermoelectrical response functions (Section 2.3). Section 3 presents exemplary results by hands of $LiRh_2O_4$ which are reproduced from Ref. [26]. Finally, Section 3.1 gives a summary and an outlook.

2 LDA+DMFT Method

The aim of this section is to give the reader a brief, elementary introduction to the LDA+DMFT approach; for more details see the reviews [22–24].

Starting point is the general ab-initio Hamiltonian for every material which, without relativistic corrections, reads in the Born–Oppenheimer approximation

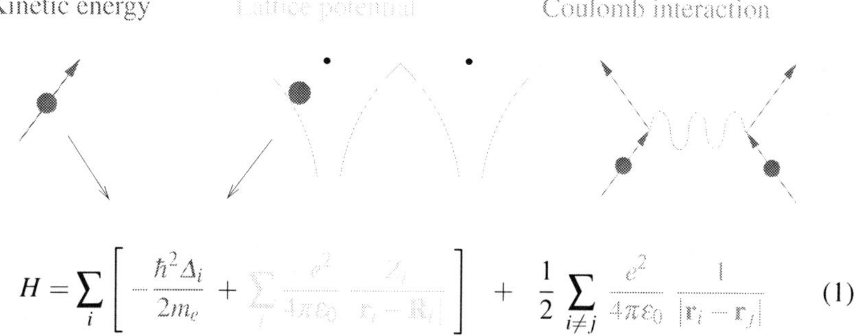

$$H = \sum_i \left[-\frac{\hbar^2 \Delta_i}{2m_e} + \sum_l \frac{e^2}{4\pi\varepsilon_0} \frac{-Z_l}{|\mathbf{r}_i - \mathbf{R}_l|} \right] + \frac{1}{2} \sum_{i \neq j} \frac{e^2}{4\pi\varepsilon_0} \frac{1}{|\mathbf{r}_i - \mathbf{r}_j|} \tag{1}$$

It consists of three terms: (1) The kinetic energy given by the Laplace operator Δ_i, Planck constant $\hbar$, and mass m_e for every electron i. (2) The lattice potential given by the Coulomb interaction between (static) ions at position $\mathbf{R}_l$ with charge $Z_l e$ and electrons at position $\mathbf{r}_i$ with charge $-e$. (3) Finally, the Coulomb interaction between each pair of electrons i and j [note the factor 1/2 is needed since each pair is counted twice in Eq. (1)]. Input for the LDA+DMFT calculation is usually the experimental crystal structure, i.e., the positions R_i as an adequate relaxation procedure to determine the R_i's from theory still needs to be developed.

While Hamiltonian (1) is easy to write down, it is impossible to solve, even numerically, for more than $\mathcal{O}(10)$ electrons, since the movement of every electron is *correlated* with that of every other electron through the last term: the Coulomb interaction between the electrons. These electronic correlations play a particularly important role if electrons are confined in or d or f-electrons or in artificial nanostructures. For such systems the typical distance $|\mathbf{r}_i - \mathbf{r}_j|$ between two such electrons on the same lattice site (i.e., two electrons in the set of d- or f-orbitals around the same ion) is small so that the Coulomb interaction and, hence, also the electronic correlations are strong.

2.1 LDA Step

Since it is impossible to solve Hamiltonian (1), we have to develop approximations, and arguably the most successful approximation so far are those developed within the DFT framework, particularly the LDA [20]. Strictly speaking, DFT only allows to calculate ground state energies and its derivatives but *not* bandstructures and thermoelectric transport functions. However, it turned out that the auxiliary Kohn-Sham Lagrange parameters ε_k often also describe bandstructures very accurately, making bandstructure calculations one of the major applications of LDA. Interpreting the LDA Lagrange parameters ε_k as the physical (one-electron) excitation energies, i.e., the bandstructure, corresponds to replace Hamiltonian (1) by the Kohn–Sham [27] LDA Hamiltonian

$$H_{\rm LDA} = \sum_{i} \left[-\frac{\hbar^2 \Delta_i}{2m_e} + \sum_{l} \frac{-e^2}{4\pi\varepsilon_0} \frac{1}{|\mathbf{r}_i - \mathbf{R}_l|} + \int d^3 r \, \frac{e^2}{4\pi\varepsilon_0} \frac{1}{|\mathbf{r}_i - \mathbf{r}|} \rho(\mathbf{r}) + V_{xc}^{\rm LDA}(\rho(\mathbf{r}_i)) \right] \quad (2)$$

This Hamiltonian shows that the complicated electron-electron interaction causing the complicated electronic correlations has been replaced by two simpler terms: The Hartree term describing the Coulomb interaction of electron $\mathbf{r}_i$ with the time-averaged mean density $\rho(\mathbf{r})$ of all electrons and an additional term $V_{xc}^{\rm LDA}$ which aims at including the effects of correlations and interactions.

However, the exact form of this term is unknown and certainly it is not local in $\mathbf{r}$ as approximated in the LDA. One can take the V_{xc} of the jellium model [28] which has a constant electron density and is only weakly correlated. Hence, it is not surprising that LDA bandstructure calculations fail for SCES [20]. For such materials, which are at the focus here, we need to take electronic correlations into account more profoundly.

A possibility to do so is to take the LDA bandstructure of the less correlated orbitals but to supplement that of the more correlated d- or f-orbitals by explicitly taking into account the most, important local Coulomb interaction. This leads to the Hamiltonian

$$\hat{\mathscr{H}} = \underbrace{\sum_{\mathbf{k}lm\sigma} \epsilon_{\mathbf{k}lm}^{\rm LDA} \hat{c}^{\dagger}_{\mathbf{k}l\sigma} \hat{c}_{\mathbf{k}m\sigma}}_{H_{\rm LDA}} + \frac{1}{2} \sum_{il\sigma m\sigma'} U_{lm}^{\sigma\sigma'} \hat{n}_{il\sigma} \hat{n}_{im\sigma'} - \Delta\varepsilon \sum_{im\sigma} \hat{n}_{im\sigma}, \quad (3)$$

where the first part is the same as the LDA Hamiltonian (2) but in second (instead of first) quantization and in $\mathbf{k}$ and orbital space (with l and m denoting two different orbitals) with creation and annihilation operators $\hat{c}^{\dagger}_{\mathbf{k}l\sigma}$ and $\hat{c}_{\mathbf{k}m\sigma}$, respectively.

The second term explicitly takes the local Coulomb interaction on the same ion site i into account. Typically only the Coulomb interactions for d (or f) l and m orbitals are considered here. These interactions are spin and orbital dependent because of the exchange matrix elements leading to Hund's rules, see Fig. 1 for an

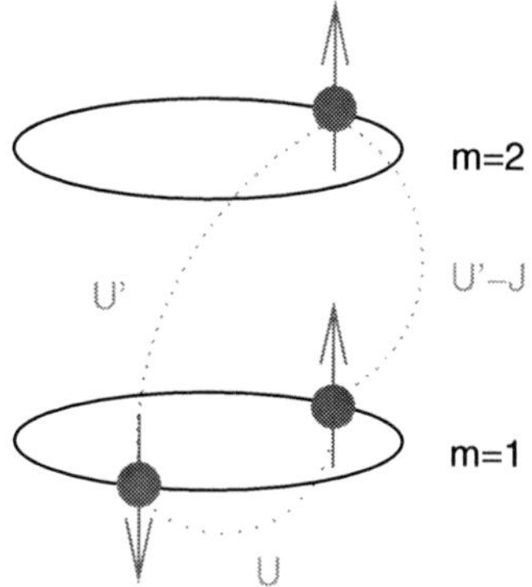

Fig. 1 Illustration of the different elements of the Coulomb interaction matrix of Hamiltonian (3). There is an inter-orbital Coulomb repulsion U', which is reduced by Hund's exchange J for a ferromagnetic spin alignment, and an intra-orbital interaction U. Orbital rotational symmetry relates these quantities as $U = U' + 2J$.

illustration. Let us note that in Hamiltonian (3) only the density-density terms are included since the inclusion of the spin-flip terms of Hund's exchange became only possible in quantum Monte Carlo (QMC) simulations [29] with recent improvements [30–34].

Finally the third $\Delta\varepsilon$ term subtracts those contributions of U already taken into account in the LDA to avoid a double counting. For a truly ab-initio calculation, U', J, and $\Delta\varepsilon$ still need to be determined. To this end, screening has to be taken into account; and a possibility within the LDA framework is to employ constrained LDA, for details see Ref. [24].

2.2 DMFT Step

Having derived a multi-orbital many-body Hamiltonian (3) from the ab-initio Hamiltonian (1), we still need to solve it. A possible way to do so is to use Hartree-Fock, allowing for symmetry breaking with respect to the spin and orbital elements, i.e.,

$$\frac{1}{2}\sum_{il\sigma m\sigma'} U_{lm}^{\sigma\sigma'}\, \hat{n}_{il\sigma}\, \hat{n}_{im\sigma'} \rightarrow \sum_{il\sigma m\sigma'} U_{lm}^{\sigma\sigma'}\, \hat{n}_{il\sigma}\, \langle \hat{n}_{im\sigma'} \rangle - \frac{1}{2}\sum_{il\sigma m\sigma'} U_{lm}^{\sigma\sigma'}\, \langle \hat{n}_{il\sigma} \rangle \langle \hat{n}_{im\sigma'} \rangle, \quad (4)$$

where $\langle \hat{n}_{im\sigma'} \rangle$ is the average occupation of the orbital m on site i with spin σ'. However, in this LDA $+\,U$ [35] approach electronic correlations are neglected through Eq. (4); and the only chance to reduce the Coulomb interaction energy is by a strong symmetry breaking. Hence, tendencies to magnetic or orbitally ordered phases are grossly overestimated, as is the tendency to open gaps. Even within these ordered phases many-body aspects such as spin-polarons are neglected as was shown in [36].

A reliable approximation to include the *local* correlations induced by the *local* Coulomb interaction of Hamiltonian (3) is possible with DMFT [11–13]. We cannot derive this approach in full detail here and refer the interested reader to [12, 24]. The basic idea is visualized in Fig. 2: We replace the local interaction on all sites but one

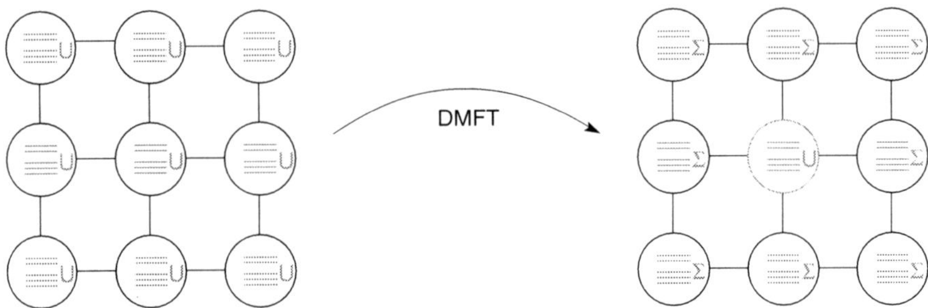

Fig. 2 In DMFT, we approximate the material specific lattice Hamiltonian (3) by a problem where the interaction is replaced by a self energy on all sites except for one. This DMFT single-site problem is equivalent to an Anderson impurity model which has to be solved self-consistently together with the **k**-integrated Dyson (Equation (6)).

by a self-energy $\Sigma(\omega)$. This gives rise to an Anderson impurity model of a single interacting site in a medium $\mathcal{G}_0(\omega)$ given by the self energy and the interacting Green function $G(\omega)$:

$$\mathcal{G}_0(\omega)^{-1} = G(\omega)^{-1} + \Sigma(\omega) \tag{5}$$

This Anderson impurity model, defined by its non-interacting Green function $\mathcal{G}_0$ has to be solved self-consistently together with the **k**-integrated Dyson equation, where the LDA bandstructure $\varepsilon^{\text{LDA}}_{l,m}(\mathbf{k})$ enters as a matrix in the orbital indices (V_{BZ} denotes the volume of the Brillouin zone):

$$G^{\sigma}_{lm}(\omega) = \int \frac{\mathrm{d}^3k}{V_{\text{BZ}}} \left[\omega+\mu-[\varepsilon^{\text{LDA}} - \Delta\varepsilon]_{lm}(\mathbf{k})-\Sigma^{\sigma}_{lm}(\omega)\right]^{-1}. \tag{6}$$

From a diagrammatic point of view, DMFT corresponds to all (topologically distinct) Feynman diagrams of which, however, only the local contribution for the self energy is taken into account. Hence, it is non-perturbative in the Coulomb interaction but neglects non-local correlations between sites. Recent improvements of DMFT include such non-local correlations by taking a cluster of interacting sites instead of a single one in Fig. 2 [37–39] or by extending the diagrammatic contributions in the dynamical vertex approximation (DΓA) [40], also see [41–43].

What we still need to do is to solve the Anderson impurity model self-consistently, which for realistic multi-orbital calculations is typically done by quantum Monte Carlo simulations, different approaches are discussed in [24]. The standard result of such a DMFT(QMC) calculation is the interacting local Green function $G(i\omega_\nu)$ for imaginary (Matsubara) frequencies $i\omega_\nu$ or its Fourier transform, the imaginary time Green function $G(\tau)$. But also various correlation functions and susceptibilities can be calculated.

2.3 Calculation of Thermoelectrical Response Functions

Starting point for calculating transport properties is the Kubo formula. For thermoelectric materials the Seebeck coefficient

$$S = -\frac{k_B}{|e|} \frac{A_1}{A_0} \tag{7}$$

is of particular importance. It is given by the constants Boltzmann k_B, unit charge e and the ratio of two correlation functions, the current–current and the current–heat-current correlation function

$$A_0 = \lim_{i\nu \to 0} \frac{i\hbar k_B T}{i\nu} \int_0^\beta d\tau \, e^{i\nu\tau} \langle T_\tau j(\tau) j(0) \rangle \tag{8}$$

$$A_1 = \lim_{i\nu \to 0} \frac{i\hbar}{i\nu} \int_0^\beta d\tau \, e^{i\nu\tau} \langle T_\tau j(\tau) j_Q(0) \rangle \tag{9}$$

in the static limit, i.e., frequency $i\nu \to 0$. Here, T_τ is Wick's time ordering operator; $j(\tau)$ and $j_Q(\tau)$ are the current and heat-current operators respectively. Also relevant is the heat-current–heat-current correlation function

$$A_2 = \lim_{i\nu \to 0} \frac{i\hbar}{i\nu k_B T} \int_0^\beta d\tau \, e^{i\nu\tau} \langle T_\tau j_Q(\tau) j_Q(0) \rangle \tag{10}$$

which yields the electronic contribution to the thermal conductivity κ similar as A_0 does for the electrical conductivity σ. Since the phononic contribution to the thermal conductivity is however typically much larger at room temperature and can be reduced by phonon engineering, we will not consider κ in the following. Instead we will concentrate of the purely electronic contributions to the power factor

$$Z = \frac{S^2 \sigma}{\kappa}, \tag{11}$$

i.e., on S and σ.

Diagrammatically, the correlation functions Eqs. (8)–(10) correspond to Fig. 3. As indicated, the vertex Γ is usually not taken into account. In case of full orbital degeneracy (of the low energy orbitals), this holds exactly since one can show by a simple argument that vertex contributions are, for the local DMFT vertex, odd in $\mathbf{k}$ and hence their integrated contribution vanishes, see [44]. In the case of $LiRh_2O_4$ where the three low energy orbitals are very similar in energy and occupation, neglecting the vertex is still justified approximately. This allows us to calculate the bubble diagram for the (heat-)current–(heat-)current correlation functions A_m from the spectral function $\rho(\mathbf{k}, \omega) = -1/\pi \, \mathrm{Im} G(\mathbf{k}, \omega)$. In the x-direction, we obtain for diagram Fig. 3

$$A_m = 2\pi\hbar \int_{-\infty}^{\infty} d\omega \, \frac{1}{V} \sum_{\mathbf{k}} \mathrm{Tr}\left[v^x(\mathbf{k}) \rho(\mathbf{k}, \omega) v^x(\mathbf{k}) \rho(\mathbf{k}, \omega) \right] f(\omega) f(-\omega) (\beta\omega)^m.$$

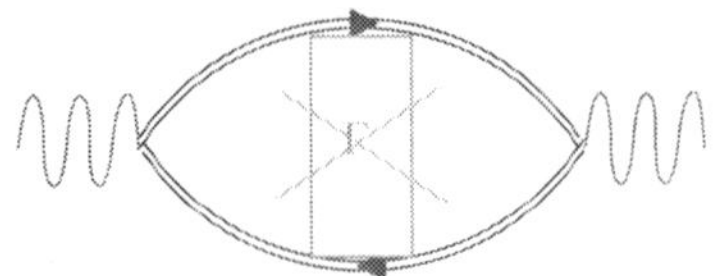

Fig. 3 Diagrammatic representation of the (heat-)current–(heat-)current correlation functions Eqs. (8)–(10) with an incoming frequency (wiggled line) $i\nu \to 0$. For the current operator the wiggled line yields a factor $v^x(\mathbf{k})$, for the heat-current operator a factor $\varepsilon_{\mathbf{k}}^{\text{LDA}} v^x(\mathbf{k})$. The vertex Γ is typically neglected as indicated so that the calculation of the correlation function reduces to the simple bubble diagram of two (interacting) Green function, i.e., a factor $G(k,\omega)$ for each of the two double lines.

Here $v_{\mathbf{k}}$ are in the general formalism the dipole matrix elements which we replaced approximately by the simpler group velocity obtained through the derivative of the dispersion relation:

$$v_{\mathbf{k}} = \frac{\partial \varepsilon_{\mathbf{k}}^{\text{LDA}}}{\partial \mathbf{k}}. \tag{12}$$

These are relatively easy to calculate from the LDA bandstructure. Note, here the quantities $v_{\mathbf{k}}$, $\varepsilon_{\mathbf{k}}^{\text{LDA}}$, and $\rho(\mathbf{k},\omega)$ are all matrices in the orbital indices.

What we still need for calculating the DMFT (heat-)current–(heat-)current correlation functions, is the $\mathbf{k}$-dependence of $\rho(\mathbf{k},\omega)$. In the DMFT self-consistency cycle, one calculates however only the local Green function

$$G(i\omega_\nu) = \frac{1}{V}\sum_k G(\mathbf{k}, i\omega_\nu) \tag{13}$$

at Matsubara frequencies $i\omega_\nu$. From the DMFT $G(i\omega_\nu)$ or its Fourier transform the imaginary time $G(\tau)$, we can determine the optical and thermal conductivity as well as the Seebeck coefficient in some post processing steps:

First, we need the self-energy for (real) frequencies. The standard procedure [46] to this end is first to analytically continue the Green function to real frequencies. This is done by the maximum entropy method [45] yielding $\text{Im}G(\omega)$ at real frequencies ω from $G(\tau)$. From this, in a second step, the full $G(\omega)$ is constructed by Kramers–Kronig transformation. Third, that self energy is determined which, if plugged into the $\mathbf{k}$-integrated Dyson (Equation. (6)), gives the Green function which is closest to the QMC-determined $G(\omega)$. Finally from the self energy $\Sigma(\omega)$ for real frequencies, we can determine

$$G^{\sigma}_{lm}(\omega) = \left[\omega + \mu - [\varepsilon^{\text{LDA}}_{\mathbf{k}} - \Delta\varepsilon]_{lm}(\mathbf{k}) - \Sigma^{\sigma}_{lm}(\omega)\right]^{-1},$$

or its imaginary part $\rho(\mathbf{k},\omega)$.

For the $LiRh_2O_4$ calculations presented below it turned out that this standard approach does not work so well because the high spectral weight close to the Fermi level makes the analytical calculation very sensitive to the statistical QMC error. On the other hand, we only need $\Sigma(\omega)$ at small frequencies, of the order $k_B T$, if we are interested in thermodynamical responses to static fields (as S and σ). Hence in [26], we did a Padé fit [47] to $\Sigma(i\omega_\nu)$ which works rather well for not too large frequencies. Comparing it with a polynomial fit allowed us to estimate the error in $\Sigma(\omega)$, see Fig. 6 in Section 3.

Let us note the connection to the Boltzmann approach. This is obtained for non-interacting electrons and a constant-τ approximation, i.e., calculation (Eq. (12)) for a self energy $\Sigma(\omega) = -i/\tau$. This reduces (Eq. (12)) to

$$A_m = \sum_k \tau \, v^x(\mathbf{k}) v^x(\mathbf{k}) \left[-\frac{\partial f(\varepsilon)}{\partial \varepsilon} \right] \left(\frac{\varepsilon^{\mathrm{LDA}}(\mathbf{k})}{k_B T} \right)^m . \tag{14}$$

Note that, in contrast to the thermal and electrical conductivity, τ cancels in the Seebeck coefficient since we divide A_1 by A_0. Hence, the exact value of the difficult to determine relaxation time is not relevant, as long as it is constant.

For a better understanding of the microscopic origin of a large thermopower, at least as far as the bandstructure effects are concerned, we can approximate the Boltzmann (Equation (14)) by summing only the states in a window $\pm k_B T$ around the Fermi energy (indicated by the tilde below):

$$A_0 \approx \tau \tilde{\sum}_k v_A^2 + v_B^2 \; ; \; A_1 \approx \tau \tilde{\sum}_k v_A^2 - v_B^2 . \tag{15}$$

Here v_A^2 and v_B^2 are the typical (averaged) velocities above and below the Fermi level, respectively. For the current–current correlation function A_0 these two contributions have to be added, whereas they have to be subtracted for the heat-current–current correlation function A_1. The reason for the latter is that a quasiparticle above the Fermi level carries a positive energy contribution relative to the Fermi energy, while we have a negative energy-contribution for quasi-hole excitations below the Fermi level. For getting an (absolutely) large Seebeck coefficient we need a large A_1 relative to A_0. Since A_1 is the difference of the same (positive) contributions which are added A_0, this requires the minuend to be much smaller than the subtrahend in Eq. (15) or vice versa. This is possible if either (i) there are many more states below the Fermi level than above (or vice versa for a large negative S) or (ii) the group velocity v_A^2 above the Fermi level is much larger than v_B^2 (or vice versa). Optimal would indeed be a combination of both. The route (i) can be heavily affected by electronic correlations, e.g., if we have a sharp Kondo peak directly above or below the Fermi level, but also bandstructure effects play a role. In contrast for mechanism (ii) the LDA group velocities (or dipole matrix elements) enter so that electronic correlations are not directly relevant.

3 An Example: $LiRh_2O_4$

Let us here briefly review the calculation of the Seebeck coefficient for $LiRh_2O_4$ of [26]. This mixed-valent spinel, see Fig. 4, was most recently synthesized by Okamoto et al. [48]. It shows two structural phase transitions: cubic-to-tetragonal transition at 230 K and tetragonal-to-orthorhombic transition at 170 K. For the high-temperature cubic phase Okamoto *et al.* reported a thermopower as large as 80 μV/K at 800 K, which for a metallic system, is quite exceptional. Together with Na_xCoO_2 [49], it shows that transition-metal-oxides are promising candidates for thermoelectric application since, even in the metallic phase, large power factors ($S^2\sigma$) are possible. Concerning Na_xCoO_2, these experimental findings led to some "heated discussion" on the origin of the large Seebeck coefficient on the theoretical side [17, 50, 51]. This makes a reliable ab-initio calculation which can put the theoretical ideas on a more solid fundament mandatory.

Starting from the experimental crystal structure Fig. 4, the first LDA+DMFT step is the calculation of the LDA bandstructure. Our results, using linearized muffin tin orbitals (LMTOs) [52], are shown in Fig. 5 (left panel; dashed line). We further simplified the LDA bandstructure by a Wannier projection [53] of the LMTO wave functions onto the subspace of Bloch waves, which were in turn Fourier transformed to Wannier functions, see [54] for details. Here, we even model the LDA bandstructure by a two-band model (solid line).

The next step is a self-consistent DMFT calculation. To this end, quantum Monte-Carlo simulations were used as an impurity solver [29]. The Coulomb interaction parameters were estimated as $(U,U',J) = (3.1, 1.7, 0.7)$eV from [55] and temperatures $\beta = 1/k_BT = 30$, 34, 40 eV^{-1} were considered. From the imaginary QMC self energy we obtained the self energy on the real axis (Fig. 6) through a Padé and polynomial (Taylor) fit. As one can see from a comparison of the two fits there is some uncertainty, but absolute differences are small, i.e., of $O(0.01)$ eV. Nonetheless, we proceeded with both self energies to have an estimate of the error.

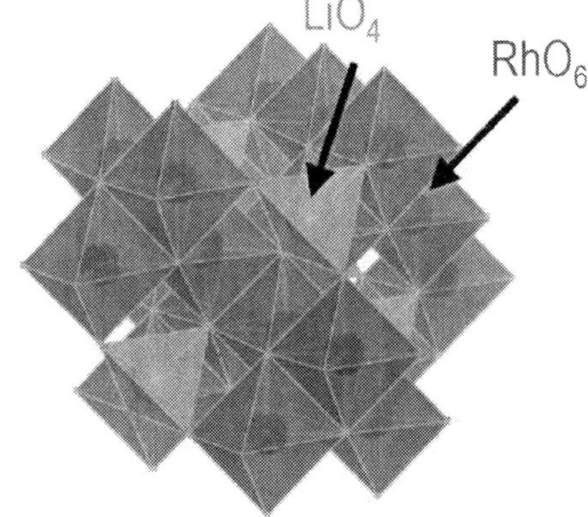

Fig. 4 Crystal structure of $LiRh_2O_4$, made up from LiO_4 tetrahedra and the RhO_5 octahedra. Reprinted with permission from [48]. Copyright (2008) by the American Physical Society.

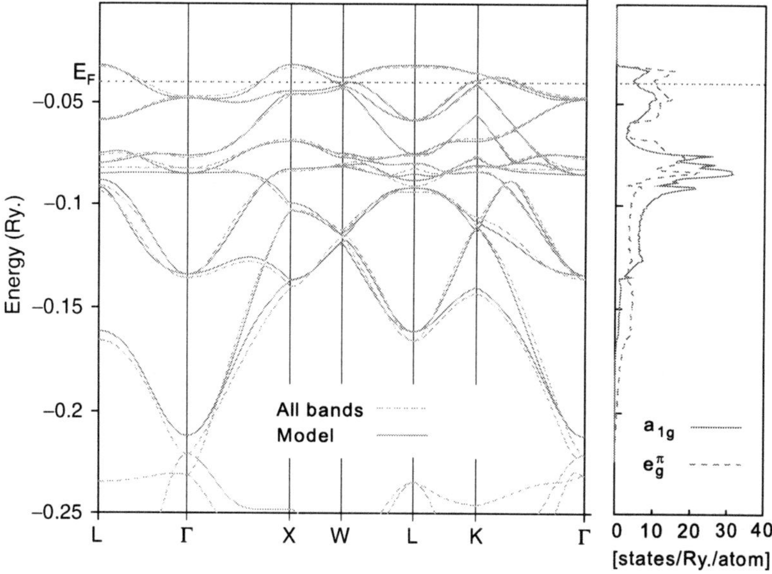

Fig. 5 Left panel: Band dispersion of the effective 3-orbital Hamiltonian (*solid line*) and total LMTO band structure (*dashed line*) of $LiRh_2O_4$. Right panel: partial a_{1g} and e_g^π density of states for the model. LDA. Reprinted with permission from [26]. Copyright (2008) by the American Physical Society.

From the self energy, Fig. 6, we can estimate the quasiparticle weight $Z = (1 - \partial \mathrm{Re}\Sigma/\partial\omega)^{-1}$ and the effective mass enhancement $m^*/m = 1/Z$. This effective mass enhancement is actually not very strong, i.e., $\approx 40\%$ for the e_g^π band and $\approx 30\%$ for the a_{1g} band. This indicates that electronic correlations are only intermediately strong for this compound, even though it is a transition metal oxide. The reason for this is the mixed-valent nature of $LiRh_2O_4$ which puts the orbital occupation far away from a (more strongly correlated) integer filling. A second noteworthy aspect, we can extract from the self energy is the strong frequency dependence and asymmetry of the imaginary part of the self energy. This poses the question whether a constant-ImΣ, i.e., a constant relaxation time τ approach as in the much less involved Boltzmann approach, works.

From the (two) self energy of Fig. 6, we calculated the Seebeck coefficient using the formulas of Section 2.3. As one can see there are some differences in the Seebeck coefficient for the Padé ($\times$ symbol) and polynomial fit ($*$ symbol), giving us an estimate of the accuracy of our calculation. Both are in good agreement with the experimental values [48].

Besides the LDA+DMFT study, we also performed calculations (i) putting a constant-τ self energy into the equations of Section 2.3 ($+$ symbol) and (ii) using directly the Boltzmann equation (solid line). As one can see in Fig. 6 both agree, as one can expect from theoretical considerations; but it is a good test in an actual implementation as two completely different programs based on different equations were employed. The Boltzmann equation yields a slightly too large Seebeck

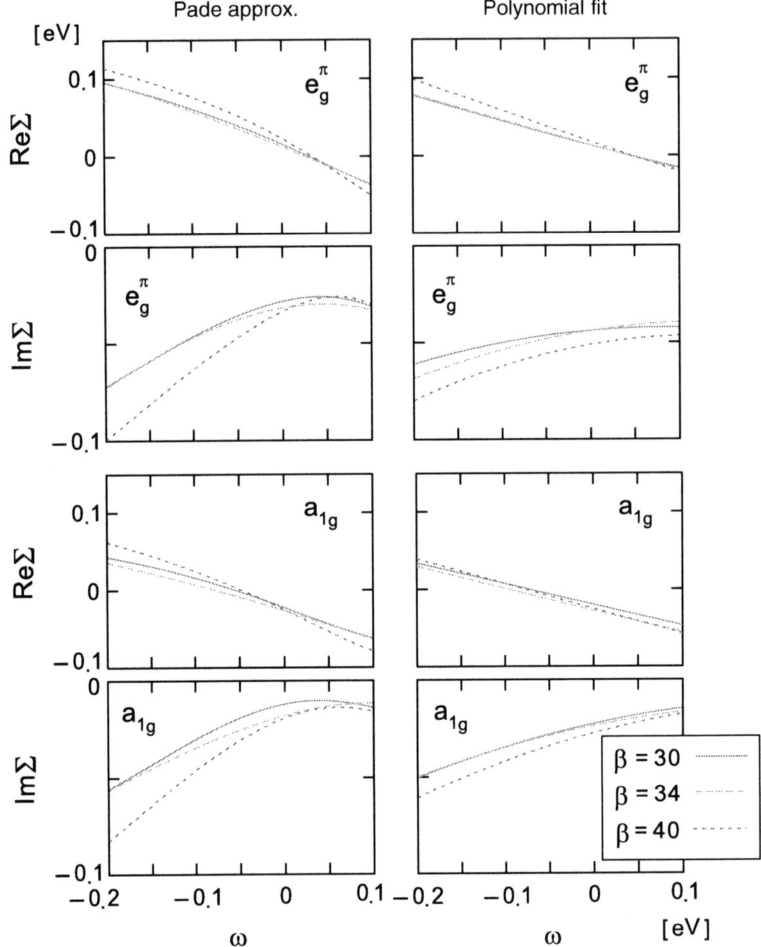

Fig. 6 DMFT(QMC) self energy calculated by the Padé approximation (*left*) and a polynomial fit (*right*). Reprinted with permission from [26]. Copyright (2008) by the American Physical Society.

coefficient S, albeit it still agrees surprisingly well with experiment. The reason for this is that $LiRh_2O_4$ is not strongly correlated. Besides, the two e_g^π and a_{1g} bands are not strongly shifted with respect to each other by electronic correlations and have a not too different self energy. Hence, we are not too far from a situation were the self energy is orbital independent. In this case, the DMFT spectral function is just a more narrow (quasiparticle renormalized) version of the LDA DOS with the same height at the Fermi level. Because of this, relatively weak electronic correlations do not strongly affect the Seebeck coefficient. We hence attribute the differences between Boltzmann approach and LDA+DMFT to the non-constant and strongly asymmetric $\mathrm{Im}\Sigma$. This means that, in contrast to the constant-τ approximation, the actual life time of quasi-holes is longer than that for quasi-particles. Let us emphasize that

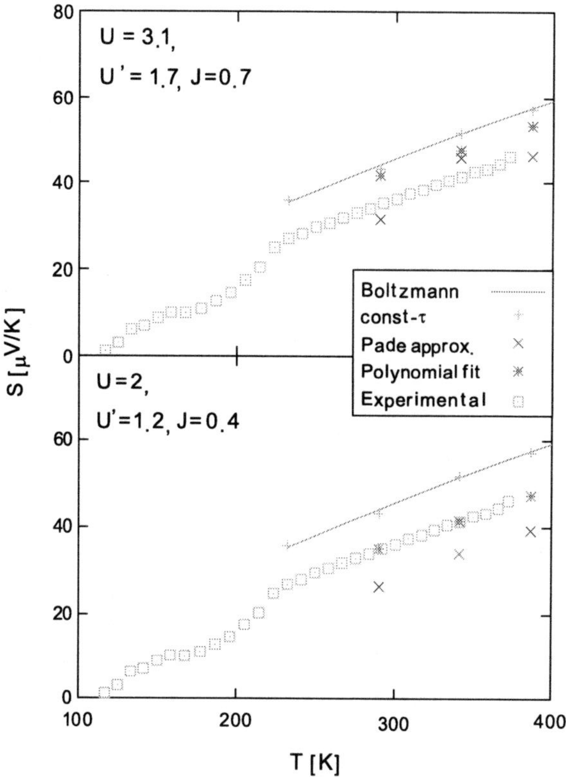

Fig. 7 Thermopower calculated by the Boltzmann equation approach and the constant-τ method as well as by LDA+DMFT, using both the Padé approximation and a polynomial fit for the self energy. Reprinted with permission from [26]. Copyright (2008) by the American Physical Society.

electronic correlations play a much more prominent role in other transition metal oxides, so that for these the Boltzmann approach will fail.

Being confident, that the Boltzmann approach roughly describes the Seebeck coefficient of $LiRh_2O_4$, we analyze Eq. (15). To this end, we plot in Fig. 8 the group velocity along the indicated paths through the Brillouin zone, within the energy window of $|\varepsilon - E_F| < 3k_BT$ at $T \simeq 300$ K. Figure 8 (upper panel) shows that v_B^2 is considerably larger than v_B^2 in large parts of the Brillouin zone, particularly around the K and W point. The reason for this difference is a particular shape of the bandstructure very similar to the ideas proposed in [17] for Na_xCoO_2. This pudding-mold type of shape is sketched in the inset of Fig. 8, for the full bandstructure see Fig. 5. In contrast to the one band situation in Na_xCoO_2 [17], we have however a double pudding mold. For the lower band, the pudding-mold shape leads to a very flat bandstructure above the Fermi level, flatter than a simple maximum because of additional turning points and minima. Consequently the group velocity above the Fermi level is very small and the Seebeck coefficient largely positive.

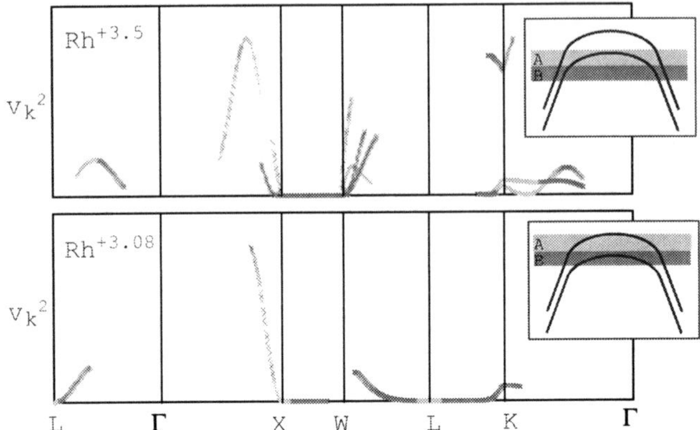

Fig. 8 Group velocity squared (v_k^2) along different directions of the first Brillouin zone for $Rh^{+3.5}$ ($LiRh_2O_4$; *upper panel*) and $Rh^{+3.08}$ (electron-doped $LiRh_2O_4$; *lower panel*). k point above the Fermi energy E_F are shown in yellow, those below E_F in red. Reprinted with permission from [26]. Copyright (2008) by the American Physical Society.

Having understood the bandstructure origin of the large thermopower in $LiRh_2O_4$, we are now in a position to identify routes to increase the thermopower even further. As the sketch in Fig. 8 suggests the upper pudding-mold band does not strongly contribute to the Seebeck coefficient or thermopower. Here, the situation is more like in a standard metal with group velocities being large above and below the Fermi level (this is the region between Γ and X and between Γ and L point in the main panel). Therefore, positive and negative contributions to the Seebeck coefficient roughly cancel for this upper pudding mold band.

We can improve the situation however by electron doping which shifts the Fermi level to higher energy. Then the situation becomes very much the same as for the lower pudding-mold band before and, at the same time, the lower pudding-mold band is still contributing with the same sign because there are states below the Fermi level but no states above. As one can see in the lower main panel of Fig. 8, doping by 0.42 electrons, i.e., for a valence $Rh^{+3.08}$, indeed leads to a situation where only the squared group velocity below the Fermi energy is large.

We further studied this idea by calculating the thermopower and the power factor for various Rh valences, using the Boltzmann equation approach. Note that we assumed the electron doping not to affect the LDA bandstructure except for a shift of the Fermi level. We also neglected the energy and filling dependence of τ, which should be present and affect ρ and, hence, the power factor (albeit not S). Depending on how the electron-doping is realized τ might change because of disorder effects. Nonetheless, we expect the tendencies to hold also for the power factor in experiments electron-doping $LiRh_2O_4$.

As one can see in Fig. 9 (inset) the Seebeck coefficient strongly increases with electron doping, i.e., with reducing the Rh valence towards 3. However if the Rh valence is 3, the band is completely occupied so that while the Seebeck coefficient is large, the conductivity $\sigma = 1/\rho$ becomes small. Hence for the power factor $S^2\sigma$

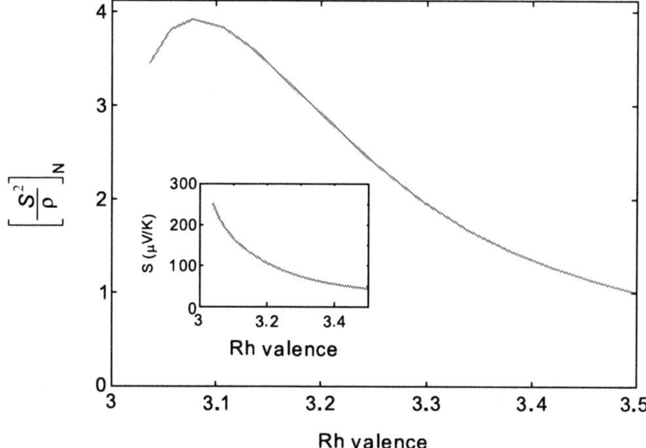

Fig. 9 Power factor (normalized by its value at Rh valence = +3.5) and thermopower (*inset*) as a function of the valence of Rh, calculated by the Boltzmann equation.

(Fig. 9 main panel), there is a trade-off between a larger Seebeck factor and a large resistivity if the valence goes towards 3. This trade-off leads to a maximum for the valence $Rh^{+3.08}$, afore shown in Fig. 8.

3.1 Summary and Outlook

We presented a brief introduction to the LDA+DMFT approach for the realistic calculation of thermoelectric properties, including bandstructure and electronic correlation effects. We have shown that the LDA+DMFT results for $LiRh_2O_4$ well agree with experiment. Furthermore, we identified the origin of the large thermopower in this material to be a particular shape of the bandstructure of the form of a (double) pudding mold. Even larger thermopowers can be obtained if the material is electron-doped, according to our prediction. For the particular material $LiRh_2O_4$ the microscopic mechanism for the large thermopower is foremost the bandstructure since electronic correlations are not very strong (the effective mass enhancement is only 40% and even less for the a_{1g} band). This shows the strength of LDA+DMFT to unbiasedly identify bandstructure effects as the origin of large thermopowers where this is appropriate and electronic correlations where these prevail.

For getting the optimal thermoelectric material, hetero- or nanostructure we likely need both ingredients. First, a good bandstructure such as the pudding-mold form discussed in the present paper which due to dramatically different group velocities above and below the Fermi energy yields an extraordinarily large Seebeck coefficient. And second, correlation effects which result in asymmetrical, sharply peaked renormalized spectra in the vicinity of the Fermi level which enhance the Seebeck coefficient as well. With LDA+DMFT, we have an ideal tool to scan and design a wide range of potential SCES materials on a computer, providing experimental physicists and chemists with valuable hints on how to improve the thermoelectric figure of merit. In the exemplary case of $LiRh_2O_4$ this would be through electron-doping the material.

Acknowledgments We would like to thank H. Takagi and Y. Okamoto for fruitful discussions; numerical calculations were performed at the facilities of the Supercomputer center, ISSP, University of Tokyo. This work was supported by Grants-in-Aid for Scientific Research (MEXT Japan) grant 19019012,19014022, 19051016 and Russian Foundation for Basic Research (RFBR) grant 07-02-00041.

References

1. T. Seebeck, Abhandlungen der Preussischen Akademie der Wissenschaften, Berlin, pp. 265–373 (1823).
2. G. D. Mahan, Solid State Phys. **51**, 81 (1997). G. D. Mahan B. Sales, and J. Sharp, Phys. Today, March 1997, p. 42.
3. F. Ritz and C. E. Peterson, IEEE, Aerospace Conference, 2004. Proceedings, **5**, 2957 (2004).
4. J. Fairbanks, "Thermoelectric applications in vehicles status 2008" in ECT 2008 – On line proceedings. Available at ect 2008.icmpe.cnrs.fr.
5. B. Poudel, Q. Hao, Y. Ma, Y. Lan, A. Minnich, B. Yu, X. Yan, D. Wang, A. Muto, D. Vashaee, X. Chen, J. Liu, M. S. Dresselhaus, G. Chen, and Z. Ren, Science **320**, 634 (2008).
6. J. P. Heremans, V. Jovovic, E. S. Toberer, A. Saramat, K. Kurosaki, A. Charoenphakdee, S. Yamanaka, and G. J. Snyder, Science **321**, 554 (2008).
7. S. Paschen, Thermoelectric aspects of strongly correlated electron systems, in CRC Handbook of Thermoelectrics, Ch. 15, (ed. D. M. Rowe), CRC Press, Boca Raton, FL, 2005.
8. A. Bentien, S. Johnsen, G. K. H. Madsen, B. B. Iversen, and F. Steglich Europhys. Lett. **80**, 39901 (2007).
9. J. K. Freericks, V. Zlatić, and A. M. Shvaika Phys. Rev. B **75**, 035133 (2007).
10. V. Zlatić, R. Monnier, and J. K. Freericks, Phys. Rev. B **78**, 045113 (2008).
11. W. Metzner and D. Vollhardt, Phys. Rev. Lett. **62**, 324 (1989).
12. A. Georges and G. Kotliar, Phys. Rev. B, **45**, 6479 (1992).
13. A. Georges, G. Kotliar, W. Krauth and M. Rozenberg, Rev. Mod. Phys. **68**, 13 (1996).
14. P. Hohenberg and W. Kohn, Phys. Rev. **136** B864 (1964).
15. D.J. Singh, Phys. Rev. B **61**, 13397 (2000).
16. G.B. Wilson-Short, D. J. Singh, M. Fornari, and M. Suewattana, Phys. Rev. B **75**, 035121 (2007).
17. K. Kuroki and R. Arita, J. Phys. Soc. Jpn., **76**, 083707 (2007).
18. V. I. Anisimov, A. I. Poteryaev, M. A. Korotin, A. O. Anokhin, and G. Kotliar, J. Phys. Condens. Matter **9**, 7359 (1997).
19. A. I. Lichtenstein and M. I. Katsnelson, Phys. Rev. B **57**, 6884 (1998).
20. For reviews see, e.g., R. O. Jones and O. Gunnarsson, Rev. Mod. Phys. **61**, 689 (1989); R. M. Martin, Electronic Structure: Basic Theory and Practical Methods, Cambridge University Press, 2004.
21. I. A. Nekrasov, K. Held, N. Blümer, A. I. Poteryaev, V. I. Anisimov and D. Vollhardt, Eur. Phys. J. B **18** 55 (2000).
22. K. Held, I. A. Nekrasov, G. Keller, V. Eyert, N. Blümer, A. McMahan, R. Scalettar, T. Pruschke, V. I. Anisimov and D. Vollhardt, Phys. Stat. Sol. (b) **243**, 2599 (2006).
23. G. Kotliar, S. Y. Savrasov, K. Haule, V. S. Oudovenko, O. Parcollet, and C. A. Marianetti, Rev. Mod. Phys. **78**, 865 (2006).

24. K. Held, Adv. Phys. **56**, 829 (2007).
25. V. S. Oudovenko, G. Pálsson, K. Haule, G. Kotliar, and S. Y. Savrasov, Phys. Rev. B **73**, 035120 (2006).
26. R. Arita, K. Kuroki, K. Held, A. V. Lukoyanov, S. Skornyakov, and V.I. Anisimov, Phys. Rev. B **78**, 115121 (2008).
27. W. Kohn and L. J. Sham, Phys. Rev. **140**, A1133 (1965).
28. D. M. Ceperley and B. J. Alder, Phys. Rev. Lett. **45**, 566 (1980).
29. J. E. Hirsch and R. M. Fye, Phys. Rev. Lett. **56**, 2521 (1986).
30. A. N. Rubtsov and A. I. Lichtenstein, JETP Lett. **80**, 61 (2004).
31. A. N. Rubtsov, V. V. Savkin and A. I. Lichtenstein, Phys. Rev. B **72**, 035122 (2005).
32. P. Werner, A. Comanac, L. De Medici, M. Troyer and A. J. Millis, Phys. Rev. Lett. **97**, 076405 (2006).
33. P. Werner and A. J. Millis, Phys. Rev. B **74**, 155107 (2006).
34. S. Sakai, R. Arita, K. Held and K. Aoki, Phys. Rev. B **74**, 155102 (2006).
35. V. I. Anisimov, J. Zaanen and O. K. Andersen, Phys. Rev. B **44**, 943 (1991).
36. G. Sangiovanni, A. Toschi, E. Koch, K. Held, M. Capone, C. Castellani, O. Gunnarsson, S.-K. Mo, J. W. Allen, H.-D. Kim, A. Sekiyama, A. Yamasaki, S. Suga and P. Metcalf, Phys. Rev. B **73**, 205121 (2006).
37. T. Maier, M. Jarrell, T. Pruschke and M. H. Hettler, Rev. Mod. Phys. **77** 1027 (2005).
38. G. Kotliar, S. Y. Savrasov, G. Pálsson and G. Biroli, Phys. Rev. Lett. **87** 186401 (2001).
39. A. I. Lichtenstein and M. I. Katsnelson, Phys. Rev. B **62** 9283 (R) (2000).
40. A. Toschi, A. A. Katanin and K. Held, Phys. Rev. B 75, 045118 (2007).
41. H. Kusunose, J. Phys. Soc. Jpn. **75**, 054713 (2006).
42. C. Slezak, M. Jarrell, T. Maier and J. Deisz (2006), cond-mat/0603421.
43. K. Held, A. A. Katanin, A. Toschi, arXiv:0807.1860.
44. T. Pruschke, D. L. Cox and M. Jarrell, Phys. Rev. B **47** 3553 (1993).
45. M. Jarrell and J. E. Gubernatis, Phys. Rep. **269** 133 (1996).
46. I. Nekrasov et al, Phys. Rev. B **73**, 155112 (2006).
47. H. J. Vidberg and J. W. Serene, J. Low Temp. Phys. **29**, 179 (1977).
48. Y. Okamoto, S. Niitaka, M. Uchida, T. Waki, M. Takigawa, Y. Nakatsu, A. Sekiyama, S. Suga, R. Arita, and H. Takagi, Phys. Rev. Lett. **101**, 086404 (2008).
49. I. Terasaki, Y. Sasago and K. Uchinokura, Phys. Rev. B **56**, R12685(1997).
50. W. Koshibae, K. Tsutsui, and S. Maekawa, Phys. Rev. B **62** 6869 (2000).
51. I. Terasaki, JPSJ Online-News and Comments [Oct. 10, 2007].
52. O. K. Andersen, Phys. Rev. B **12**, 3060 (1975); O. Gunnarsson, O. Jepsen, and O. K. Andersen, Phys. Rev. B **27**, 7144 (1983).
53. N. Marzari and D. Vanderbilt, Phys. Rev. B **56**, 12847 (1997).
54. V. I. Anisimov, D. E. Kondakov, A. V. Kozhevnikov, I. A. Nekrasov, Z. V. Pchelkina, J. W. Allen, S.-K. Mo, H.-D. Kim, P. Metcalf, S. Suga, A. Sekiyama, G. Keller, I. Leonov, X. Ren, and D. Vollhardt, Phys. Rev. B **71**, 125119 (2005).
55. Z. V. Pchelkina, I. A. Nekrasov, Th. Pruschke, A. Sekiyama, S. Suga, V. I. Anisimov, and D. Vollhardt, Phys. Rev. B **75**, 035122 (2007).

Theory of Electronic Transport and Thermoelectricity in Ordered and Disordered Heavy Fermion Systems

C. Grenzebach, F. B. Anders, and G. Czycholl

Abstract A theory for electronic transport in heavy fermion systems has been developed and applied to the calculation of the temperature dependence of the resistivity and of the thermoelectric power. The dynamical mean-field theory has been used for the periodic Anderson model (PAM) together with the numerical renormalization group (NRG) as the impurity solver. To investigate also the influence of impurities and disorder, the method has been combined with the coherent potential approximation (CPA) for disordered systems. Considering two distinct local environments of a binary alloy with arbitrary concentration, two types of disorder have been investigated: on the f-sites and on the ligand sites. The temperature and concentration dependence of the thermoelectric power is calculated. The characteristic concentration dependence as well as the order of magnitude of the thermopower are reproduced for metallic heavy-fermion systems and for Kondo insulators. In particular, sign changes of the Seebeck coefficient as function of temperature and concentration are observed.

1 Introduction

Characteristic information on heavy-fermion systems (HFS) [1, 2] is obtained by measurements of the temperature (T) dependence of the transport coefficients. The resistivity $\rho(T)$ of HFS usually shows the following characteristic behavior: For high T, the resistivity $\rho(T)$ is determined by a negative temperature coefficient, and one usually observes a logarithmic, "Kondo"-like increase of the resistivity $\rho(T)$ with decreasing T. In metallic heavy-fermion systems – for example, $CePd_3$ [3], $CeAl_3$ [4] or $CeCu_6$ [5] – a maximum is observed in $\rho(T)$ at a characteristic

---

C. Grenzebach, F. B. Anders, and G. Czycholl
Institute for Theoretical Physics, University of Bremen, D-28334 Bremen, Germany
e-mail: cgrenzebach@itp.uni-bremen.de, czycholl@itp.uni-bremen.de

F. B. Anders
Present address: Lehrstuhl für Theoretische Physik II, Technische Universität Dortmund, Otto-Hahn Straße 4, 44221 Dortmund, Germany
e-mail: frithjof.anders@tu-dortmund.de

V. Zlatić and A. C. Hewson (eds.), *Properties and Applications of Thermoelectric Materials*, 159
NATO Science for Peace and Security Series B: Physics and Biophysics,

temperature $T_{\max}$ before the resistivity approaches a small residual value for $T \to 0$. At low temperatures T, $\rho(T)$ often obeys a T^2 law in such materials. The observation of a T^2 behavior well below the characteristic temperature scale $T_{\max}$ (in combination with a strongly enhanced γ coefficient of the specific heat) indicates the formation of a Landau Fermi-liquid with heavy quasiparticles. In Kondo insulators (KI), $\rho(T)$ crosses over from low values to an activation behavior for smaller T, indicating the opening of a gap and an insulating ground state. Another interesting transport quantity is the Seebeck coefficient $S(T)$, or the thermoelectric power. At temperatures comparable to $T_{\max}$, the thermoelectric power can reach absolute values of about 50 $\mu V/K$ in metallic HFS, often accompanied with sign changes at intermediate temperatures [1,2]. In Kondo insulators, even larger values of $S(T)$ up to 300 $\mu V/K$ have been observed [6–13]. Much of the recent interest in the heavy-fermion thermoelectricity results from these large absolute values at low temperatures which might be useful for applications, for instance in solid-state cooling devices [14].

Disorder has a strong impact on the transport properties such as $\rho(T)$ and $S(T)$. In substitutional alloys like $La_{1-x}Ce_xPd_3$ [3], $Ce_xLa_{1-x}Cu_6$ [5], $Ce_xLa_{1-x}Cu_{2.05}Si_2$ [15, 16], $U_{1-x}Th_xPd_2Al_3$ [17], etc., the residual resistivity $\rho(0)$ rapidly increases with increasing concentration x of the nonmagnetic impurities, and one obtains a crossover from the metallic $\rho(T)$ behavior with a maximum to a monotonic curve, where $\rho(T)$ decreases with increasing T. The thermoelectric power $S(T)$ for $CeCu_{2.05}Si_2$ shows a crossover from one case with a negative minimum, a sign change, and a positive maximum to a behavior with two positive maxima and no sign change when substituting Ce by La [15, 16]. For materials such as $Ce_3Cu_xPt_{3-x}Sb_4$, ligand alloying introduces a transition from a Kondo insulator to a dirty metal [7].

The basic model for a description of the electronic properties of HFSs is the periodic Anderson model (PAM) [18]. We study the PAM within the dynamical mean-field theory (DMFT) [19,20] and show that the characteristic transport properties of HFSs can be understood within this framework. Within the DMFT, the lattice model is mapped on an effective single-impurity Anderson model (SIAM) [21] by a self-consistency condition. We use the numerical renormalization group [22–24] (NRG) as impurity solver for this effective SIAM. The NRG is a nonperturbative method applicable at very low temperatures which reproduces the correct characteristic low-temperature scale (Kondo temperature) [25].

To study also the influence of disorder, we have extended this DMFT-NRG treatment to the disordered PAM. Thereby the disorder is studied within the coherent-potential approximation [26, 27] (CPA), which was originally developed for noninteracting disordered systems. In the CPA, lattice coherence is restored by introducing a (complex) "coherent" potential (local site-diagonal selfenergy). Janiš and Vollhardt have pointed out that the local nature allows one to embed the CPA into the DMFT framework [28]. Therefore, within the DMFT a unique, straightforward generalization of the CPA to disordered, correlated systems is possible [28–32]. We have applied this approach to the disordered PAM and calculate the transport quantities $\rho(T)$ and $S(T)$ for disorder on the f site as well as for disorder on the ligand sites. We treat the whole range of impurity concentrations $c \in [0;1]$

for different choices of the PAM parameters, starting with either a metallic HFS or a Kondo insulator for $c = 0$. We find that this CPA-NRG treatment of the PAM is able to explain the strong disorder dependence of the transport quantities of HFSs.

This paper gives an overview or review of results obtained within a research project over the last three years. Most of the results presented here have been published already in two original papers [33,34] and in Claas Grenzebach's thesis [35]. We would also like to mention some older related work [36], in which not the selfconsistent DMFT treatment of the PAM was used but an effective SIAM was studied. In another earlier paper [37] also the role of disorder on the transport properties of the PAM was investigated within a DMFT/CPA combination, however, not yet the NRG was used as the impurity solver. Furthermore, we would like to draw the attention to some very recent related work [38], also investigating the thermopower and thermal conductivity of the PAM within the DMFT starting from Fermi-liquid laws and relations.

The paper is organized as follows: We start with a short description of the model in Section 2, namely the PAM for periodic HFS and the "disordered" PAM, i.e. a straightforward extension of the PAM to describe also disordered (substitutional alloys of) HFS. Section 3 contains a short description of the different approximations applied, namely the DMFT mapping on an effective single impurity Anderson model (SIAM), the NRG treatment of this SIAM, and the combination of DMFT and CPA to study disordered, correlated systems. In Section 4 the (standard) starting relations and equations for a calculation of the transport coefficients based on the linear response theory and the simplifications arising within the DMFT/CPA are summarized. The results obtained for the resistivity and the thermopower of periodic metallic HFS and Kondo insulators and of disordered HFS with Kondo hole and ligand disorder are presented in Section 5 before the paper ends with a short summary and conclusion in Section 6.

2 Model

The basic model for the electronic properties of periodic HFS is the periodic Anderson model (PAM) given by

$$\hat{H} = \sum_{\mathbf{k}\sigma} \varepsilon_{\mathbf{k}\sigma} c^{\dagger}_{\mathbf{k}\sigma} c_{\mathbf{k}\sigma} + \sum_{i\sigma} \varepsilon_{f\sigma} \hat{n}^{f}_{i\sigma} + \frac{U}{2} \sum_{i\sigma} \hat{n}^{f}_{i\sigma} \hat{n}^{f}_{i-\sigma} + V \sum_{i\sigma} (f^{\dagger}_{i\sigma} c_{i\sigma} + c^{\dagger}_{i\sigma} f_{i\sigma}). \quad (1)$$

where $c_{\mathbf{k}\sigma}$ ($c^{\dagger}_{\mathbf{k}\sigma}$) destroys (creates) a conduction electron with spin σ, momentum $\mathbf{k}$ and energy $\varepsilon_{\mathbf{k}\sigma}$. The energy $\varepsilon_{f\sigma}$ denotes the spin-dependent single particle f-level energy at lattice site i, $\hat{n}^{f}_{i\sigma} = f^{\dagger}_{i\sigma} f_{i\sigma}$ is the f-electron occupation operator (per site and spin), $f_{i\sigma}$ ($f^{\dagger}_{i\sigma}$) destroys (creates) an f electron with spin σ at site i, and U denotes the on-site Coulomb repulsion between two f electrons on the same site i. The uncorrelated conduction electrons hybridize locally with the f electrons via the matrix element V. In this simplified model, only a single band is considered.

Moreover, the restriction to a single twofold degenerate f-orbital is only correct if all other crystal field states are energetically well separated from this magnetic ground state doublet.

The total filling per site, $n_{\text{tot}} = \sum_\sigma(\langle\hat{n}^c_{i\sigma}\rangle + \langle\hat{n}^f_{i\sigma}\rangle)$, is kept constant by a temperature-dependent chemical potential $\mu(T)$. We absorb the energy shifts (due to μ) into the band center ε_c of the conduction band, $\varepsilon_{\mathbf{k}} = \varepsilon_c + \tilde{\varepsilon}_{\mathbf{k}}$, as well as the f level ε_f (i.e. energies are measured relative to μ). For $n_{\text{tot}} = 2$ and $U = 0$, the uncorrelated system is an insulator at $T = 0$, since the lower of the two hybridized bands is completely filled. According to Luttinger's theorem a finite U of arbitrary strength does not change the Fermi volume which includes the full first Brillouin zone. As long as the ground state does not change symmetry due to a phase transition, the system remains an insulator at arbitrarily large Coulomb repulsion. Therefore, the nonmetallic ground state of Kondo insulators is not correlation induced, but it is already present for the noninteracting system and a consequence of Luttinger's theorem. For nonintegral values of n_{tot}, the paramagnetic phase of the system must be metallic.

We want to investigate, in particular, also the influence of substitutional disorder, i.e. not only pure periodic systems will be considered and modelled but also alloys of different materials with arbitrary concentration of the constituents. The model (1) is easily extended to such disordered HFS:

$$\hat{H} = \sum_{\mathbf{k}\sigma} \varepsilon_{\mathbf{k}\sigma} c^\dagger_{\mathbf{k}\sigma} c_{\mathbf{k}\sigma} + \frac{U}{2} \sum_{i\sigma} \hat{n}^f_{i\sigma} \hat{n}^f_{i-\sigma} + \sum_{i\sigma} \left(f^\dagger_{i\sigma}\; c^\dagger_{i\sigma} \right) \mathcal{V}_{i\sigma} \begin{pmatrix} f_{i\sigma} \\ c_{i\sigma} \end{pmatrix} \tag{2}$$

$$\text{with } \mathcal{V}_{i\sigma} = \begin{pmatrix} \varepsilon_{if\sigma} & V_{i\sigma} \\ V_{i\sigma} & \varepsilon_{ic\sigma} \end{pmatrix}. \tag{3}$$

To describe an $A_c B_{1-c}$ alloy we set:

$$\mathcal{V}_{i\sigma} = \begin{cases} \mathcal{V}^A_\sigma & \text{with probability } c \\ \mathcal{V}^B_\sigma & \text{with probability } 1-c. \end{cases} \tag{4}$$

where $\mathcal{V}^{A/B}$ contains the local on-site matrix elements of the Hamiltonian of the pure $A(B)$-system.

We will, in particular, consider two types of disorder, namely the substitution of the correlated, f-electron (lanthanide or actinide) ions by (chemically similar) non-f-electron ions, e.g. the substitution of Ce by La as in $Ce_{1-x}La_xPd_3$. We call this type of disordered (substituted) HFS *Kondo holes* and model it by allowing two different values for the f-electron energy level:

$$\varepsilon_{if} = \begin{cases} \varepsilon^A_f & \text{with probability} \quad c \\ \varepsilon^B_f \to \infty & \text{with probability} \quad 1-c \end{cases}. \tag{5}$$

By shifting ε^B_f to ∞ the f-level of this ion remains unfilled because of which it is a suitable model for a non-f-electron ion.

The second type of disorder to be considered is the substitution of the non-correlated component (ions) by other ions, for instance in $Ce_3Cu_xPt_{3-x}Sb_4$, in which the Pt-ions are substituted by Cu. We call this type of disorder *ligand disorder* and model it by assuming that the f-electron level is the same at all sites but that the band electron level and the hybridization vary depending whether an A- or B-ion is at the corresponding site. Therefore, for ligand disorder one has:

$$\varepsilon_{ic} = \begin{cases} \varepsilon_c^A \text{ with probability } c \\ \varepsilon_c^B \text{ with probability } 1-c \end{cases}, \quad V_i = \begin{cases} V^A \text{ with probability } c \\ V^B \text{ with probability } 1-c \end{cases}. \tag{6}$$

3 Approximations

The transport properties depend on the one-particle Green function, which is a 2×2-matrix for a two-band model like the PAM. For the periodic PAM (1) without disorder the Green function can be written as

$$G(\mathbf{k},z) = \begin{bmatrix} G_{ff}(\mathbf{k},z) & G_{fc}(\mathbf{k},z) \\ G_{cf}(\mathbf{k},z) & G_{cc}(\mathbf{k},z) \end{bmatrix} = \left\{ \begin{pmatrix} z & 0 \\ 0 & z-\varepsilon_{\mathbf{k}} \end{pmatrix} - \begin{pmatrix} \varepsilon_f + \Sigma_f(\mathbf{k},z) & V \\ V & \varepsilon_c \end{pmatrix} \right\}^{-1}, \tag{7}$$

where $\Sigma_f(\mathbf{k},z)$ denotes the f-electron self-energy, which is $\mathbf{k}$-dependent, in general. Our first approximation will be the assumption of a local, $\mathbf{k}$-independent self-energy, i.e. $\Sigma_f(\mathbf{k},z) = \Sigma_f(z)$ which corresponds to the dynamical mean-field theory (DMFT), which becomes correct in the limit of infinite spatial dimension $d \to \infty$. Just for the PAM this DMFT-assumption should be quite good also for finite d and small hybridization V, because it can be shown that corrections to the local approximation occur only in order V^6.

Within the DMFT the f-electron Green function of the PAM can be mapped on that of an effective single-impurity Anderson model (SIAM) via the self-consistency condition

$$G_{ff}(z) = \frac{1}{N}\sum_{\mathbf{k}} G_{ff}(z,\mathbf{k}) = \frac{1}{N}\sum_{\mathbf{k}} \frac{1}{z-\varepsilon_f-\Sigma_f(z)-\frac{V^2}{z-\varepsilon_{\mathbf{k}}}} \tag{8}$$

$$= G_f^{\mathrm{SIAM}}(z) = \frac{1}{z-\varepsilon_f-\Sigma_f(z)-\Delta(z)} \tag{9}$$

Here $\Delta(z)$ describes an effective bath (of conduction electrons) to which the single, local, correlated impurity is coupled, and this $\Delta(z)$ has to be determined self-consistently.

We accurately solve this effective SIAM using Wilson's numerical renormalization group (NRG) [22, 23]. The key ingredient in the NRG is a logarithmic discretization of the continuous bath, controlled by the parameter [22] $\Lambda > 1$. The Hamiltonian is mapped onto a semi-infinite chain, where the Nth link represents an exponentially decreasing energy scale $D_N \sim \Lambda^{-N/2}$. Using this hierarchy of scales the sequence of finite-size Hamiltonians $\mathscr{H}_N$ for the N-site chain is solved iteratively, truncating the high-energy states at each step to maintain a manageable number of states. The reduced basis set of $\mathscr{H}_N$ thus obtained is expected to faithfully describe the spectrum of the full Hamiltonian on the scale of D_N, corresponding [22] to a temperature $T_N \sim D_N$ from which all thermodynamic expectation values are calculated. The energy-dependent hybridization function $\Delta(z)$ determines the coefficients of the semi-infinite chain [39].

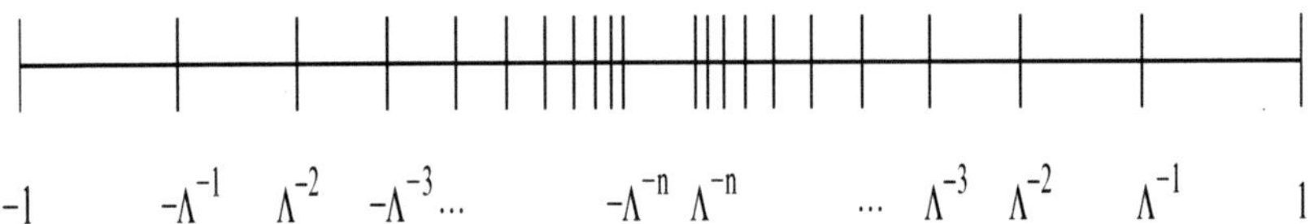

To treat the disorder problem we apply the coherent potential approximation (CPA) for the configurationally averaged Green function (matrix)

$$\mathscr{G}(z) = \begin{pmatrix} \mathscr{G}_{ff}(z) & \mathscr{G}_{fc}(z) \\ \mathscr{G}_{cf}(z) & \mathscr{G}_{cc}(z) \end{pmatrix} = cG^A(z) + (1-c)G^B(z) \tag{10}$$

$$= \frac{1}{N}\sum_{\mathbf{k}} \left[\begin{pmatrix} z & 0 \\ 0 & z-\varepsilon_{\mathbf{k}} \end{pmatrix} - \Sigma^{\mathrm{CPA}}(z) \right]^{-1} \tag{11}$$

Here the latter equation defines the "coherent potential" $\Sigma^{\mathrm{CPA}}(z)$, i.e. the (site-diagonal, complex) CPA-self-energy, which is a 2×2-matrix for a two-band model like the disordered PAM (2). The CPA is the best single-site approximation for the self-energy, which is again local (site-diagonal, i.e. **k**-independent). Therefore, the CPA also becomes correct in the limit of large dimension or coordination number, and a combination of DMFT and CPA is consistent for a description of disordered, correlated systems. The G^A and G^B are the local Green function matrices of the effective single-impurity problems for the two subsystems A and B, determined by the effective media $\Delta_A(z)$ and $\Delta_B(z)$, respectively:

$$G_{ff}^{A/B}(z) = \left[z - \varepsilon_f^{A/B} - \Delta_{A/B}(z) - \Sigma_f^{A/B}(z) \right]^{-1}, \tag{12}$$

$$G_{fc}^{A/B}(z) = \frac{\Sigma_c^{A/B}(z)}{V_{A/B}} G_{cc}^{A/B}(z) = \frac{\Delta_{A/B}(z)}{V_{A/B}} G_{ff}^{A/B}(z). \tag{13}$$

Therefore, to solve the problem of correlated, disordered systems within a suitable combination of DMFT and CPA, one has first to determine the correlation self-energies $\Sigma_f^{A/B}(z)$ for the pure A- and B-system within the DMFT using the NRG as impurity solver; then one has to use these Green function matrices $G^{A/B}(z)$ of the A- and B-system in the CPA-equation (Eq. 11) to determine the disorder-self-energy $\Sigma^{CPA}(z)$.

4 Transport Quantities

We start from the standard relations [40] for the generalized transport coefficients, according to which the electrical current density $\mathbf{J}$ and the heat current density $\mathbf{q}$ depend linearly on the electric field $\mathbf{E}$ and the temperature gradient ∇T:

$$\mathbf{J} = L_{11}\mathbf{E} + L_{12}\left(-\frac{1}{T}\nabla T\right),$$
$$\mathbf{q} = L_{21}\mathbf{E} + L_{22}\left(-\frac{1}{T}\nabla T\right). \tag{14}$$

All coefficients are calculated within the linear response approach, starting from similar Kubo formulas [40, 41]; for symmetry reasons, $L_{12} = L_{21}$ must hold.

According to the standard linear response theory, the real part of the frequency-dependent (optical) conductivity tensor [40, 42, 43] $\sigma(\omega) = L_{11}(\omega)$ is given by the current–current correlation function as

$$\sigma_{\alpha\beta}(\omega) = -\frac{1}{\omega N V_0}\mathrm{Im}\langle\langle j_\alpha | j_\beta^\dagger \rangle\rangle(\omega + i0^+), \tag{15}$$

where $V_0 = a^3$ is the volume of the unit cell and N counts the number of unit cells. For a $\mathbf{k}$-independent hybridization the current operator consistent with the PAM is given by [44]:

$$\mathbf{j} = e\sum_{\mathbf{k}\sigma} \mathbf{v_k} c^\dagger_{\mathbf{k}\sigma} c_{\mathbf{k}\sigma}, \tag{16}$$

where $\mathbf{v_k} = \frac{1}{\hbar}\nabla_\mathbf{k}\varepsilon_\mathbf{k}$ is the group velocity. Then the current-susceptibility tensor $\langle\langle \mathbf{j}|\mathbf{j}^\dagger\rangle\rangle(z)$ is connected to the particle-hole Green function

$$\langle\langle \mathbf{j}|\mathbf{j}^\dagger\rangle\rangle(z) = e^2 \sum_{\sigma\sigma'\mathbf{k}\mathbf{k}'} \mathbf{v_k}\mathbf{v}^T_{\mathbf{k}'} \langle\langle c^\dagger_{\mathbf{k}\sigma} c_{\mathbf{k}\sigma} | c^\dagger_{\mathbf{k}'\sigma'} c_{\mathbf{k}'\sigma'}\rangle\rangle(z). \tag{17}$$

In general, the current-current correlation function involves vertex corrections [40]. However, for a k-independent hybridization, the current operator vertex corrections vanish [45–48]. This prevails even in the disordered system within a CPA treatment, since the two-particle vertex remains local. Then the current-current correlation function is completely determined by the free particle-hole propagator, and all transport coefficients depend only on the single-particle Green function and its self-energy.

Then in the limit $\omega \to 0$, one obtains for the generalized transport coefficients:

$$L_{11} = \frac{e^2}{\hbar a}\int_{-\infty}^{\infty} [-f'(\omega)]\ \tau(\omega)\, d\omega, \tag{18}$$

$$L_{12} = \frac{e}{\hbar a}\int_{-\infty}^{\infty} [-f'(\omega)]\ \omega\ \tau(\omega)\, d\omega, \tag{19}$$

$$L_{22} = \frac{1}{\hbar a}\int_{-\infty}^{\infty} [-f'(\omega)]\ \omega^2\ \tau(\omega)\, d\omega. \tag{20}$$

where f' is the derivative of the Fermi function and $\tau(\omega)$ represents a generalized relaxation time defined as

$$\begin{aligned}\tau(\omega) &= \frac{2\pi}{d}(t^*)^2 \int_{-\infty}^{\infty} \rho_0(\varepsilon)\rho_c^2(\varepsilon,\omega)d\varepsilon \\ &= \frac{(t^*)^2}{\pi d}\left[\frac{\partial D(z)}{\partial z} - \frac{\mathrm{Im}D(z)}{\mathrm{Im}z}\right],\end{aligned} \tag{21}$$

which is to be evaluated at $z = x + iy$

$$z = \omega + i0^+ - \varepsilon_c - \Sigma_c(\omega + i0^+)$$

Above, we have used the notations

$$\rho_c(\varepsilon,\omega) = -\frac{1}{\pi}\mathrm{Im}G_c(\omega + i0,\varepsilon),\ G_c(z,\varepsilon_{\mathbf{k}}) = \frac{1}{z - \varepsilon_{\mathbf{k}} - \Sigma_c(z)} \tag{22}$$

$$\Sigma_c(z) = \frac{V^2}{z - \Sigma_f(z)} + \Sigma_c^{\mathrm{CPA}}(z),\ D(z) = \int d\varepsilon \frac{\rho_0(\varepsilon)}{z - \varepsilon}. \tag{23}$$

Here $\rho_0(\varepsilon)$ is the unperturbed conduction band density of states for which we have chosen a Gaussian model density of states, which is just the density of states of the d-dimensional simple hypercubic tight-binding system in the limit $d \to \infty$:

$$\rho_0(\varepsilon) = \frac{1}{t^*\sqrt{2\pi}}\exp(-\frac{\varepsilon^2}{2t^{*2}}). \tag{24}$$

From the generalized transport coefficients (18–20) one finally obtains the measurable quantities resistivity, thermopower and thermal conductivity as [40]

$$\rho = \sigma^{-1} = L_{11}^{-1},\ S = \frac{1}{T}\frac{L_{12}}{L_{11}},\ \kappa = \frac{1}{T}\left(L_{22} - \frac{L_{12}^2}{L_{11}}\right). \tag{25}$$

The thermopower S is directly obtained in natural units $k_B/|e| = 86\,\mu\mathrm{V/K}$, but the unit of the resistivity depends on the lattice constant a. If we assume $a \approx 10^{-10}\,\mathrm{m}$, the resistivity has the unit $\hbar a/e^2 \approx 41\,\mu\Omega\,\mathrm{cm}$.

The dimensionless "figure of merit" defined by

$$ZT = \frac{T\sigma S^2}{\kappa}, \tag{26}$$

measures the efficiency of a thermoelectric material, which is not only determined by a high thermopower S, since a simultaneously large thermal conductivity κ may easily compensate the temperature gradient generated by the large S. An efficient thermoelectric cooler requires a large figure of merit which should be $ZT > 1$.

5 Results

5.1 Periodic Systems

5.1.1 Metallic HFS

Besides the parameters of the PAM ($\varepsilon_f - \varepsilon_c, U, V$ and the model assumption concerning the band structure of the unperturbed conduction band) one has to fix the total number of electrons n_{tot} per site. To achieve a heavy fermion situation ε_f must be below the chemical potential μ and $\varepsilon_f + U$ above μ so that (for small V) the f-level at each site is singly occupied and thus carries a magnetic moment. The chemical potential and the f-electron occupation numbers have to be determined selfconsistently for a given n_{tot} which ranges between 0 and 4. For $1 < n_{\text{tot}} < 2$ μ falls automatically into a region of a finite density of states and one has, therefore, a metallic system. For the results presented in the following figures energies (and temperatures) are measured in units of $\Gamma_0 = \pi V^2 \rho_0(0)$ (V hybridization, $\rho_0(0)$ unperturbed conduction band density of states at the band center). Γ_0 corresponds to the "Anderson width" of the (original) SIAM and measures the charge fluctuation scale. For HFS one can expect Γ_0 to be of the magnitude 100 meV.

For a metal with $n_{tot} = 1.6$, and $\varepsilon_f - \varepsilon_c = -U/2$, $\rho(T)$ is plotted for different values of the Coulomb repulsion U in Fig. 1. We observe a small residual resistivity $\rho(0)$, a maximum at a characteristic temperature T_{max}, and $\rho(T)$ (Kondo like) decreasing with increasing T for $T > T_{\text{max}}$. The fact that $\rho(T = 0)$ is finite (and not 0 though the system is periodic and does not contain any residual scattering anymore) is due to a finite broadening, in particular a finite energy imaginary part δ,

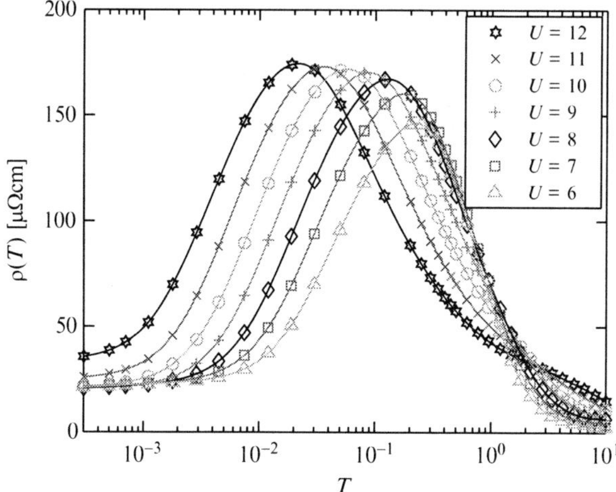

Fig. 1 Resistivity $\rho(T)$ as function of T for different U, $\varepsilon_f - \varepsilon_c = -U/2$, and fixed total filling $n_{\text{tot}} = 1.6$.

which has to be introduced to obtain a continuous density of states from the discrete NRG spectrum and for numerical reasons to achieve a better convergence of the self-consistency equations. This broadening may be interpreted physically as simulating the effects of residual impurity scattering. $T_{\max}$ is shifted to lower values for increasing U; within DMFT-NRG one obtains an exponential dependency of $T_{\max}$ on U in agreement with the (non-analytic) dependence of the Kondo temperature on U according to the Schrieffer–Wolff transformation. For sufficiently strong U the peak height at $T_{\max}$ is nearly U independent.

For the same parameters the T-dependence of the thermopower $S(T)$ is shown in Fig. 2. One observes a rich two-peak structure. Obviously, the low-T peak again moves to lower temperature with increasing U whereas the upper peak moves to higher T. The characteristic temperature, at which the first extremum of $S(T)$ is obtained, scales similarly as the $T_{\max}$ at which the resistivity $\rho(T)$ has its maximum and as the Kondo-temperature with increasing U. Therefore, this temperature of the first (low-T) extremum is essentially just the characteristic low T (Kondo temperature) scale of the system. The two maxima are separated by a minimum in $S(T)$, and for $U > 8$ $S(T)$ is negative at this minimum, i.e. $S(T)$ changes sign and has a negative minimum. Several extrema and similar sign changes were found experimentally for several HFS materials, in particular below 80 K for $CeCu_{2.2}Si_2$ [49, 50]. Note furthermore the large absolute values of $S(T)$ of the magnitude $100\mu V/K$.

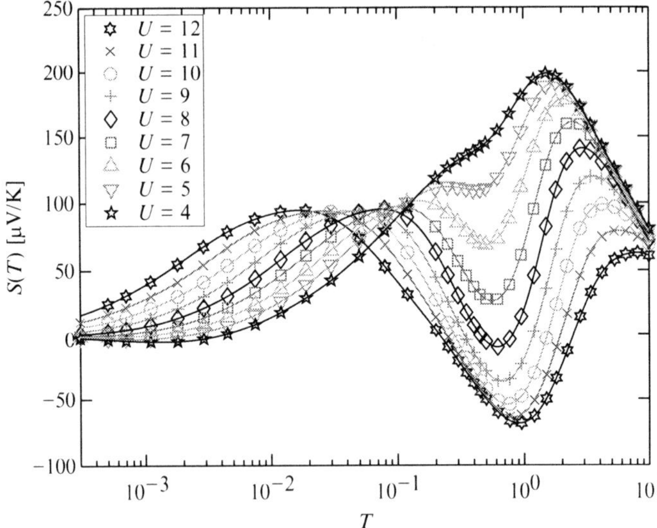

Fig. 2 Thermoelectric power for various values of U and fixed total occupation $n_{\mathrm{tot}} = 1.6$ and $\varepsilon_f - \varepsilon_c = -U/2$.

5.1.2 Kondo Insulators

For $n_{\text{tot}} = 2$ the lower hybridized bands are completely filled in the (exactly solvable) uncorrelated case $U = 0$. Therefore, the chemical potential μ falls into the hybridization gap and one obtains an insulating ground state. According to the Luttinger theorem this statement holds also for the correlated system with finite or large U: for $n_{\text{tot}} = 2$ μ always falls into a small hybridization gap and the system is, therefore, an insulator for $T \to 0$. For such a filling, different $\varepsilon_f - \varepsilon_c$ and $U = 10$ the T-dependence of $\rho(T)$ is shown in Fig. 3. One obtains the $\rho(T)$-behavior typical for Kondo insulators, i.e. a crossover from a Kondo like (logarithmic) increase of $\rho(T)$ to an activation behavior for smaller T. The (artificial) finite $\rho(0)$ depends hardly on $\varepsilon_f - \varepsilon_c$ but only on the finite residual imaginary part (broadening) δ.

For the same parameters the thermoelectric power $S(T)$ is shown in Fig. 4. In the symmetric case $\varepsilon_f - \varepsilon_c = -U/2 = -5$, $S(T)$ vanishes due to particle-hole symmetry; away from the symmetric case, the peaks reach extremely large absolute values of S up to 200 μV/K. This shows that Kondo insulators could be excellent candidates for thermoelectric applications.

5.1.3 Approaching the Kondo Insulator from the Metallic Side

We investigate the crossover from the metallic HFS to the Kondo insulator by varying n_{tot}. For fixed U $= 10$ and $\varepsilon_f - \varepsilon_c = -U/2$, Fig. 5 shows the results of the temperature dependence of the resistivity $\rho(T)$. One observes the crossover from the typical metallic HF behavior with a finite residual resistance and a maximum in $\rho(T)$ to the KI activation behavior.

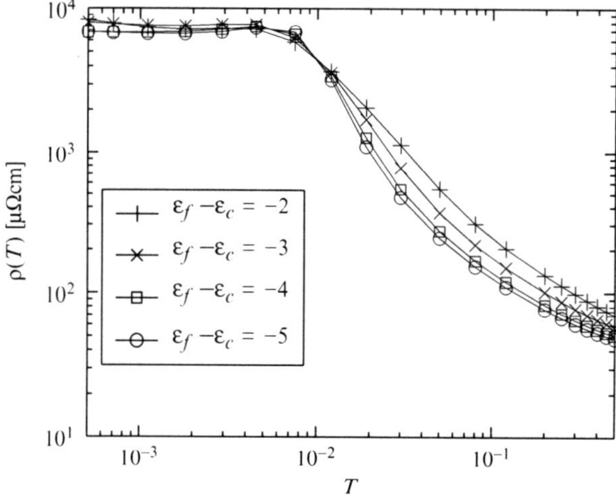

Fig. 3 Resistivity $\rho(T)$ for different $\varepsilon_f - \varepsilon_c$, $U = 10$, $n_{\text{tot}} = 2$ (Kondo insulators).

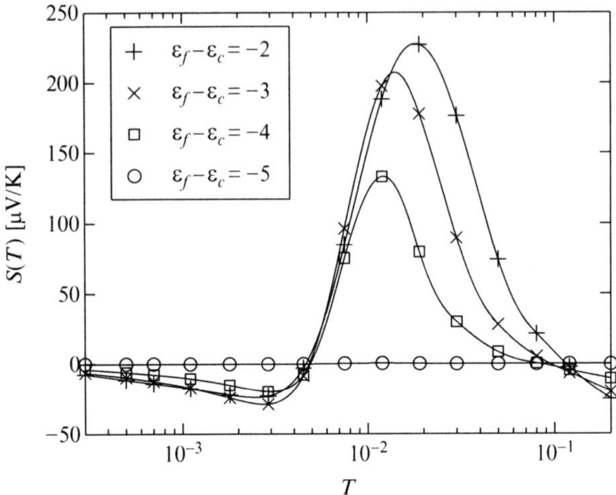

Fig. 4 Thermoelectric power for $U = 10$, $n_{\text{tot}} = 2$ (Kondo insulator regime) and different values for ε_f (or $\varepsilon_f - \varepsilon_c$).

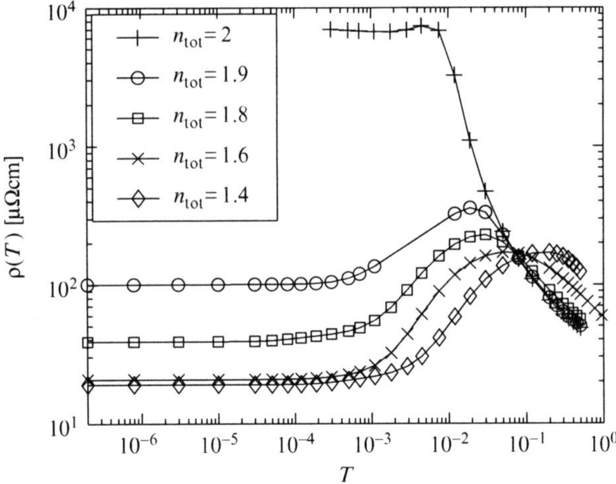

Fig. 5 Resistivity $\rho(T)$ for different total fillings n_{tot}, $U = 10$, $\varepsilon_f - \varepsilon_c = -U/2$, $\delta = 10^{-3}$.

Corresponding results for the temperature dependence of the thermopower are shown in Fig. 6. This choice of parameters corresponds to the (particle-hole) symmetric PAM exactly in the KI situation $n_{\text{tot}} = 2$; therefore, the thermopower vanishes identically for $n_{\text{tot}} = 2$. But very close to the KI behavior a very large $S(T)$ is obtained. The closer one is to the KI regime the larger becomes $S(T)$. Therefore, very large absolute values of $S(T)$ cannot only be obtained for KI but also for metallic HFS which are close to the KI regime.

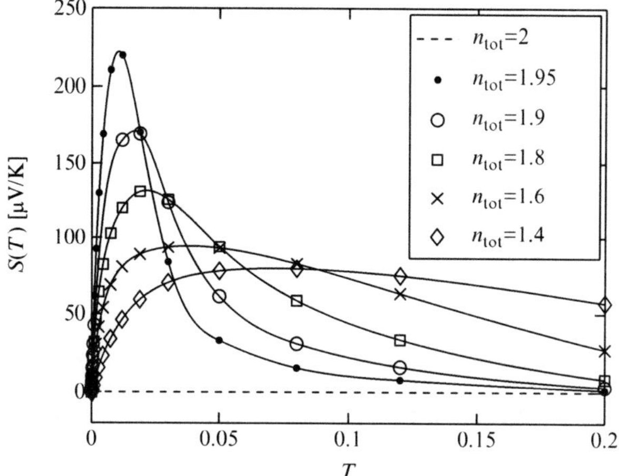

Fig. 6 Thermoelectric power for a fixed $U/\Gamma_0 = 10$ and various occupations $n_{\text{tot}} = 1.4$, $1.6, 1.9, 1.95, 2$ and $\varepsilon_f - \varepsilon_c = -U/2$.

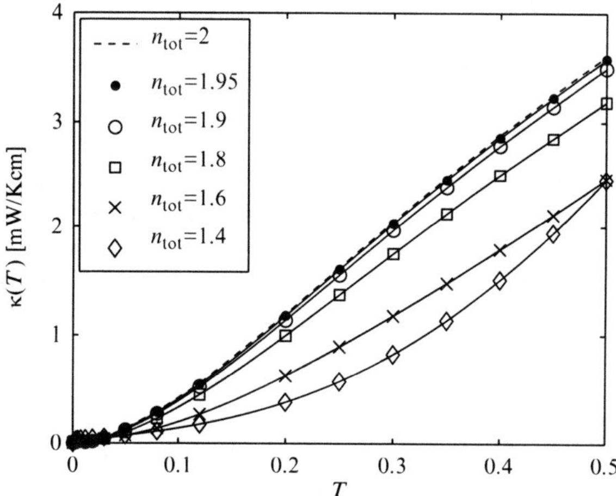

Fig. 7 Electronic part of thermal conductivity for $U = 10$, $\varepsilon_f - \varepsilon_c = -U/2$. All parameters: as in Figs. 5 and 6.

For this set of parameters we have also calculated the electronic contribution to the thermal conductivity shown in Fig. 7. Combining this with the electrical conductivity and the thermopower we get the figure of merit shown in Fig. 8. Obviously, just in the situation close to the KI case a large figure of merit $ZT \approx 2 > 1$ is possible. However, additional phonon contributions not included in the PAM will increase the thermal conductance and, therefore, reduce the figure of merit. Hence, a low phonon thermal conductance is needed to optimize ZT. One possible root is disorder, because phonon scattering at the impurity atoms may reduce the phonon mean free path and therefore the phonon contribution to the thermal conductivity.

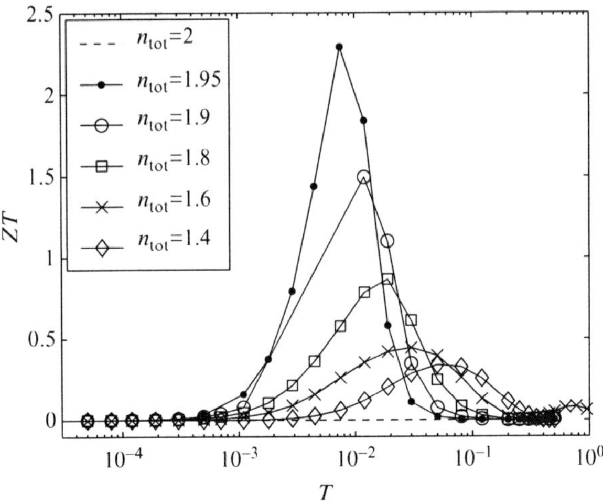

Fig. 8 Figure of merit for $U = 10$, $\varepsilon_f - \varepsilon_c = -U/2$. All parameters: as in Figs. 5 and 6.

5.2 *Disordered Systems*

5.2.1 Kondo Holes

First we consider the case of a metallic, substitutional A_cB_{1-c} alloy, where A denotes a (correlated) heavy-fermion system and B a (less correlated) system with empty f-shells far above the chemical potential. Such a situation is realized, for instance, in $Ce_cLa_{1-c}Al_3$, where the rare earth Ce-ion is substituted by La. Results for the temperature dependence of the resistivity $\rho(T)$ are shown in Fig. 9 for the whole concentration regime $0 < c \leq 1$. For the pure HFS ($c = 1$) we obtain once more the typical behavior with a small residual resistance $\rho(0)$, a rapid increase with increasing T and a maximum at a characteristic (Kondo) temperature $T_{\max}$ and a $\rho(T)$ decreasing with increasing T for $T > T_{\max}$. When substituting A by the non-magnetic B-system, the residual resistance $\rho(0)$ quickly increases and the characteristic temperature $T_{\max}$, at which the maximum is reached, moves to lower values. At $c = 0.6$, $\rho(0)$ assumes its largest value and there is no longer a maximum but the resistivity $\rho(T)$ monotonically decreases with increasing T and saturates for $T \to 0$ to a large, finite value. The resistivity behaves as that of a typical Kondo impurity system. For still smaller c a qualitatively similar behavior is obtained but the value $\rho(0)$ decreases again with increasing c, and finally in the opposite limit $c \to 0$ of an uncorrelated metal a very small resistivity is obtained.

Nevertheless, for small finite concentration of the magnetic f-electron ions, i.e. here for $c = 10^{-4}$, the typical Kondo impurity behavior is obtained. This can be seen by plotting the resistivity normalized to the concentration, $\rho(T)/c$ shown in Fig. 10. We obtain the typical resistivity behavior of a Kondo impurity system for all $c \leq 0.6$, but the normalized resistivity $\rho(T)/c$ does not saturate but has a maximum at $c = 0.5$

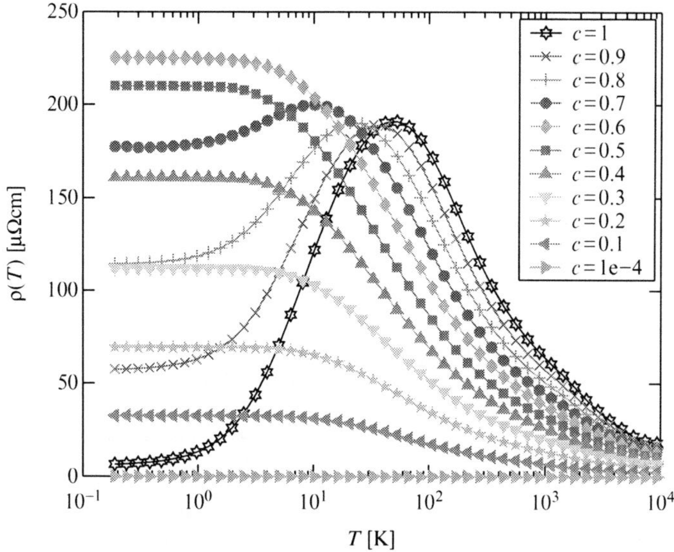

Fig. 9 Resistivity $\rho(T)$ as a function of T for different concentrations c of system A, calculated with DMFT/CPA-NRG; $U = 10$, $\varepsilon_f^A - \varepsilon_c^A = -U/2$, $\varepsilon_f^B = \infty$, and $n_{\text{tot}} = 1.6 - (1 - c)$.

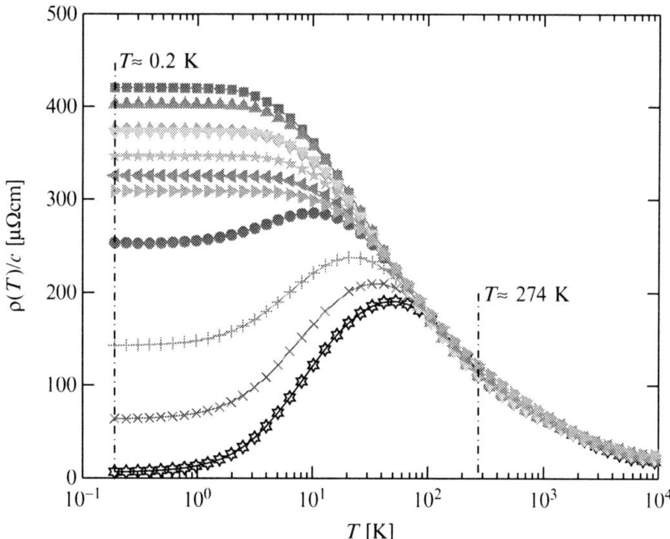

Fig. 10 The same data as in Fig. 9 plotted as $\rho(T)/c$.

and decreases with decreasing c approaching a finite value in the limit $c \to 0$. To demonstrate this in more detail, we show the concentration (c-) dependence of the residual resistance $\rho(T=0)$ and $\rho(T=0)/c$ in Fig. 11. Obviously a behavior

$$\rho(T=0,c) \sim \frac{c}{1-0.6c} \tag{27}$$

is obtained.

Results for the thermopower are depicted in Fig. 12. Again we find very large absolute values for $S(T)$ (of the magnitude $100\,\mu\mathrm{V/K}$). Similar to the resistivity, the thermoelectric power exhibits a low-temperature peak which is correlated with the maximum of the resistivity. The position of this low-temperature peak and the peak

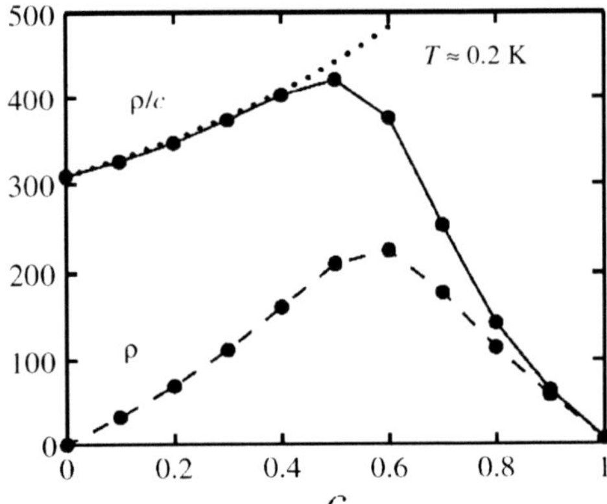

Fig. 11 Resistivity as a function of the concentration c at $T \approx 0.2$ K $\rho(c)$ (*dashed line*), $\rho(c)/c$ (*full line*), and approximation $\propto 1/(1-0.6c)$ (*dotted line*).

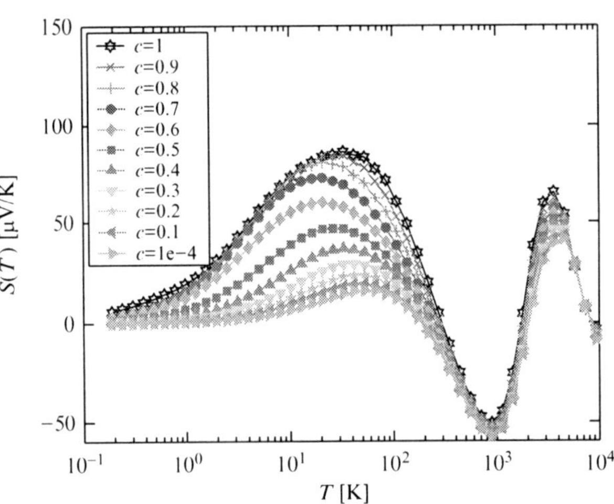

Fig. 12 Thermoelectric power $S(T)$ as a function of T for fixed $U = 10$ and different concentrations c of system A; all parameters as in Fig. 9, among others $n_{\mathrm{tot}} = 1.6 - (1-c)$.

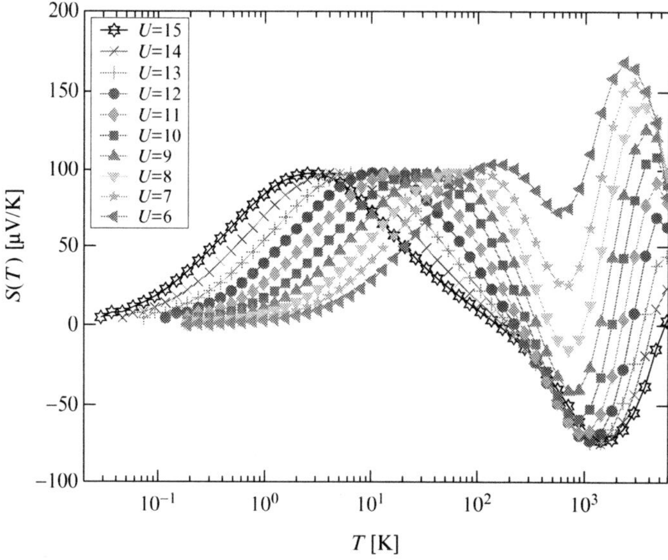

Fig. 13 Thermoelectric power $S(T)$ as a function of T for fixed concentration $c = 0.8$ of system A and different U; $\varepsilon_f^A - \varepsilon_c^A = -U/2$, $\varepsilon_f^B = \infty$, chemical potential $\mu = 0$, and a filling $n_{\mathrm{tot}} = 1.4$.

height depend on the concentration c. In addition, we observe a second extremum at a very high temperature independent of the concentration and the low-temperature scale which results from the charge fluctuations on the energy scale $\varepsilon_f - \mu$. This maximum moves to higher temperatures with increasing U, as displayed in Fig. 13.

5.2.2 Ligand Disorder

In the case of ligand disorder, i.e. disorder in the non-correlated (non f-electron) part of the HFS, we have first investigated a situation, in which the HFS is driven from a metallic HFS to a KI by alloying. In Fig. 14, the resistivity and the thermoelectric power are displayed for $\varepsilon_f^A - \varepsilon_c^A = -U/2$, $\varepsilon_f^B - \varepsilon_c^B = -1$, and various values of the concentration c of system A. The curves for $c = 1$ are identical to the $c = 1$ curves in Figs. 9 and 12. We plot the resistivity on a log-log scale in Fig. 14a to cover the almost three orders of magnitude of resistivity change.

In contrast to the previous case of f disorder introduced by Kondo holes, the number of Kondo scatterers remains constant as a function of the concentration c. Therefore, the resistivity does not vanish for $c \to 0$. For temperatures $T > T_{\max}$, the absolute values of the resistivity remain nearly the same; above this characteristic low-temperature scale $T_{\max}$, lattice and Kondo disorder scattering become indistinguishable. For temperatures $T < T_{\max}$, an increase in the resistivity with decreasing concentration c is observed; Fig. 14a demonstrates such a crossover from low resistivity to an insulating behavior for $T \to 0$ and $c \to 0$. The disor-

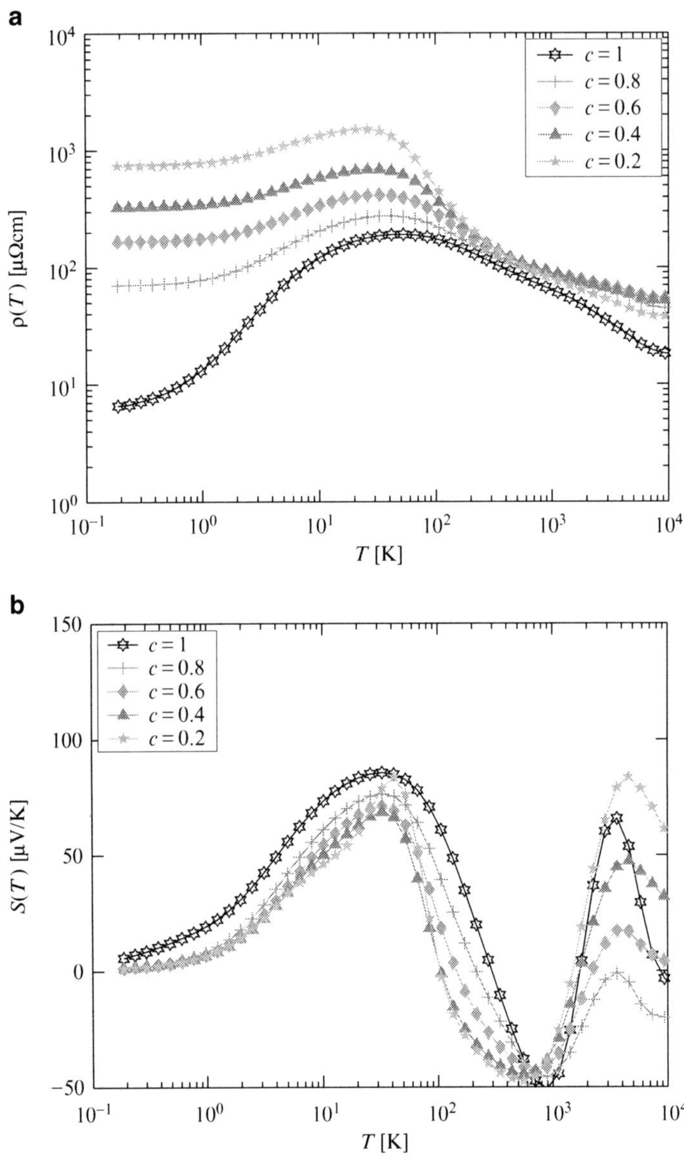

Fig. 14 (**a**) Resistivity $\rho(T)$ and (**b**) thermoelectric power $S(T)$ as functions of T for different concentrations c of system A, calculated with CPA-NRG for $V_A = V_B = V_0$, $U = 10$, $\varepsilon_f^A - \varepsilon_c^A = -U/2$, $\varepsilon_f^B - \varepsilon_c^B = -\Gamma_0$, $\varepsilon_f^A = \varepsilon_f^B$, chemical potential $\mu = 0$, and a filling $n_{\mathrm{tot}} = 2 - 0.4c$ and $\delta/\Gamma_0 = 10^{-3}$.

der introduced in the conduction band destroys the lattice coherence of the heavy quasiparticles at low temperatures. Even though translational invariance is restored for $c \to 0$, the residual resistivity remains increasing for $c < 0.5$ due to the crossover to an insulating behavior.

The thermoelectric power does not show much variation with the concentration c as can be seen in Fig. 14b. In contrast to the resistivity, the thermoelectric power depends on the asymmetry of the spectrum [see Eqs. (25) and (19)]. This means that ligand disorder, which has strong influence on the value of the resistivity (and also on the thermal conductivity), has only weak influence on the thermopower. Therefore, a large figure of merit should become possible by suitable choices of ligand disorder, because the thermopower may be nearly unaffected (and large), whereas the thermal conductivity is reduced.

Next, we investigate the transport properties as a function of disorder in the case in which ligand substitution keeps the system in a metallic regime for all values of c. In Fig. 15a, the resistivity is displayed for $\varepsilon_f^A - \varepsilon_c^A = -6$ and $\varepsilon_f^B - \varepsilon_c^B = -4$. In this case, we have a disordered conduction band where both subsystems A and B are asymmetric. By choosing a filling of $n_{\text{tot}} = 1.8 - 0.2c$ and $\varepsilon_f^A - \varepsilon_f^B = 0.2$, a nearly c-independent f occupation $n_f \approx 0.92$ is achieved, significantly departed from the Kondo insulator regime. This is clearly visible in the figure: in the limits $c = 0$ and $c = 1$, the residual resistivity $\rho(T \to 0)$ is small as expected for metallic systems. For $0 < c < 1$, the disorder in the conduction band leads to an enhanced resistivity. The residual resistivity peaks between $c = 0.4$ and 0.6. At high temperatures, the resistivity becomes independent of the concentration c: the incoherent Kondo scattering dominates the scattering processes over the lattice disorder.

In Fig. 15b, the thermoelectric power is shown for the same parameter set. The overall appearance is equal to the case with Kondo holes, but, in addition, we observe a sign change at small temperatures. This sign change originates from the change of asymmetry in the relaxation time $\tau(\omega)$ as defined in Eq. (21).

6 Conclusion

We end by shortly summarizing and emphasizing the most important conclusions to be drawn from our investigations and calculations:

1. The DMFT/CPA treatment of the PAM in combination with an NRG impurity solver for the effective SIAM is able and in our opinion the most suitable method to study the influence of correlations and of disorder on the electronic properties, in particular the transport properties resistivity and thermopower, of metallic HFS and Kondo insulators.
2. The method is able to describe the crossover from the coherent behavior typical for periodic HFS to the behavior typical for diluted Kondo impurity systems.
3. A simple scaling of the residual resistivity with the concentration of the Kondo "impurities" ($\rho(T = 0, c) \sim c$) is usually *not* obtained.
4. The large absolute value of the thermopower observed in HFS (Kondo systems) can be reproduced and understood within the PAM. The absolute value becomes even larger when approaching the Kondo insulator state.
5. The thermopower of HFS has several extrema, the first extremum at the characteristic low (Kondo) temperature.

a

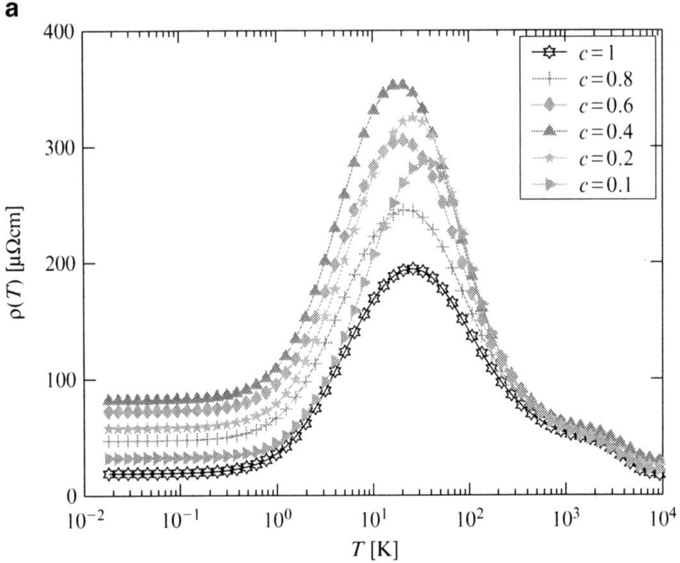

b

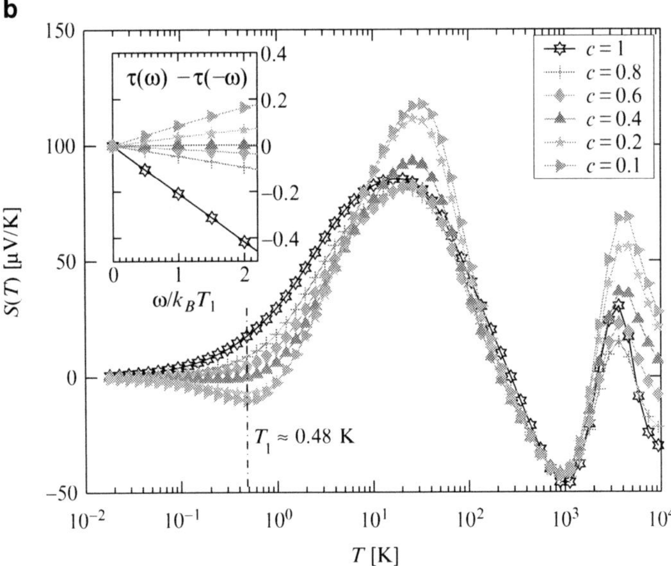

Fig. 15 (**a**) Resistivity $\rho(T)$ and (**b**) thermoelectric power $S(T)$ as functions of T for different concentrations c of system A, calculated with CPA-NRG for $V_A = V_B = V_0$, $U = 10$, $\varepsilon_f^A - \varepsilon_c^A = -6$, $\varepsilon_f^B - \varepsilon_c^B = -4$, $\varepsilon_f^A - \varepsilon_f^B = 0.2$, chemical potential $\mu = 0$, and a filling $n_{\text{tot}} = 1.8 - 0.2c$. The inset of (b) shows the antisymmetrized relaxation time $\tau(\omega) - \tau(-\omega)$ [Eq. (21)] in the vicinity of the chemical potential for the temperature $T_1 = 0.48$ K. NRG parameters: number of retained NRG states, $N_s = 800$, $\Lambda = 1.6$, and $\delta/\Gamma_0 = 10^{-3}$.

6. The thermopower (and its absolute value) is nearly unaffected by ligand disorder.
7. A figure of merit larger than 1 (and thus the possibility of applications as thermoelectric cooling devices) should be possible for HFS close to the Kondo insulator state.

Acknowledgements We acknowledge useful discussions and partly collaborations on the subject of this paper with Bernard Coqblin, Jim Freericks, Norbert Grewe, Thomas Pruschke and Veljko Zlatic. This research project has been supported by the German Science Foundation (Deutsche Forschungsgemeinschaft, DFG), project AN 275/5-1, and by a grant for computation time from the Neumann Institute of Computation (NIC, Jülich), project HHB000.

References

1. N. Grewe and F. Steglich, in *Handbook on the Physics and Chemistry of Rare Earths*, edited by K. A. Gschneidner, Jr. and L. Eyring (North-Holland, Amsterdam, 1991), Vol. 14, p. 343.
2. G. R. Stewart, Rev. Mod. Phys. **73**, 797 (2001).
3. P. Scoboria, J. E. Crow, and T. Mihalisin, J. Appl. Phys. **50**, 1895 (1979).
4. K. Andres, J. E. Graebner, and H. R. Ott, Phys. Rev. Lett. **35**, 1779 (1975); H. R. Ott, H. Rudigier, Z. Fisk, J. L. Smith, Physica B **127**, 359 (1984).
5. Y. Onuki and T. Komatsubara, J. Magn. Magn. Mat. **63–64**, 281 (1987).
6. C. D. W. Jones, K. A.Regan, F. J. DiSalvo, Phys. Rev. B **58**, 16057 (1998).
7. C. D. W. Jones, K. A.Regan, F. J. DiSalvo, Phys. Rev. B **60**, 5282 (1999).
8. F. Steglich, in *Festkörperprobleme: Advances in Solid State Physics*, edited by J. Treusch (Vieweg, Braunschweig, 1977), Vol. 17, p. 319.
9. H. Schneider, Z. Kletowski, F. Oster, and D. Wohlleben, Solid State Commun. **48**, 1093 (1983).
10. D. Jaccard, J. M. Mignot, B. Bellarbi, A. Benoit, H. F. Braun, and J. Sierro, J. Magn. Magn. Mat. **47–48**, 23 (1985).
11. D. Jaccard, A. Basset, J. Sierro, and J. Pierre, J. Low Temp. Phys. **80**, 285 (1990).
12. M. Očko, B. Bushinger, C. Geibel, and F. Steglich, Physica B **259–261**, 87 (1999).
13. S. Paschen, B. Wand, G. Sparn, F. Steglich, Y. Echizen, and T. Takabatake, Phys. Rev. B **62**, 14912 (2000).
14. G. Mahan, B. Sales, and J. Sharp, Phys. Today **50**(3), 42 (1997).
15. M. Očko, B. Buschinger, C. Geibel, F. Steglich, Physica B **259–261**, 87 (1999).
16. M. Očko, D. Drobac, B. Buschinger, C. Geibel, F. Steglich, Phys. Rev. B **64**, 195106 (2001).
17. P. de V. du Plessis, A. M. Strydom, R. Troć, T. Cichorek, Cz. Marucha, R. P. Gers, J. Phys.: Condens. Matter **11**, 9775 (1999).
18. C. M. Varma, Y. Yafet, Phys. Rev. B **13**, 2950 (1976).
19. Th. Pruschke, M. Jarrell, and J. K. Freericks, Adv. Phys. **42**, 187 (1995).
20. A. Georges, G. Kotliar, W. Krauth, and M. J. Rozenberg, Rev. Mod. Phys. **68**, 13 (1996), for a review on the DMFT.
21. P. W. Anderson, Phys. Rev. **124**, 41 (1961).
22. K. G. Wilson, Rev. Mod. Phys. **47**, 773 (1975).
23. H. R. Krishna-murthy, J. W. Wilkins, K. G. Wilson, Phys. Rev. B **21**, 1003 (1980); **21**, 1044 (1980).
24. R. Bulla, T. A. Costi, T. Pruschke, Rev. Mod. Phys **80**, 395 (2008).
25. T. Pruschke, R. Bulla, M. Jarrell, Phys. Rev. B **61**, 12799 (2000).
26. R. J. Elliott, J. A. Krumhansl, P. L. Leath, Rev. Mod. Phys. **46**, 465 (1974).
27. F. Yonezawa, in: *The Structure and Properties of Matter*, edited by T. Matsubara, Springer (Berlin, Heidelberg, New York), 1982, Vol. 28, Springer Series in Solid-State Sciences, chap. 11, pp. 383–432.

28. V. Janiš, D. Vollhardt, Phys. Rev. B **46**, 15712 (1992).
29. V. Janiš, M. Ulmke, D. Vollhardt, Europhys. Lett. **24**, 287 (1993).
30. M. Ulmke, V. Janiš, D. Vollhardt, Phys. Rev. B **51**, 10411 (1995).
31. P. J. H. Denteneer, M. Ulmke, R. T. Scalettar, G. T. Zimanyi, Physica A **251**, 162 (1998).
32. K. Byczuk, W. Hofstetter, D. Vollhardt, Phys. Rev. B **69**, 045112 (2004).
33. C. Grenzebach, F. B. Anders, G. Czycholl and Th. Pruschke, Phys. Rev. B **74**, 195119 (2006).
34. C. Grenzebach, F. B. Anders, G. Czycholl and Th. Pruschke, Phys. Rev. B **77**, 115125 (2008).
35. C. Grenzebach: *Transporttheorie für geordnete und ungeordnete Systeme schwerer Fermionen*, Thesis (Univ. Bremen), Mensch-und-Buch-Verlag Berlin 2008, ISBN 978-3-86664-504-2.
36. V. Zlatic, I. Milat, B. Coqblin, G. Czycholl, C. Grenzbach, Phys. Rev. B **68**, 104432 (2003).
37. C. Grenzebach, G. Czycholl, Physica B **359–361**, 732 (2005).
38. V. Zlatic, R. Monnier, J. K. Freericks, Phys. Rev. B **78**, 045113 (2008).
39. R. Bulla, Th. Pruschke, and A. C. Hewson, J. Phys.: Condens. Matter **9**, 10463 (1997).
40. G. Mahan, *Many-Particle Physics* (Plenum Press, New York, 1981).
41. J. M. Luttinger, Phys. Rev. **135**, A1505 (1964).
42. P. Voruganti, A. Golubentsev, and S. John, Phys. Rev. B **45**, 13945 (1992).
43. G. Czycholl, *Theoretische Festkörperphysik*, 3rd edition (Springer, Heidelberg, 2007).
44. G. Czycholl and H. J. Leder, Z. Phys. B: Condens. Matter **44**, 59 (1981).
45. A. Khurana, Phys. Rev. Lett. **64**, 1990 (1990).
46. H. Schweitzer and G. Czycholl, Phys. Rev. Lett. **67**, 3724 (1991).
47. A. Lorek, N. Grewe and F. B. Anders, Solid State Commun. **78**, 167 (1991).
48. F. B. Anders and D. L. Cox, Physica B **230–232**, 441 (1997).
49. F. Steglich, B. Buschinger, P. Gegenwart, M. Lohmann, R. Helfrich, C. Langhammer, P. Hellmann, L. Donnevert, S. Thomas, A. Link, et al., J. Phys.: Condens. Matter **8**, 9909 (1996).
50. F. B. Anders and M. Huth, Eur. Phys. J. B **19**, 491 (2001).

Role of Multiple Subband Renormalization in the Electronic Transport of Correlated Oxide Superlattices

A. Rüegg and M. Sigrist

Abstract Metallic behavior of band-insulator/ Mott-insulator interfaces was observed in artificial perovskite superlattices such as in nanoscale $SrTiO_3/LaTiO_3$ multilayers. Applying a semiclassical perspective to the parallel electronic transport we identify two major ingredients relevant for such systems: (i) the quantum confinement of the conduction electrons (superlattice modulation) leads to a complex, quasi-two dimensional subband structure with both hole- and electron-like Fermi surfaces. (ii) strong electron-electron interaction requires a substantial renormalization of the quasi-particle dispersion. We characterize this renormalization by two sets of parameters, namely, the quasi-particle weight and the induced particle-hole asymmetry of each partially filled subband. In our study, the quasi-particle dispersion is calculated self-consistently as function of microscopic parameters using the slave-boson mean-field approximation introduced by Kotliar and Ruckenstein. We discuss the consequences of strong local correlations on the normal-state free-carrier response in the optical conductivity and on the thermoelectric effects.

1 Introduction

Recent experiments [1] have shown that a metallic state can be stabilized at the interface between the Mott insulator $LaTiO_3$ and the band insulator $SrTiO_3$. There is strong evidence that in such systems electronic charge is redistributed between the Mott insulator (MI) and the band insulator (BI) in order to compensate for the mismatch of the work functions and to avoid the so-called polar catastrophe [2]. The electronic charge reconstruction [3] at the interface leads to metallic behavior.

The experimental data [1, 4–6] are consistent with Fermi liquid behavior and a single-particle perspective, where transport properties are studied by the semiclassical transport equations, offers a natural starting point. However, it is necessary to clarify how the single-particle picture for weakly interacting electrons, which

A. Rüegg and M. Sigrist
Theoretische Physik, ETH Zürich, 8093 Zürich, Switzerland
e-mail: rueegga@phys.ethz.ch, sigrist@phys.ethz.ch

V. Zlatić and A. C. Hewson (eds.), *Properties and Applications of Thermoelectric Materials*, 181
NATO Science for Peace and Security Series B: Physics and Biophysics,

is successfully applied in the study of semiconductor nano-structures, is modified by strong electronic correlations. In particular, understanding the renormalization of the quasi-particle dispersion as function of microscopic parameters is crucial. Let us in the following ignore complicating aspects related to the orbital degrees of freedom [7] or to possible symmetry-broken phases [3]. Then, from quite general considerations, we can expect two major ingredients determining the electronic structure:

(i) The $LaTiO_3$ bulk system has a Ti-$3d^1$ configuration whereas the $SrTiO_3$ compound has a Ti-$3d^0$ configuration. Therefore, from a single-particle point of view, the BI/MI/BI sandwich acts as a quantum well confining the conduction electrons to the MI region, cf. Fig. 1. The bound states of the quantum well form *quasi-two dimensional subbands* with dispersion $E_{\mathbf{k}\nu}$ labeled by the in-plane momentum $\mathbf{k}$ and the subband index ν.[1] The Fermi surface (FS) defined by the $\mathbf{k}$ points satisfying $E_{\mathbf{k}\nu} = 0$ contains in general both open and closed sheets which brings about that electron-like and hole-like contributions can lead to partial compensation [10].

(ii) *Strong electron–electron interaction* introduces novel electronic physics at the band-insulator/ Mott-insulator interface. When the local self-energy corrections are dominant, we can assume that $E_{\mathbf{k}\nu} = E_\nu(\varepsilon_{\mathbf{k}})$, where $\varepsilon_{\mathbf{k}}$ is the non-interacting in-plane dispersion [8, 9]. In this case, the renormalization of the quasi-particle dispersion is characterized by two sets of parameters. On the one hand, the on-site repulsion leads to a reduction of the Fermi velocity of the subband ν by a factor

$$Z_\nu = \left.\frac{\partial E_{\mathbf{k}\nu}}{\partial \varepsilon_{\mathbf{k}}}\right|_{FS} \tag{1}$$

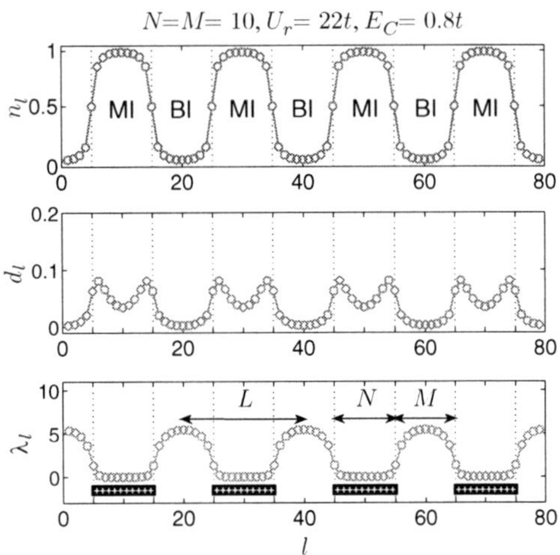

Fig. 1 The charge density n_l, the fraction of doubly occupied sites d_l^2 and the Lagrange multiplier λ_l as obtained by the present mean-field approach for a band-insulator (BI)/Mott-insulator (MI) superlattice. $L = N + M$ denotes the number of unit cells of the superlattice modulation where N is the number of MI-layers and M the number of BI-layers. The set of microscopic parameters is given in the figure.

[1] Our discussion is restricted to large superlattice periods where the dependence on the perpendicular momentum Q can be neglected when studying the parallel transport.

which is equal to the quasi-particle weight of the subband ν. On the other hand, at the interface, the hybridization of (almost) localized with itinerant degrees of freedom induces an enhanced particle-hole asymmetry. We quantify this asymmetry by the dimensionless parameter

$$\alpha_\nu = \left(\Delta\varepsilon_{\mathbf{k}} \frac{\partial^2 E_{\mathbf{k}\nu}}{\partial\varepsilon_{\mathbf{k}}^2}\right) \Big/ \left(\frac{\partial E_{\mathbf{k}\nu}}{\partial\varepsilon_{\mathbf{k}}}\right)\Bigg|_{\mathrm{FS}}, \tag{2}$$

where we have defined $\Delta\varepsilon_{\mathbf{k}} = \varepsilon_{\mathbf{k}} - \varepsilon_b$ with ε_b the energy of the lower band edge. Starting from a microscopic model, we calculate the quasi-particle dispersion in a self-consistent way to obtain the dependence on microscopic parameters. We obtain an *interfacial heavy-fermion state* and discuss correlation effects as characterized by Eqs. (1) and (2), cf. Fig. 2. Implications for transport are illustrated by calculating the free carrier response and the thermoelectric effects, cf. Fig. 3. We find that in both cases important contributions arise from the interface.

2 Microscopic Model

For the microscopic description we assume perfect lattice match between the two materials, thereby neglecting aspects related to the lattice relaxation [11]. The microscopic model is given by an extended single-orbital Hubbard model on a cubic lattice (introduced in Ref. [13])

$$H = H_t + H_U + H_{ee} + H_{ei} + H_{ii}. \tag{3}$$

Here, the kinetic energy is given by a nearest-neighbor tight-binding model and the on-site repulsion is modeled by a Hubbard interaction,

$$H_t = -t \sum_{\langle ij\rangle,\sigma} c^\dagger_{i\sigma} c_{j\sigma} + \text{h.c.} \quad \text{and} \quad H_U = U \sum_i n_{i\uparrow} n_{i\downarrow}, \tag{4}$$

where $n_{i\sigma} = c^\dagger_{i\sigma} c_{i\sigma}$. The nanoscale structure is defined by the superlattice period $L = N + M$ and the number N of counter-ion layers (see also Fig. 1). They simulate the difference between Sr^{2+} and La^{3+} and sit in the center between the electronic sites [3] interacting with the electrons through the long-range electron–ion interaction

$$H_{ei} = -E_C \sum_{i,j} \frac{n_i}{|\mathbf{r}_i - \mathbf{r}_j^{\text{ion}}|} \tag{5}$$

where $\mathbf{r}_j^{\text{ion}}$ denotes the position of the ions, $n_i = n_{i\uparrow} + n_{i\downarrow}$, and we have introduced the parameter E_C controlling the screening length. Furthermore, the long-range electron–electron and ion–ion interaction energies are given by

$$H_{ee} = \frac{E_C}{2} \sum_{i\neq j} \frac{n_i n_j}{|\mathbf{r}_i - \mathbf{r}_j|}, \quad H_{ii} = \frac{E_C}{2} \sum_{i\neq j} \frac{1}{|\mathbf{r}_i^{\text{ion}} - \mathbf{r}_j^{\text{ion}}|}, \tag{6}$$

respectively. The number of electrons is fixed by the charge-neutrality condition. Notice that we can formally relate the parameter E_C to an effective dielectric constant $\varepsilon = e^2/E_C a$ where a is the lattice constant and $e > 0$ the elementary charge. However, for a more realistic description of the screening at the interface a single parameter for the long-range electron–electron interaction is too crude. In fact, the polarization of the lattice dominates the dielectric constant in the considered transition metal oxides and the effect of the relaxation of the lattice near the interface introduces additional parameters in an effective model description [11, 12]. For simplicity, such effects are not considered here.

3 Slave-Boson Mean Field Approximation

To discuss the low-energy behavior of the model (3) in the normal state we apply the four-boson mean-field approximation of Kotliar and Ruckenstein [14]. This approach allows one to discuss the effect of local self-energy corrections by introducing auxiliary bosons representing the local charge and spin degrees of freedom together with pseudo fermions. The effective low-energy theory is then obtained by the saddle-point approximation for the slave bosons which can be controlled by a $1/N$-expansion of a suitable generalization of the slave-boson action [15, 16]. The remaining fermionic degrees of freedom are interpreted as the Landau quasiparticles of a Fermi liquid which are dressed by the interactions and therefore have modified single-particle properties.

3.1 *Superlattice Geometry*

For a quantum well system, the effective low-energy model was derived in Ref. [8]. In the following, we will briefly discuss the case of a superlattice. Assuming a translational invariant state in the in-plane direction, the problem of finding the eigenvalues of the effective low-energy Hamiltonian reduces to a one-dimensional problem which is parameterized by the non-interacting dispersion $\varepsilon_{\mathbf{k}} = -2t(\cos k_x + \cos k_y)$:

$$(z_l^2 \varepsilon_{\mathbf{k}} + \lambda_l)\psi(l) - t \sum_{\gamma=\pm 1} z_l z_{l+\gamma}\psi(l+\gamma) = E\psi(l). \tag{7}$$

Here, l labels the layers along the direction of the superlattice modulation. The hopping renormalization amplitude z_l depends on the charge density n_l and the fraction of doubly occupied sites d_l^2 in layer l and is given by the standard Gutzwiller expression [17]

$$z_l = \frac{\sqrt{(1-n_l+d_l^2)(n_l-2d_l^2)} + d_l\sqrt{n_l-2d_l^2}}{\sqrt{n_l(1-n_l/2)}}. \tag{8}$$

The mean fields n_l and d_l are determined by the minimum of an appropriate free energy, as discussed below. The Lagrange multiplier λ_l acts as a single-particle potential and enforces the self-consistency of the electronic charge distribution.

Decomposing ψ according to $\psi(l) = \psi_{\mathbf{K}\nu}(l)e^{iQl}$, where $\mathbf{K} = (\mathbf{k}, Q)$, $-\pi/La \leq Q < \pi/La$ with $\psi_{\mathbf{K}\nu}(l+L) = \psi_{\mathbf{K}\nu}(l)$, the problem reduces to diagonalizing the following matrix

$$\hat{K}(Q) = \begin{pmatrix} z_1^2\varepsilon_{\mathbf{k}} + \lambda_1 & -tz_1z_2e^{iQ} & \cdots & -tz_1z_Le^{-iQ} \\ -tz_2z_1e^{-iQ} & z_2^2\varepsilon_{\mathbf{k}} + \lambda_2 & -tz_2z_3e^{iQ} & \cdots \\ \cdots & \cdots & \cdots & \cdots \\ -tz_Lz_1e^{iQ} & \cdots & -tz_{L-1}z_Le^{-iQ} & z_L^2\varepsilon_{\mathbf{k}} + \lambda_L \end{pmatrix}. \tag{9}$$

A further simplification is obtained by restricting to superlattices with a large period $L \gg 1$. In this case, the Q-dependence can be safely neglected for the parallel transport and it is sufficient to consider only $\hat{K}(0)$. In this case, the quasi-particle dispersion has the form stated in the introduction, $E = E_\nu(\varepsilon_{\mathbf{k}})$, $\nu = 1, \ldots, L$.

3.1.1 Long-Range Coulomb Interaction

The long-range Coulomb interaction is treated in the Hartree-type of mean-field calculation [8]. In order to find the interaction matrix $W_{ll'}$ between electrons in layer l and l' of the superlattice unit cell we explicitly take into account the periodicity of the charge distribution. In the same way, we also determine the resulting (screened) potential V_l of the counterions. In the end, these potentials have to be found self-consistently by simultaneously solving the mean-field equations and the Poisson equation, $\Delta\phi(z) = -4\pi\frac{\rho(z)}{\varepsilon}$, where $\phi(z)$ is the electrostatic potential and $\rho(z)$ the charge distribution.

We start with the approximation commonly found in the literature [18], namely, we replace each layer by a uniformly charged plane, thereby respecting the polar nature of the Mott insulator. Corrections due to the discrete nature of the charge distribution (lattice) are calculated numerically, but are only significant very close to (or within) the considered layer [19]. It is convenient to use the following elementary solution ϕ_o determined by a periodical array of uniformly charged layers with period La

$$\rho_o(z) = \sigma \sum_{m \in Z} \delta(z + mLa) - \bar{\rho} \tag{10}$$

where $\sigma = e/a^2$ is an elementary surface-charge density and $\bar{\rho} = \sigma/La$ is a uniform background charge to keep the total system charge neutral. The solution can be written in a compact form by use of the polylogarithm $\mathrm{Li}_b(z) = \sum_{n>0} \frac{z^n}{n^b}$:

$$\phi_o(z) = \frac{\sigma La}{\varepsilon\pi} \left[\mathrm{Li}_2(e^{2\pi iz/La}) + \mathrm{Li}_2(e^{-2\pi iz/La})\right]. \tag{11}$$

Between two neighboring layers at nLa and $(n+1)La$, the resulting potential is simply given by the parabola

$$\phi_o(z) = \frac{2\pi\sigma}{L\varepsilon}(z - nLa)(nLa + La - z). \tag{12}$$

From the elementary solution Eq. (11) it is straightforward to determine $W_{ll'}$ and V_l by summing up the contributions from the different layers within the superlattice unit cell and adding the numerically determined correction terms due to the discreteness of the charge distribution.

3.1.2 Free Energy

Eventually, we give the expression for the free energy per lattice site ($\beta^{-1} = k_B T$)

$$f(n,d,\lambda) = \frac{-2}{\beta N_{||} L}\sum_{\mathbf{k}\nu}\ln(1+e^{-\beta E_{\mathbf{k}\nu}}) + \frac{U_r}{L}\sum_l {d_l}^2 + \frac{1}{2L}\sum_{ll'} n_l W_{ll'} n_{l'} - \frac{1}{L}\sum_l (\lambda_l - V_l) n_l. \tag{13}$$

Sums over layers are restricted to a single superlattice unit cell. The self-consistency equations are solved by maximizing Eq. (13) with respect to λ_l and minimizing the resulting function with respect to the mean fields n_l and d_l under the constraint of charge neutrality, $\sum_l n_l = N$.

A typical solution of the self-consistency equations at $T = 0$ is shown in Fig. 1 for a superlattice with $N = M = 10$. The charge distribution n_l allows one to distinguish naturally between three different regions, $n_l \approx 1$, $n_l \approx 0.5$ and $n_l \approx 0$. The layers with filling $n_l \approx 0.5$ separate the "Mott-insulating" (MI) regions ($n_l \approx 1$) from the "band-insulating" (BI) regions with $n_l \approx 0$ (notice that the whole system is actually metallic). The presence of the interface (IF) layers with filling $n_l \approx 0.5$ is a consequence of the polar nature of the Mott insulator.

4 Transport Properties

From the self-consistent solution of the mean-field equations we also obtain the quasi-particle dispersion $E_{\mathbf{k}\nu}$ and the envelope wave-function $\psi_{\mathbf{k}\nu}(l)$. At low temperatures, it is expected that the transport properties can be understood from the properties of the quasi-particles [9]. We therefore start by discussing the electronic structure.

4.1 Characterization of the Generic Electronic Structure

The quasi-particle dispersion $E_{\mathbf{k}\nu}$ is shown in Fig. 2a for the $N = 15$, $M = 5$ superlattice. The value of the on-site repulsion, $U_r = 22t$, is well above the critical interaction strength of the Mott transition in the half-filled bulk system [20], $U_c \approx 16t$. Thus, the quasi-particle dispersion shown in Fig. 2 corresponds to the strongly correlated regime. In panels (b) to (d) we show the subband filling n_ν, the quasi-particle weight Z_ν [Eq. (1)] and the induced particle-hole asymmetry α_ν [Eq. (2)] of the subbands shown in (a). Although the spatial weight of the envelope wave-function $\psi_{\mathbf{k}\nu}(l)$ extends over the whole super unit cell and also depends on the value of $\varepsilon_{\mathbf{k}}$ [8], it is possible to group the different subband states according to the regions where most of their spatial weight is located. We thus define $\nu_{\mathrm{MI}} = 1,\ldots,N-1$, $\nu_{\mathrm{IF}} = N, N+1$ and $\nu_{\mathrm{BI}} = N+2,\ldots,L$. Notice that due to strong local correlations, the quasi-particle weight $Z_{\nu_{\mathrm{MI}}}$ and $Z_{\nu_{\mathrm{IF}}}$ is strongly reduced for the subbands of the MI and IF region [panel (c)], whereas the particle-hole asymmetry $\alpha_{\nu_{\mathrm{IF}}}$ is enhanced most dominantly for the subbands of the IF region [panel (d)]. These are the basic characteristics of the *interfacial heavy-fermion state* obtained by the slave-boson mean-field approximation at low temperatures. The coherent hybridization of the itinerant degrees of freedom in the BI and IF region with the almost localized degrees of freedom in the MI region is mediated by the intra-layer hopping and leads to heavy-fermion behavior of the interfacial subbands. The situation is reminiscent of heavy-fermion systems as described for example by the periodic Anderson model [21]. However, in the present case, localized and itinerant degrees of freedom have the same orbital character but are separated spatially – in contrast to the classical heavy-fermion systems where the localized f-electrons hybridize with the states of the conduction band.

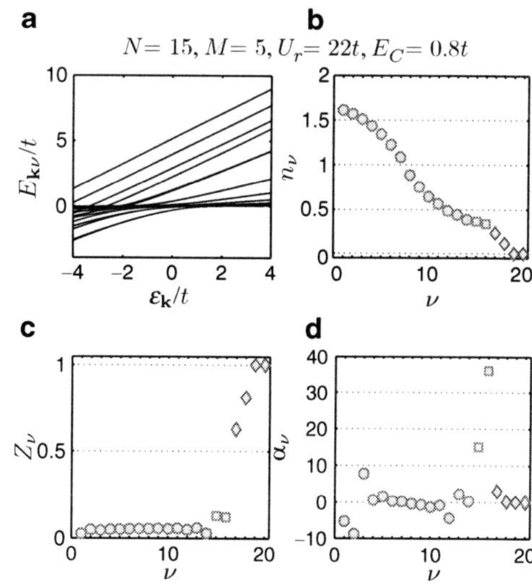

Fig. 2 (**a**) The quasi-particle dispersion $E_{\mathbf{k}\nu}$ as function of the non-interacting in-plane dispersion $\varepsilon_{\mathbf{k}}$. (**b**) The subband filling n_ν, (**c**) the quasi-particle weight Z_ν and (**d**) the induced particle-hole asymmetry α_ν for the individual subbands which correspond to the dispersion in (**a**). *Circles* are associated with the subbands of the MI region, *squares* with the IF and *diamonds* with the BI region.

4.2 Drude Weight

The *Drude weight* in the optical conductivity is obtained as described in [8,9]. One finds the familiar expression for a quasi-particle response,

$$D = \sum_{\nu} D_{\nu}, \quad D_{\nu} = \frac{e^2}{4\pi L a} A_{\nu} \bar{v}_{\nu}, \tag{14}$$

where A_{ν} is the Fermi surface volume of the sheet ν and $\bar{v}_{\nu} = Z_{\nu} \langle |\nabla_{\mathbf{k}} \varepsilon_{\mathbf{k}}| \rangle_{\mathrm{FS}} / \hbar$ is the Fermi velocity averaged over the Fermi surface. In Fig. 3a we illustrate how D evolves from the band insulator ($N = 0$) to the Mott insulator ($N = L$), which both have a vanishing D. The maximal D as a function of the averaged electronic density N/L depends on the value of E_C/t and shifts to lower N's for increasing E_C because the screening length is reduced.

4.3 Seebeck Coefficient

Thermoelectric effects are characterized by the Seebeck coefficient S. In terms of the subband contributions it is written as

$$S = \sum_{\nu} \frac{S_{\nu} \sigma_{\nu}}{\sigma}, \quad \sigma = \sum_{\nu} \sigma_{\nu}. \tag{15}$$

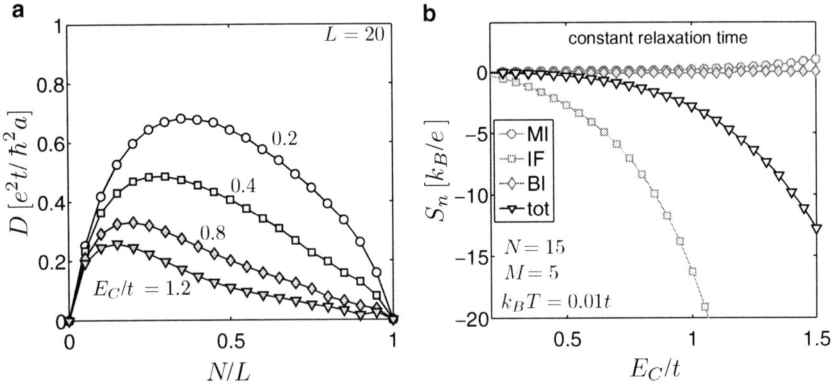

Fig. 3 (**a**) The Drude weight D as function of N/L for different values of E_C for a superlattice with period $L = 20$. (**b**) The total Seebeck coefficient S (*triangles*) and the different contributions S_n associated with the MI (*circles*), IF (*squares*) and BI (*diamonds*) regions as a function of E_C evaluated for a constant relaxation time. In both panels, the value of the on-site interaction is fixed at $U_r = 22t$.

Here, σ is the total electrical conductivity and S_ν the Seebeck coefficient associated with the subband ν. In lowest order in the temperature we have [9]

$$S_\nu = -\frac{\pi^2}{3}\frac{k_B}{e}\frac{k_B T}{Z_\nu \Delta\varepsilon_\nu^*}\left[\alpha_\nu + \Delta\varepsilon_\nu^*\left(\frac{\tau_\nu'}{\tau_\nu} + \frac{\mathcal{N}_\nu'}{\mathcal{N}_\nu}\right)\right] \tag{16}$$

where $\Delta\varepsilon_\nu^* = \varepsilon_\nu^* - \varepsilon_b$ is measured from the band edge and defined through $E_\nu(\varepsilon_\nu^*) = 0$, $\tau_\nu(\varepsilon)$ is the relaxation time of the subband ν and

$$\mathcal{N}_\nu(\varepsilon) = \int \frac{d^2k}{(2\pi)^2}|\nabla_{\mathbf{k}}\varepsilon_{\mathbf{k}}|^2\delta(\varepsilon - \varepsilon_{\mathbf{k}}). \tag{17}$$

The prime ($'$) in Eq. (16) denotes the derivative with respect to ε at the Fermi surface. Attempts to calculate τ_ν from a microscopic model offers a challenging task in correlated, disordered and inhomogeneous systems (see also Ref. [9]). For simplicity, we discuss here the case of a constant relaxation time. We assume an energy-independent relaxation time τ_n for the subbands associated with the different regions, $n =$ MI, IF, BI, and therefore obtain

$$S = \sum_n \frac{S_n\sigma_n}{\sigma}, \quad S_n = \sum_{\nu_n}\frac{S_\nu D_\nu}{D_n}, \quad D_n = \sum_{\nu_n} D_\nu, \quad \sigma_n = D_n\tau_n. \tag{18}$$

Figure 3b shows the different contributions S_n to the total Seebeck coefficient for a superlattice with $N = 15$ and $M = 5$ as a function of the parameter E_C. The contribution S_{IF} from the subbands associated with the interface is most dominant and $|S_{\mathrm{IF}}|$ increases for increasing E_C (remember that a large value of E_C yields a sharp charge distribution). This can be understood by the fact that the particle-hole asymmetry α_ν induced by the correlations is largest for the subbands of the interface region and increases for a sharper interface due to a reduction in the hybridization (intra-layer hopping). However, the total (absolute) Seebeck coefficient is at best equal to its largest subband contribution, $|S| \leq \max_\nu |S_\nu|$, but is in general smaller, as shown in Fig. 3b by assuming $\tau_{\mathrm{MI}} = \tau_{\mathrm{IF}} = \tau_{\mathrm{BI}}$. Nevertheless, we can state the conditions for which the interface contribution becomes large in the constant-relaxation-time approximation: (i) strong electronic correlations, $U_r > U_c$, (ii) large values of N, such that bulk-like properties in the center of the MI are obtained, and (iii) a sharp interface ($E_C > t$).

4.4 Comparison to the Atomic Limit Result

Note that the above discussed enhancement of $|S_{\mathrm{IF}}|$ at low temperatures is a non-local effect. It is therefore expected that any reduction of spatial coherence across the interface will suppress and eventually destroy this mechanism. The influence of the reduction of spatial coherence due to thermal fluctuations is clearly seen by analyzing S in the high-temperature (atomic) limit. Let us in the following discuss the situation $t \ll k_B T \ll U$ such that doubly occupied sites are completely suppressed. In the *homogeneous* system at density n, the S is given by the entropic contribution alone [22,23]. This yields Heikes formula

$$S = -\frac{k_{\mathrm{B}}}{e} \log \left[\frac{2(1-n)}{n} \right] \tag{19}$$

where the $\log 2$ contribution arises from the entropy of the spin degree of freedom. As pointed out in Ref. [24], in transition metal oxides also the inclusion of orbital degrees of freedom is necessary. This is in principle straight forward but we will not discuss it further here. For the *inhomogeneous* system an appropriate generalization of Eq. (19) is given by

$$S = \frac{\sum_l S_l \sigma_l}{\sigma} = -\frac{k_B}{e} \frac{\sum_l \log \left[\frac{2(1-n_l)}{n_l} \right] n_l(1-n_l)}{\sum_l n_l(1-n_l)}. \tag{20}$$

Here we have used the *local* quantities

$$S_l = -\frac{k_{\mathrm{B}}}{e} \log \left[\frac{2(1-n_l)}{n_l} \right], \quad \sigma_l = \frac{e^2 A \beta}{2} n_l(1-n_l), \tag{21}$$

with a temperature and doping independent constant A [25]. The weighted sum in Eq. (20) clearly shows that an inhomogeneous system is not favorable as long as a *local* description is appropriate. In fact, if one seeks to optimize the powerfactor $\mathrm{PF} = S^2 \sigma$ in this limit for a spatially varying density profile, the optimal solution is found to be the homogeneous solution with optimal density $n \approx 0.12$, c.f. Ref. [25]. This is exactly the opposite behavior than found in the low-temperature limit. Restricting to purely electronic contribution, we therefore conclude that a spatially non-uniform system can only be favorable if spatial coherence is sustained.

5 Conclusions

In summary, we have studied aspects of the parallel transport at low-temperatures in strongly-correlated superlattices from the semiclassical point of view. The generic electronic structure is discussed and self-consistently computed from microscopic parameters using the four-boson approach of Kotilar and Ruckenstein to deal with strong local correlations. Implications for the parallel transport are illustrated by the free-carrier response and the thermoelectric effects. The presence of the interface introduces new aspects not feasible in the bulk systems. Here, we have discussed the scenario of an interfacial heavy-fermion state where the coherent hybridization of itinerant and almost localized degrees of freedom leads to a large particle-hole asymmetry which can be responsible for a high value of the Seebeck coefficient. We find that this mechanism is a non-local effect and that spatial coherence is crucial. If such mechanisms are also relevant for the giant value observed in $SrTiO_3/SrTi_{0.8}Nb_{0.2}O_3$ superlattices [26] need to be clarified by further studies of more realistic models.

Acknowledgements We would like to thank S. Pilgram, M. Ossadnik, R. Asahi and T. M. Rice for valuable discussions. We acknowledge financial support of the Toyota Central R&D Laboratories, Nagakute, Japan and the NCCR MaNEP of the Swiss Nationalfonds.

References

1. A. Ohtomo, D.A. Muller, J.L. Grazul, and H.Y. Hwang, Nature **419**, 378 (2002).
2. N. Nakagawa, H.Y. Hwang, and D.A. Muller, Nature Mat. **5**, 204 (2006).
3. S. Okamoto and A.J. Millis, Nature **428**, 630 (2004).
4. S.S.A. Seo et al., Phys. Rev. Lett. **99**, 266801 (2007).
5. M. Takizawa et al., Phys. Rev. Lett. **97**, 057601 (2006).
6. K. Shibuya et al., Jpn. J. Appl. Phys. **43**, L1178 (2004).
7. J. Chakhalian et al., Science **318**, 1114 (2007).
8. A. Rüegg, S. Pilgram, and M. Sigrist, Phys. Rev. B **75**, 195117 (2007).
9. A. Rüegg, S. Pilgram, and M. Sigrist, Phy. Rev. B **77**, 245118 (2008).
10. A.B. Pippard *Magnetotransport in metals* (Cambridge University Press, 1989).
11. S. Okamoto, A.J. Millis, and N.A. Spaldin, Phys. Rev. Lett. **97**, 056802 (2006).
12. D.R. Hamann, D.A. Muller, and H.Y. Hwang, Phys. Rev. B **73**, 195403 (2006).
13. S. Okamoto and A.J. Millis, Phys. Rev. B **70**, 241104(R) (2004).
14. G. Kotliar and A.E. Ruckenstein, Phys. Rev. Lett. **57**, 1362 (1986).
15. R. Frésard and P. Wölfle, Int. J. Mod. Phys. B **6**, 237 (1992).
16. E. Arrigoni and G.C. Strinati, Phy. Rev. B **52** (1995).
17. M.C. Gutzwiller, Phys. Rev. Lett. **10**, 159 (1963).
18. J.K. Freericks, *Transport in multilayered nanostructures: the dynamical mean-field approach* (Imperial College Press, London, 2006).
19. S. Wehrli, D. Poilblanc, and T.M. Rice, Eur. Phys. J. B **23**, 345 (2001).
20. W.F. Brinkman and T.M. Rice, Phys. Rev. B **2**, 4302 (1970).
21. T.M. Rice and K. Ueda, Phys. Rev. Lett. **55**, 995 (1985).
22. G. Beni, Phys. Rev. B **10**, 2186 (1974).
23. P.M. Chaikin and G. Beni, Phys. Rev. B **13**, 647 (1976).
24. W. Koshibae, K. Tsutsui, and S. Maekawa, Phys. Rev. B **62**, 6869 (2000).
25. S. Mukerjee and J.E. Moore Appl. Phys. Lett **90**, 112107 (2007).
26. H. Ohta et al., Nat. Mat. **6**, 129 (2007).

Thermoelectric Properties of Junctions Between Metal and Models of Strongly Correlated Semiconductors

M. Rontani and L. J. Sham

Abstract We study the thermopower of a junction between a metal and a strongly correlated semiconductor. Both in the electronic ferroelectric regime and in the Kondo insulator regime the thermoelectric figures of merit, ZT, of these junctions are compared with that of the ordinary semiconductor. By inserting at the interface one or two monolayers of atoms different from the bulk, with a suitable choice of rare-earth elements, very high values of thermopower may be reached at low temperatures.

1 Introduction

New thermoelectric coolers and power generators are under massive investigation [1]. The quality of the material needed in a thermoelectric device is defined by the dimensionless figure of merit ZT, where T is the absolute temperature, and Z is expressed in Eq. (1) in terms of transport coefficients. Currently, the highest value of ZT, which is ~ 1 at room temperature, is found in Bi–Te alloys [2]. New devices would be competitive with traditional refrigerators if ZT were about $3 \sim 4$: the major lack of high-ZT materials is at temperatures below 300 K.

In this paper [3] we propose a junction of metal and strongly correlated semiconductor as the basis for a possible efficient low-temperature thermoelectric device. This system embodies previous intuitions that were recognized as fruitful [4–16] in different materials/devices such as rare-earth compounds, superlattices, and metal/superconductor junctions. We now outline these concepts and briefly review the related literature.

M. Rontani
CNR-INFM Research Center S3, Via Campi 213/A, 41100 Modena MO, Italy
e-mail: rontani@unimore.it

L. J. Sham
Department of Physics, University of California San Diego, Gilman Drive 9500, La Jolla 92093-0319, California
e-mail: lsham@ucsd.edu

V. Zlatić and A. C. Hewson (eds.), *Properties and Applications of Thermoelectric Materials*, 193
NATO Science for Peace and Security Series B: Physics and Biophysics,

From the definition

$$Z = \frac{Q^2\sigma}{(\kappa_e + \kappa_l)}, \tag{1}$$

where Q is the absolute thermopower (Seebeck coefficient), σ the electrical conductivity, and κ_e and κ_l the electronic and lattice part, respectively, of the thermal conductivity, it follows that an ideal thermoelectric material should have high thermopower, high electrical conductivity, and low thermal conductivity. Semiconductors seemed to have the optimum collection of these properties, in contrast with metals, which have high σ but low Q, and insulators, which have high Q but low σ. However, in the last thirty years no substantial enhancement of ZT beyond ~ 1 has been obtained.

A breakthrough came with the synthesis of new materials such as filled skutterudites, which have nearly bound rare-earth atoms, closed in an atomic cage, whose "rattling" under thermal excitation scatters phonons, dramatically reducing κ_l [4]. More generally, heavy atoms in compounds help with lowering κ_l. Mahan and Sofo [5] also showed that the best bulk band structure for high Q is one with a sharp singularity in the density of states very close to the Fermi energy. These results provide the first idea in the search for the best thermoelectric, namely to look at rare–earth compounds as major candidates. In fact, mixed valence metallic compounds (e.g. $CePd_3$, $YbAl_3$) show high values of Q, but at the present time no useful value of ZT has been reported [17].

The second idea is that the best thermoelectric must have an energy gap. Because in ordinary semiconductors the optimum band gap is predicted to be about 10 k_BT (k_B being the Boltzmann constant) [6], one is led to consider small-gap semiconductors for low-temperature applications. If the chemical composition of semiconductors includes transition metals or rare earths, conduction and valence bands are frequently strongly renormalized by correlation effects, forming a temperature-dependent gap (see Fig. 1): this is the case for mixed-valent semiconductors, usually cubic, whose relevant electronic properties may be modeled by a f-flat band and a broad conduction band, with two electrons per unit cell [18]. This class of materials consists of two subclasses: The first is the Kondo insulator, characterized by a very strong Coulomb interaction between electrons on the same rare-earth site, usually described by the slave boson solution of the Anderson lattice Hamiltonian [19, 20]. Mao and Bedell predicted a high value of ZT for bulk Kondo insulators (the lower the dimension, the higher the value) [7]: however, some experimental reports seem to exclude this possibility [21]. The second, called the electronic ferroelectric (FE) [22], consists of semiconductors with high dielectric constants, such as SmB_6 and Sm_2Se_3, and it is modeled by the self-consistent mean-field (MF) solution of the Falicov–Kimball Hamiltonian [23]. The ground state of the insulating phase is found to be a coherent condensate of d-electron/f-hole pairs, giving a net built-in macroscopic polarization which breaks the crystal inversion symmetry and makes the material ferroelectric [22].

Another useful observation is that ZT is expected to increase in quantum-well superlattices, due to the modification of the density of states [8–10]. Moreover,

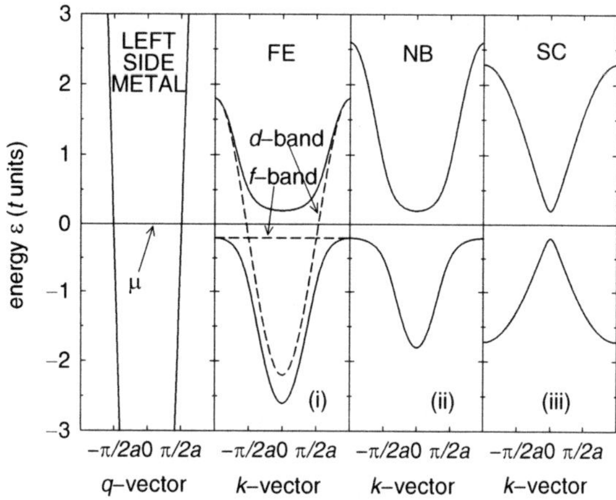

Fig. 1 Bulk energy bands of materials on both sides of the junction vs wave vector, for the three cases under study. The energy unit is t, and $\varepsilon_d = \varepsilon'_d = \varepsilon_f = \mu = 0$. The second panel [case (i)] shows also the "bare" band structure (*dashed lines*, $V_{df} = \Delta = 0$). The metal (semiconductor) bare bandwidth is $4t'$ ($4t$), with $t' = 5t$. The chosen parameters are: (i) $\Delta(T = 0) = 0.49t$, $V_{df} = 0$. (ii) $\Delta = 0$, $V_{df} = 0.49t$. (iii) $t'' = 0.141t$, $V_e = 0.071t$.

superlattices with large thermal impedance mismatch between layers seem very efficient at reducing κ_l because interfaces scatter phonons very effectively [11] (see also Ref. [24]).

In addition to the above literature, this work was stimulated by some recent advances in thermoelectric applications of junctions. Nahum and coworkers [12] built an electronic microrefrigerator based on a metal–insulator–superconductor (NIS) junction. Subsequent experimental [25] and theoretical work [13, 26, 27] confirmed this new idea. Edwards and coworkers [13] showed that tunneling through structures with sharp energy features in the density of states, like quantum dots and NIS junctions, can be used for cryogenic cooling. It seems then natural to us to study the junction between a metal and a strongly correlated semiconductor.

There were recent proposals for devices such as semiconductor/metal superlattices with transport perpendicular to interfaces. Mahan and Woods [14] suggested a multilayer geometry with the thickness of the metallic layer smaller than the electronic mean free path and the semiconductor acting as a potential barrier (thermionic refrigeration). Independently, Moyzhes and Nemchinsky [15] proposed a similar configuration with the metallic layer thickness comparable to the energy relaxation length. We cite also Min and Rowe's idea [16] of using Fermi-gas/liquid interfaces. While the analysis of electronic transport of Refs. [14, 15] is not directly applicable to our study, nevertheless this experimental geometry, with the strongly correlated semiconductor replacing the barrier layer, could be actually implemented.

In summary, we build on two key ideas from the above literature for increasing ZT. One is to utilize the sharp energy features in the density of states of bulk materials as in strongly correlated semiconductors. The other is to exploit the good thermoelectric characteristics of a junction. In this paper, we combine these ideas in exploring the thermopower behavior of a junction between a metal and different classes of semiconductors. We describe the gapped material on one side of the junction as the solution of the Falicov–Kimball Hamiltonian [23] in different regimes. In particular, we consider the case of: (i) electronic ferroelectric (FE), where, because of the Coulomb interaction between f-holes and d-electrons, the MF insulating ground state is a condensate of excitons [22, 28]; (ii) narrow band (NB) semiconductor, characterized by the d-f band hybridization: it would be a Kondo insulator had we taken into account the f-f electron repulsion; (iii) broad band semiconductor (SC), for comparison. In these three cases, we solve the electronic motion across the junction by means of a two-band model in a finite-difference form. In addition to the clean interface, we consider an "impurity" overlayer made either of rare-earth atoms, with relevant atomic orbitals of f-type, or of atoms with d-type orbitals, like transition metals. In the latter case, electrons can hop from these "d-impurity" sites to adjacent neighbor atoms, while in the former f-impurity case hopping is assumed to be negligible. This scenario is motivated by recent advances in atomic layer fabrication. We compute Q and ZT for the interface via a linear response. We find that Q can be greatly enhanced by the presence of a suitable f-impurity layer at the interface. In these regimes, bulk thermal conductivity would be dominated by phonons which would reduce ZT. However, one can fabricate the junction with two materials with large thermal impedance mismatch, so that phonon scatterings at the interface decrease the thermal conductivity. Thus, phonon conductivity would not diminish the high ZT found [3].

The structure of the paper is as follows: in Section 2 we describe the model Hamiltonian, in Section 3 we solve the electronic motion across the junction, and in Section 4 we compute transport coefficients and ZT. In Section 5 we present and discuss our results, for the clean interface (5.1), and for the d- (5.2) and f- (5.3) impurity layer.

2 The Model

We introduce the one-dimensional spinless Hamiltonian $\mathscr{H}$ to model the motion across the junction along the z direction perpendicular to the interface between a metal and different types of semiconductor. $\mathscr{H}$ is given by the sum of three terms:

$$\mathscr{H} = \mathscr{H}_{\text{metal}} + \mathscr{H}_{\text{interface}} + \mathscr{H}_{\text{FK}}. \tag{2}$$

$\mathscr{H}_{\text{metal}}$ is a tight-binding Hamiltonian describing the metal on the left side of the junction:

$$\mathscr{H}_{\text{metal}} = \left(\varepsilon_d' + eV\right) \sum_{j<0} d_j^{\dagger} d_j - t' \sum_{j<0} d_j^{\dagger} d_{j+1} + \text{H.c.} \tag{3}$$

Here d_j destroys an electron at the lattice site with energy ε_d' and position $z = aj$, a is the lattice constant, t' is the hopping parameter for nearest-neighbor sites, and V is the electrostatic potential across the junction by applying an external bias. In our model, V is constant in the two bulk regions with a discontinuous step at the interface, although the actual profile should be determined self-consistently together with the electronic density. The third term $\mathscr{H}_{\mathrm{FK}}$ in Eq. (2) is the Falicov–Kimball Hamiltonian [23] referring to the insulator on the right side of the interface. In addition to the Coulomb interaction U_{df} between the d-electrons at sites $z = aj$ and the f-electrons localized on atoms at $z = a(j+1/2)$, we add a hybridization term (V_{df} is the modulus of the hybridization integral) between d and f orbitals. Because the crystal has inversion symmetry, this term has to be odd:

$$\mathscr{H}_{\mathrm{FK}} = \mathscr{H}_{\mathrm{FK}}^{(1)} + \mathscr{H}_{\mathrm{FK}}^{(2)}, \tag{4}$$

$$\mathscr{H}_{\mathrm{FK}}^{(1)} = \tilde{\varepsilon}_d \sum_{j>0} d_j^{\dagger} d_j + \tilde{\varepsilon}_f \sum_{j>0} f_{j+1/2}^{\dagger} f_{j+1/2} - t \sum_{j\geq 0} d_j^{\dagger} d_{j+1} \tag{5}$$

$$-V_{df} \sum_{j\geq 0} f_{j+1/2}^{\dagger} d_j + V_{df} \sum_{j\geq 1} f_{j-1/2}^{\dagger} d_j + \text{H.c.}, \tag{6}$$

$$\mathscr{H}_{\mathrm{FK}}^{(2)} = U_{df} \sum_{j\geq 0} d_j^{\dagger} d_j f_{j+1/2}^{\dagger} f_{j+1/2} + U_{df} \sum_{j>1} d_j^{\dagger} d_j f_{j-1/2}^{\dagger} f_{j-1/2}, \tag{7}$$

where t is the hopping coefficient, and $\tilde{\varepsilon}_f$ and $\tilde{\varepsilon}_d$ are the f- and d-site energies, respectively. The second term $\mathscr{H}_{\mathrm{interface}}$ in Eq. (2) is the Hamiltonian at the interface, which describes the overlayer made of d-sites at $z = 0$ and f-sites at $z = a/2$:

$$\mathscr{H}_{\mathrm{interface}} = \tilde{\varepsilon}_{d0} d_0^{\dagger} d_0 + \tilde{\varepsilon}_{f1/2} f_{1/2}^{\dagger} f_{1/2}. \tag{8}$$

Here we included the possibility of "impurity" atoms at the interface, namely one with energy $\tilde{\varepsilon}_{d0}$ at $z = 0$ and another one with energy $\tilde{\varepsilon}_{f1/2}$ at $z = a/2$.

To provide the MF solution of the Hamiltonian $\mathscr{H}$ of Eq. (2), we assume that $\left\langle d_j^{\dagger} d_j \right\rangle = n_d$, $\left\langle f_{j+1/2}^{\dagger} f_{j+1/2} \right\rangle = n_f$, for $j \geq 1$, $U_{df} \left\langle d_j^{\dagger} f_{j+1/2} \right\rangle = U_{df} \left\langle d_j^{\dagger} f_{j-1/2} \right\rangle = \Delta$, and $\left\langle d_0^{\dagger} d_0 \right\rangle = n_{d0}$, $\left\langle f_{1/2}^{\dagger} f_{1/2} \right\rangle = n_{f1/2}$. Here $\langle \ldots \rangle$ is the symbol for the quantum statistical average. Note that, in addition to the usual mean orbital occupations n_d and n_f of standard Hartree-Fock theory ($0 \leq n_d, n_f \leq 1$), we also introduce the non-vanishing pairing potential Δ, characteristic built-in coherence of the d-electron/ f-hole condensate.

The MF Hamiltonian $\mathscr{H}^{\mathrm{MF}}$, to be computed self-consistently together with the energy spectrum, is:

$$\begin{aligned}\mathscr{H}^{\mathrm{MF}} = {} & eV \sum_{j<0} d_j^{\dagger} d_j + \varepsilon_{d0} d_0^{\dagger} d_0 + \varepsilon_{f1/2} f_{1/2}^{\dagger} f_{1/2} - t' \sum_{j<0} d_j^{\dagger} d_{j+1} - t \sum_{j\geq 0} d_j^{\dagger} d_{j+1} \\ & -(\Delta + V_{df}) \sum_{j\geq 0} f_{j+1/2}^{\dagger} d_j - (\Delta - V_{df}) \sum_{j\geq 1} f_{j-1/2}^{\dagger} d_j + \text{H.c.},\end{aligned} \tag{9}$$

where we introduced the renormalized quantities ε_{d0}, $\varepsilon_{f1/2}$, ε_f, and ε_d, treated as material parameters. For the sake of simplicity, we assumed $t' > t$ and $\varepsilon_d' = \varepsilon_d = 0$, i.e., the middle of the d-band on both sides of the junction is aligned, and the metal bandwidth ($4t'$) is larger than the semiconductor bandwidth ($4t$). We consider only the case $\varepsilon_f = 0$, that is the flat band is in the middle of the d-band in the right-side material (see Fig. 1) [29]. Note the different parity of Δ- and V_{df}-terms: while the V_{df}-term is odd, the Δ-term is even [20].

The values of orbital occupancies as well as Δ are spatially inhomogeneous and should be determined self-consistently together with the equations of motions. We make the approximation that those are the bulk averages appropriate to each side of the interface layer. In particular, we take the order parameter Δ as constant on the right side of the interface, and zero on the left side.

The temperature dependence of Δ is much stronger than that of the average orbital occupancy. We determine numerically Δ, together with the chemical potential μ, as the bulk value deep inside the right material in a self-consistent way. We assume a bulk electronic occupation of one electron per unit cell (a cell contains one d-site and one f-site), namely

$$\frac{1}{N_s}\sum_j \left\langle d_j^\dagger d_j + f_{j+1/2}^\dagger f_{j+1/2} \right\rangle = 1, \tag{10}$$

where N_s is the number of cells. Equation (10) allows us to calculate the chemical potential μ for the bulk. Because the MF Hamiltonian is strongly temperature-dependent, so μ is. Since μ refers to the bulk value inside the semiconductor, the metal on the left side acts as an electron reservoir.

We will consider three specific cases for the actual values of the parameters in the Hamiltonian $\mathscr{H}^{\mathrm{MF}}$ of Eq. (9): (i) $V_{df} \rightarrow 0$ and non-zero U_{df}, i.e., the case FE of the electronic ferroelectric with negligible hybridization [30]. (ii) $U_{df} \rightarrow 0$ and non-zero V_{df}, i.e., the case NB of a f-band hybridized with a d-band, that is a narrow band semiconductor which would be a Kondo insulator if it were driven by the condensation of slave bosons. (iii) A ordinary, "broad band" semiconductor (SC), without center of inversion. In the latter case we still use $\mathscr{H}^{\mathrm{MF}}$ but we regard it simply as a one particle Hamiltonian where the "d" and "f" indices are pure labels, and we rename V_{df} as t, considering it as the odd part of a hopping coefficient between nearest neighbor sites, Δ as V_{e}, the even part, and t as t'', a second nearest neighbor hopping coefficient. We make the choice $t \gg V_{\mathrm{e}}$, $V_{\mathrm{e}} \sim t''$, leading to broad conduction and valence bands with different effective masses.

3 The Equations of Motion

We now find the canonical transformation that diagonalizes $\mathscr{H}^{\mathrm{MF}}$ in a self-consistent way. It is convenient to refer all the excitation energies to the chemical potential μ. Thus, we replace the Hamiltonian $\mathscr{H}^{\mathrm{MF}}$ with $\mathscr{H}^{\mathrm{MF}} - \mu N$, where the number operator N is $N = \sum_j d_j^\dagger d_j + f_{j+1/2}^\dagger f_{j+1/2}$. We then introduce the unitary transformation for electrons

$$\gamma_{ke} = \frac{1}{\sqrt{N_s}}\sum_j [u_k^*(j)\, d_j + v_k^*(j+1/2)\, f_{j+1/2}], \tag{11}$$

and similarly for holes, where k is a quantum index, only equivalent to the crystal momentum in the bulk case. The idea is that the operators $\gamma^{\dagger}_{ke}$, $\gamma^{\dagger}_{-kh}$ diagonalize $\mathscr{H}^{\mathrm{MF}} - \mu N$ and create elementary quasi-particle excitations of energy $\omega(k)$ (electrons) and $\bar{\omega}(-k)$ (holes), respectively, if applied to the ground state. Therefore the equations of motion are derived from

$$\mathrm{i}\hbar\dot{\gamma}_{ke} = \left[\gamma_{ke}, \mathscr{H}^{\mathrm{MF}} - \mu N\right] = \omega(k)\,\gamma_{ke}, \tag{12}$$

and the Hamiltonian $\mathscr{H}^{\mathrm{MF}} - \mu N$ acquires the form

$$\mathscr{H}^{\mathrm{MF}} - \mu N = \sum_{k} \omega(k)\,\gamma^{\dagger}_{ke}\gamma_{ke} + \bar{\omega}(-k)\,\gamma^{\dagger}_{-kh}\gamma_{-kh}. \tag{13}$$

Note that in our definition the operator $\gamma^{\dagger}_{-kh}$ is a fermionic creation operator which *excites* a hole, hence the hole energy is positive, $\bar{\omega}(-k) > 0$ ["excitation representation" (ER)]. Since $\omega(k)$ depends on μ and Δ, it depends on the temperature T as well. From Eq. (12) we obtain finite-difference equations of motion for the site-coefficients u and v of the junction. For the bulk semiconductor the equations are:

$$\begin{aligned}
[\omega(k)+\mu]\,u_k(j) &= -tu_k(j-1) - tu_k(j+1) - \left(\Delta^* - V^*_{df}\right)v_k(j-1/2) \\
&\quad - \left(\Delta^* + V^*_{df}\right)v_k(j+1/2), \\
[\omega(k)+\mu]\,v_k(j+1/2) &= -\left(\Delta + V_{df}\right)u_k(j) - \left(\Delta - V_{df}\right)u_k(j+1) \qquad j \geq 1.
\end{aligned} \tag{14}$$

To solve the system of Eqs. (14), we write down the trial two-component wavefunction

$$\begin{pmatrix} u_k(j) \\ v_k(j+1/2) \end{pmatrix} = \begin{pmatrix} u_k \\ v_k \mathrm{e}^{\mathrm{i}ka/2} \end{pmatrix} \mathrm{e}^{\mathrm{i}kaj}. \tag{15}$$

In the following, we take without loss of generality $u_k > 0$, as well as Δ, V_{df} real. Equation (15) is compatible with Eq. (14) only if $\omega(k) = \xi_k + E_k - \mu$, where $\xi_k = \varepsilon_k/2$, $\varepsilon_k = -2t\cos(ka)$, $E_k = \sqrt{\xi_k^2 + |\Delta_k - V_k|^2}$, $\Delta_k = 2\Delta\cos(ka/2)$, $V_k = 2\mathrm{i}V_{df}\sin(ka/2)$. Consequently, the amplitudes (u_k, v_k) are given by $(V_k - \Delta_k)\,u_k = (\xi_k + E_k)\,v_k$, plus the normalization condition $u_k^2 + |v_k|^2 = 1$, that is $u_k^2 = (1 + \xi_k/E_k)/2$, $|v_k|^2 = (1 - \xi_k/E_k)/2$. Figure 1 shows typical quasi-particle band structures on both sides of the junction for the three different cases under study. In this picture the energy branches are drawn according to the "semiconductor representation" (SCR), where the quasi-particle energy ε for holes is negative, namely $\varepsilon = -\bar{\omega}$ for holes. In SCR the ground state (vacuum of γ-operators in ER) is obtained by filling with electrons the valence band and leaving empty the conduction band. Note that the asymmetry of conduction and valence bands close to the gap is remarkable in case (i) and (ii), contrary to case (iii). For FE (NB) the gap is indirect, and the bottom (top) of conduction (valence) band is much flatter than the top (bottom) of valence (conduction) band, while the curvature of the two SC bands close to the direct gap is comparable.

It is easy to derive an expression for the probability current density J_{keN} associated with the quasi-particle wavefunction (u,v) (see Ref. [33]). Clearly the bulk current is site-independent. It can be shown that $J_{keN} = (a\hbar)^{-1}\partial\,\omega(k)\,/\partial k$, i.e., the quasi-particle velocity has the same expression as the semi-classical one. According to the latter formula, the solution (15) represents a two-component electron wavefunction with wave vector k travelling from left to right if $k > 0$.

Let us focus on case (i). The temperature dependence of the order parameter Δ is given by

$$\Delta(T) = \frac{U_{df}}{N_s}\sum_k \left\{ \frac{\Delta_k}{2E_k}\cos(ka/2)\Big[1 - f(\bar{\omega}(-k)) - f(\omega(k))\Big]\right\}, \tag{16}$$

where $f(\omega) = \left(e^{\beta\omega}+1\right)^{-1}$ is the Fermi function ($\beta = 1/k_{\mathrm{B}}T$). Equation (16) is the analogous of the BCS gap equation, implicitly defining $\Delta(T)$. The constraint (10) on the electron number is an implicit definition of $\mu(T)$; in terms of γ operators it turns into:

$$\frac{1}{N_s}\sum_k f(\omega(k)) = \frac{1}{N_s}\sum_k f(\bar{\omega}(-k)). \tag{17}$$

We solve both Eqs. (16) and (17) simultaneously to obtain the values of $\Delta(T)$ and $\mu(T)$ deep inside the FE bulk. A critical temperature T_C exists at which $\Delta = \mu = 0$, i.e., the energy gap of the material vanishes.

Since we know the bulk solution on the semiconductor side of the junction, we now study the motion of quasi-particles along the whole junction. The equations of motion for the bulk metal on the left hand side,

$$\begin{aligned}
&[\omega(k)+\mu]\,u_k(j) = eVu_k(j) - t'u_k(j-1) - t'u_k(j+1)\,,\\
&[\omega(k)+\mu]\,v_k(j+1/2) = 0 \quad j<0,
\end{aligned} \tag{18}$$

have the Bloch solution

$$\begin{pmatrix} u_k(j) \\ v_k(j+1/2)\end{pmatrix} = \begin{pmatrix}1\\0\end{pmatrix} e^{iqaj},$$

with energy $\omega'(q) = eV - 2t'\cos(qa) - \mu$, and probability current density $J_{keN}(j) = 2t'\hbar^{-1}\sin(qa)$, or, equivalently, $J_{keN} = (a\hbar)^{-1}\partial\,\omega'(q)\,/\partial q$.

The idea is to match the bulk solutions on both sides of the junction. The physical boundary condition is that an electron, travelling e.g. from $z=-\infty$ towards positive values of z, is partly transmitted through the junction and partly reflected. Note that the same energy ω corresponds to two different bulk wave vectors q and k. We always use k to label the coherent electronic state through the whole space, with the convention that both k and q, wave vectors in their respective bulks, correspond to the same energy ω. It is convenient to define the wavefunctions

$$\begin{pmatrix} \Psi_{1L}(j) \\ \Psi_{2L}(j+1/2) \end{pmatrix} = \begin{pmatrix} 1 \\ 0 \end{pmatrix} \mathrm{e}^{\mathrm{i}qaj} - R_k \begin{pmatrix} 1 \\ 0 \end{pmatrix} \mathrm{e}^{-\mathrm{i}qaj} \quad \forall j, \tag{19}$$

$$\begin{pmatrix} \Psi_{1R}(j) \\ \Psi_{2R}(j+1/2) \end{pmatrix} = T_k \begin{pmatrix} u_k \\ v_k \mathrm{e}^{\mathrm{i}ka/2} \end{pmatrix} \mathrm{e}^{\mathrm{i}kaj} \quad \forall j, \tag{20}$$

with the elastic scattering condition $\omega'(q) = \omega(k)$. If we define

$$\begin{pmatrix} u_k(j) \\ v_k(j+1/2) \end{pmatrix}$$

as the solution of the motion across the whole junction, we immediately have from Eq. (18) that

$$\begin{pmatrix} u_k(j) \\ v_k(j+1/2) \end{pmatrix} = \begin{pmatrix} \Psi_{1L}(j) \\ \Psi_{2L}(j+1/2) \end{pmatrix} \quad j < 0, \tag{21}$$

$u_k(0) = \Psi_{1L}(0)$, and from Eq. (14) that

$$\begin{pmatrix} u_k(j) \\ v_k(j+1/2) \end{pmatrix} = \begin{pmatrix} \Psi_{1R}(j) \\ \Psi_{2R}(j+1/2) \end{pmatrix} \quad j \geq 1, \tag{22}$$

$v_k(1/2) = \Psi_{2R}(1/2)$, because Eqs. (14) and (18) are linear and homogeneous. We have still to determine the two unknown parameters T_k and R_k, so we need the two interface equations not yet employed:

$$[\omega(k) + \mu]\, u_k(0) = \varepsilon_{d0} u_k(0) - t' u_k(-1) - t u_k(1) - \left(\Delta^* + V_{df}^*\right) v_k(1/2), \tag{23}$$

$$[\omega(k) + \mu]\, v_k(1/2) = \varepsilon_{f1/2} v_k(1/2) - \left(\Delta + V_{df}\right) u_k(0) - \left(\Delta - V_{df}\right) u_k(1). \tag{24}$$

We replace $u_k(-1)$, $u_k(0)$, $u_k(1)$, $v_k(1/2)$, in Eqs. (23) and (24) with $\Psi_{1L}(-1)$, $\Psi_{1L}(0)$, $\Psi_{1R}(1)$, $\Psi_{2R}(1/2)$, respectively, obtaining a linear system for the two unknowns R_k and T_k:

$$-\left(\Delta + V_{df}\right) R_k + \left[\left(\omega - \varepsilon_{f1/2}\right) v_k \mathrm{e}^{\mathrm{i}ka/2} + \left(\Delta - V_{df}\right) u_k \mathrm{e}^{\mathrm{i}ka}\right] T_k = -\left(\Delta + V_{df}\right)$$

$$\left(-\omega + \varepsilon_{d0} - t'\mathrm{e}^{\mathrm{i}qa}\right) R_k + [(\Delta^* + V_{df}^*) v_k \mathrm{e}^{\mathrm{i}ka/2} + t u_k \mathrm{e}^{\mathrm{i}ka}] T_k = -\omega + \varepsilon_{d0} - t'\mathrm{e}^{-\mathrm{i}qa}. \tag{25}$$

To solve system (25) we fix ω, then we obtain k and q, and hence u_k and v_k: once we input as parameters Δ, V_{df}, ε_{d0}, and $\varepsilon_{f1/2}$, the whole coefficient matrix of system (25) is known and T_k and R_k can be eventually obtained. In Eq. (19) we chose a normalization such that the flux of the incident wave is $J_{keN}^{\mathrm{inc}} = 2t'\hbar^{-1} \sin(qa)$, and that of the reflected wave is $J_{keN}^{\mathrm{refl}} = -|R_k|^2 2t'\hbar^{-1} \sin(qa)$, hence, by definition, the reflection coefficient is given by $\mathscr{R}(\omega) = \left|J_{keN}^{\mathrm{refl}}\right| / \left|J_{keN}^{\mathrm{inc}}\right| = |R_k|^2$. The transmission coefficient $\mathscr{T}(\omega)$ may be calculated most simply by probability conservation, $\mathscr{T}(\omega) = 1 - \mathscr{R}(\omega)$. $\mathscr{T}(\omega)$ is the key quantity to compute the thermopower.

Because of the two-fold degeneracy of the energy ω, we have to consider also the quasi-particle associated with $-k$ (and $-q$), with $k > 0$. It is clear now how to rewrite Eqs. (19) and (20), because an electron coming from the right is partly reflected into the right-hand side of the junction and partly transmitted into the left-hand side. Therefore, we can proceed as above. By time-reversal symmetry, one has $\mathscr{T}(\omega(k)) = \mathscr{T}(\omega(-k))$.

4 Transport Coefficients and *ZT*

If we apply an electric field E or a temperature gradient ∇T across the junction, we produce electric and thermal currents J_E and J_T, respectively. In a stationary state the relation between fluxes and driving forces is linear, if fields are small enough [34]:

$$J_E = L_{EE}E + L_{ET}\nabla T \tag{26}$$

$$J_T = L_{TE}E + L_{TT}\nabla T. \tag{27}$$

Since the computation of the electrostatic and thermal field across the junction is complicated and not essential to the junction thermopower, we simply assume that the voltage V and the temperature T are constant on both sides of the junction and have a sharp step at the interface. Thus, instead of Eq. (27), we write

$$J_E = K_{EE}(-\delta V) + K_{ET}\,\delta T \tag{28}$$

$$J_T = K_{TE}(-\delta V) + K_{TT}\,\delta T, \tag{29}$$

where we have replaced the gradients $E = -\nabla V$ and ∇T with $(-\delta V)$ and δT, respectively, where $\delta V = V\,(\text{right}) - V\,(\text{left})$ and $\delta T = T - T_n$, being T the temperature associated to the right-hand side, and T_n to the normal metal on the left-hand side. Since we have assumed $V\,(\text{right}) = 0$, we have $-\delta V = V\,(\text{left}) = V$. The interface conductance is $G = K_{EE}$, the thermopower is $Q = -K_{ET}/K_{EE}$, the thermal conductance is $G_T = -(K_{TT} - K_{TE}K_{ET}/K_{EE})$, and the figure of merit is $Z = Q^2G/G_T$. Moreover, the Onsager relation holds, $K_{ET} = -K_{TE}/T$. From Eq. (29) the formulae for the K transport coefficients follow: $K_{EE} = [\partial J_E/\partial(-\delta V)]_{\delta T, \delta V = 0}$, and similarly for the other coefficients.

Here we closely follow the approach of Blonder et al. [35]. We assume that the two sides of the junction are in contact with perfect electron reservoirs, at different temperatures and chemical potentials, and that quasi-particle wavefunctions keep their phase coherence across the whole system except at $z = \pm\infty$, where they completely lose their phase, thermalize and relax in energy due to inelastic scattering processes of the Fermi sea. Namely, we regard the contacts as perfect emitters or adsorbers. The electric current density, J_E, will be given by the sum of all contributions of the quasi-particle excitations to the current, each one weighted by the correct Fermi distribution function, depending on whether quasi-particles originate from the left or right reservoir. Because J_E is stationary and conserved, we can calculate it in every point of the space, so we choose a lattice site $j < 0$ in the bulk on the left-hand side. In particular, the contribution to the density current $J_E^{L\to R}$ produced by electrons going from left to right is

$$J_E^{L\to R} = 2\frac{1}{N_s}\sum_{k>0} J_{ke}^{\text{inc}}(j)\, f(\omega(k) - eV) \quad j < 0, \tag{30}$$

where the Fermi function refers to the temperature T_n of the metal on the left-hand side, and the electrochemical potential differs from that on the right-hand side by eV. The prefactor 2 accounts for the spin degeneracy. Note that Eq. (30) gives the only electron flux running from left to right on the bulk metal side. On the other hand, the contribution to the electron current $J_E^{L\leftarrow R}$ from right to left, always on the bulk metal side, is given by two distinct terms, one for electrons reflected at the interface and coming from the left side, hence in equilibrium with the left reservoir, and one for electrons transmitted through the interface and coming from the right reservoir:

$$J_E^{L\leftarrow R} = 2\frac{1}{N_s}\sum_{k>0} J_{ke}^{\text{refl}}(j)\, f(\omega(k) - eV) + 2\frac{1}{N_s}\sum_{k<0} J_{ke}^{\text{trans}}(j)\, f(\omega(k)) \quad j < 0. \tag{31}$$

Note that the second sum in Eq. (31) runs over negative k values and the related electron wavefunctions extend all over the junction, being therefore well defined at $j < 0$. Now we add Eqs. (30) and (31),

$$\begin{aligned} J_E^{L\to R} + J_E^{L\leftarrow R} &= 2\frac{1}{N_s}\sum_{k>0} J_{ke}^{\text{inc}}(j<0)\,[1 - \mathscr{R}(\omega(k))]\, f(\omega(k) - eV) \\ &+ 2\frac{1}{N_s}\sum_{k<0} J_{ke}^{\text{inc}}(j>0)\, \mathscr{T}(\omega(k))\, f(\omega(k)). \end{aligned} \tag{32}$$

Here the notation $J_{ke}^{\text{inc}}(j > 0)$ refers to electrons, incident on the interface, coming from $z = +\infty$. We have

$$\begin{aligned} J_E^{L\to R} + J_E^{L\leftarrow R} &= 2\frac{1}{N_s}\sum_{q>0} \frac{e}{a\hbar}\frac{\partial\, \omega'(q)}{\partial q}\left[1 - \mathscr{R}(\omega'(q))\right] f(\omega'(q) - eV) \\ &+ 2\frac{1}{N_s}\sum_{k<0} \frac{e}{a\hbar}\frac{\partial\, \omega(k)}{\partial k}\, \mathscr{T}(\omega(k))\, f(\omega(k)), \end{aligned} \tag{33}$$

and going to the continuum limit we obtain

$$J_E^{L\to R} + J_E^{L\leftarrow R} = \frac{2e}{h}\int_0^\infty \mathrm{d}\,\omega\, \mathscr{T}(\omega)\,[f(\omega - eV) - f(\omega)],$$

with the convention that $\mathscr{T}(\omega) = 0$ if ω is not in the range of excitation energies allowed, and $h = 2\pi\hbar$. Similarly, the hole contribution to J_E is given by

$$-\frac{2e}{h}\int_0^\infty \mathrm{d}\,\bar{\omega}\, \bar{\mathscr{T}}(\bar{\omega})\,[f(\bar{\omega} + eV) - f(\bar{\omega})]. \tag{34}$$

Therefore, passing to SCR and using the equivalence $f(-\varepsilon) = 1 - f(\varepsilon)$, the total current J_E is:

$$J_E = \frac{2e}{h} \int_{-\infty}^{\infty} \mathrm{d}\varepsilon \left[\mathscr{T}(\varepsilon) + \bar{\mathscr{T}}(-\varepsilon) \right] \left[f(\varepsilon - eV) - f(\varepsilon) \right].$$

The electron heat current density $J_{keT}(j)$ is defined by $J_{keT}(j) = \omega(k) J_{keN}(j)$. Reasoning in the same way as for J_E, we obtain the total heat current J_T:

$$J_T = \frac{2}{h} \int_{-\infty}^{\infty} \mathrm{d}\varepsilon \left[\mathscr{T}(\varepsilon) + \bar{\mathscr{T}}(-\varepsilon) \right] \varepsilon \left[f(\varepsilon - eV) - f(\varepsilon) \right].$$

Now it is easy to compute K-transport coefficients from their definitions. We only summarize the results: $K_{EE} = 2e^2 L_0 / h$, $K_{ET} = -2eL_1/Th$, $K_{TE} = 2eL_1/h$, $K_{TT} = -2L_2/Th$, with

$$L_n = \int_{-\infty}^{\infty} \mathrm{d}\varepsilon \left[\mathscr{T}(\varepsilon) + \bar{\mathscr{T}}(-\varepsilon) \right] \varepsilon^n \left(-\frac{\partial f(\varepsilon)}{\partial \varepsilon} \right). \tag{35}$$

From the transmission coefficient momenta (35) we may also compute the alternative figure of merit $L_1^2 / L_0 L_2 = ZT/(ZT+1)$.

5 Results and Discussion

We present the results for the thermopower Q and the figure of merit ZT for the different types of junction under study. Bulk parameters are chosen to describe a large class of materials and are the same as in Fig. 1, with the metal band much broader than the semiconductor one. With this choice of parameters, the three semiconductors have the same gap at $T = 0$ ($E_{\mathrm{gap}} = 0.4t$) and approximately the same bandwidth ($\approx 4t$).

5.1 Clean Interface

First we study the junction with a clean interface ($\varepsilon_{d0} = \varepsilon_{f1/2} = 0$). In Fig. 2a we plot the absolute value of thermopower $|Q|$ vs T. In all three cases $|Q|$ goes to infinity as $T \to 0$, as expected for both indirect-gap narrow-band semiconductors [36] and direct-gap ordinary semiconductors [34]. As T approaches the critical temperature T_c of the ferroelectric ($k_{\mathrm{B}} T_c = 0.195t$ for our choice of parameters), the thermopower $|Q|$ of FE goes to zero, while $|Q|$ for NB and SC decreases in a exponential-like manner as the temperature rises. At $T = T_C$ the FE gap vanishes and the semiconductor turns into a metal with symmetric bands, therefore $Q = 0$. In the low temperature region, instead, for $k_{\mathrm{B}} T \leq 0.05t$ ($T \leq 0.25 T_c$), i.e., in the region where

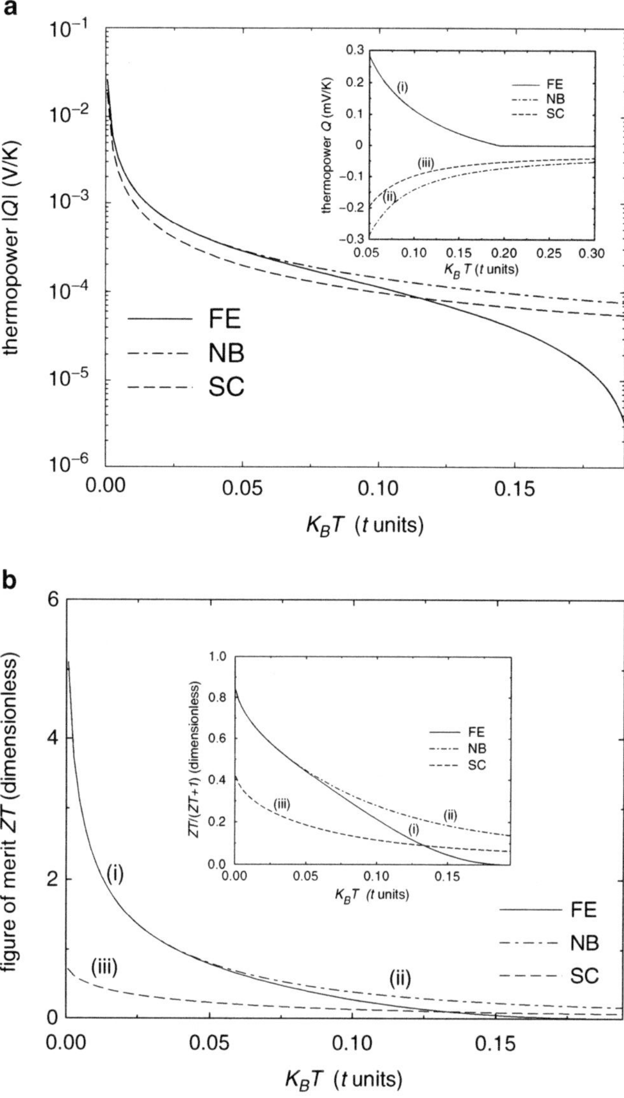

Fig. 2 (**a**) Absolute value of the thermopower $|Q|$ vs T, for the three cases under study. Inset: Magnification of the same plot, in linear scale, in the neighborhood of T_c for FE. (**b**) Plot of the figure of merit ZT vs T. Inset: same plot for the alternative figure of merit $ZT/(ZT+1)$.

$\Delta(T)$ is almost constant, the absolute value of Q for FE and NB is nearly identical. For SC $|Q|$ is about 60–70% of the NB value in the whole range of temperature. To compare our results with reported bulk data, we can assume a typical value of $4t = 1$ eV for FE or NB, i.e. a gap around 0.1 eV at $T = 0$, with room temperature $\sim 0.1t$. With these numbers, we have $|Q| = 0.11 - 0.14$ mV/K at room temperature, values comparable with those of some rare-earth metals with very high thermopower.

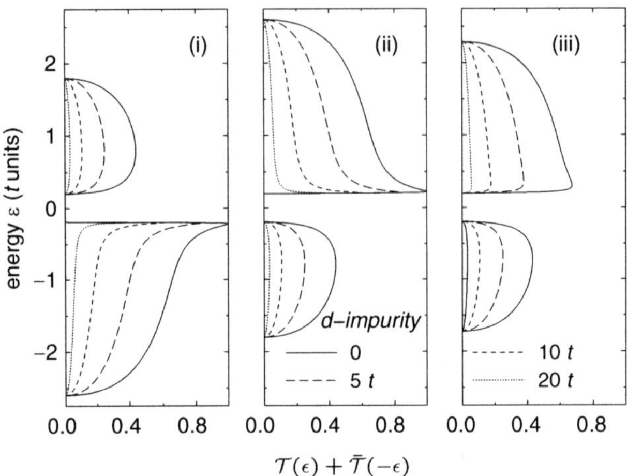

Fig. 3 Total transmission coefficient $\mathcal{T}(\varepsilon)+\bar{\mathcal{T}}(-\varepsilon)$ as a function of the energy ε (vertical axis). From left to right panels correspond to FE, NB, and SC, respectively. Curves for different values of the d-impurity energy ε_{d0} are plotted. The parameters used are the same as in Fig. 1.

If instead we assume for SC a typical gap of 0.5 eV, then $|Q|$ is around 0.5 mV/K at room temperature, consistently with characteristic bulk data. The inset of Fig. 2a is a magnification of the plot (in linear scale) in the neighborhood of T_c. The discontinuity of the derivative of Q for the FE curve is a signature of the second-order ferroelectric transition. Similar kinks are found also for the other transport coefficients G and G_T. One sees that NB and SC have electron transport character ($Q<0$) while FE has hole character ($Q>0$).

To understand the above features, in Fig. 3 we plot the total (electron plus hole) transmission coefficient $\mathcal{T}(\varepsilon)+\bar{\mathcal{T}}(-\varepsilon)$ vs energy in SCR for the three cases under study (at $T=0$). Solid lines refer to the clean interface case. Since the sign and magnitude of Q is established by Eq. (35) for L_1, i.e. by the competition between the different weights of electron and hole transmission functions $\mathcal{T}$ and $\bar{\mathcal{T}}$, respectively, it is clear that the stronger the electron/hole asymmetry of the transmission coefficient, the higher the thermopower. Panels (i) and (ii) of Fig. 3 present the remarkable electron/hole asymmetry close to the gap, hence Q for FE has a strong hole character ($Q>0$) while NB has a dominant electron character ($Q<0$). The situation is different for SC [see panel (iii)], because electron and hole transmission coefficients are more symmetric, and thus $|Q|$ has a lower value. In the limit of electron/hole symmetry [$\mathcal{T}(\varepsilon)=\bar{\mathcal{T}}(-\varepsilon)$, as it is the case for FE when $T\to T_c$], the thermopower is zero.

The other transport parameters, apart from Q, are the electrical conductance G and the thermal conductance G_T. We find that, for all the three cases under study, there exists an activation temperature around $0.03t$, below which G and G_T, as functions of T, exponentially drop to zero, because the number of thermally excited carriers becomes too small. This behavior, characteristic of a gapped material, shows that at low temperature the main contribution to the thermal conductance G_T is given by the lattice, which is not included, thus dramatically decreasing the actual value of ZT. This contribution, and hence the minimum working temperature of the junction, depends on the thermal impedance mismatch of the interface.

Once all transport coefficients are known, the most relevant quantity to be computed for practical applications is the figure of merit ZT, plotted in Fig. 2b as a function of the temperature. In all three cases ZT is a monotonic decreasing function of T. However, ZT is much bigger for FE and NB than for SC (except that for FE $ZT \to 0$ if $T \to T_c$). At room temperature ZT turns out to be $\sim$0.3 for FE and $\sim$0.4 for NB, but already at $T = 100$ K ZT is $\sim$1.1 for FE and NB and only $\sim$0.5 for SC. In the inset of Fig. 2b, we redraw the same plot in term of the alternative figure of merit $ZT/(ZT+1)$. While ZT has no theoretical upper bound, the maximum of $ZT/(ZT+1)$ is 1, corresponding to $ZT = \infty$.

It is likely that the one-dimensional model artificially enhances ZT. Besides, we do not know how the order parameter $\Delta(T)$ actually varies at the interface, and all the effects of charge polarization are neglected: this does not necessarily imply a reduction of ZT. The only scattering mechanism we have considered is the interface: in particular, the contribution of phonons to the thermal conductance is neglected.

5.2 d-Impurity Layer

An overlayer of d-impurity atoms at the interface may be built by epitaxial growth techniques: we find that this configuration does not improve considerably the figure of merit ZT. In our model, the overlayer is obtained by putting one atom at $z = 0$ with site energy ε_{d0}. Since the hopping coefficient t is left unchanged with respect to the clean-interface case, the atomic orbital of the impurity is still of d-type. The essential point here is that the atom at $z = 0$ substantially participates in the electronic motion, contrary to a f-site. In case (iii) this distinction between f- and d-sites is no longer relevant: actually we look at SC only for comparison.

To understand the role of the d-impurity layer in the transport, consider the transmission coefficient vs ε in Fig. 3. Here, in addition to the clean interface case (solid lines), we have also plotted curves corresponding to increasing values of ε_{d0}. As ε_{d0} increases, the transmission is uniformly depressed over the whole range of energies. In cases (i) and (ii), results depend only on the absolute value of ε_{d0}, while (iii) is more complex, due to the presence of a second nearest neighbor hopping coefficient.

While the effect of the impurity seems important for both NB and FE semiconductors, some caution is needed in accepting these results. First, one cannot make $|\varepsilon_{d0}|$ arbitrarily large, because its value is physically limited and cannot differ too much from typical bulk energies, otherwise the impurity would be screened. Second, in our computation we have taken all parameters but ε_{d0} unchanged with respect to the clean-interface case, and it is clear that this approximation becomes worse as long as $|\varepsilon_{d0}|$ increases.

The conclusion is that the enhancement of the figure of merit $ZT/(ZT+1)$ for reasonable values of ε_{d0} is quite limited. In general, as long as we increase the magnitude of ε_{d0}, we slightly enhance ZT uniformly over the whole range of temperatures. It is remarkable that FE and NB values of ZT are always much higher than SC values. For example, at room temperature, with the usual choice of numerical parameters and $\varepsilon_{d0} = 5t$, $ZT \sim 0.4$ for FE, ~ 0.6 for NB, and ~ 0.2 for SC, but already at 100 K $ZT \sim 1.8$ for FE and NB, while it is only ~ 0.4 for SC.

5.3 f-Impurity Layer

In order to improve dramatically the figure of merit, we propose the insertion of a rare-earth overlayer at $z = a/2$, with energy $\varepsilon_{f1/2}$. Contrary to the results of the previous section, here it is essential that the localized impurity level is of f-type, i.e., hardly sharing the electronic conduction. This is not the case for SC, where tunneling from this "f-site" to adjacent neighbors is allowed: in fact, for SC, the situation is basically similar to the previous d-impurity case. In case (iii), hopping to adjacent neighbors is allowed both from the "f-site" and from the "d"-site. In fact, SC results for the figure of merit ZT are not dissimilar from the values obtained in the d-impurity case. In order to gain some insight into the thermopower behavior, in Fig. 4 we have plotted the total transmission coefficient $\mathscr{T}(\varepsilon) + \bar{\mathscr{T}}(-\varepsilon)$ vs ε for different values of $\varepsilon_{f1/2}$, as in Fig. 3. FE and NB curves [panels (i) and (ii), respectively] are qualitatively different from those of SC [panel (iii)], as we set $\varepsilon_{f1/2}$ to negative values in the energy band range. For FE and NB, $\bar{\mathscr{T}}$ goes to zero in the neighborhood of $\varepsilon_{f1/2}$, as if the hole were completely backscattered from the interface at energies resonant with $\varepsilon_{f1/2}$. On the contrary, in case (iii) the effect is opposite, with $\bar{\mathscr{T}}$ gaining weight for energies close to $\varepsilon_{f1/2}$. The trend is similar for

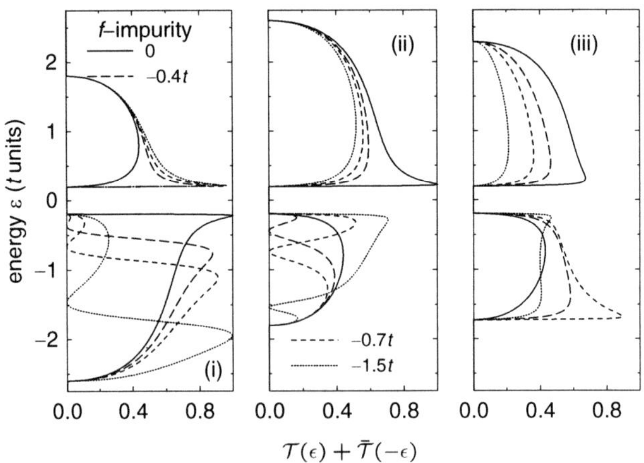

Fig. 4 Total transmission coefficient $\mathscr{T}(\varepsilon) + \bar{\mathscr{T}}(-\varepsilon)$ as a function of the energy ε (*vertical axis*). From left to right vertical panels correspond to FE, NB, and SC, respectively. Curves for different values of the f-impurity energy $\varepsilon_{f1/2}$ are plotted. The parameters used are the same as in Fig. 1.

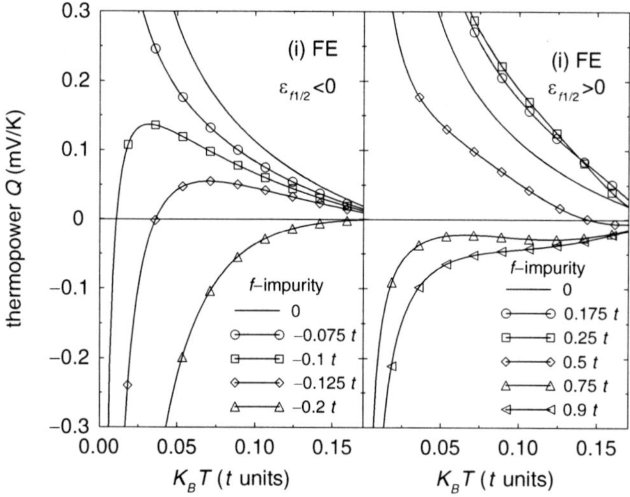

Fig. 5 Thermopower Q vs T for FE. *Left panel*: Curves for different values of $\varepsilon_{f1/2} < 0$. *Right panel*: The same with $\varepsilon_{f1/2} > 0$. Bulk parameters as in Fig. 1.

$\varepsilon_{f1/2} > 0$, showing total reflection in the neighborhood of positive values of $\varepsilon_{f1/2}$, in cases (i) and (ii). These results demonstrate that while the f-impurity level of FE and NB does not share the electronic conduction, due to localization, it strongly affects the transmission, because either in case (i) the quasi-particle excitation is a coherent superposition of d- and f-states, or in case (ii) the "bare" bands are hybridized.

The overall effect of the f-impurity layer on transport is so strong that it even changes the dominant (electron or hole) character of Q, i.e., its sign. For the sake of simplicity, we now consider only FE. In Fig. 5 we plot Q vs T for different values of $\varepsilon_{f1/2}$. Left panel refers to $\varepsilon_{f1/2} < 0$. We see that, as we set $\varepsilon_{f1/2}$ from zero to negative values, Q changes sign: already at $\varepsilon_{f1/2} = -0.1t$ Q has electron character ($Q < 0$) as $T \to 0$, while for $k_B T > 0.02t$ Q has hole character ($Q > 0$). This behavior can be understood by examining Fig. 4: the impurity level drastically diminishes the weight of the transmission coefficient at the top of the valence band, while increasing it at the bottom of the conduction band, so that the sign of Q is reversed at low temperatures, where the only excited carriers are those whose energies are close to the gap. Note that Q is *extremely* sensitive to the position of $\varepsilon_{f1/2}$, in contrast to the case of the d-impurity layer. The right panel of Fig. 5 presents a similar situation for $\varepsilon_{f1/2} > 0$. Here the situation is a little less obvious: as we raise the value of $\varepsilon_{f1/2}$ first Q rises then drops and changes sign. This is due to the asymmetry of electron and hole bands. At first $\varepsilon_{f1/2}$, now positive, is close to the bottom of the conduction band, favoring the hole transport because it depresses the thermally activated electronic channels. Then, as $\varepsilon_{f1/2}$ is increased, weight is added to the transmission coefficient at the bottom of the conduction band. This weight is swept away from the energy neighborhood resonant with $\varepsilon_{f1/2}$, that now is higher in energy with respect to the band bottom. This mechanism also makes Q change sign. A full analysis of the behavior of ZT, which may be reach very high values, is reported in Refs. [3,33].

6 Conclusions

We have made a qualitative theoretical study of the possibility that a junction of metal and FE or NB (as opposed to bulk materials or to a junction metal/SC) produces high thermopower. This is possible if a δ-layer of suitable rare-earth impurity atoms substitutes the original layer at the interface. The localized character of the impurity f-orbital has a strong effect on the transmission of carriers across the junction.

Acknowledgement This work is supported by the CNR Short Term Mobility Program 2008.

References

1. G. D. Mahan, Solid State Phys. **51**, 81 (1998); F. Giazotto et al., Rev. Mod. Phys. **78**, 217 (2006).
2. A value of $ZT = 2.4$ at room temperature was reported for p-type Bi_2Te_3/Sb_2Te_3 superlattice devices, see R. Venkatasubramanian et al., Nature **413**, 597 (2001).
3. For a preliminary report of this work see M. Rontani and L. J. Sham, Appl. Phys. Lett. **77**, 3033 (2000).
4. See B. C. Sales, D. Mandrus, and R. K. Williams, Science **272**, 1325 (1996), and references therein.
5. G. D. Mahan and J. O. Sofo, Proc. Natl. Acad. Sci. USA, **93**, 7436 (1996).
6. G. D. Mahan, J. Appl. Phys. **65**, 1578 (1989); J. O. Sofo and G. D. Mahan, Phys. Rev. B **49**, 4565 (1994).
7. W. Mao and K. S. Bedell, Phys. Rev. B **59**, R15590 (1999).
8. L. D. Hicks and M. S. Dresselhaus, Phys. Rev. B **47**, 12727 (1993); L. D. Hicks, T. C. Harman, and M. S. Dresselhaus, Appl. Phys. Lett. **63**, 3230 (1993).
9. J. O. Sofo and G. D. Mahan, Appl. Phys. Lett. **65**, 2690 (1994).
10. L. D. Hicks et al., Phys. Rev. B **53**, R10493 (1996); T. C. Harman, D. L. Spears, and M. P. Walsh, J. Electr. Mater. **28**, L1 (1999).
11. T. Yao, Appl. Phys. Lett. **51**, 1798 (1987); S. M. Lee, D. G. Cahill, and R. Ventakasubramanian, Appl. Phys. Lett. **70**, 2957 (1997); G. Chen and M. Neagu, Appl. Phys. Lett. **71**, 2761 (1997); P. Hyldgaard and G. D. Mahan, Phys. Rev. B **56**, 10754 (1997); G. Chen, Phys. Rev. B **57**, 14958 (1998); M. V. Simkin and G. D. Mahan, Phys. Rev. Lett. **84**, 927 (2000).
12. M. Nahum, T. M. Eiles, and J. M. Martinis, Appl. Phys. Lett. **65**, 3123 (1994).
13. H. L. Edwards et al., Phys. Rev. B **52**, 5714 (1995).
14. G. D. Mahan and L. M. Woods, Phys. Rev. Lett. **80**, 4016 (1998); G. D. Mahan, J. O. Sofo, and M. Bartkowiak, J. Appl. Phys. **83**, 4683 (1998).
15. B. Moyzhes and V. Nemchinsky, Appl. Phys. Lett. **73**, 1895 (1998).
16. G. Min and D. M. Rowe, J. Phys. D: Appl. Phys. **32**, L1, L26 (1999).
17. F. J. Blatt et al., *Thermoelectric Power of Metals*, (Plenum, New York, 1976).
18. G. Aeppli and Z. Fisk, Comments Cond. Mat. Phys. **16**, 155 (1992).

19. Piers Coleman, Phys. Rev. B **29**, 3035 (1984); A. J. Millis and P. A. Lee, Phys. Rev. B **35**, 3394 (1987).
20. Ji-Min Duan, Daniel P. Arovas, and L. J. Sham, Phys. Rev. Lett. **79**, 2097 (1997).
21. B. C. Sales et al., Phys. Rev. B **50**, 8207 (1994); C. D. W. Jones, K. A. Regan, and F. J. DiSalvo, *ibid.* **58**, 16057 (1998).
22. T. Portengen, Th. Östreich, and L. J. Sham, Phys. Rev. Lett. **76**, 3384 (1996); Phys. Rev. B **54**, 17452 (1996).
23. L. M. Falicov and J. C. Kimball, Phys. Rev. Lett. **22**, 997 (1969); R. Ramirez, L. M. Falicov, and J. C. Kimball, Phys. Rev. B **2**, 3383 (1970).
24. D. G. Cahill et al., J. Appl. Phys. **93**, 793 (2003).
25. M. M. Leivo, J. P. Pekola, and D. V. Averin, Appl. Phys. Lett. **68**, 1996 (1996); A. J. Manninen, M. M. Leivo, and J. P. Pekola, Appl. Phys. Lett. **70**, 1885 (1997).
26. A. Bardas and D. Averin, Phys. Rev. B **52**, 12873 (1995).
27. J. E. Hirsch, Phys. Rev. B **58**, 8727 (1998).
28. For the peculiar features of a junction between semimetal and excitonic condensate cf. M. Rontani and L. J. Sham, Phys. Rev. Lett. **94**, 186404 (2005); Solid State Commun. **134**, 89 (2005).
29. The possibility of a charge density wave [G. Czycholl, Phys. Rev. B **59**, 2642 (1999)] is removed either by introducing the spin degrees of freedom to the electrons and strong on-site interaction to the f-electrons or by including the Coulomb energy cost of the charge density wave.
30. When V_{df} is exactly zero, the local gauge invariance of the Falicov-Kimball Hamiltonian renders the f electron site occupation number classical. See J. K. Freericks and V. Zlatić [31] and references therein. However, a small V_{df} breaks the local gauge invariance and the long-range order of FE is possible [22]. The broken local gauge invariance by hopping f-electrons is shown beyond mean-field approximation to lead to a rich phase diagram including the FE phase [32].
31. J. K. Freericks and V. Zlatić, Phys. Rev. B **58**, 322 (1998).
32. C. D. Batista, Phys. Rev. Lett. **89**, 166403 (2002).
33. An extended account of this work is posted on the web at arXiv:cond-mat/0309687.
34. J. M. Ziman, *Electrons and Phonons*, (Oxford, London, 1960).
35. G. E. Blonder, M. Tinkham, and T. M. Klapwijk, Phys. Rev. B **25**, 4515 (1982).
36. C. Sanchez-Castro, K. S. Bedell, and B. R. Cooper, Phys. Rev. B **47**, 6879 (1993).

Theory of the Nernst Effect Caused by Fluctuations of the Superconducting Order Parameter

K. Michaeli and A. M. Finkel'stein

Abstract We show that the strong Nernst effect observed recently in amorphous superconducting films far above the critical temperature is caused by the fluctuations of the superconducting order parameter. We present here the main steps of the calculation and discuss some subtle issues in the theoretical study of thermal phenomena that we have encountered while calculating the Nernst coefficient. Unlike the calculation of the electric conductivity, the use of the Kubo formula for the thermal and thermoelectric transport coefficients meets with certain complications in the presence of interactions – the interaction enters the heat current operator. In practice a simplified expression for the heat current operator ignoring interactions is commonly used. Since in the presence of superconducting fluctuations there is no justification for the simplified form of the Kubo formula, we preferred to derive the expression for the Nernst coefficient in the quantum kinetic equation approach. The Nernst effect provides an excellent opportunity to test the use of the quantum kinetic equation in the description of thermoelectric transport phenomena. We show how the third law of thermodynamics constrains the magnitude of the Nernst signal. In particular, we demonstrate the cancellation of the non-vanishing quantum contributions in the limit $T \to 0$ by the magnetization current. As a result, we obtained a striking agreement between our theoretical calculations and the experimental data in a broad region of temperatures and magnetic fields.

K. Michaeli
Department of Condensed Matter Physics, The Weizmann Institute of Science, Rehovot 76100, Israel
e-mail: karen.michaeli@weizmann.ac.il

A. M. Finkel'stein
Department of Condensed Matter Physics, The Weizmann Institute of Science, Rehovot 76100, Israel
and
Department of Physics, Texas A&M University, College Station, TX 77843 – 4242, USA
e-mail: alexander.finkelstein@weizmann.ac.il

V. Zlatić and A. C. Hewson (eds.), *Properties and Applications of Thermoelectric Materials*, 213
NATO Science for Peace and Security Series B: Physics and Biophysics,

1 Introduction

Consideration of the transverse thermoelectric signal (the Nernst effect) in the field of condensed matter has been a neglected topic for many years. Currently, however, the Nernst effect is receiving much attention, extending even to fields of research far from condensed matter such as the theory of gravitation [1]. The "rediscovery" of the Nernst effect by the condensed matter community happened after the measurement of the effect in high-T_c materials above the superconducting transition temperatures [2, 3]. Since then the Nernst effect has been also observed in conventional amorphous superconducting films far above T_c [4, 5]. The Nernst effect in high-T_c superconductors [2, 3] has been attributed to the motion of vortices [6–8] existing even above T_c (the vortex-liquid regime). In conventional amorphous superconducting films the strong Nernst signal observed deep in the normal state [4,5] cannot be explained by the vortex-like fluctuations. The authors of Refs. [4, 5] have suggested that the effect is caused by fluctuations of the superconducting order parameter. Here we present a comprehensive analysis of this mechanism using the quantum kinetic approach and demonstrate a quantitative agreement between the theoretical expressions and experiment [5]. No fitting parameters have been used; the values of T_c and the diffusion coefficient have been taken from independent measurements (see Refs. [4,5]). In particular, we succeeded in reproducing the non-trivial dependence of the signal on the magnetic field. Our results imply that in the quest for understanding the thermoelectric phenomena in high-T_c materials the fluctuations of the order parameter should not be ignored.

The Nernst effect and its counterpart, the Ettingshausen effect, are effective tool for studying the superconducting fluctuations because in metallic conductors the contribution of the quasi-particle excitations is negligible. Under the approximation of a constant density of states at the Fermi energy, which is a standard approximation for the Fermi liquid theory, this contribution vanishes completely [9]. On the other hand, the collective modes describing all kinds of fluctuations can in general generate significant contributions to the Nernst effect. Since the neutral modes are not deflected by the Lorentz force they do not contribute to the transverse thermoelectric current. The charged modes, such as fluctuations of superconducting order parameter, are a possible source for the giant Nernst effect even far from the superconducting transition. The fact that the main contribution to the Nernst signal originates from the superconducting fluctuations is in contrast to other transport phenomena such as the electric conductivity. The contributions to the electric conductivity caused by the superconducting fluctuations (paraconductivity [10–12]) can be obtained only close enough to the superconducting transition, where the paraconductivity increases rapidly and may even overcome the Drude conductivity. Far from the transition the superconducting fluctuations produce only one among many corrections to the conductivity and are therefore hard to identify. Owing to the fact that in the absence of fluctuations the Nernst effect is negligible, measurements of the Nernst signal provide a unique opportunity to study the superconducting fluctuations deep inside the normal state.

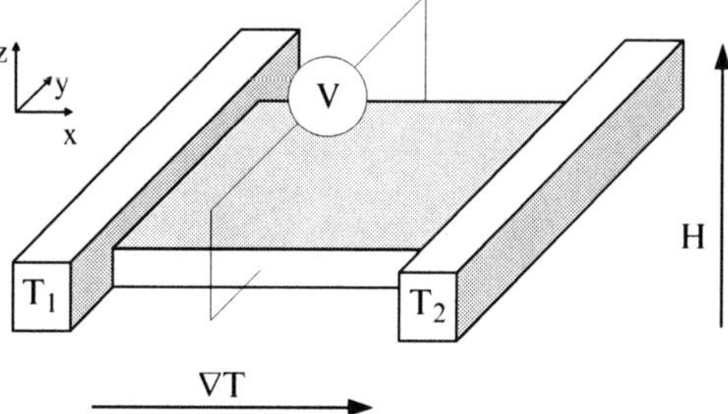

Fig. 1 The setup of the Nernst effect measurement. The sample is placed between two thermal baths of different temperatures. The temperature gradient is in the x-direction, the magnetic field is along the z-direction and the electric field is induced in the y-direction.

The transport coefficients for the electric and thermal currents are defined via the standard conductivity tensor:

$$\begin{pmatrix} \mathbf{j}_e \\ \mathbf{j}_h \end{pmatrix} = \begin{pmatrix} \hat{\sigma} & \hat{\alpha} \\ \hat{\tilde{\alpha}} & \hat{\kappa} \end{pmatrix} \begin{pmatrix} \mathbf{E} \\ -\nabla T \end{pmatrix}. \tag{1}$$

When the thermo-magnetic phenomena are studied in films (or layered conductors) the magnetic field is conventionally directed perpendicularly to the conducting plane, see Fig. 1. Then each element of the conductivity tensor corresponds to a 2×2 matrix describing the conductivity components in the $x - y$ plane (see Fig. 1). The Onsager relations imply that $\sigma_{ij}(\mathbf{H}) = \sigma_{ji}(-\mathbf{H})$ and $\tilde{\alpha}_{ij}(\mathbf{H}) = T\alpha_{ji}(-\mathbf{H})$. In an open circuit setup, from the condition $\mathbf{j}_e = 0$ one gets that the Nernst coefficient is:

$$e_N = \frac{E_y}{-\nabla_x T} = \frac{\sigma_{xx}\alpha_{xy} - \sigma_{xy}\alpha_{xx}}{\sigma_{xx}^2 + \sigma_{xy}^2}. \tag{2}$$

We checked that the second term in the numerator is negligible in comparison to the first one (see the comment below Eq. (27)). This observation has been verified by experiment as one can conclude from the results presented in Fig. 2a of Ref. [4]. Therefore, the leading order term for the Nernst coefficient is $e_N \approx \alpha_{xy}/\sigma_{xx}$ and our goal is to find the transverse Peltier coefficient, α_{xy}.

The electric current generated as a response to an external force, such as the electric field, can be found in the linear regime by the Kubo formula [13] which expresses the response in terms of a corresponding correlation function. Extending the Kubo formalism to the calculation of the response to the temperature gradient is not trivial because this gradient is not directly connected to any mechanical force. Following the scheme used in the derivation of the Einstein relation, Luttinger [14] made a connection between the responses to the temperature gradient and to an artificial gravitational field. As a result, Luttinger succeeded to relate all transport coefficients with various current–current correlation functions. A main ingredient in using the Kubo formula is to derive the quantum mechanical expression for the current operators that enter the correlation function. The correlation functions that describe either the thermoelectric or the thermal current involve the heat current. Luttinger derived the operator for the heat current for a system of electrons that are scattered by their mutual interaction and by disorder. When the electron–electron interactions are neglected, Luttinger's expression for the thermal current operator in the second quantization becomes:

$$\mathbf{j}_h(\tau) = \sum_{\mathbf{p}} \frac{\partial \varepsilon_{\mathbf{p}}}{\partial \mathbf{p}} \varepsilon_{\mathbf{p}} c^{\dagger}_{\mathbf{p}}(\tau) c_{\mathbf{p}}(\tau) + \sum_{\mathbf{p},\mathbf{p}'} \frac{\partial \varepsilon_{\mathbf{p}}}{\partial \mathbf{p}} V_{imp}(\mathbf{p},\mathbf{p}') c^{\dagger}_{\mathbf{p}}(\tau) c_{\mathbf{p}'}(\tau) \tag{3}$$

where $c^{\dagger}_{\mathbf{p}}(\tau)$ ($c_{\mathbf{p}}(\tau)$) is the creation (annihilation) operator of a single electron in imaginary time with energy $\varepsilon_{\mathbf{p}}$ and $V_{imp}(\mathbf{p},\mathbf{p}')$ is the potential created by the disorder. Using the equations of motion for the creation and annihilation operators one may replace the Hamiltonian density by the time derivative which in the terms of the Matsubara frequency becomes:

$$\mathbf{j}_h(\omega_n) = \sum_{\mathbf{p},\varepsilon_n} \frac{\partial \varepsilon_{\mathbf{p}}}{\partial \mathbf{p}} \frac{2i\varepsilon_n - i\omega_n}{2} c^{\dagger}_{\mathbf{p}}(\varepsilon_n) c_{\mathbf{p}}(\varepsilon_n - \omega_n). \tag{4}$$

When the mutual interactions between the electrons or the interaction of the electrons with dynamical fields (such as phonons) are considered, the expression for the heat current becomes a non-trivial function of the interaction. In general, the resulting expression for the current is more complicated than just the frequency multiplied by the velocity as it appears in Eq. (4). There are some exceptional cases in which the heat current reduces to be the same as for a free electron gas; for example, in the case of an electron–phonon interaction under the adiabatic approximation [15]. Unfortunately, the simplified expression for the heat current is used in many cases where there is no real justification for it. Moreover, one can check that this form of the Kubo formula fails to reproduce the thermal conductivity even for Fermi liquids. According to the Fermi liquid theory the thermal conductivity of the quasiparticles obeys the Wiedemann–Franz law. However, when the simplified form of the Kubo formula is used, the resulting expression for the current does not satisfy the Wiedemann–Franz law (see Ref. [16]). The incorrect result that emerges from Eq. (4) does not imply that the use of the Kubo formula for the thermal transport coefficients is generally wrong. The weak point is in replacing the full expression for the heat current by the one in Eq. (4). The problem with using the full expression for the heat current is its complexity.

In addition, the Kubo formalism meets with some difficulties when the thermoelectric currents are considered in the presence of a magnetic field. Obraztsov [17] has pointed out that when a magnetic field is applied, the heat current describing the change in the entropy must include a contribution from the magnetization. This is because the thermodynamic expression for the heat contains the magnetization term. Thus, the additional problem of this approach is that the current cannot be expressed entirely by a correlation function. In order to determine the transverse thermoelectric currents one needs to combine the quantum mechanical response to the external field with the magnetization, which is a thermodynamic quantity [17–19].

In the derivation of the thermoelectric currents we have decided, instead of applying the Kubo formula, to employ a different approach and to use the quantum kinetic equation [20–22]. One main advantage of the quantum kinetic equation approach is that the problem of the magnetization current has been solved straightforwardly. We have directly obtained the expression for the thermoelectric current which includes the magnetization current. In this way, the electric current generated by the temperature gradient can be related to the flow of entropy. Therefore, according to the third law of thermodynamics the Nernst signal must vanish at $T \to 0$ [23]. As we will see, this argument imposes a strict constraint on the structure of the different contributions to the Peltier coefficient. Note that the calculation of the thermoelectric transport using the kinetic equation permits a direct verification of the Onsager relations between the off-diagonal components of the conductivity tensor (see Eq. (1)).

The paper is organized as follows: in Section 2 we present the main steps in the derivation of the electric current as a response to a gradient of the temperature in the presence of fluctuations of the superconducting order parameter using the quantum kinetic equation. Then, in Sections 3 and 4 we give details of the calculation that are specific to the transverse current. We devote Section 5 to the magnetization current and discuss its role in cancelling the quantum contributions. This cancellation makes the Nernst signal compatible with the third law of thermodynamics. The results of the calculation of the Nernst signal for various values of the temperature and the magnetic field are presented in Section 6. In view of the frequently used argument that the particle-hole symmetry limits the magnitude of the Nernst effect (see e.g. Ref [24]) we discuss this issue in the Appendix. We demonstrate that the value of the Nernst coefficient is not constrained by the particle-hole symmetry. Rather, the contribution from the quasi-particle excitations is zero when their density of states is taken to be constant, which is often confused with the particle-hole symmetry.

In separate publications we have presented the complete scheme for studying the thermoelectric currents using the quantum kinetic approach in the presence of electron-electron interactions. In Ref. [16] we considered the Coulomb interaction as an example for an interaction with uncharged collective modes, while in Ref. [25] we derive the expressions for the thermoelectric currents for interactions via charged superconducting fluctuations. Here, we sketch the main steps of the derivation and present the final results.

2 The Quantum Kinetic Equation Above T_c in the Presence of a Temperature Gradient

In the presence of superconducting fluctuations we describe the system using two fields. One is the quasi-particle field ψ, while the other represents the fluctuations of the superconducting order parameter Δ. The matrix functions $\hat{G}(\mathbf{r},t;\mathbf{r}',t')$ and $\hat{L}(\mathbf{r},t;\mathbf{r}',t')$ written in the Keldysh form [20–22] describe the propagation of these two fields, respectively. [Notice that we use the term propagators when referring to both these functions, while separately we name $\hat{G}(\mathbf{r},t;\mathbf{r}',t')$ the quasi-particle Green function and $\hat{L}(\mathbf{r},t;\mathbf{r}',t')$ the propagator of the superconducting fluctuations.] The derivation of the transport coefficients in the quantum kinetic approach is divided into two steps. In the first step, the expression for the current in terms of the propagators is obtained. Then, the propagators are found using the quantum kinetic equations.

To calculate the Nernst coefficient we derive the expression for the electric current as a response to a temperature gradient. In the presence of a magnetic field, the electric current is a sum of two terms:

$$\mathbf{j}_e = \mathbf{j}_e^{con} + \mathbf{j}_e^{mag}. \tag{5}$$

The first one, $\mathbf{j}_e^{con}$, is derived using the continuity equation for the electric charge. In the presence of superconducting fluctuations, it is obvious that the charge of the quasi-particles, $-e|\psi(\mathbf{r})|^2$, is not conserved unless the current carried by the superconducting fluctuations is also included. Then, we get from the continuity equation that:

$$\mathbf{j}_e^{con} = \frac{ie}{2}\int d\mathbf{r}'dt'2\left[\underline{\underline{\hat{\mathbf{v}}}}(\mathbf{r},t;\mathbf{r}',t')\underline{\underline{\hat{G}}}(\mathbf{r}',t';\mathbf{r},t) + \underline{\underline{\hat{\mathscr{V}}}}(\mathbf{r},t;\mathbf{r}',t')\underline{\underline{\hat{L}}}(\mathbf{r}',t';\mathbf{r},t)\right]^{<} + h.c,$$

where we redefined all quantities by the following transformation: $\underline{\underline{\hat{Y}}}(\mathbf{r},t;\mathbf{r}',t') = (1+\mathbf{r}\nabla T/T_0)^{1/2}\hat{Y}(\mathbf{r},t;\mathbf{r}',t')(1+\mathbf{r}'\nabla T/T_0)^{1/2}$, with $\hat{Y}$ denoting $\hat{G}$, $\hat{L}$, $\hat{\mathbf{v}}$ or $\hat{\mathscr{V}}$. This transformation eliminates the explicit dependence of the current on ∇T. The remaining dependence of the current on the gradient of the temperature is through the propagators and velocities. We use the notation $[...]^{<}$ as a reminder that the expression inside the square brackets is a product of matrices and to indicate that the current corresponds to the lesser component of the resulting matrix. In general, the Green function $\hat{G}$ contains spin indices over which we sum. Since we do not consider scattering mechanisms that flip the spins and ignore the Zeeman splitting, this sum yields a factor of 2 which eliminates the $1/2$ arising from the symmetric form of the current. Similarly, in the second term the factor $1/2$ is cancelled by the charge $2e$ of the superconducting fluctuations. In the following we omit the spin indices from the Green functions.

The matrix $\underline{\underline{\hat{\mathbf{v}}}}(\mathbf{r},t;\mathbf{r}',t')$ is the velocity of the quasi-particles renormalized by the self-energy $\underline{\underline{\hat{\Sigma}}}(\mathbf{r},t;\mathbf{r}',t')$. In the presence of a magnetic field the velocity also includes the vector potential $\mathbf{A}(\mathbf{r})$:

$$\underline{\underline{\hat{\mathbf{v}}}}(\mathbf{r},t;\mathbf{r}',t') = -\frac{i}{2m}\lim_{\substack{\mathbf{r}'\to\mathbf{r} \\ t'\to t}}\left(\nabla - \frac{ie}{c}\mathbf{A}(\mathbf{r}) - \nabla' - \frac{ie}{c}\mathbf{A}(\mathbf{r}')\right) - i(\mathbf{r}-\mathbf{r}')\underline{\underline{\hat{\Sigma}}}(\mathbf{r},t;\mathbf{r}',t'). \tag{6}$$

Similarly, $\underline{\underline{\hat{\mathcal{V}}}}(\mathbf{r},t;\mathbf{r}',t')$ is equivalent to the "renormalized velocity" of the collective mode describing the superconducting fluctuations (with a contact s-wave interaction):

$$\hat{\mathcal{V}}(\mathbf{r},\mathbf{r}',) = -i(\mathbf{r}-\mathbf{r}')\hat{\Pi}(\mathbf{r},\mathbf{r}',\omega), \tag{7}$$

where $\underline{\underline{\hat{\Pi}}}(\mathbf{r},t;\mathbf{r}',t')$ is the self-energy of the superconducting fluctuations. (Note that in fact $\underline{\underline{\hat{\mathcal{V}}}}$ does not have the dimension of a velocity.)

The second contribution to the electric current is from the magnetization current. Since the magnetization current is divergenceless it cannot be obtained using the continuity equation:

$$\mathbf{j}_e^{mag} = -2ic\nabla\times\mathbf{M}(\mathbf{r}) \lim_{\substack{\mathbf{r}'\to\mathbf{r}\\ t'\to t}} \left[\underline{\underline{\hat{G}}}(\mathbf{r},t;\mathbf{r}',t')\right]^<, \tag{8}$$

where $\mathbf{M}(\mathbf{r})$ denotes the magnetization and the factor of 2 is due to the summation over the spin index.

We write all the components in the expression for the current in coordinate space because $\underline{\underline{\hat{G}}}$ and $\underline{\underline{\hat{L}}}$ that enter the expression for the current are not translational invariant. There are three reasons for the lack of translational invariance. The first one is because the propagators and velocities are not gauge invariant. They depend on the magnetic field through the vector potential which is a function of both the center of mass and the relative coordinates. The second reason is due to the fact that we did not yet performed the averaging over the disorder. Finally, and most important, in the presence of a gradient of the temperature (even in the absence of magnetic field) the propagators become functions of the center of mass coordinate.

Since we postpone the averaging over the impurities, the Green function of the quasi-particles contains open impurity lines as illustrated in Fig. 2:

$$\hat{G}(\mathbf{r},t;\mathbf{r}',t') = \hat{G}_{int}(\mathbf{r},t;\mathbf{r}',t') + \int d\mathbf{r}_1 dt_1 \hat{G}_{int}(\mathbf{r},t;\mathbf{r}_1,t_1)V_{imp}(\mathbf{r}_1)\hat{G}(\mathbf{r}_1,t_1;\mathbf{r}',t'), \tag{9a}$$

$$\begin{aligned}\hat{G}_{int}(\mathbf{r},t;\mathbf{r}',t') &= \hat{G}_b(\mathbf{r},t;\mathbf{r}',t')\\ &+ \int d\mathbf{r}_1 dt_1 d\mathbf{r}_2 dt_2 \hat{G}_b(\mathbf{r},t;\mathbf{r}_1,t_1)\hat{\Sigma}(\mathbf{r}_1,t_1;\mathbf{r}_2,t_2)\hat{G}_{int}(\mathbf{r}_2,t_2;\mathbf{r}',t').\end{aligned} \tag{9b}$$

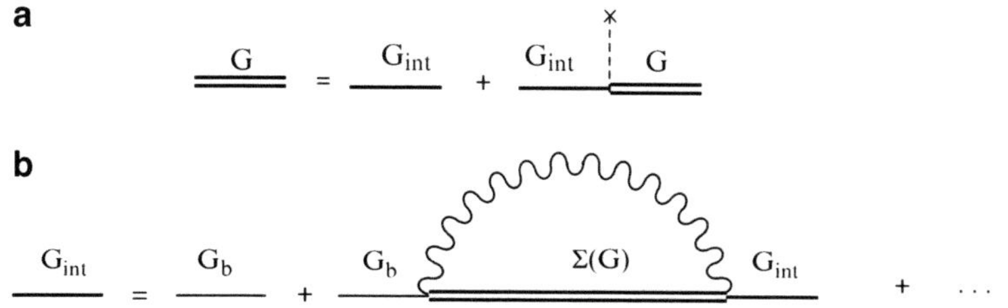

a

G = G_{int} + G_{int} G

b

G_{int} = G_b + G_b Σ(G) G_{int} + . . .

Fig. 2 (**a**) Illustration of Eq. (9a) for the full Green function $\hat{G}$. (**b**) The Dyson equation for $\hat{G}_{int}$ (see Eq. (**??**)). Note that $\hat{G}_{int}$ includes scattering by impurities only through $\hat{\Sigma}(G)$ which is a function of the full Green function $\hat{G}$. The bare Green function is denoted by $\hat{G}_b$.

The lack of translational invariance of the propagator of superconducting fluctuations before averaging over the disorder is a consequence of the dependence of $\hat{L}$ on the quasi-particle Green functions.

We now derive the quantum kinetic equations for the two propagators, $\underline{\underline{\hat{G}}}$ and $\underline{\underline{\hat{L}}}$, in the presence of a temperature gradient. Inspired by Luttinger [14], we introduced an artificial gravitational field in the derivation of the quantum kinetic equation; for more details see Ref. [16]. We considered a system in which both a temperature gradient and a gravitational field are applied at $t=-\infty$ in such a way that the system is at equilibrium. Then, starting at $t=-\infty$, the gravitational field is adiabatically switched off. From the response to the switching off the gravitational field, we can learn about the effect of a temperature gradient on the system. In the regime of linear response the quantum kinetic equation for the Green function of the quasi-particle is:

$$\left[i\left(1-\frac{\mathbf{r}\nabla T}{T_0}\right)\frac{\partial}{\partial t}+\frac{1}{2m}\left(\nabla-\frac{ie}{c}\mathbf{A}(\mathbf{r},t)\right)^2-V_{imp}(\mathbf{r})+\mu\right]\underline{\underline{\hat{G}}}(\mathbf{r},t;\mathbf{r}',t')$$
$$=\delta(\mathbf{r}-\mathbf{r}')\delta(t-t')+\int dt_1 d\mathbf{r}_1\underline{\underline{\hat{\Sigma}}}(\mathbf{r},t;\mathbf{r}_1,t_1)\underline{\underline{\hat{G}}}(\mathbf{r}_1,t_1;\mathbf{r}',t'). \quad (10)$$

The dependence of this equation on the gradient of the temperature is much simplified by the transformation to $\underline{\underline{\hat{G}}}$ and $\underline{\underline{\hat{\Sigma}}}$ because ∇T has been eliminated from all the terms in the equation except the derivative with respect to time. The success of this transformation in simplifying the quantum kinetic equation is not surprising if one recalls that the non-uniform temperature is related to the gravitational field. In field theories which include a non-trivial space–time metric $\hat{g}$ the transformation of the kind $\underline{Y}(\mathbf{x},\mathbf{x}')=\sqrt{-\det\hat{g}(\mathbf{x})}Y(\mathbf{x},\mathbf{x}')\sqrt{-\det\hat{g}(\mathbf{x}')}$ (where $\mathbf{x}$ is a 4-vector) is standard. We write the quantum kinetic equation using the center of mass coordinates, $\mathbf{R}=(\mathbf{r}+\mathbf{r}')/2$, $\mathscr{T}=(t+t')/2$ and the relative coordinates, $\rho=\mathbf{r}-\mathbf{r}'$, $\tau=t-t'$. Since the temperature gradient is independent of time and we are interested in the steady state solution, the Green function will be taken to be independent of $\mathscr{T}$. On the other hand, the dependence on $\mathbf{R}$ remains because the gradient of the temperature enters the equation as the product $\mathbf{r}\nabla T$. This dependence on the center of mass coordinate is the main difference between the response to a temperature gradient and the response to an electric field. The point is that in the presence of an electric field the quantum kinetic equation can be formulated in such a way that the electric field enters only as a product with the relative coordinate, $(\mathbf{r}-\mathbf{r}')\mathbf{E}$. Therefore, after averaging over the disorder the electric field dependent Green function becomes translationally invariant.

In order to find the expression for the ∇T-dependent Green function using the quantum kinetic equation, we separate the Green function into three parts:

$$\underline{\underline{\hat{G}}}=\hat{g}_{eq}+\hat{G}_{loc-eq}+\hat{G}_{\nabla T}. \quad (11)$$

The first part describes the propagation at equilibrium:

$$g_{eq}^{R,A}(\rho,\varepsilon;\mathbf{A},imp)=\left[\varepsilon+\frac{1}{2m}\left(\nabla-\frac{ie}{c}\mathbf{A}(\mathbf{r},t)\right)^2+\mu-V_{imp}-\sigma^{R,A}(\rho,\varepsilon;\mathbf{A},imp)\right]^{-1};$$

$$g_{eq}^{K}(\rho,\varepsilon;\mathbf{A},imp)=(1-2n_F(\varepsilon))\left[g_{eq}^{R}(\rho,\varepsilon;\mathbf{A},imp)-g_{eq}^{A}(\rho,\varepsilon;\mathbf{A},imp)\right]. \quad (12)$$

Here, we perform a Fourier transform of the relative time τ. The Green function $\hat{g}_{eq}$ depends on the center of mass coordinate through the potential V_{imp} created by the impurities at a specific realization and the vector potential. The Laplacian in the operatorial expression for $\hat{g}_{eq}$ contains the derivatives with respect to $\mathbf{R}$ and ρ.

The expressions given in Eq. (12) can be rewritten as a product of the phase

$$\exp\left\{i\frac{e}{c}\int_{\mathbf{r}'}^{\mathbf{r}}\mathbf{A}(\mathbf{r}_1)d\mathbf{r}_1\right\}=\exp\left\{-i\frac{e\mathbf{B}}{4c}(\mathbf{r}-\mathbf{r}')\times(\mathbf{r}+\mathbf{r}')\right\}, \quad (13)$$

and the gauge invariant Green functions (see Ref. [26] and references therein):

$$\tilde{g}_{eq}^{R,A}(\rho,\varepsilon;imp)=\left[\varepsilon+\frac{1}{2m}\left(\nabla-i\frac{e\mathbf{B}\times\rho}{2c}\right)^2-V_{imp}-\sigma^{R,A}(\rho,\varepsilon;imp)\right]^{-1}. \quad (14)$$

In the following the permeability is taken to be 1, and correspondingly we will not distinguish between $\mathbf{B}$ (the magnetic flux density) and the magnetic field $\mathbf{H}$. After averaging over the disorder, the gauge invariant Green functions at equilibrium become translationally invariant functions of the relative coordinate ρ alone:

$$\tilde{g}_{eq}^{R,A}(\rho,\varepsilon)=\left[\varepsilon+\frac{1}{2m}\left(\frac{\partial^2}{\partial\rho^2}-\frac{e^2H^2\rho^2}{4c^2}\right)+\mu\pm i/2\tau-\sigma^{R,A}(\rho,\varepsilon)\right]^{-1}. \quad (15)$$

As we have already discussed, when we turn from the equilibrium Green function to the ∇T-dependent Green function, an additional dependence on the center of mass coordinate appears. We wish to isolate this dependence on $\mathbf{R}$ from the others. Similar to $\hat{g}_{eq}(\rho,\varepsilon;\mathbf{A},imp)$, we denote the dependencies of $\underline{\underline{\hat{G}}}$ on the center of mass coordinate caused by the impurity potential and the vector potential by imp and $\mathbf{A}$, respectively. Then, the remaining explicit dependence on $\mathbf{R}$ in $\underline{\underline{\hat{G}}}(\mathbf{R};\rho,\varepsilon;\mathbf{A},imp)$ arises only due to the gradient of the temperature. Therefore, in the process of linearizing the equation with respect to $\nabla T/T_0$, we expand $\underline{\underline{\hat{G}}}$ and $\underline{\underline{\hat{\Sigma}}}$ in the collision integral with respect to this explicit dependence on $\mathbf{R}$ alone. In other words, we may write

$$\int dr_1\underline{\underline{\hat{\Sigma}}}(\mathbf{R}+\mathbf{r}_1/2;\rho-\mathbf{r}_1,\varepsilon)\underline{\underline{\hat{G}}}(\mathbf{R}-(\rho-\mathbf{r}_1)/2;\mathbf{r}_1,\varepsilon)$$
$$\approx\int dr_1\underline{\underline{\hat{\Sigma}}}(\mathbf{R};\rho-\mathbf{r}_1,\varepsilon)\underline{\underline{\hat{G}}}(\mathbf{R};\mathbf{r}_1,\varepsilon)+\int dr_1\frac{\mathbf{r}_1}{2}\frac{\partial\underline{\underline{\hat{\Sigma}}}(\mathbf{R};\rho-\mathbf{r}_1,\varepsilon)}{\partial\mathbf{R}}\underline{\underline{\hat{G}}}(\mathbf{R};\mathbf{r}_1,\varepsilon)$$
$$-\int dr_1\underline{\underline{\hat{\Sigma}}}(\mathbf{R};\rho-\mathbf{r}_1,\varepsilon)\frac{\rho-\mathbf{r}_1}{2}\frac{\partial\underline{\underline{\hat{G}}}(\mathbf{R};\mathbf{r}_1,\varepsilon)}{\partial\mathbf{R}}. \quad (16)$$

As we shall see below, the last two terms in the expansion are actually proportional to $\nabla T/T_0$.

Collecting all the terms in Eq. (10) that explicitly contain $\mathbf{R}$, one gets the equation for the local equilibrium Green function:

$$\int d\mathbf{r}_1 \hat{g}_{eq}^{-1}(\rho-\mathbf{r}_1,\varepsilon;\mathbf{A},imp)\,\hat{G}_{loc-eq}(\mathbf{R};\mathbf{r}_1,\varepsilon)-\frac{\mathbf{R}\nabla T}{T_0}\varepsilon\hat{g}_{eq}(\rho,\varepsilon;\mathbf{A},imp)$$
$$=\int d\mathbf{r}_1\hat{\Sigma}_{loc-eq}(\mathbf{R};\rho-\mathbf{r}_1,\varepsilon)\,\hat{g}_{eq}(\mathbf{r}_1,\varepsilon;\mathbf{A},imp). \tag{17}$$

This equation is solved by:

$$\hat{G}_{loc-eq}(\mathbf{R};\rho,\varepsilon)=-\frac{\mathbf{R}\nabla T}{T_0}\varepsilon\frac{\partial\hat{g}_{eq}(\rho,\varepsilon;\mathbf{A},imp)}{\partial\varepsilon}, \tag{18}$$

where the corresponding self-energy should be taken as $\hat{\Sigma}_{loc-eq}(\mathbf{R};\rho,\varepsilon)=-(\mathbf{R}\nabla T/T_0)\,\varepsilon\partial\hat{\sigma}_{eq}(\rho,\varepsilon;\mathbf{A},imp)/\partial\varepsilon$. We see that the local equilibrium Green function is a straightforward extension of the equilibrium Green function for a non-uniform temperature. Since the same holds for $\hat{\Sigma}_{loc-eq}$, the equation for $\hat{G}_{loc-eq}$ is a closed equation determined by the equilibrium properties of the system.

The Green function $\hat{G}_{loc-eq}$ describes the readjustment of quasi-particles to the non-uniform temperature when the system is trying to maintain a local equilibrium. This response of the electrons to the gradient of the temperature attempts to induce modulation of the density. Since for charged particles it is impossible to have a large scale charge modulation, the gradient of the temperature transfers into a gradient of the electro-chemical potential. Therefore, $\mathbf{j}_e=\hat{\sigma}(\mathbf{E}+\nabla\mu/e)=\hat{\sigma}\mathbf{E}^*$ where the effective field $\mathbf{E}^*$ is the one measured in experiments.

The role of the local-equilibrium Green function is most peculiar when the response to the temperature gradient is considered in the presence of a magnetic field. Under these conditions, as we show below, $\hat{G}_{loc-eq}(\mathbf{R};\rho,\varepsilon)$ is responsible for the non-vanishing contribution to $\mathbf{j}_e$ from the magnetization current [27] which is given in Eq. (8).

All the remaining terms in the quantum kinetic equation determine the last term of the Green function, $\hat{G}_{\nabla T}$:

$$\int d\mathbf{r}_1 \hat{g}_{eq}^{-1}(\rho-\mathbf{r}_1,\varepsilon;\mathbf{A},imp)\,\hat{G}_{\nabla T}(\mathbf{R};\mathbf{r}_1,\varepsilon)-\frac{\rho\nabla T}{2T_0}\varepsilon\hat{g}_{eq}(\rho,\varepsilon;\mathbf{A},imp)$$
$$+\frac{1}{2m}\left(\frac{\partial}{\partial\rho}-\frac{ie}{c}\mathbf{A}(\mathbf{R}+\rho/2)\right)\frac{\partial\hat{G}_{loc-eq}(\mathbf{R};\rho,\varepsilon)}{\partial\mathbf{R}}$$
$$=\int d\mathbf{r}_1\hat{\Sigma}_{\nabla T}(\mathbf{R};\rho-\mathbf{r}_1,\varepsilon)\,\hat{g}_{eq}(\mathbf{r}_1,\varepsilon;\mathbf{A},imp)+\int d\mathbf{r}_1\frac{\mathbf{r}_1}{2}\frac{\partial\hat{\Sigma}_{loc-eq}(\mathbf{R};\rho-\mathbf{r}_1,\varepsilon)}{\partial\mathbf{R}}$$
$$\times\,\hat{g}_{eq}(\mathbf{r}_1,\varepsilon;\mathbf{A},imp)-\int d\mathbf{r}_1\hat{\sigma}_{eq}(\rho-\mathbf{r}_1,\varepsilon;\mathbf{A},imp)\frac{\rho-\mathbf{r}_1}{2}\frac{\partial\hat{G}_{loc-eq}(\mathbf{R};\mathbf{r}_1,\varepsilon)}{\partial\mathbf{R}}. \tag{19}$$

In the above equation, the derivatives with respect to the center of mass coordinate act only on the explicit dependence of $\hat{G}_{loc-eq}(\mathbf{R},\rho,\varepsilon;\mathbf{A},imp)$ and $\hat{\Sigma}_{loc-eq}$ $(\mathbf{R},\rho,\varepsilon;\mathbf{A},imp)$ on $\mathbf{R}$ (i.e., through the spatial dependent temperature). Recall that the derivatives with respect to the center of mass coordinate which act on V_{imp} and $\mathbf{A}$ in the local equilibrium Green function have been already included in g_{eq}^{-1} that appears in Eq. (17).

Once the explicit expressions for $\hat{G}_{loc-eq}$ and $\hat{\Sigma}_{loc-eq}$ are inserted, the equation becomes much simpler:

$$\hat{G}_{\nabla T}(\rho,\varepsilon;\mathbf{A},imp) = \hat{g}_{eq}(\varepsilon)\,\hat{\Sigma}_{\nabla T}(\varepsilon)\,\hat{g}_{eq}(\varepsilon) - \frac{i\nabla T}{2T_0}\varepsilon\left[\frac{\partial \hat{g}_{eq}(\varepsilon)}{\partial\varepsilon}\hat{\mathbf{v}}_{eq}(\varepsilon)\hat{g}_{eq}(\varepsilon) - \hat{g}_{eq}(\varepsilon)\,\hat{\mathbf{v}}_{eq}(\varepsilon)\frac{\partial \hat{g}_{eq}(\varepsilon)}{\partial\varepsilon}\right]. \tag{20}$$

The product of matrices should be understood as a convolution in real space. The velocity $\hat{\mathbf{v}}_{eq}$ is the renormalized velocity at equilibrium, i.e., it is independent of the gradient of the temperature. Let us point out an important difference between the two parts of the Green function depending on the temperature gradient, $\hat{G}_{loc-eq}$ and $\hat{G}_{\nabla T}$. As has been already mentioned, $\hat{G}_{loc-eq}$ and $\hat{\Sigma}_{loc-eq}$ are a straightforward extension of the equilibrium Green function and self-energy for a non-uniform temperature. On the other hand, the equation for $\hat{G}_{\nabla T}$ contains the ∇T-dependent self-energy which by itself is a function of $\hat{G}_{\nabla T}$. Thus, this is a self consistent equation, and in order to find a closed expression for $\hat{G}_{\nabla T}$, one has to determine the structure of the self-energy. Once the form of the self-energy is known, one should take into consideration in the coarse linearization with respect to ∇T that all the Green functions in $\hat{\Sigma}_{\nabla T}$ may depend on the temperature gradient.

To complete the derivation of the electric current as a response to a gradient of the temperature, we must also find the dependence of the propagator of the superconducting fluctuations $\hat{L}(\mathbf{r},t;\mathbf{r}',t')$ on ∇T. The explicit dependence on the temperature gradient in the regime of linear response can be eliminated from the kinetic equation for $\hat{L}$ by transforming to the propagator $\hat{\underline{\underline{L}}}$:

$$-\lambda^{-1}\hat{\underline{\underline{L}}}(\mathbf{r},t;\mathbf{r}',t') = \delta(\mathbf{r}-\mathbf{r}') - \int d\mathbf{r}_1 dt_1 \hat{\underline{\underline{\Pi}}}(\mathbf{r},t;\mathbf{r}_1,t_1)\hat{\underline{\underline{L}}}(\mathbf{r}_1,t_1;\mathbf{r}',t'). \tag{21}$$

Here $\lambda > 0$ is the coupling constant of the attractive s-wave coupling. The entire dependence of the propagator on the temperature gradient is through the self-energy term $\hat{\underline{\underline{\Pi}}}$ which in turn depends on the quasi-particle Green functions.

Let us separate the solution of Eq. (21) into the equilibrium and ∇T-dependent propagators, $\hat{\underline{\underline{L}}} = \hat{L}_{eq} + \hat{L}_{loc-eq} + \hat{L}_{\nabla T}$. The propagator at equilibrium can be written symbolically as:

$$\hat{L}_{eq}(\omega) = \left[-\lambda^{-1} + \hat{\Pi}_{eq}(\omega)\right]^{-1}. \tag{22}$$

The RHS should be real as a spatial convolution of the inverted matrix.

The entire dependence of $\hat{L}_{eq}(\mathbf{R};\rho,\omega)$ on the frequency is due to dressing of the bare propagator by its self-energy $\hat{\Pi}_{eq}(\mathbf{R};\rho,\omega)$. The equations for the ∇T-dependent propagators resemble the first term in Eq. (20) for $\hat{G}_{\nabla T}$:

$$\hat{L}_{loc-eq}(\mathbf{R};\rho,\omega) = -\hat{L}_{eq}(\omega)\hat{\Pi}_{loc-eq}(\omega)\hat{L}_{eq}(\omega), \tag{23}$$

and

$$\hat{L}_{\nabla T}(\mathbf{R};\rho,\omega) = -\hat{L}_{eq}(\omega)\hat{\Pi}_{\nabla T}(\omega)\hat{L}_{eq}(\omega). \tag{24}$$

Once again, one should understand the product as a convolution of the coordinates.

3 Derivation of the Transverse Component of $\mathbf{j}_e^{con}$

At this stage of the derivation we shall consider only $\mathbf{j}_e^{con}$ keeping for a while the magnetization current aside. Inserting the expressions for the ∇T-dependent propagators given in Eqs. (20), (18) and (24) into Eq. (6) and extracting the lesser component, we obtain $\mathbf{j}_e^{con}$ as a response to a temperature gradient. First of all, one may observe that the contributions of the local equilibrium functions $\hat{G}_{loc-eq}$ and $\hat{L}_{loc-eq}$ to $\mathbf{j}_e^{con}$ vanish under averaging over the volume of the sample. Since we are not interested in terms that vanish after averaging over the volume the non-zero part of $\mathbf{j}_e^{con}$ becomes:

$$\begin{aligned} j^{con}_{e\ i} = &-\frac{e\nabla_j T}{2T_0}\int\frac{d\varepsilon}{2\pi}\varepsilon\frac{\partial n_F(\varepsilon)}{\partial\varepsilon}\left[v_i^R(\varepsilon)g^R(\varepsilon)v_j^A(\varepsilon)g^A(\varepsilon)+v_i^R(\varepsilon)g^R(\varepsilon)v_j^R(\varepsilon)g^A(\varepsilon)\right.\\ &\left.-v_i^R(\varepsilon)g^R(\varepsilon)v_j^R(\varepsilon)g^R(\varepsilon)-g^R(\varepsilon)v_j^R(\varepsilon)g^R(\varepsilon)v_i^A(\varepsilon)\right]\\ &-\frac{e\nabla_j T}{T_0}\int\frac{d\varepsilon}{2\pi}\varepsilon n_F(\varepsilon)\left[v_i^R(\varepsilon)\frac{\partial g^R(\varepsilon)}{\partial\varepsilon}v_j^R(\varepsilon)g^R(\varepsilon)-v_i^R(\varepsilon)g^R(\varepsilon)v_j^R(\varepsilon)\frac{\partial g^R(\varepsilon)}{\partial\varepsilon}\right]\\ &-ie\int\frac{d\varepsilon}{2\pi}v_i^R(\varepsilon)g^R(\varepsilon)\left[\Sigma^<_{\nabla T}(\varepsilon)(1-n_F(\varepsilon))+\Sigma^>_{\nabla T}(\varepsilon)n_F(\varepsilon)\right](g^R(\varepsilon)-g^A(\varepsilon))\\ &+ie\int\frac{d\omega}{2\pi}\mathcal{V}_i^R(\omega)L^R(\omega)\left[\Pi^<_{\nabla T}(\omega)(1+n_P(\omega))-\Pi^>_{\nabla T}(\omega)n_P(\omega)\right](L^R(\omega)\\ &-L^A(\omega))+c.c. \end{aligned} \tag{25}$$

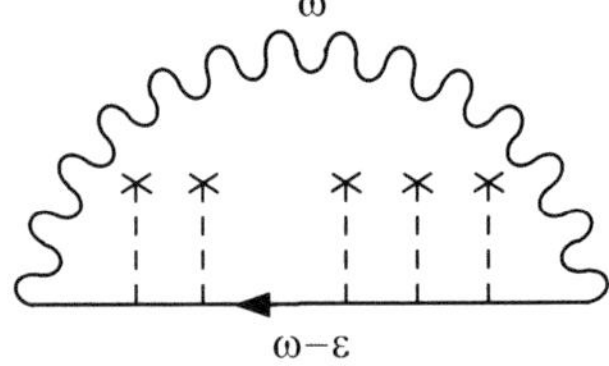

Fig. 3 The self-energy in the first order with respect to the propagator superconducting fluctuations before averaging over the disorder.

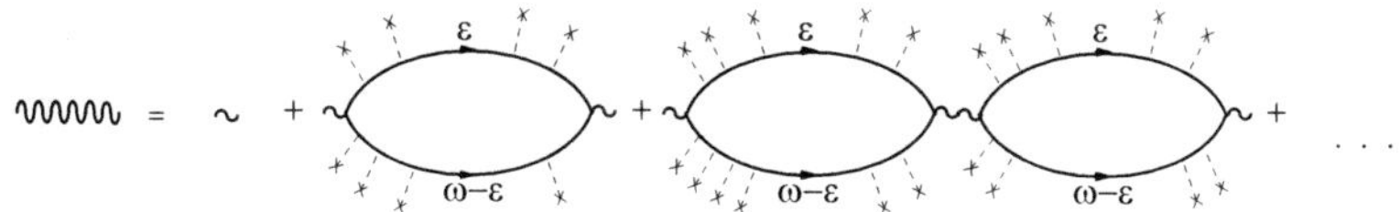

Fig. 4 The geometrical series describing the fluctuations propagator in the Cooper channel.

Here and from now on we omit the notation *eq* from the equilibrium quantities such as the propagators, self-energies and velocities.

As we are interested only in the Gaussian fluctuations, we shall expand the expression for the current with respect to the interaction. For this reason, we replace the equilibrium Green function by $\hat{g}(\mathbf{r},\mathbf{r}',\varepsilon) = \hat{g}_0(\mathbf{r},\mathbf{r}',\varepsilon) + \int d\mathbf{r}_1 d\mathbf{r}_2 \hat{g}_0(\mathbf{r},\mathbf{r}_1,\varepsilon)$ $\hat{\sigma}(\mathbf{r}_1,\mathbf{r}_2,\varepsilon)\ \hat{g}_0(\mathbf{r}_2,\mathbf{r}',\varepsilon)$ and consider a self-energy with only one propagator of the superconducting fluctuations as illustrated in Fig. 3. Furthermore, the propagator of the superconducting fluctuations is determined by the standard geometrical series $\hat{L}(\omega) = [-\lambda^{-1}) + \hat{\Pi}(\omega)]^{-1}$, where $\hat{\Pi}(\omega)$ is approximated by the particle-particle polarization operator as it is shown in Fig. 4. Applying the regular rules of the Keldysh technique, one may check that at equilibrium $\Pi^K_{eq} = (1+2n_P(\omega))(\Pi^R_{eq} - \Pi^A_{eq})$ where $n_P(\omega)$ is the Bose distribution function and Π^R and Π^A, after averaging over the disorder, are given by the standard expressions.

We may now obtain the leading order corrections in the interaction to the electric current as a response to a gradient of the temperature in the linear regime. We should consider all possibilities for linearizing the expressions for $\hat{\Sigma}_{\nabla T}$ and $\hat{\Pi}_{\nabla T}$ with respect to ∇T in Eq. (25). The diagrammatic interpretation for the different contributions to the transverse electric current, obtained in the quantum kinetic approach, corresponds to the three diagrams shown in Fig. 5. After averaging over the disorder, the leading order contributions to the Nernst signal in the diffusive regime are obtained from the two diagrams with three Cooperons [12] presented in Fig. 6a and b, and the Aslamazov–Larkin diagram [10] shown in Fig. 6c. [The Cooperon is a singular diffusion propagator which describes the rescattering on impurities in the particle-particle channel.] Since we use the quantum kinetic equation in order to generate these terms, the analytic structure of the diagrams is given by the equation.

To get the explicit expression for the current we return to the gauge invariant equilibrium Green functions $\tilde{g}$ given in Eq. (15). Since we restrict our calculation to the limit $\omega_c\tau \ll 1$ (where $\omega_c = eH/m^*c$ is the cyclotron frequency of the quasi-particles), we may neglect the dependence of $\tilde{g}$ on the magnetic field entering through the Landau quantization of the quasi-particles states. Therefore, the entire dependence of the quasi-particle Green functions on the magnetic field is through the phase. Unlike the quasi-particles, the Landau quantization of the collective modes (both the Cooperons and the fluctuations of the superconducting order parameter) cannot be neglected because the quantization condition for these modes is $\Omega_c/T_0 > 1$, where in the diffusive regime $\Omega_c = 4eHD/c$ is the cyclotron frequency in the Cooper channel. Note that $\Omega_c \propto \omega_c(\varepsilon_F\tau) \gg \omega_c$, because the product of the Fermi energy and the mean free time is assumed to be a large parameter. [In Ω_c the effective charge is equal to $2e$ and the diffusion coefficient D replaces $1/2m$

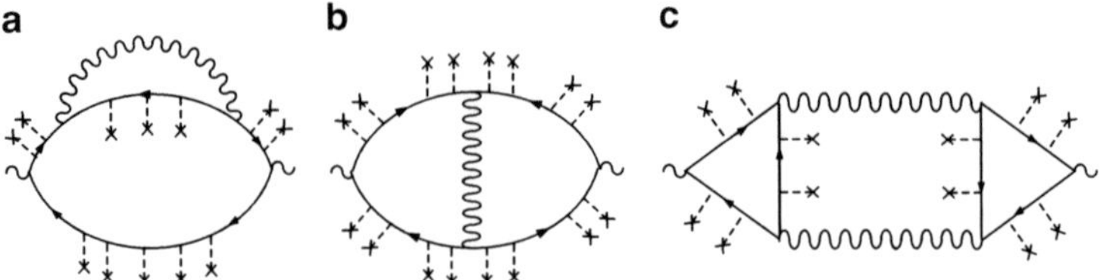

Fig. 5 The diagrammatic contributions the transverse component of $\mathbf{j}_e^{con}$ before averaging over the disorder. (The obvious counterpart diagram for (**a**) is not shown.)

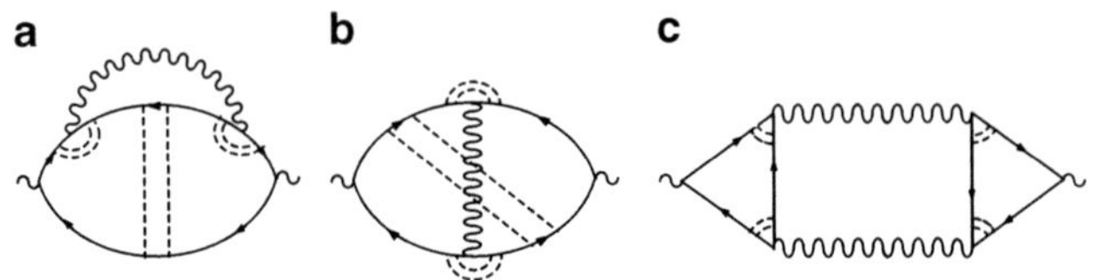

Fig. 6 The diagrammatic contributions to the transverse component of $\mathbf{j}_e^{con}$. Diagrams (**a**) and (**b**) describe the fluctuation of the superconducting order parameter decorated by three Cooperons and (**c**) is the Aslamazov–Larkin diagram. (The obvious counterpart diagrams for (**a**) and (**b**) are not shown.)

because in the Cooperons and the fluctuations propagators, the term Dq^2 substitutes the kinetic energy $p^2/2m$.]

In the limit of small magnetic fields, the Cooperons, like the quasi-particle Green functions, can be also separated into the phase $\exp\{2ie\int_{\mathbf{r}'}^{\mathbf{r}}\mathbf{A}(\mathbf{r}_1)d\mathbf{r}_1/c\}$ and the gauge invariant part at $\mathbf{H}=0$, $C^{R,A}(\rho,\varepsilon,\omega-\varepsilon)=\left[\mp i(2\varepsilon-\omega)\tau-D\nabla_\rho^2\right]^{-1}$, see Appendix in Ref. [26]. At a finite magnetic field, one may express the gauge invariant part of the Cooperon propagator using the Landau level quantization:

$$C_N^{R,A}(\varepsilon,\omega-\varepsilon)=[\mp i(2\varepsilon-\omega)\tau+\Omega_c\tau(N+1/2)]^{-1}, \tag{26}$$

where N is the index of the Landau level. Similarly, the propagator of the superconducting fluctuations written in terms of the Landau levels becomes:

$$L_N^{R,A}(\omega)=\frac{-1}{\nu}\left[\ln\left(\frac{T}{T_c}\right)+\psi\left(\frac{\mp i\omega}{4\pi T}+\frac{\Omega_c(N+1/2)}{4\pi T}\right)-\psi\left(\frac{1}{2}\right)+\varsigma\omega\right]^{-1}. \tag{27}$$

The primary goal of this calculation is to analyze the measurements of the Nernst effect in superconducting films [4,5]. In such films the electron states are not quantized and therefore ν is the density of states of three dimensional electrons (as well as D). The parameter $\varsigma\propto 1/(\nu\lambda\varepsilon_F)$ is important for understanding the difference in magnitude between the longitudinal and transverse Peltier coefficients. The longitudinal Peltier coefficient, α_{xx}, contains an integral over the frequency that vanishes when $\varsigma=0$ while the integrand determining α_{xy} remains finite even in the absence of ς. As a result, in the expression for the Nernst coefficient given in Eq. (2) the second term in the numerator is smaller than the first one by a factor of the order $T/(\nu\lambda\varepsilon_F)$ [28].

Using the expressions for the quasi-particle Green functions, the Cooperons and the propagators of the superconducting fluctuations in the equilibrium state we may investigate the contributions of $\hat{G}_{\nabla T}$ and $\hat{L}_{\nabla T}$ to the current. Recall that we are interested in the transverse current. Let us show how to find the transverse current using for a representative example, one specific term out of the few contributions to the Aslamazov–Larkin diagram:

$$j_{e\ x}^{con} = \frac{e\nabla_y T}{2T_0}\int\frac{d\varepsilon d\varepsilon' d\omega}{(2\pi)^3}\int d\mathbf{r}_2...d\mathbf{r}_{12}\lim_{\mathbf{r}_{12}\to\mathbf{r}_1}\left(\frac{\nabla_1^x}{2m}+\frac{ieHy_1}{4m}-\frac{\nabla_{12}^x}{2m}+\frac{ieHy_{12}}{4m}\right)$$

$$\lim_{\mathbf{r}_6\to\mathbf{r}_7}\left(\frac{\nabla_7^y}{2m}-\frac{ieHx_7}{4m}-\frac{\nabla_6^y}{2m}-\frac{ieHx_6}{4m}\right)g_0^R(\mathbf{r}_1,\mathbf{r}_2,\varepsilon)g_0^A(\mathbf{r}_{11},\mathbf{r}_2,\omega-\varepsilon)g_0^R(\mathbf{r}_{11},\mathbf{r}_{12},\varepsilon)$$

$$C_R(\mathbf{r}_2,\mathbf{r}_3,\varepsilon,\omega-\varepsilon)C_R(\mathbf{r}_{10},\mathbf{r}_{11},\varepsilon,\omega-\varepsilon)L^R(\mathbf{r}_3,\mathbf{r}_4,\omega)L^A(\mathbf{r}_9,\mathbf{r}_{10},\omega)g_0^R(\mathbf{r}_5,\mathbf{r}_6,\varepsilon')$$

$$g_0^A(\mathbf{r}_5,\mathbf{r}_8,\varepsilon')g_0^R(\mathbf{r}_7,\mathbf{r}_8,\varepsilon')C_R(\mathbf{r}_4,\mathbf{r}_5,\varepsilon',\omega-\varepsilon')C_R(\mathbf{r}_8,\mathbf{r}_9,\varepsilon',\omega-\varepsilon')F(\varepsilon,\varepsilon',\omega). \quad (28)$$

In Fig. 7 we show the spatial coordinates corresponding to the expression given above. Since in this part of the calculation we concentrate on the integration over the spatial coordinates, we collect all the frequency dependent factors into the function $F(\varepsilon,\varepsilon',\omega)$ given by

$$F(\varepsilon,\varepsilon',\omega)=\varepsilon\left[\tanh\left(\varepsilon/2T\right)-\tanh\left((\varepsilon-\omega)/2T\right)\right]\tanh\left((\omega-\varepsilon')/2T\right)\partial n_P(\omega)/\partial\omega, \quad (29)$$

and leave them aside for a while.

Next, we rewrite the Cooperons and the propagators of the superconducting fluctuations using the basis of the Landau levels states, $\varphi_{N,n}(\mathbf{r}) = R_{N,n}(r)e^{in\phi}/\sqrt{2\pi}$ (where $R_{N,n}(r)$ are the generalized Laguerre polynomials). In addition, we separate the quasi-particles Green functions into the phases and the gauge invariant Green functions. Then, following the flux technique introduced in Ref. [26], we rearrange Eq. (28) as:

$$j_{e\ x}^{con} = \frac{e\nabla_y T}{4\pi T_0\ell_H^2}\int\frac{d\varepsilon d\varepsilon' d\omega}{(2\pi)^3}\sum_{N,M}\int d\mathbf{r}_2...d\mathbf{r}_{12}e^{-ie\mathbf{H}(\mathbf{r}_{11}-\mathbf{r}_1)\times(\mathbf{r}_1-\mathbf{r}_2)/2c} \quad (30)$$

$$\lim_{\mathbf{r}_{12}\to\mathbf{r}_1}\left(\frac{\nabla_1^x}{2m}+\frac{ieH(y_1-y_2)}{4m}-\frac{\nabla_{12}^x}{2m}-\frac{ieH(y_{11}-y_{12})}{4m}\right)\tilde{g}_0^R(\mathbf{r}_1-\mathbf{r}_2,\varepsilon)$$

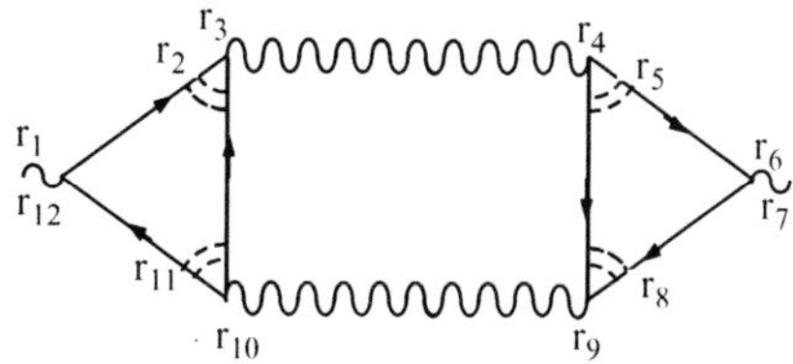

Fig. 7 The Aslamazov–Larkin diagram.

$$\tilde{g}_0^A(\mathbf{r}_{11}-\mathbf{r}_2,\omega-\varepsilon)\tilde{g}_0^R(\mathbf{r}_{11}-\mathbf{r}_{12},\varepsilon)e^{-ie\mathbf{H}(\mathbf{r}_5-\mathbf{r}_6)\times(\mathbf{r}_6-\mathbf{r}_8)/2c}\tilde{g}_0^A(\mathbf{r}_5-\mathbf{r}_8,\varepsilon')$$
$$\lim_{\mathbf{r}_6\to\mathbf{r}_7}\left(\frac{\nabla_7^y}{2m}-\frac{ieH(x_7-x_8)}{4m}-\frac{\nabla_6^y}{2m}+\frac{ieH(x_5-x_6)}{4m}\right)\tilde{g}_0^R(\mathbf{r}_5-\mathbf{r}_6,\varepsilon')\tilde{g}_0^R(\mathbf{r}_7-\mathbf{r}_8,\varepsilon')$$
$$e^{-ie\mathbf{H}(\mathbf{r}_8-\mathbf{r}_{11})\times(\mathbf{r}_{11}-\mathbf{r}_2)/c-ie\mathbf{H}(\mathbf{r}_2-\mathbf{r}_5)\times(\mathbf{r}_5-\mathbf{r}_8)/c}\varphi_{N,0}(\mathbf{r}_2-\mathbf{r}_5)\varphi_{M,0}(\mathbf{r}_8-\mathbf{r}_{11})$$
$$C_N^R(\varepsilon,\omega-\varepsilon)C_M^2(\varepsilon,\omega-\varepsilon)L_N^R(\omega)L_M^A(\omega)C_N^R(\varepsilon',\omega-\varepsilon')C_M^R(\varepsilon',\omega-\varepsilon')F(\varepsilon,\varepsilon',\omega),$$

where $\ell_H=\sqrt{c/2eH}$ is the magnetic length in the Cooper channel. In the last step we have used the orthogonality of the generalized Laguerre polynomials (an example for the treatment of the propagators in this basis can be found in Ref. [29]). The first two exponents in Eq. (30) contain the magnetic fluxes accumulated in the triangles $(\mathbf{r}_1,\mathbf{r}_2,\mathbf{r}_{11})$ and $(\mathbf{r}_5,\mathbf{r}_6,\mathbf{r}_8)$, respectively. One way to get the transverse current, is to extract the magnetic field from these two fluxes or from the diamagnetic terms. As a result the transverse current appears with a coefficient $\omega_c\tau$. We neglect these terms; we will see that when the magnetic field responsible for turning the current to the transverse direction is extracted from the Cooperons and the propagators of the superconducting fluctuations one gets a much larger factor of the order Ω_C/T. Therefore, the integration over the coordinates of the two triangles can be done with the quasi-particle Green functions taken at $\mathbf{H}=0$:

$$j_{e\ x}^{con}=-\frac{e\nabla_yT}{4\pi^2T_0\ell_H^2}v^2\tau^4\int d\varepsilon d\varepsilon' d\omega\int d\mathbf{r}\sum_{N,M}\left[2D\left(\frac{\partial}{\partial x}+\frac{ieHy}{c}\right)\varphi_{N,0}(\mathbf{r})\right] \tag{31}$$
$$\left[2D\left(\frac{\partial}{\partial y}-\frac{ieHx}{c}\right)\varphi_{M,0}(\mathbf{r})\right]C_N^R(\varepsilon,\omega-\varepsilon)C_M^R(\varepsilon,\omega-\varepsilon)L_N^R(\omega)L_M^A(\omega)$$
$$C_N^R(\varepsilon',\omega-\varepsilon')C_M^R(\varepsilon',\omega-\varepsilon')F(\varepsilon,\varepsilon',\omega). \tag{32}$$

The integral over the coordinate corresponds to the matrix element of the velocity operators $\langle N,0|V_xV_y|M,0\rangle$, where $|M,0\rangle=\varphi_{M,0}$ is the quantum state of a particle with a mass equal $1/2D$ in the M Landau level and zero angular momentum in the z-direction. Using the known properties of the Laguerre polynomials, the matrix element can be written as $\langle N,0|V_xV_y|M,0\rangle=2iD^2[(N+1)\delta_{N,M-1}-(M+1)\delta_{M,N-1}]/\ell_H^2$. Finally, the contribution to the current becomes:

$$j_{e\ x}^{con}=-i\frac{e\nabla_yT}{2\pi^2T_0\ell_H^4}v^2D^2\tau^4\int d\varepsilon d\varepsilon' d\omega\sum_{N=0}^{\infty}(N+1)C_N^R(\varepsilon,\omega-\varepsilon)C_{N+1}^R(\varepsilon,\omega-\varepsilon)$$
$$\left[L_N^R(\omega)L_{N+1}^A(\omega)-L_{N+1}^R(\omega)L_N^A(\omega)\right]C_N^R(\varepsilon',\omega-\varepsilon')C_{N+1}^R(\varepsilon',\omega-\varepsilon')F(\varepsilon,\varepsilon',\omega). \tag{33}$$

In the limit $\mathbf{H}\to 0$ when the quantization of the collective modes can be neglected, one may replace in Eq. (28) the Cooperons and the propagators of the superconducting fluctuations by the product of the phase terms (with charge $2e$) and the corresponding propagators in the absence of magnetic field. Then, by using the flux technique [26] the contribution to the current at vanishingly small magnetic field can be found. One may check that the same result is obtained when the transformation from the discrete sum into an integral over a continuous variable is performed in Eq. (33).

Let us conclude with a remark regarding the diagrammatic interpretation of the different contributions to $\mathbf{j}_e^{con}$. As has been already mentioned, the analytic structure and the expressions for the vertices of these diagrams have been found from the quantum kinetic equation. In principal, the same diagrams can be calculated using the Kubo formula. However, if for simplicity one uses in the Kubo formula the heat current operator of non-interacting electrons described in Eq. (4), the resulting expressions for these diagrams differ from those obtained using the quantum kinetic equation. Most important, as one can see from Eq. (25), in the quantum kinetic approach the frequency accompanies the renormalized velocity, so that the expression for the electric current is generally of the form $eg(\varepsilon)v_i(\varepsilon)g(\varepsilon)\varepsilon v_j(\varepsilon)\nabla_j T/T$, as Eq. (25) shows. In other words, the frequency appears together with the velocity that has been already renormalized by the interaction. On the other hand, owing to the fact that the frequency in the simplified version of the Kubo formula is attached to the external vertex before the renormalization of the velocity, the expression for the current has a totally different structure.

This is also the proper place to explain what is so unique in the superconducting fluctuations in the diffusive limit that leads to the giant Nernst effect. As any transverse current coefficient, α_{xy} contains a difference of two almost equal terms. In addition, like all thermoelectric coefficients the integral over the frequency in α contains a factor of the quasi-particle frequency. Consequently, as discussed in the Appendix, the contribution of the quasi-particles to the transverse Peltier coefficient includes two small parameters. The first, is the usual $\omega_c\tau$ that appears in all transverse currents. The second is a reminiscent of the fact that the frequency factor (that in the Boltzmann equation converts into the energy) is responsible for the vanishing of the Peltier coefficient under the approximation of a constant density of states. When a non-constant density of states is considered, the integration over the energy yields another small parameter proportional to T/ε_F. Now we turn to the contribution of the superconducting fluctuations to the transverse component of $\mathbf{j}_e^{con}$ and consider Eq. (33) as a representative example. The difference between the two almost identical terms results in an odd integrand with respect to the frequency of the superconducting fluctuations, ω, which potentially may lead to the vanishing of α_{xy}. So, how can the superconducting fluctuations induce a strong Nernst signal? The explanation lies in the fact that the Cooperons accompanying the superconducting fluctuation depend on the frequency of the incoming/outgoing quasi-particles and not only on the frequency ω carried by the fluctuations (Eq. (26)). The combination of the quasi-particles frequency factor and the Cooperons in the integral over ε saves the situation. This is because after the integration over ε the resulting integrand becomes an even function of ω and hence there is no longer danger that the transverse Peltier coefficient vanishes. We shall see that instead of the two small parameters obtained for the quasi-particles, the contribution of the superconducting fluctuations includes only one. Because of the extra-sensitivity of these fluctuations to the magnetic field this parameter is Ω_c/T_0.

4 Final Expressions for the Transverse Component of $\mathbf{j}_e^{con}$

If one examines Eq. (25) which presents the general expression for the contributions to the electric current from $\hat{G}_{\nabla T}$ and $\hat{L}_{\nabla T}$, one may notice that not all the terms contain the derivative of the Fermi distribution function. As one may expect, the terms in which the Fermi distribution function is not differentiated contribute only to the transverse component of $\mathbf{j}_e^{con}$ and not to the longitudinal one. After integration over the Fermion degrees of freedom (the frequency ε and the coordinates of the quasi-particles Green functions), the terms proportional to $\partial n_F(\varepsilon)/\partial\varepsilon$ give two non-vanishing contributions. The first one corresponds to the Aslamazov–Larkin diagram presented in Fig. 6c:

$$
\begin{aligned}
(j^{con}_{e\ i})^1 &= \varepsilon_{ij}\frac{e\nabla_j T}{16\pi^2 T_0}\nu^2\int d\omega \sum_{N=0}^{\infty}(N+1)\frac{\partial n_P(\omega)}{\partial\omega}\left[L_N^R(\omega)L_{N+1}^A(\omega)-L_{N+1}^R(\omega)L_N^A(\omega)\right]\\
&\left[\psi\left(\frac{1}{2}-\frac{i\omega}{4\pi T_0}+\frac{\Omega_c(N+1/2)}{4\pi T_0}\right)-\psi\left(\frac{1}{2}-\frac{i\omega}{4\pi T_0}+\frac{\Omega_c(N+3/2)}{4\pi T_0}\right)\right.\\
&\left.+\psi\left(\frac{1}{2}+\frac{i\omega}{4\pi T_0}+\frac{\Omega_c(N+1/2)}{4\pi T_0}\right)-\psi\left(\frac{1}{2}+\frac{i\omega}{4\pi T_0}+\frac{\Omega_c(N+3/2)}{4\pi T_0}\right)\right]\\
&\left[\Omega_c(N+1/2)\left(\psi\left(\frac{1}{2}-\frac{i\omega}{4\pi T_0}+\frac{\Omega_c(N+1/2)}{4\pi T_0}\right)\right.\right.\\
&\quad\left.-\psi\left(\frac{1}{2}+\frac{i\omega}{4\pi T_0}+\frac{\Omega_c(N+1/2)}{4\pi T_0}\right)\right)\\
&\quad-\Omega_c(N+3/2)\left(\psi\left(\frac{1}{2}-\frac{i\omega}{4\pi T_0}+\frac{\Omega_c(N+3/2)}{4\pi T_0}\right)\right.\\
&\quad\left.\left.-\psi\left(\frac{1}{2}+\frac{i\omega}{4\pi T_0}+\frac{\Omega_c(N+3/2)}{4\pi T_0}\right)\right)\right],
\end{aligned}
\tag{34}
$$

where ε_{ij} is the anti-symmetric tensor. The second contribution generated by the terms with the derivative $\partial n_F(\varepsilon)/\partial\varepsilon$ corresponds to the diagram with three Cooperons shown in Fig. 6a:

$$
\begin{aligned}
(j^{con}_{e\ i})^2 &= -\varepsilon_{ij}\frac{e\nabla_j T}{4\pi^2 T_0}\nu\int d\omega\sum_{N=0}^{\infty}\Omega_c(N+1)\left\{\frac{1}{4}L_N^R(\omega)\frac{\partial n_P(\omega)}{\partial\omega}\left[\frac{i\omega+\Omega_c(N+1/2)}{4\pi T_0}\right.\right.\\
&\times\left(\psi'\left(\frac{1}{2}+\frac{i\omega}{4\pi T_0}+\frac{\Omega_c(N+1/2)}{4\pi T_0}\right)-\psi'\left(\frac{1}{2}-\frac{i\omega}{4\pi T_0}+\frac{\Omega_c(N+1/2)}{4\pi T_0}\right)\right)
\end{aligned}
$$

$$+\frac{i\omega+\Omega_c(N+3/2)}{\Omega_c}\left(\psi\left(\frac{1}{2}+\frac{i\omega}{4\pi T_0}+\frac{\Omega_c(N+1/2)}{4\pi T_0}\right)-\psi\left(\frac{1}{2}-\frac{i\omega}{4\pi T_0}+\frac{\Omega_c(N+1/2)}{4\pi T_0}\right)\right.$$
$$\left.\left.-\psi\left(\frac{1}{2}+\frac{i\omega}{4\pi T_0}+\frac{\Omega_c(N+3/2)}{4\pi T_0}\right)+\psi\left(\frac{1}{2}-\frac{i\omega}{4\pi T_0}+\frac{\Omega_c(N+3/2)}{4\pi T_0}\right)\right)\right]$$
$$+\frac{i}{2}L_N^A n_P(\omega)\left[\frac{i\omega+\Omega_c(N+1/2)}{(4\pi T_0)^2}\psi^{(2)}\left(\frac{1}{2}+\frac{i\omega}{4\pi T_0}+\frac{\Omega_c(N+1/2)}{4\pi T_0}\right)+\frac{i\omega+\Omega_c(N+3/2)}{4\pi T_0\Omega_c}\right.$$
$$\left.\times\left(\psi'\left(\frac{1}{2}+\frac{i\omega}{4\pi T_0}+\frac{\Omega_c(N+1/2)}{4\pi T_0}\right)-\psi'\left(\frac{1}{2}+\frac{i\omega}{4\pi T_0}+\frac{\Omega_c(N+3/2)}{4\pi T_0}\right)\right)\right]$$
$$\left.+(N+1/2)\leftrightarrow(N+3/2)\right\}+c.c. \tag{35}$$

The notation $(N+1/2)\leftrightarrow(N+3/2)$ means that L_N is replaced by L_{N+1} and in all the terms $N+1/2$ should be replaced by $N+3/2$ and the other way around. Notice that there are no contributions proportional to the derivative of the distribution function which can be attributed to the diagram shown in Fig 6b.

Next, we discuss the group of terms that are proportional to $n_F(\varepsilon)$. The diagrammatic interpretation of these terms, which are generated by Eq. (25), includes all three diagrams presented in Fig. 6. However, one may check that the contributions from the diagrams shown in Fig. 6b and c are cancelled by a part of the contribution from the diagram given in Fig. 6a. The remaining contribution is:

$$(j_{e\ i}^{con})^3=-i\varepsilon_{ij}\frac{e\nabla_j T}{4\pi^2 T_0}\nu\int d\omega\sum_{N=0}^{\infty}(N+1)n_P(\omega)\left\{\left[L_N^R(\omega)+L_{N+1}^R(\omega)\right]\right.$$
$$\times\left[\psi\left(\frac{1}{2}-\frac{i\omega}{4\pi T_0}+\frac{\Omega_c(N+1/2)}{4\pi T_0}\right)-\psi\left(\frac{1}{2}-\frac{i\omega}{4\pi T_0}+\frac{\Omega_c(N+3/2)}{4\pi T_0}\right)\right]$$
$$+\frac{\Omega_c}{4\pi T}L_N^R(\omega)\psi'\left(\frac{1}{2}-\frac{i\omega}{4\pi T_0}+\frac{\Omega_c(N+1/2)}{4\pi T_0}\right)$$
$$\left.+\frac{\Omega_c}{4\pi T_0}L_{N+1}^R(\omega)\psi'\left(\frac{1}{2}-\frac{i\omega}{4\pi T_0}+\frac{\Omega_c(N+3/2)}{4\pi T_0}\right)\right\} \tag{36}$$

Further analysis of $\mathbf{j}_e^{con}$ at arbitrary temperatures and magnetic fields shows that in $(j_{e\ i}^{con})^1$ and $(j_{e\ i}^{con})^2$ the integration over the frequency accumulates at $\omega\sim T\ll 1/\tau$ where τ is the mean free time of the electrons. As a consequence of the narrow range of the integration, the final expressions for these two contributions vanish in the limit $T\to 0$. On the other hand, in $(j_{e\ i}^{con})^3$ the integration over the frequency is not limited to small frequencies and hence the outcome of the integration depends logarithmically on the scattering rate $1/\tau$ which acts as an ultraviolet cutoff. In addition, as the temperature goes to zero there is even a more serious problem with this term, because its pre-factor is proportional to Ω_c/T_0. Such a dependence on the temperature violates the third law of thermodynamics. [The connection between the third law of thermodynamics and the Nernst effect has been discussed in the Introduction.] We shall see that the dangerous parts in $(j_{e\ i}^{con})^3$ are cancelled out by the magnetization current that up to now we have not yet considered.

5 The Magnetization Current and the Third Law of Thermodynamics

In this section we examine the magnetization current given in Eq. (8). In general, we need to insert the ∇T-dependent part of the Green function, $\hat{G}_{loc-eq} + \hat{G}_{\nabla T}$, into Eq. (8). Since after the averaging over the disorder $\hat{G}_{\nabla T}(\mathbf{r} \to \mathbf{r}', \varepsilon)$ is translational invariant, it is clear that this part of the Green function does not contribute to the magnetization current. On the other hand, the explicit dependence of the local equilibrium Green function on the center of mass coordinate leads to a non-zero contribution to the magnetization current:

$$\mathbf{j}_e^{mag} = 2ic\nabla_{\mathbf{R}} \times \mathbf{M}(\mathbf{R}) \int \frac{d\varepsilon}{2\pi} \lim_{\rho \to 0} \varepsilon \frac{\mathbf{R}\nabla T}{T_0} \frac{\partial g^<(\rho, \varepsilon; \mathbf{A}, imp)}{\partial \varepsilon} \tag{37}$$

Thus, $\hat{G}_{\nabla T}$ and $\hat{G}_{loc-eq}$ are complementary to each other; while the first contributes only to $\mathbf{j}_e^{con}$, the other one fully determines $\mathbf{j}_e^{mag}$. One should recall that we are looking for a current that does not vanish after spatial averaging, i.e., after integration with respect to the center of mass coordinate $\mathbf{R}$. Since in the process of averaging over $\mathbf{R}$ we may integrate by parts, the magnetization current can be written as

$$\mathbf{j}_{e\ i}^{mag} = 2i\varepsilon_{ij}cM_z \lim_{\rho \to 0} \int \frac{d\varepsilon}{2\pi} \frac{\nabla_j T}{T_0} \tilde{g}^<(\rho, \varepsilon). \tag{38}$$

Here we integrated by parts over the frequency as well. One may recognize that $\mathbf{j}_e^{mag}$ is directly related to the magnetization density at equilibrium:

$$\mathbf{j}_{e\ i}^{mag} = -\varepsilon_{ij} c \langle M_z \rangle \frac{\nabla_j T}{T_0}. \tag{39}$$

The result demonstrates the strength of the quantum kinetic approach. This method provides a way to derive both components of the current quantum mechanically without engaging any thermodynamical arguments.

Actually, at this point one may employ in Eq. (39) the known expression for the magnetization in the presence of superconducting fluctuation. Still, since we are interested in the interplay between the quasi-particle excitations and the fluctuations of the superconducting order parameter, let us derive the expression for the first order correction to the magnetization induced by the fluctuations starting with Eq. (38). Using the standard identities for the Keldysh Green function at equilibrium, one gets:

$$\begin{aligned}\mathbf{j}_{e\ i}^{mag} = ic\varepsilon_{ij} \frac{\nabla_j T}{T_0} \lim_{\mathbf{r}' \to \mathbf{r}} \int \frac{d\varepsilon}{2\pi} d\mathbf{r}_1 d\mathbf{r}_2 \left[M_z(\mathbf{r}) + M_z(\mathbf{r}')\right] n_F(\varepsilon) \\ \times \left[g_0^A(\mathbf{r} - \mathbf{r}_1, \varepsilon; \mathbf{A}, imp)\sigma^A(\mathbf{r}_1 - \mathbf{r}_2, \varepsilon; \mathbf{A}, imp) g_0^A(\mathbf{r}_2 - \mathbf{r}', \varepsilon; \mathbf{A}, imp)\right. \\ \left. - g_0^R(\mathbf{r} - \mathbf{r}_1, \varepsilon; \mathbf{A}, imp)\sigma^R(\mathbf{r}_1 - \mathbf{r}_2, \varepsilon; \mathbf{A}, imp) g_0^R(\mathbf{r}_2 - \mathbf{r}', \varepsilon; \mathbf{A}, imp)\right].\end{aligned} \tag{40}$$

Here, for convenience, we returned to the initial coordinates. Next, we use the fact that the equilibrium Green function in the absence of fluctuations satisfies the following operator identity $-\partial g_0^{R,A}/\partial \mathbf{H} = g_0^{R,A}\mathbf{M}g_0^{R,A}$. Therefore, the expression for the magnetization current can be rewritten as

$$\mathbf{j}^{mag}_{e\ i} = -2ic\varepsilon_{ij}\frac{\nabla_j T}{T_0}\lim_{\mathbf{r}'\to\mathbf{r}}\int\frac{d\varepsilon}{2\pi}d\mathbf{r}_1 n_F(\varepsilon)\left[\frac{\partial g_0^A(\mathbf{r}-\mathbf{r}_1,\varepsilon;\mathbf{A},imp)}{\partial H_z}\right.$$
$$\left.\times\sigma^A(\mathbf{r}_1-\mathbf{r}',\varepsilon;\mathbf{A},imp)-\frac{\partial g_0^R(\mathbf{r}-\mathbf{r}_1,\varepsilon;\mathbf{A},imp)}{\partial H_z}\sigma^R(\mathbf{r}_1-\mathbf{r}_2,\varepsilon;\mathbf{A},imp)\right]. \quad (41)$$

Finally, using the explicit expression for the self-energy and rearranging all the terms we reformulate the expression for the magnetization current in terms of the propagator of the superconducting fluctuations:

$$\mathbf{j}^{mag}_{e\ i} = -ic\varepsilon_{ij}\frac{\nabla_j T}{T_0}\lim_{\mathbf{r}'\to\mathbf{r}}\int\frac{d\omega}{2\pi}d\mathbf{r}_1 n_P(\omega)\left[\frac{\partial \Pi^R(\mathbf{r}-\mathbf{r}_1,\omega;\mathbf{A},imp)}{\partial H_z}\right.$$
$$\left.\times Ł^R(\mathbf{r}_1-\mathbf{r}',\omega;\mathbf{A},imp)-\frac{\partial \Pi^A(\mathbf{r}-\mathbf{r}_1,\omega;\mathbf{A},imp)}{\partial H_z}L^A(\mathbf{r}_1-\mathbf{r}_2,\omega;\mathbf{A},imp)\right]$$
$$= -ic\varepsilon_{ij}\frac{\nabla_j T}{T_0}\frac{\partial}{\partial H_z}\lim_{\rho\to 0}\int\frac{d\omega}{2\pi}n_P(\omega)\left[\ln L_R^{-1}(\rho,\omega;\mathbf{A},imp)-\ln L_A^{-1}(\rho,\omega;\mathbf{A},imp)\right]. \quad (42)$$

The transition from Eq. (40) to the last line in Eq. (42) is illustrated in Fig. 8. Averaging over the disorder and transforming from the expression for the propagator as a function of the coordinates to the basis of Landau levels, one obtains the known expression for the correction to the magnetization in the lowest order with respect to the fluctuations [12, 30]:

$$\mathbf{j}^{mag}_{e\ i} = \varepsilon_{ij}\frac{\nabla_j T}{T_0}\frac{\partial}{\partial H}\frac{eH}{\pi}\int\frac{d\omega}{2\pi i}\sum_{N=0}^{\infty}n_P(\omega)\left[\ln\left(L_N^R(\omega)\right)^{-1}-\ln\left(L_N^A(\omega)\right)^{-1}\right]. \quad (43)$$

The discussion of the higher order corrections to the magnetization has been given in Ref. [30].

Similar to $(\mathbf{j}_e^{con})^3$ in Eq. (36), the integration over the frequency in the magnetization current is restricted by the scattering rate, and at low temperature $\mathbf{j}_e^{mag}$ also diverges as Ω_c/T. The opposite sign of the magnetization current relative to $(\mathbf{j}_e^{con})^3$ suggests that these dangerous parts may cancel each other making the Nernst signal

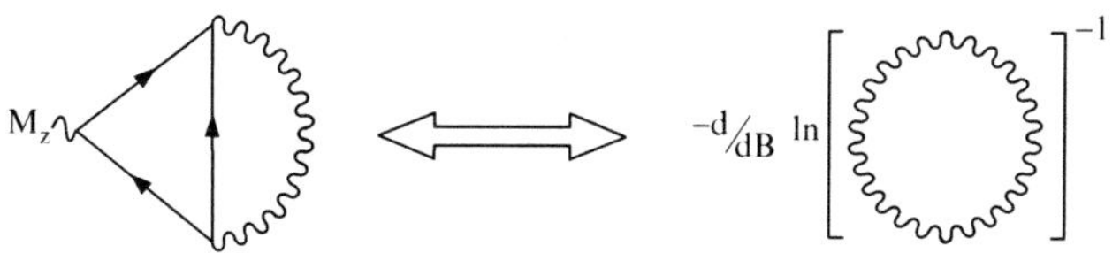

Fig. 8 An illustration of the relation between the magnetization current term that is obtained from the local equilibrium Green function and the thermodynamic diagram for the magnetization.

compatible with the third law of thermodynamics. Another hint for this cancellation is the similar analytical structure of $\mathbf{j}_e^{mag}$ and $(\mathbf{j}_e^{con})^3$. [All the terms in Eq. (36) and in Eq. (43) are a product of either retarded or advanced functions only.] We have found a way [25] to verify that the diverging parts of the magnetization current indeed identically cancel out the diverging parts of $(\mathbf{j}_e^{con})^3$. We have obtained that the total current is independent of τ in the whole temperature range $T \ll 1/\tau$. As a result, the Nernst signal is regular at $T \to 0$. Moreover, the contributions which are constant with respect to the temperature also vanish and the remaining terms are linear in T.

6 The Phase Diagram for the Nernst Effect – Comparison Between the Theoretical Results and the Experiment

In the following section we present the theoretical expressions for the transverse Peltier coefficient for a superconducting film in the normal state for various regions of the temperature and the magnetic field. The phase diagram for the Peltier coefficient is plotted in Fig. 9. In the area below the line $\ln(T/T_c(H)) = \Omega_c/4\pi T$ the Landau level quantization of the superconducting fluctuations becomes essential. The line $\ln(H/H_{c_2}(T)) = 4\pi T/\Omega_c$ separates the regions of classical and quantum fluctuations. From now on T_0 is replaced by T which represents the spatially-averaged temperature.

For a small magnetic field, $\Omega_c \ll T$, close to the transition temperature ($T \approx T_c$) the leading contribution to α_{xy} is given by the Aslamazov–Larkin term (see Fig. 6 c) and the magnetization current:

$$\alpha_{xy} \approx \frac{e\Omega_c}{192T \ln T/T_c(H)}. \tag{44}$$

In the previous section we have discussed in detail the importance of the magnetization current in cancelling the quantum contributions to the Nernst signal. In the vicinity of T_c one can interpret the expression in Eq. (44) in terms of the classical picture in which the Cooper pairs with a finite lifetime are responsible

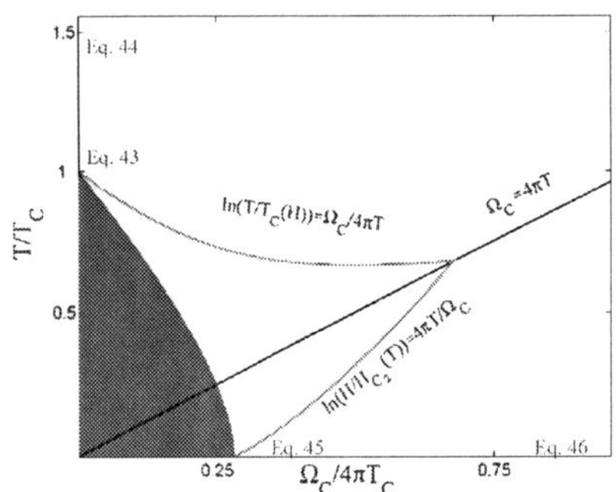

Fig. 9 The phase diagram for the Peltier coefficient α_{xy}. We indicate the equations in the text which give the corresponding expressions for α_{xy} in the different limits. $\Omega_c = 4eHD/c$ is the cyclotron frequency for the fluctuations of the superconducting order parameter in the diffusive regime.

for the thermoelectric current. The magnetization current is just equal to $-2/3$ of the leading order contribution from the Aslamazov–Larkin term. Note that unlike the electric conductivity, σ_{xx}, for which the anomalous Maki-Thompson [11] and the Aslamazov–Larkin terms yield comparable corrections, the contribution from the anomalous Maki-Thompson term to the Nernst signal is $\sim (T/\varepsilon_F)^2 \ll 1$ smaller than the one given by Eq. (44). Therefore, it is natural that in the vicinity of T_c our result coincides with the expression [12, 31] obtained phenomenologically from the time dependent Ginzburg–Landau equation (TDGL).

When temperature is increased further away from the critical temperature, the sum of the contributions to the transverse Peltier coefficient from all the diagrams and the magnetization current yields:

$$\alpha_{xy} \approx \frac{e\Omega_c}{24\pi^2 T \ln T/T_c}. \tag{45}$$

A comparison between the transverse Peltier coefficient in the vicinity of T_c (Eq. (44)) and far from the transition (Eq. (45)) reveals that the two expressions differ by a numerical coefficient. The similarity between α_{xy} in the two different limits is not seen in paraconductivity and it is a consequence of the cancellation of the quantum corrections by the magnetization current. Away from the critical region $T \gg T_c$, the quantum nature of the fluctuations reveals itself in contributions to $(\mathbf{j}_e^{con})^3$ and $\mathbf{j}_e^{mag}$ that contain an integration over a wide interval of frequencies between T and $1/\tau$. As a result, these terms become of the order $\ln(\ln 1/T\tau) - \ln(\ln T/T_c)$. However, we have obtained that these τ-dependent terms in $(\mathbf{j}_e^{con})^3$ and $\mathbf{j}_e^{mag}$ cancel each other. The Peltier coefficient far from T_c demonstrates how the third law of thermodynamics constrains the magnitude of the Nernst signal not only at $T \to 0$ but also at high temperatures, $T \gg T_c$.

The comparison of our result with the experimental observation of Ref. [5] for two $Nb_{0.15}Si_{0.85}$ films of thicknesses 35 and 12.5 nm (with critical temperatures $T_c = 380$ and 165 mK, correspondingly) is given in Fig. 10. The Peltier coefficient depends on the mean field temperature of the superconducting transition, T_c^{MF}, and on the diffusion coefficient through Ω_c. Throughout the paper we fit the data using the same diffusion coefficient $D = 0.187\,\mathrm{cm}^2/\mathrm{s}$ which is within the measurement accuracy of the value that has been extracted from the experiment in Ref. [5]. We take $T_c^{MF} = 385$ mK for the first film which is slightly higher than the measured critical temperature anticipating a small suppression of the temperature of the transition by fluctuations while for the second film we use $T_c^{MF} = T_c$.

The cancellation of the terms proportional to Ω_c/T in the limit $T \ll \Omega_c$ has been described in the previous section. (Without this cancellation we would get a finite Nernst effect in the limit $T \to 0$ and the third law of thermodynamics would be violated.) After the cancellation, the remaining contributions to α_{xy} in the limit $T \to 0$ are linear in the temperature:

$$\alpha_{xy} \approx -\frac{eT \ln 3}{3\Omega_c (\ln H/H_{c_2}(T))^2} \qquad \text{for } H \approx H_{c_2}, \tag{46}$$

and

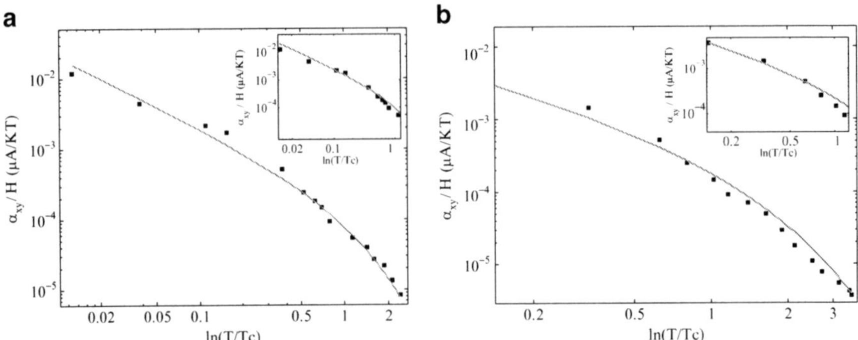

Fig. 10 The transverse Peltier coefficient α_{xy} divided by the magnetic field H as a function of $\ln T/T_c$ in the limit $\mathbf{H} \to 0$ for films of thicknesses (**a**) 35 nm and (**b**) 12.5 nm. The experimental data of Ref. [5] is presented by the *black squares* and the *solid line* corresponds to the theoretical curve given by Eq. (45). The inset presents the fitting of the data in the vicinity of T_c with Eq. (44).

$$\alpha_{xy} \approx \frac{2eT}{3\Omega_c \ln H/H_{c_2}} \qquad \text{for } H \gg H_{c_2}. \tag{47}$$

Similar to the limit $\Omega_c < T$ the integrals determining the final expression for α accumulate at low frequency. This situation is not typical for fluctuations induced by a quantum phase transition. Notice that α_{xy} changes its sign in this region. Since the transverse signal is non-dissipative, the sign of the effect is not fixed. As it has been already mentioned, in the vicinity of T_c for $\Omega_c \ll T$, the main contribution to the Peltier coefficient is from the Aslamazov Larkin term and the magnetization current. The magnetization current is opposite in sign to the Aslamazov Larkin terms and equals 2/3 of it. When crossing to the region $\ln(T/T_c(H)) < \Omega_c/T$ (see the phase diagram in Fig. 9) the contribution from the magnetization current grows. To the first order in Ω_c/T the magnetization current cancels the Aslamazov Larkin term and the Peltier coefficient turns out to be proportional to $O[(\Omega_c/T)^2]$. Lowering further the temperature and increasing the magnetic field one reaches the region $\Omega_c > T$ and $\ln(H/H_{c_2}(T)) < T/\Omega_c$. In this region the magnetization current becomes dominant. Since the magnetization current gives a contribution that is opposite in sign to the Aslamazov Larkin term, we obtain that the Peltier coefficient is negative.

In Fig. 11 we plot the curve for the Peltier coefficient as a function of the magnetic field at a temperature higher than T_c. Figure 11 demonstrates the agreement between the theoretical expressions and the experimental observation for a broad range of magnetic fields. In addition, we show that the experimental data is well described by Eq. (44) in the limit of vanishing magnetic field (see inset of Fig. 11). Since Eq. (44) is valid in the limit $\Omega_c \ll T$, it can describe only the first few points in the measurement. In order to fit the entire range of the magnetic field we had to include higher order terms of Ω_c/T. For that we needed to sum the contributions from all diagrams and the magnetization current. We performed the calculation assuming that $\ln(T/T_c(H)) \ll 1$; therefore the theoretical curve starts to deviate from the measured data when $\ln(T/T_c(H))$ is no longer small ($H \approx 1$ T).

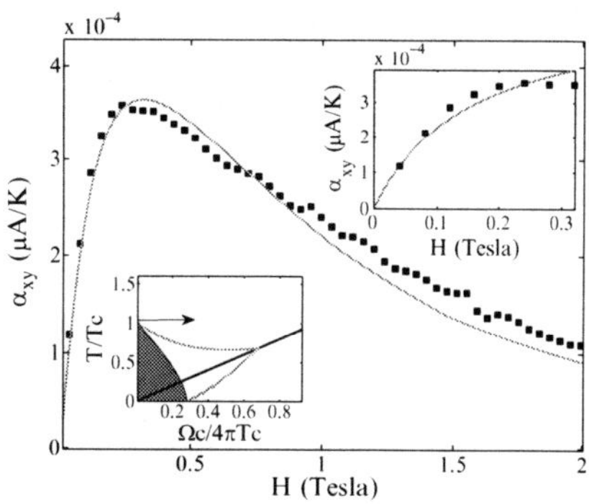

Fig. 11 The transverse Peltier coefficient α_{xy} as a function of the magnetic field measured at $T = 410$ mK. The *black squares* correspond to the experimental data of Ref. [8] while the *solid line* describes the theoretical result. The *arrow* on the phase diagram illustrates the direction of the measurement. In the inset the low magnetic field data is fitted with the theoretical curve given by Eq. (44).

7 Summary

We have demonstrated that the contribution from the fluctuations of the superconducting order parameter to the Nernst effect in disordered films is dominant and can be observed far away from the transition. We have shown that the important role of the magnetization current is in cancelling the quantum contributions, thus making the Nernst signal compatible with the third law of thermodynamics. The third law of thermodynamics constrains the magnitude of the Nernst signal not only at low temperatures, but also far from T_c. As a consequence of this constraint, the phase diagram is less rich and diverse than one expects in the vicinity of a quantum phase transition.

The Nernst effect provides an excellent opportunity to test the use of the quantum kinetic equation in the description of thermoelectric transport phenomena. We have described the main steps in the derivation of the Nernst effect using the quantum kinetic equation. In this method one gets automatically all contributions to the Nernst coefficient as a response to the temperature gradient, in particular the one from the magnetization current. This is an advantage of the quantum kinetic approach but not the only one. This method allows us to verify the Onsager relations and also to demonstrate in a simple way the validity of the Wiedemann–Franz law in the Fermi liquid theory [16, 25]. We should remark that our results are different in few aspects from the expressions for the Peltier coefficient recently obtained using the Kubo formula in Ref. [32]. The striking agreement between our results and the experiment in the different limits (and the fact that we have reproduced the phenomenological result of the TDGL [31]) reinforces our confidence in the correctness of our method.

Appendix: The Role of Particle-hole Asymmetry and the Constant Density of States Approximation in Determining the Thermoelectric Transport Coefficients

In metallic conductors the quasi-particle excitations yield a negligible contribution to the Nernst effect and to its counterpart, the Ettingshausen effect. Let us consider a system with an electron and hole conducting bands that have a particle-hole symmetry, i.e., the bands are identical and the two species of particles differ only in their charge. We shall describe the deviation of the distribution functions of the two species (δf_e and δf_h) from their equilibrium value in the linear response to a temperature gradient. For that we write the classical Boltzmann equation in the relaxation time approximation:

$$\frac{\delta f_{e,h}(\varepsilon_{\mathbf{k}})}{\tau} = \frac{\partial f_0(\varepsilon_{\mathbf{k}})}{\partial T}\mathbf{v}_{\mathbf{k}}\nabla T \mp \frac{e\mathbf{v}_{\mathbf{k}} \times \mathbf{H}}{c}\frac{\partial f_0(\varepsilon_{\mathbf{k}})}{\partial \mathbf{k}}. \tag{48}$$

Here the equilibrium (Fermi–Dirac) distribution function is denoted by $f_0(\varepsilon_k)$, $\mathbf{v}_k$ is the velocity of the particles.

The electric current is the sum of the electric currents due to the electrons and the holes:

$$\mathbf{j}_e^{total} = -2e\int\frac{d\mathbf{k}}{(2\pi)^d}\mathbf{v}_{\mathbf{k}}\delta f_e(\varepsilon_{\mathbf{k}}) + 2e\int\frac{d\mathbf{k}}{(2\pi)^d}\mathbf{v}_{\mathbf{k}}\delta f_h(\varepsilon_{\mathbf{k}}). \tag{49}$$

Notice that the factor of 2 results from the sum over the two spin directions. For simplicity we only examine the limit of vanishingly small magnetic field. However the conclusions apply also for an arbitrary value of the magnetic field. In order to determine whether a current vanishes in the particle-hole symmetric system we just need to count the powers of the electric charge; an odd power means cancellation of the two contributions to the current.

We start from the longitudinal electric current induced by the gradient of the temperature. In the limit $\mathbf{H} \to 0$, the longitudinal current is independent of the magnetic field:

$$\mathbf{j}_e^x = 2e\int\frac{d\mathbf{k}}{(2\pi)^d}\frac{\partial f_0(\varepsilon_{\mathbf{k}})}{\partial \varepsilon_{\mathbf{k}}}\left[\varepsilon_{\mathbf{k}}D_e - \varepsilon_{\mathbf{k}}D_h\right]\frac{\nabla_x T}{T_0} = 0, \tag{50}$$

where $D_e = D_h \equiv D = v_k^2\tau/d$ with d the dimension of the system. Since the expression includes only one power of the charge, there is no longitudinal electric current unless particle-hole asymmetry is introduced.

The transverse current is obtained when the Lorentz force in the Boltzmann equation acts on the distribution function. Therefore, the expression for the transverse current contains an additional power of the charge:

$$\mathbf{j}_e^y = 2e\int\frac{d\mathbf{k}}{(2\pi)^d}\frac{\partial f_0(\varepsilon_{\mathbf{k}})}{\partial \varepsilon_{\mathbf{k}}}\varepsilon_{\mathbf{k}}D\left[\omega_c\tau - (-\omega_c\tau)\right]\frac{\nabla_x T}{T_0} \neq 0. \tag{51}$$

The additional charge enters through the cyclotron frequency $\omega_c = eH/m^*c$. The even power of the charge means that the particle-hole symmetry does not constrain the Nernst effect.

Next, we shall look at the contribution for the transverse electric current of the electrons. We shall use the approximation of a constant density of states. Under this approximation the expression for the transverse current is:

$$\mathbf{j}_e^y = 2ev_0 D(\omega_c\tau)\frac{\nabla T}{T_0}\int d\varepsilon_{\mathbf{k}}\frac{\partial f_0(\varepsilon_{\mathbf{k}})}{\partial\varepsilon_{\mathbf{k}}}\varepsilon_{\mathbf{k}}. \tag{52}$$

Since near the Fermi energy the integrand is an odd function of the energy, the resulting current is zero. Therefore, under the approximation of a constant density of states at the Fermi energy, which is a standard approximation for the Fermi liquid theory, this contribution vanishes completely [9]. One may conclude that in metallic systems with high Fermi energy the contribution of the quasi-particles to the Nernst signal includes a small factor related to the deviation from the constant density of states which is of the order T/ε_F. In semi-metals like *Bi* where the constant density of states approximation cannot be used, a large Nernst signal has been measured [33].

Let us compare the magnitudes of the transverse Peltier coefficient generated by the quasi-particles and by the superconducting fluctuations in a film of thickness a. The first is of the order $\sim(\omega_c\tau)evDaT/\varepsilon_F$ for $\omega_c\tau \ll 1$ while the second one is of the order $\sim e\Omega_c/T$ for $\Omega_c/T \ll 1$ and $\sim eT/\Omega_c$ for the opposite limit. Thus in the limit of vanishing small magnetic field the ratio between the contribution of the quasi-particles and the fluctuations is $\alpha_{xy}^{qp}/\alpha_{xy}^{fl} \sim (k_F a)T^2\tau/\varepsilon_F$. At higher magnetic fields (but still in the limit $\omega_c\tau \ll 1$) this ratio becomes $\alpha_{xy}^{qp}/\alpha_{xy}^{fl} \sim (k_F a)\varepsilon_F\tau(\omega_c\tau)^2 \ll 1$. Under the condition of the experiment [4, 5], the ratio $\alpha_{xy}^{qp}/\alpha_{xy}^{fl} \ll 1$ up to $T \lesssim 100T_c$ and $H \lesssim 100H_{c_2}$. The reason why the Nernst signal generated by the superconducting fluctuations dominates the one produced by the quasi-particles has been explained in the end of Section 3.

Acknowledgements We thank H. Aubin, K. Behnia, J. Sinova, M. A. Skvortsov and A. A. Varlamov for valuable discussions. The research was supported by the US-Israel BSF and the Minerva Foundation.

References

1. Zaanen, J., A black hole full of answers, *Nature* **448**, 1000–1001 (2007).
2. Xu, Z. A., et al., Vortex-like excitations and the onset of superconducting phase fluctuation in underdoped $La_{2-x}Sr_xCuO_4$, *Nature* **406**, 486–488 (2000).
3. Wang, Y., Li, L., Ong. N. P., Nernst effect in high-Tc superconductors, *Phys. Rev. B* **73**, 024510 (2006).

4. Pourret, A. et al., Observation of the Nernst signal generated by fluctuating Cooper pairs, *Nat. Phys.* **2**, 683–686 (2006).
5. Pourret, A. et al., Length scale for the superconducting Nernst signal above Tc in $Nb_{0.15}Si_{0.85}$, *Phys. Rev. B* **76**, 214504 (2007).
6. Anderson, P. W., Two new vortex liquids, *Nat. Phys.* **3**, 160–162 (2007).
7. Podolsky, D., Raghu, S., Vishwanath, A., Nernst effect and diamagnetism in phase fluctuating superconductors, *Phys. Rev. Lett.* **99**, 117004 (2007).
8. Mukerjee, S., Huse, D. A., Nernst effect in the vortex-liquid regime of a type-II superconductor, *Phys. Rev. B* **70**, 014506 (2004).
9. Sondheimer, E., The theory of the galvanomagnetic and thermomagnetic effect in metals, *Proc. R. Soc. (London)* **A193**, 484–512 (1948).
10. Aslamazov, L. G., Larkin, A. I., Effect of fluctuations on the properties of a superconductor above the critical temperature, *Fiz. Tverd. Tela* **10**, 1104 (1968) [*Sov. Phys. Solid State* **10**, 875 (1968)].
11. Maki, K., Critical fluctuation of the order parameter in a superconductor. I, *Prog. Theor. Phys.* **40**, 193–200 (1968).
12. Larkin, A. I., Varlamov, A. A., *Theory of fluctuations in superconductors*, (Carendon, Oxford, 2005).
13. Kubo, R., Statistical-mechanical theory of irreversible processes. I. General theory and simple applications to magnetic and conduction problems, *J. Phys. Soc. Jpn.* **12**, 570–586 (1957).
14. Luttinger, J. M., Theory of thermal transport coefficients, *Phys. Rev.* **135**, A1505–1514 (1964).
15. Jonson, M., Mahan, G. D., Mott's formula for the thermopower and the Wiedemann-Franz law, *Phys. Rev. B* **21**, 4223 (1980).
16. Michaeli, K., Finkel'stein, A. M., Quantum kinetic approach for studying thermal transport in the presence of electron-electron interactions and disorder, to be published.
17. Obraztsov, Y. N., The thermal EMF of semiconductors in a quantizing magnetic field, *Fiz. Tverd. Tela.* **7**, 573–581 (1965) [*Sov. Phys. Solid State* **7**, 455–461 (1965)].
18. Smrcka, L., Streda, P., Transport coefficients in strong magnetic fields, *J. Phys. C: Solid State Phys.* **10**, 2153–2161 (1977).
19. Cooper, N. R., Halperin, B. I., Ruzin, I. M., Thermoelectric response of an interacting two-dimensional electron gas in a quantizing magnetic field, *Phys. Rev. B* **55**, 2344–2359 (1997).
20. Keldysh, L. V., Diagram thechnique for nonequilibrium processes, *Zh. Eksp. Teor. Fiz.* **47**, 1515–1527 (1964) [*Sov. Phys. JETP* **20**, 1018–1026 (1965)].
21. Rammer, J., Smith, H., Quantum field-theoretical methods in transport theory of metals, *Rev. Mod. Phys.* **58**, 323–359 (1986).
22. Haug, H., Jauho, A.-P., *Quantum Kinetics in Transport and Optics of Semiconductors*, (Springer, Berlin, 1996).
23. Hu, C.-R., Transport entropy of vortices in superconductors with paramagnetic impurities, *Phys. Rev. B* **14**, 4834–4853 (1976).
24. Sergeev, A., Reizer, M. Y., Mitin, V., Heat current in the magnetic field: Nernst-Ettingshausen effect above the superconducting transition, *Phys. Rev. B* **77**, 064501 (2008).
25. Michaeli, K., Finkel'stein, A. M., The quantum kinetic equation approach to the calculation of the Nernst effect, to be published.
26. Khodas, M., Finkelstein, A. M., Hall coefficient in an interacting electron gas, *Phys. Rev. B.* **68**, 155114 (2003).
27. Michaeli, K., Finkel'stein, A. M., Fluctuations of the superconducting order parameter as an origin of the Nernst effect, *Europhys. Lett.* **86**, 27007 (2009).
28. Aronov, A. G., Hikami, S., Larkin, A. I., Gauge invariance and transport properties in superconductors above T_c, *Phys. Rev. B* **51**, 3880 (1995).
29. Laikhtman, B., Altshuler, E. L., Quasiclassical theory of Shubnikov-de Haas effect in 2D electron gas, *Ann. Phys.* **232**, 332 (1994).
30. Galitski, V. M., Larkin, A. I., Superconducting fluctuations at low temperature, *Phys. Rev. B* **63**, 174506 (2001).

31. Ussishkin, I., Sondhi, S. L., Huse, D. A., Gaussian superconducting fluctuations, thermal rransport, and the Nernst effect, *Phys. Rev. Lett.* **89**, 287001 (2002).
32. Serbyn, M. N., Skvortsov, M. A., Varlamov, A. A., Galitski, V., Giant Nernst effect due to fluctuating cooper pairs in superconductors, *arXiv:0806.4427* (2008).
33. Behnia, K., Measson, M.-A., Kopelevich, Y., Oscillating Nernst-Ettingshausen effect in Bismuth across the quantum limit, *Phys. Rev. Lett.* **98**, 166602 (2007).

Quantum Criticality in Heavy Electron Compounds

M. C. Bennett, D. A. Sokolov, M. S. Kim, Y. Janssen, and M. C. Aronson

Abstract Heavy electron systems provide ideal venues to study a range of issues associated with quantum criticality, including unconventional electronic phases, moment formation, and complex phase diagrams with exotic critical phenomena. In the heavy electron antiferromagnets studied so far, magnetic order occurs via a second order phase transition which can be tuned via pressure or field to a quantum critical point. Fermi liquid behavior is found beyond the quantum critical point, and the quasiparticle mass diverges at the quantum critical point, nucleating the moments required to enable magnetic order itself. We review here our experimental results on a new heavy electron system, Yb_3Pt_4, where antiferromagnetic order is weakly first order in zero field, but becomes second order at a critical endpoint with the application of magnetic field. No divergence of the quasiparticle mass is observed near the quantum critical field, and instead magnetic order is driven by the exchange enhancement of the Fermi liquid itself. These data support the thesis that there are multiple routes to quantum criticality in the heavy electron compounds.

1 Introduction

It is increasingly believed that quantum criticality provides an underlying organization and structure to the phase behavior of virtually all classes of correlated electron systems. Qualitatively new types of collective behavior are found in the vicinity of these quantum critical point, including magnetically mediated superconductivity [1, 2], quasi-ordered phases such as 'spin nematics' [3] and even more exotic phases where the fundamental nature of the electrons are themselves in transition [4]. Such quantum critical points are found in virtually all classes of strongly

M. C. Bennett and M. C. Aronson (also at BNL)
Department of Physics and Astronomy, Stony Brook University, Stony Brook, NY 11974
e-mail: mcbennet@umich.edu, maronson@bnl.gov

D. A. Sokolov, M. S. Kim, and Y. Janssen
Brookhaven National Laboratory, Upton, NY 11973
e-mail: fermiliquid@gmail.com, mskim@bnl.gov, yjanssen@bnl.gov

V. Zlatić and A. C. Hewson (eds.), *Properties and Applications of Thermoelectric Materials*, 243
NATO Science for Peace and Security Series B: Physics and Biophysics,

interacting systems, from complex oxides [5,6], to heavy electron compounds [7–9], to low dimensional conductors [10]. Despite the very different microscopic physics characteristic of these diverse systems, many of the fundamental questions are universal. What are the properties of the quantum critical phenomena, and how are they affected by the specifics of these host systems, such as low dimensionality, proximity to metal–insulator transitions, and competing instabilities? How do the critical phenomena couple to the electronic degrees of freedom, and how are they reflected in the new collective phases found only near quantum critical points? What is the role of quantum critical points in localizing electrons and their moments? Answering these questions remains compelling for experimentalists and theorists alike.

Kondo lattice compounds provide an extremely rich setting for controlled studies of these central issues, since many Kondo lattice compound systems are known, spanning a range of rare earth and actinide moment types, arranged in a number of different crystal structures, with different degeneracies. Many Kondo lattice systems order at low temperatures, either ferromagnetically or antiferromagnetically, while more exotic electronic phases such as magnetically mediated superconductivity can coincide with the suppression of magnetic order itself. Especially attractive is the hierarchy of well-separated energy scales which can be realized in Kondo lattice systems, where scales associated with local physics, such as crystal field energy levels and the Kondo temperature T_K are larger than the energy scales associated with the onset of collective instabilities such as magnetic or superconducting order. Further, these energy scales are generally much smaller in f-electron systems than in transition metal compounds, especially oxides, making it feasible to track the entire sequence of behaviors from high temperature paramagnetism, through Kondo compensation, and to the ordered ground state in a single experiment.

The Kondo lattice systems provide a much-needed flexibility as experimental hosts for investigations of the physics associated with magnetic quantum critical points. It is especially significant that there is a convenient and well tested framework for classifying these rare earth and actinide based intermetallic compounds, i.e. the Doniach phase diagram [11]. The fundamental physics captured by this simple model is that the exchange coupling J between conduction electrons and electrons localized in f-orbitals, supplemented by the conduction electron density of states ρ control both the Kondo temperature $T_K \sim \rho^{-1} \exp(-1/\rho J)$ and the onset temperature for antiferromagnetic or ferromagnetic order $T_{N,C} \sim J^2$. It is the implied competition between T_K and $T_{N,C}$ that underlies the phase diagram itself, and if the relative sizes of these two scales can be tuned by pressure, magnetic field, or composition, magnetic order can be eventually suppressed to zero temperature at a quantum critical point (QCP). It is generally found that the paramagnetic phase which occurs beyond the QCP is a Fermi liquid at low temperatures, and in some systems the quasiparticle mass diverges as the QCP is approached [12–14]. Much interest has focused on the critical phenomena associated with the QCP, which are manifest as unconventional temperature dependences in quantities such as the specific heat ($C/T \sim \ln T$), the magnetic susceptibility ($\chi \sim T^{-\delta}$, $\delta < 1$) and the electrical resistivity ($\rho \sim T^{1+\Delta}$) [7–9]. Neutron scattering measurements demonstrate unusual dynamical scaling near the QCP, distinguished by the absence of characteristic energy scales, other than temperature itself [15–17].

Accompanying the evolution of critical phenomena from conventional to quantum critical are profound changes in the underlying electronic behaviors of the heavy electron compounds. In particular, the description of the magnetic moments varies considerably across the phase diagram. In the limit of small J, $T_{N,C} \sim J^2$ is always larger than $T_K \sim \exp(-1/\rho J)$, so the f-electrons are fully localized and have no appreciable compensation by the Kondo effect. As J increases, one possibility is that the f-states become increasingly hybridized with the conduction electrons, leading to a magnetic transition which resembles a spin density wave (SDW) instability of the Fermi surface (FS). The implication here is that the electron which was previously localized on the f-orbital becomes itinerant and is included in the FS which goes from *small* (i.e. not containing the localized f-electron) to *large* containing the previously localized f-electron). A second possibility is that the delocalization of the f-electron occurs at the quantum critical point, possibly driven by the failure of the Kondo effect, evidenced by the divergence of the quasiparticle mass and the related breakdown of the Fermi liquid description itself [18–22]. Direct measurements of the Fermi surface volume indicate that it may jump discontinuously at $T = 0$ as implied by Hall effect measurements of field-tuned $YbRh_2Si_2$ [23], and by de Haas–van Alphen measurements of pressure tuned $CeRhIn_5$ [14]. Alternatively, there are systems in which the zero temperature Fermi surface volume shrinks continuously to zero at the quantum critical point, as demonstrated by quantum oscillation studies of field-tuned $CeRu_2Si_2$ [24].

2 The Yb_3Pt_4 Case

It is fair to say that the full potential of the heavy electron compounds to demonstrate a diversity of routes to quantum criticality and its transformative impact on electronic structure has yet to be tapped, either experimentally or theoretically. It is evident that there are more complex magnetic phase behaviors possible, where both first and second order transitions may compete in ferromagnetic or antiferromagnetic systems. The magnetic instabilities themselves may be local moment like, having a small Fermi surface, or may be better described as collective instabilities of electrons inhabiting a large Fermi surface. Developing new systems to serve as exemplars of these different behaviors is consequently of great importance for sustained progress. We will describe here our experiments on a new compound Yb_3Pt_4, which we believe represents a new type of quantum critical system, distinguished both by its unusual magnetic phase diagram as well as by the dominance of the itinerant electrons in determining the physics near a field-driven quantum critical point. Yb_3Pt_4 displays the same high temperature behavior found in a number of f-electron intermetallics. The dc magnetic susceptibility χ has a Curie–Weiss like temperature dependence (Fig. 1a), and for temperatures larger than $\sim$150 K the moment obtained from the Curie constant is $4.54\,\mu_B$ per Yb, very close to the Hund's rule value for Yb^{3+} [25]. At lower temperatures a modest anisotropy develops, as well as a reduction of the moment, suggesting that a crystal field split manifold of

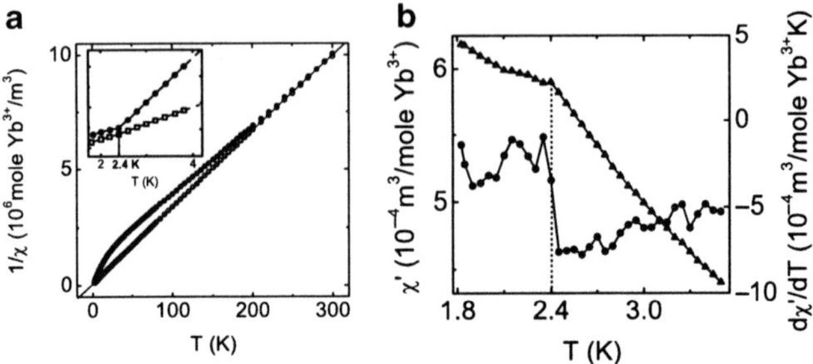

Fig. 1 (**a**) The inverse of $\chi = M/H$, measured with a 2,000 Oe field parallel to the a and c axes. Inset: Same as (**a**) for expanded range of low temperatures near T_N. (**b**) Antiferromagnetic ordering cusp in the ac susceptibility $\chi'(\triangle)$, demonstrating slope discontinuity in $d\chi'/dT$ (•) at T_N.

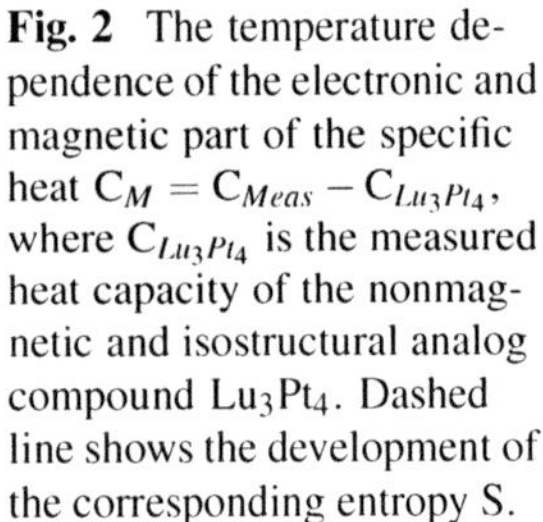

Fig. 2 The temperature dependence of the electronic and magnetic part of the specific heat $C_M = C_{Meas} - C_{Lu_3Pt_4}$, where $C_{Lu_3Pt_4}$ is the measured heat capacity of the nonmagnetic and isostructural analog compound Lu_3Pt_4. Dashed line shows the development of the corresponding entropy S.

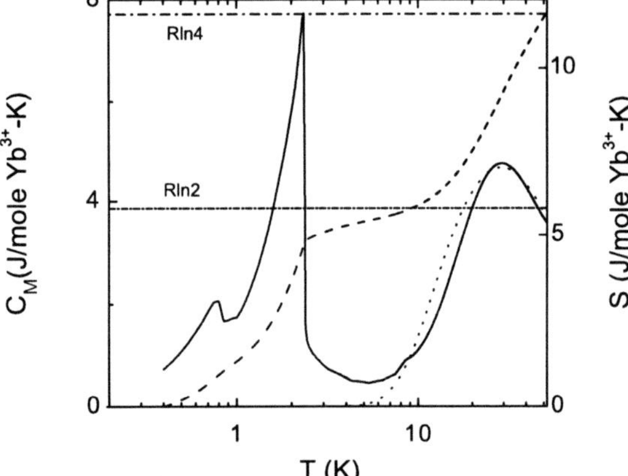

states is being depopulated. Antiferromagnetic order is found at $T_N = 2.4$K, both from the slope discontinuity in $1/\chi$ (Fig. 1a, inset) and from the pronounced cusp in the real part of the ac susceptibility χ'(Fig. 1b). As demonstrated in Fig. 1a, the temperature linearity of χ^{-1} is found all the way down to T_N itself, indicating that the antiferromagnetic order develops directly from a local moment paramagnet with minimal Kondo compensation. It is clear that the Yb-based f-electron is excluded from the Fermi surface at zero field, at least for $T \geq T_N$.

A large ordering anomaly is observed at T_N in the specific heat C(T), presented in Fig. 2. A broad peak is evident near ~25 K in Fig. 2, and our fits to the Schottky expression indicate that the crystal field manifold consists of the four doublets expected for Yb^{3+} in tetragonal symmetry, with inelastic neutron scattering measurements [26] showing that the first and second excited states are 87 and 230 K above the ground doublet. We have measured a small latent heat associated with the onset of antiferromagnetic order [25], and the weakly first order character of the transition is consistent with the mean-field like appearance of the ordering anomaly

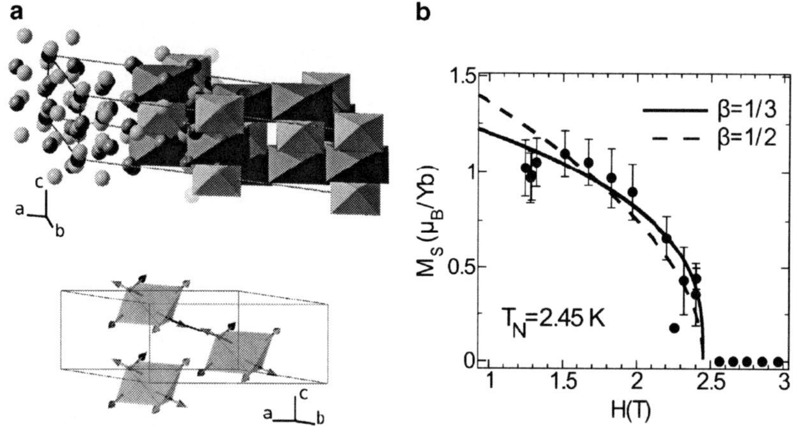

Fig. 3 (**a**) *Top*: A unit cell of Yb_3Pt_4, showing how the Yb atoms (*dark spheres*) are collected into octahedra, and the stacking of these Yb octahedra. The antiferromagnetic structure of Yb_3Pt_4 is based on octahedra of six Yb moments, each antiferromagnetically coupled to its nearest neighbor, which is in another octahedron. *Bottom*: the orientations for the six moments in the Yb octahedra. (**b**) The antiferromagnetic order parameter determined from the 110 magnetic peak. *Dashed line* is the mean field expression, solid line is $T_N \sim (H - 1.62\text{T})^{1/3}$.

in C(T). We have also plotted the temperature dependence of the entropy in Fig. 2, and see that 0.8 Rln2 is released at T_N, attesting to the quite localized character of the doublet ground state in Yb_3Pt_4.

Neutron diffraction measurements have been used to determine the magnetic structure and to study the magnetic order parameter [27]. Yb_3Pt_4 forms in the rhombohedral hR14 (No. 148) space group. As shown in Fig. 3a, the full unit cell contains 18 Yb atoms and 24 Pt atoms, although the primitive cell has a single Yb atom and only three Pt atoms. The fundamental building blocks of the crystal structure are Yb and Pt octahedra which are stacked in a staggered fashion along the c-axis. As indicated in Fig. 3a, the six spins making up the Yb octahedra split into two 3-spin units, one with a net moment component pointing up and the other pointing down. Each moment in the triad is paired antiferromagnetically with its nearest neighbor, which lies in a neighboring octahedron. Interestingly, the magnetic unit cell is the same size as the crystal unit cell in Yb_3Pt_4, and so the propagation wave vector for the antiferromagnetic order is $q_{AF} = 0$. The antiferromagnetic order parameter was measured (Fig. 3b), and displays a mean field-like temperature dependence with no measurable discontinuity at T_N.

Taken together, our data show that Yb_3Pt_4 has many of the attributes of other Yb-based intermetallic compounds. For a wide range of temperatures, the Yb moments are weakly coupled to the conduction electrons, while magnetic susceptibility measures find a well separated doublet ground state without appreciable Kondo compensation of the moments. Antiferromagnetic order is found below 2.4 K, remarkable in that there is little sign of critical fluctuations preceding the transition, which the susceptibility suggests develops directly from the local moment paramagnet. This weakly first order character to the phase transition is unique among heavy electron antiferromagnets studied so far, and the small value of the Neel temperature is suggestive that the phase transition might be suppressed to zero temperature with modest fields or pressures. We will use magnetic fields to probe the stability of antiferromagnetic order in Yb_3Pt_4, with the aim of driving the Neel temperature to zero temperature.

The phase diagram which is produced by field tuning the antiferromagnetic transition in Yb_3Pt_4 [28] is qualitatively different from those found previously in other quantum critical antiferromagnets. To begin, we present the temperature dependence of the specific heat C, measured with fields ranging from 0 T to 2 T applied along the a-axis (Fig. 4a). With increasing field, the step in the specific heat is reduced in magnitude and occurs at lower temperatures, ultimately vanishing at ~1.5 T (Fig. 4b). The transition temperature T_N taken from these data is plotted as a function of magnetic field in Fig. 5. Since T_N is still nonzero at 1.5 T (Fig. 5), we conclude that the

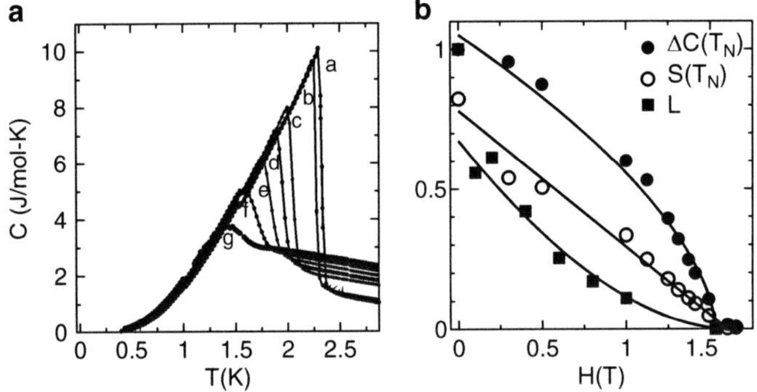

Fig. 4 (**a**) The specific heat C of Yb_3Pt_4 measured in magnetic fields H||a (a = 0.3T, b = 0.5T, c = 1.0T, d = 1.128T, e = 1.25T, f = 1.37T, g = 1.49T). (**b**) $\Delta C(T_N)$ (normalized to zero field value 8.82 J/mol-K, *filled circles*), entropy $S(T_N)$ (units of Rln2, *open circles*), latent heat L (normalized to zero field value 0.09 J/mol-Yb, *filled squares*) Lines are guides for the eye.

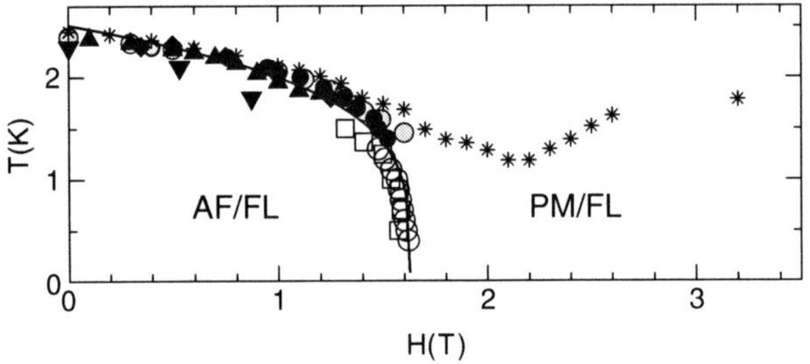

Fig. 5 The magnetic phase diagram of Yb_3Pt_4 for H||a taken from temperature sweeps of χ (*open triangles*), dM/dT (*open diamonds*), C(T) (*gray circles*) and field sweeps of C(H) (1st order: •, 2nd order: ◯) and ρ(H) (1st order: *filled triangles*, 2nd order: *open squares*). T_{FL} (∗) is the limit of $\Delta\rho \sim T^2$. The first and second order lines are jointly fit by $T_N \sim (H-1.62\,T)^{0.28}$ (*solid line*).

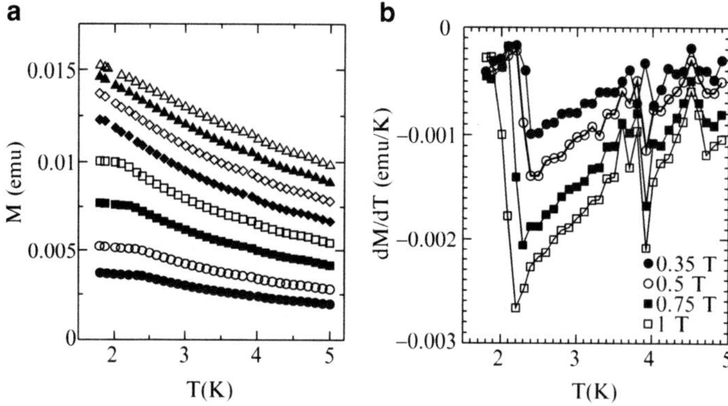

Fig. 6 (**a**) The temperature dependence of the magnetization M, measured in different fixed fields H||a (*filled circle* = 0.35 T, *open circle* = 0.5 T, *filled square* = 0.75 T, *open square* = 1 T, *filled diamond* = 1.25 T, *open diamond* = 1.5 T, *filled triangle* = 1.75 T, *open triangle* = 2 T). (**b**) The temperature dependence of dM/dT, measured in different fixed fields, as indicated.

line of first order transitions in the field-temperature plane terminates at a critical endpoint (T_{CEP}, H_{CEP}), which is (1.2 K, 1.52 T) for fields along the a-axis, and (1.2 K, 3.5 T) when the field is along the c-axis.

Signatures of the first order phase line can be detected in other measurements as well. We have plotted in Fig. 6a the temperature dependence of the dc magnetization M, measured in different magnetic fields. It is evident that M is continuous at the transition, although its temperature derivative dM/dT is not. dM/dT is plotted in Fig. 6b, demonstrating a sharp steplike discontinuity at the Neel temperature T_N. Indeed, Fig. 5 shows that the values for T_N(H) determined in this fashion overlay those from the specific heat measurements. Similar behavior is found in the field derivative of the magnetization, which is obtained from the ac susceptibility $\chi'(H,T) = dM/dH\|_H(T)$ (Fig. 7), measured as a function of temperature at various fixed fields. Again, the step in χ'(T) defines T_N(H), and Fig. 5 shows that it is in complete agreement with the values found in C and dM/dT. We note that the step in χ' is only measurable for fields larger than ∼0.5 T, and this is because its magnitude, like that of dM/dT, increases with proximity to the critical endpoint. Both observations are our first indicators that the critical endpoint gives way to a new line of second order transitions for $H \geq H_{CEP}$, requiring the full divergence of the magnetic susceptibility.

The appearance of a critical endpoint in the field-temperature phase diagram for Yb_3Pt_4 implies either that there is a coexistence region which separates the ordered phase at low fields and temperatures from the paramagnetic phase at high fields and low temperatures, or that there are two phase lines which emanate from the critical endpoint. We have measured the field dependencies of the specific heat C and the electrical resistivity ρ to distinguish between these two possibilities. The field dependence of C measured at fixed temperature above and below $T_{CEP} = 1.2$ K is presented in Fig. 8a. If the temperature is larger than T_{CEP}, this horizontal scan across the H–T phase diagram traverses the first order phase line, and a step in C(H) is de-

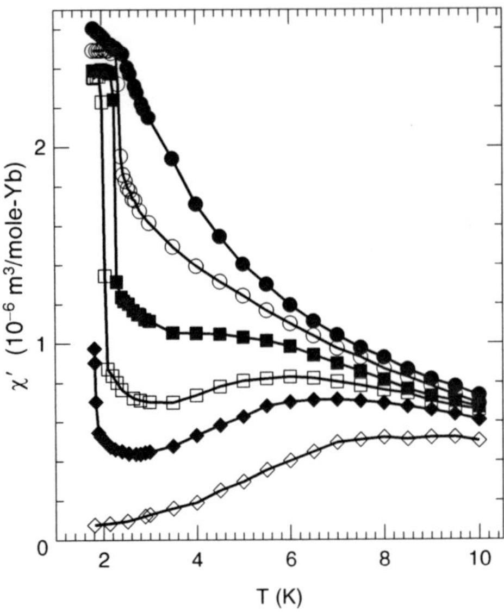

Fig. 7 The temperature dependence of the real part of the ac susceptibility χ', measured in different fixed fields H ∥ a: *filled circles* = 0 T, *open circles* = 0.5 T, *filled squares* = 0.8 T, *open squares* = 1.1 T, *filled diamonds* = 1.35 T, *open diamonds* = 2 T.

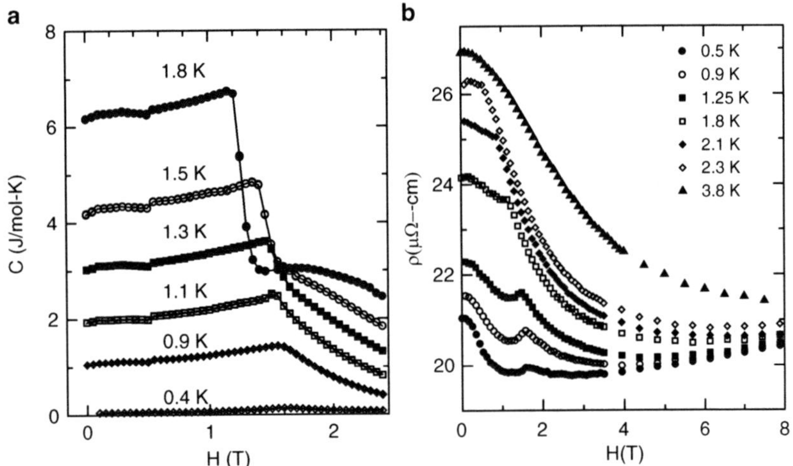

Fig. 8 (**a**) The field dependence of the specific heat C, measured at different temperatures T, as indicated. (**b**) The field dependence of the electrical resistivity ρ, measured at different temperatures.

tected, much as was observed in C(T) for vertical traversals of the same phase line. However, if the temperature is below T_{CEP} the step is absent, and instead a lambda-like peak is found in C(H), as expected for a second order phase transition. The magnitude of the peak in C(H) decreases with decreasing temperature in agreement with the constraint that the entropy associated with this second order phase transition must vanish at T = 0. The fields and temperatures at which the specific heat peak occurs have been added to the phase diagram (Fig. 5), and demonstrate that a line of second order phase transitions emerges from the critical endpoint and vanishes at a quantum critical point (QCP) with $H_{QCP} = 1.62$ T.

A very similar conclusion can be drawn from the field dependence of the electrical resistivity (Fig. 8b). While we do not present the data here, we have found that $\rho(H,T) - \rho(H=0,T=0) \propto M^2(H, T)$, in both the antiferromagnetically ordered and paramagnetic phases. We conclude that ρ is dominated at low temperatures by spin disorder scattering, and in this sense is a proxy measurement for the magnetization itself. As we saw for the specific heat, the signature at the phase transition is very different when the first and second order phase lines are crossed. Specifically, we observe a slope discontinuity in ρ(H) at the first order phase line when $T \geq T_{CEP} = 1.2$ K, reminiscent of the step in χ' reported in Fig. 7. The slope discontinuity in ρ(H) becomes a peak for $T \leq T_{CEP}$, implying that the magnetization has a peak on the second order phase line. This second order phase line is traced out by adding the fields and temperatures at which the peak in ρ(H) occurs to the phase diagram (Fig. 5). While the data do not perfectly overlay the phase line deduced from specific heat measurements, we ascribe this difference to a slight difference in the field alignment between the two measurements.

The completed field-temperature phase diagram is presented in Fig. 5. A line of weakly first order phase transitions extends from the zero field Neel temperature $T_{N,0} = 2.4$ K to a critical endpoint at 1.5 T, where the Neel temperature is reduced to $T_{CEP} = 1.2$ K. No evidence is found for a coexistence region below the critical endpoint. Instead, we find that the first order phase line is extended by a line of second order phase transitions which terminate at a quantum critical point at 1.62 T. Thermodynamics requires a second phase line to emerge from the critical endpoint. Since there is no sign of this second phase line in our measurements, we conclude that it must be associated with a second control variable, perhaps pressure. Remarkably, the first and second order phase lines found in the H–T plane appear to be well described by the same expression, i.e. $T_N \propto (H - H_{QCP})^{0.28\pm0.03}$ with $H_{QCP} = 1.62$ T. We note that this critical exponent is much smaller than 2/3, the value expected for a mean field quantum critical antiferromagnet [29], and does not reproduce the quasi-linear suppression of T_N found in other quantum critical antiferromagnets [30, 31].

The H–T phase diagram of Yb_3Pt_4 is distinct from those found in other heavy electron antiferromagnets which can be tuned to quantum criticality by either field or pressure [13,30–32]. Like these systems, antiferromagnetism in Yb_3Pt_4 occurs at the lowest temperatures via a second order transition. However, as magnetic fields tune Yb_3Pt_4 away from the quantum critical point, the phase line is driven weakly first order. Qualitatively, we can describe Yb_3Pt_4 as a system where fluctuations are, for the most part, too weak to drive antiferromagnetic order which must then occur via free energy differences between the ordered and paramagnetic states. Only near the quantum critical point are fluctuations sufficiently strong to enable a bona fide second order transition, typical of other heavy electron antiferromagnets. We note that the central feature of the H–T phase diagram of Yb_3Pt_4, the critical endpoint, is also found in quantum critical ferromagnets such as MnSi [33] and $ZrZn_2$ [34], and field theoretical techniques have as well provided a supporting phenomenology [35]. We stress, however, that these ferromagnets are apparently never truly quantum critical [36], and that first order character is associated with the vanishing of ferromagnetic order. This is, of course, completely opposite to the scenario outlined in the phase diagram which our measurements reveal for Yb_3Pt_4, where a first order transition gives way to a second order transition when the transition temperature becomes very small.

Given the unique phase diagram found for Yb_3Pt_4, it is likely that the underlying electronic structure will have quite different properties as well. Is the high field phase a Fermi liquid, and does the quasiparticle mass diverge at the quantum critical point, as found in other quantum critical antiferromagnets? Is there a Fermi surface volume change associated with the field-driven $T = 0$ transition to antiferromagnetic order? What is the mechanism for magnetic order, and is it different near the quantum critical point and far away? We will address these questions in the remainder of this paper.

A central feature of virtually all quantum critical systems studied so far is the formation of a Fermi liquid state, once pressure, composition, or magnetic field suppresses magnetic order. In some cases, the characteristic temperature for the Fermi liquid, T_{FL}, vanishes at the quantum critical point [18], and the associated divergence of the quasiparticle mass has been documented in specific heat measurements [12, 13]. Qualitatively speaking, the destruction of the Fermi liquid in these systems seems to be a necessary step for the formation of moments and the eventual stability of magnetic order. We have carried out an extensive study of the low temperature electrical resistivity, specific heat, and magnetic susceptibility in both the high field paramagnetic state and as well the low field antiferromagnetic state to test whether this scenario is also appropriate for understanding the unconventional field-temperature phase diagram of Yb_3Pt_4.

The electrical resistivity ρ shows clear Fermi liquid behavior in both the antiferromagnetic and paramagnetic phases, where $\rho(T) = \rho(T = 0) + AT^2$. $\Delta\rho/\rho = (\rho - \rho(T = 0))/\rho(T = 0)$ is plotted in Fig. 9 as a function of T^2 for fields $H \leq H_{QCP}$. Unusually, the quadratic temperature dependence is observed for $H \leq H_{QCP}$ up to the Neel temperature itself, suggesting that the long ranged and long lived critical fluctuations which usually suppress Fermi liquid behavior near phase transitions are either especially weak or even absent in Yb_3Pt_4. The resistivity coefficient A, taken from the slope of $\Delta\rho$ vs. T^2 is very nearly constant for $H \leq H_{QCP}$ (Fig. 10a). The disappearance of antiferromagnetic order at $H = H_{QCP}$ is marked by a rapid reduction of A (Fig. 10a), as well as a rapid increase in the magnitude of T_{FL}, defined as the highest temperature at which the resistivity has a quadratic temperature dependence. T_{FL} has been added to the phase diagram in Fig. 5, revealing that T_{FL} has a minimum but nonzero value in the range of fields encompassing both H_{CEP} and H_{QCP}. We note that the inevitable differences incurred in aligning the magnetic field relative to the crystal axes in the different experiments prohibits a definitive statement as to whether the minimum in T_{FL} occurs at H_{QCP}, H_{CEP}, or simply nearby.

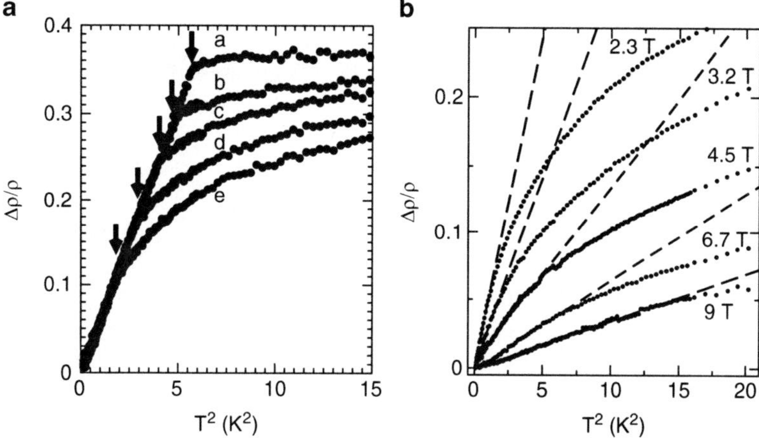

Fig. 9 (**a**) The normalized electrical resistivity $\Delta\rho/\rho$ plotted as a function of temperature squared, for different fixed fields $H \leq H_{QCP}$: a = 0T, b = 0.8T, c = 1.2T, d = 1.5T, e = 1.8T. *Arrows* indicate Neel temperature T_N. (**b**) Same, but for different fixed fields $H \geq H_{QCP}$, as indicated. *Dashed lines* are linear fits, showing temperature regimes for Fermi liquid behavior.

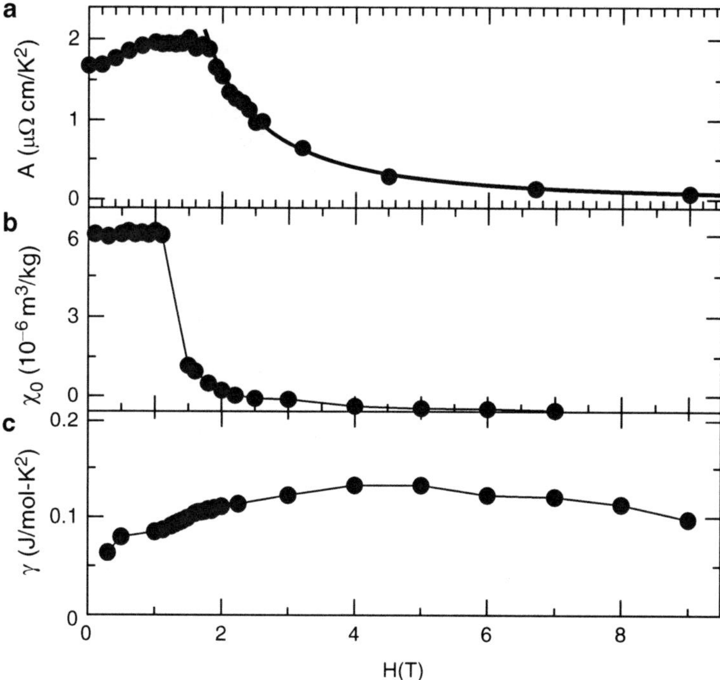

Fig. 10 A comparison of the field dependencies of the Fermi liquid parameters A, χ_0, and γ (see text). Small differences in field alignment are responsible for the different quantum critical fields found in the three quantities. *Solid lines* are guides for the eye.

Nonetheless, it is very clear that T_{FL} is never significantly smaller than T_{CEP}, and certainly never approaches zero at any field.

The temperature dependence of the ac magnetic susceptibility χ' in zero field shows that antiferromagnetic order develops suddenly from a paramagnetic state where $\chi' \sim 1/T$ (Fig. 7). The susceptibility in the ordered state is largely temperature independent, although there is a weak ordering cusp on the low temperature side of the transition. This temperature independent Pauli susceptibility χ_0 is expected for the Fermi liquid state first revealed by electrical resistivity measurements. The field dependence of χ_0 is compared to that of A in Fig. 10. In the ordered state, there is little field dependence for either A or χ_0, while both Fermi liquid parameters drop suddenly when the field increases beyond H_{QCP}.

While the field dependencies of A and χ_0 would both be consistent with the conventional view that the quasiparticle mass m* diverges near the quantum critical point, direct measurements of the electronic specific heat reveal a very different scenario. The measured specific heat C_{meas} combines a number of contributions in addition to the Fermi liquid term, γT. These include the field independent phonon contribution, which we equate to the measured specific heat of the nonmagnetic analog compound, polycrystalline Lu_3Pt_4, a Schottky peak C(T) which traverses our temperature window as the field is tuned from 0 to 9 T, and a critical contribution associated with $T_N(H)$ and absent for fields above H_{CEP}. Given this complexity, we restrict our analysis to $H \geq H_{QCP}$, where the absence of the critical component permits us to fit accurately the remaining terms (Fig. 11a). By subtracting the Schottky peak, we can plot the remaining specific heat $\Delta C/T = (C_{Meas} - C_{Schottky})/T$ as a function of T^2 in Fig. 11b. The $T = 0$ intercept is defined as γ, and is in turn plotted in Fig. 10c. We see that paramagnetic Yb_3Pt_4 is a moderately heavy Fermi liquid, with

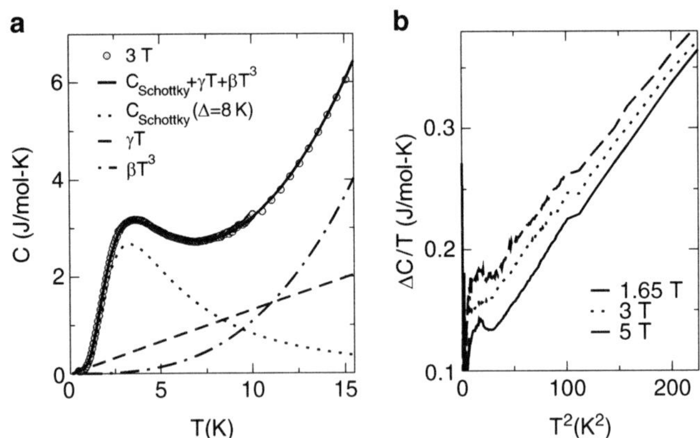

Fig. 11 (**a**) The measured specific heat C of Yb_3Pt_4 in a field of 3 T. *Solid line* shows fit to data, consisting of a Schottky term (*dotted line*), a term linear in temperature γT (*dashed line*) and our fit to the measured specific heat of Lu_3Pt_4, which is very well described on this range of temperatures as βT^3 (*dotted* and *dashed line*). (**b**) $\Delta C/T = (C - C_{Schottky})/T = \gamma + \beta T^2$, plotted as a function of T^2 to isolate the $T = 0$ intercept, γ.

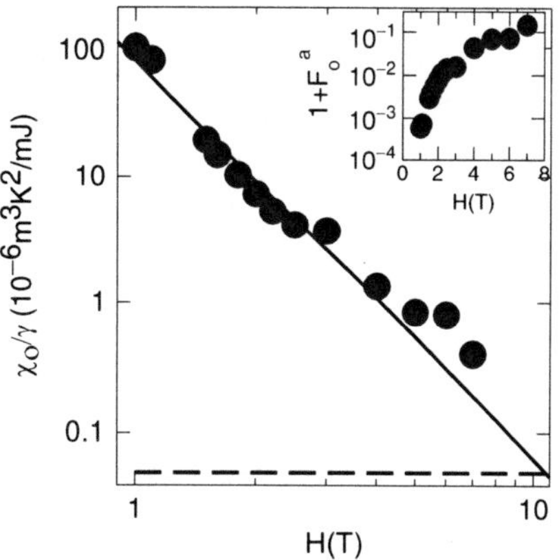

Fig. 12 The field dependence of χ_0/γ. *Solid line* is $1/H^3$, *dashed line* is typical value of χ_0/γ found in other heavy electron materials. Inset: $1+F_0^a$ goes to zero as $H \rightarrow H_{QCP}$.

γ varying from $\sim$0.1 to 0.16 J/mol K^2. There is only a small field dependence for γ, and certainly no evidence for the divergence near H_{QCP} found in systems such as $YbRh_2Si_2$ [13]. Remarkably, the onset of magnetic order in Yb_3Pt_4 does not require a divergence or even an increase in the quasiparticle mass, which is found instead to be almost independent of magnetic fields.

Given that antiferromagnetic order emerges at $T = 0$ and $H = H_{QCP}$ from a Fermi liquid state without an accompanying divergent quasiparticle mass, what instability of the Fermi liquid leads to the onset of magnetic order? To probe this question, we have plotted χ_0/γ in Fig. 12, eliminating the dependence of χ_0 on the quasiparticle mass and isolating a term which is proportional to $1/(1+F_0^a)$, where F_o^a is a Landau parameter. We see that at 9 T, i.e. far from the quantum critical point, χ_0/γ has a value of $\sim 0.2 \times 10^{-6}$ $m^3 K^2/mJ$, comparable to values found in heavy electron compounds such as URu_2Si_2 and UPt_3 [37]. As the field is reduced, χ_0/γ increases rapidly, reaching a value at H_{QCP} which is $\sim$500 times larger than at 9 T. Since $\chi_0/\gamma = (1+F_0^a)\,(\mu_0\mu_{eff}/\pi^2 k_B^2)$, we have isolated $1+F_0^a$ and plotted it in the inset of Fig. 12, showing that $1+F_0^a$ decreases from a value near 0.1 at 9 T to 0.002 at H_{QCP}. We conclude that the $T = 0$ antiferromagnetic instability in Yb_3Pt_4 is apparently driven by long-ranged interactions among the quasiparticles, and not by the enhancement of the quasiparticle mass. Significantly, this finding indicates that there is no role for localized Yb moments in stabilizing antiferromagnetic order near the quantum critical point, which we take as the instability of a 'large' Fermi surface, i.e. one which includes the very electrons which were localized on Yb-based f-orbitals in zero field.

Finally, we point out that $\chi_0/\gamma \propto H^{-3\pm0.2}$. Attempts to fit χ_0/γ to powers of $(H - H^*)$ uniformly degraded the quality of the fit for any power and any nonzero H^*. This suggests that the divergence of $1/(1+F_0^a)$ is controlled by a fixed point at zero field, and the saturation of χ_0 in the ordered state signals that the first order transition cuts off further development of these long-ranged quasiparticle interactions.

The $T = 0$ divergence of χ_0 evident in Fig. 10b demonstrates that the $q = 0$ part of the magnetic susceptibility is greatly enhanced, if not divergent, on the approach to the ordered state. Normally, the $q = 0$ part of the susceptibility is associated with the ferromagnetic part of the response. However, our neutron diffraction measurements show that the antiferromagnetic ordering vector for Yb_3Pt_4 is $q_{AF} = 0$, meaning that every antiferromagnetic Bragg peak coincides with a nuclear Bragg peak [27]. Given the inherent weakness of small angle scattering in the electrical resistivity [38], it is not a priori evident whether critical fluctuations play a decisive role in the resistivity, especially near H_{QCP}. Accordingly, we have plotted the temperature derivative of ρ together with the temperature dependence of the specific heat to demonstrate that they are proportional near the $H = 0$ antiferromagnetic instability at 2.4 K (Fig. 13). Figure 13 shows that Yb_3Pt_4 obeys the Fisher–Langer relation [38] in zero field as well as in fields which approach H_{CEP} and H_{QCP}. This indicates that for much of the ordered region of the phase diagram $H \leq H_{QCP}$, both the resistivity and the specific heat are dominated by the same antiferromagnetic critical fluctuations. The observation of the Fisher–Langer relation in a system where $q_{AF} = 0$ indicates that these same critical fluctuations are effective at scattering the quasiparticles as the quantum critical point is approached from the $H \geq H_{QCP}$ Fermi liquid. To isolate the field enhancement of the quasiparticle scattering cross-section from that of the quasiparticle mass itself, we plot the Kadowaki–Woods ratio A/γ^2 in Fig. 14 for fields from 1.4 to 9 T. A/γ^2 has a value of $10\,\mu\Omega$ cm $mol^2\,K^2/mJ^2$ at 9 T, which is typical for a heavy electron compound with $\gamma \sim 0.1$ J/mol-K^2 [39]. As the field is lowered towards the quantum critical point, A/γ^2 increases dramatically with a field dependence $1/H^{2.6\pm0.2}$ which is very similar to that found in χ_0/γ. The inset to Fig. 14 shows that A/γ^2 and χ_0/γ are approximately proportional to each other for fields between 1.4 and 9 T. This is significant, since in the case of dominantly ferromagnetic quasiparticle interactions we expect A/γ^2 to be proportional to $(\chi_0/\gamma)^2$ [8, 40, 41]. Of course, since antiferromagnetic fluctuations dominate the small angle scattering in Yb_3Pt_4, we do not expect that this relationship would hold and indeed we see a linear, and not a quadratic relationship between A/γ^2 and χ_0/γ.

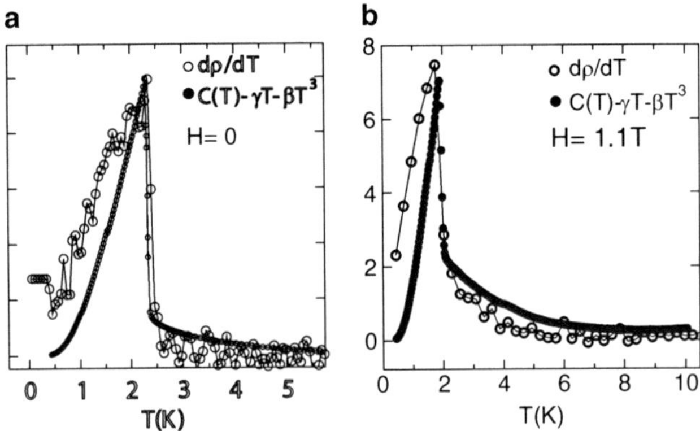

Fig. 13 A comparison of the temperature dependencies of $d\rho/dT$ and the critical part of the specific heat $C(T) = C_{Meas} - \gamma T - \beta T^3$ for $H = 0$ (**a**) and $H = 1.1$ T (**b**) (both in arbitrary units).

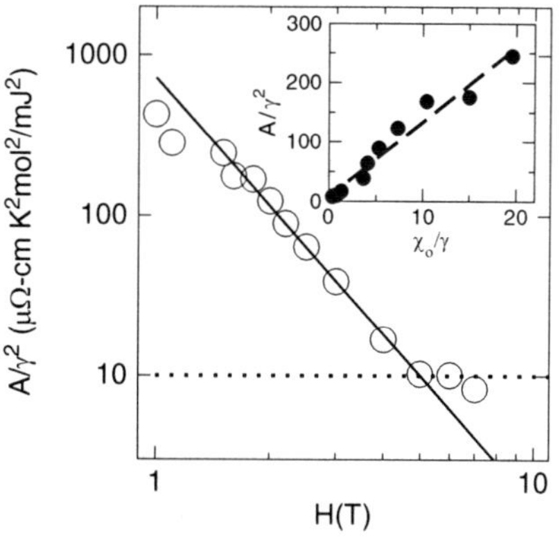

Fig. 14 The field dependence of the Kadowaki–Woods ratio, A/γ^2. *Dotted line* indicates typical value for other heavy electron compounds. Inset: A/γ^2 is approximately proportional to χ_0/γ for fields $H \geq H_{QCP}$.

The data which are reported here show that the relationship between magnetic order and the underlying electronic structure is played out in Yb_3Pt_4 in a way which is very different from that of previously studied quantum critical systems, such as $YbRh_2Si_2$ and $CeCu_{5.9}Au_{0.1}$. We stress that the starting point for Yb_3Pt_4 is very similar. In zero field, Yb_3Pt_4 is a system in which spatially localized Yb^{3+} moments are weakly coupled to conduction electrons, with a Kondo temperature T_K which is not appreciably larger than the Neel temperature itself. The magnetic susceptibility follows a Curie–Weiss law down to T_N, where it is discontinuously replaced by the temperature independent susceptibility of a Fermi liquid state. Considering as well the weakly first order character of the magnetic order, we hypothesize that the onset of order and the accompanying loss of magnetic moment in Yb_3Pt_4 might be the result of a sudden expansion of the underlying Kondo temperature, akin to the "Kondo Volume Collapse" found in other f-electron compounds, notably elemental Ce and $YbInCu_4$ [42, 43]. However, the coincidence of antiferromagnetic order with this possible Fermi surface expansion would be, as far as we know, unique to Yb_3Pt_4. Of course, direct investigations of the Fermi surface above and below the transition are necessary in order to assess the appropriateness of this explanation. However, we note that no change is observed in the crystal symmetry with the onset of antiferromagnetic order [27], and that the lattice constants above and below the transition are identical at the level of one part in 10^4, apparently shedding some doubt on this scenario.

The weakly first order character of the $H = 0$ antiferromagnetic order in Yb_3Pt_4 extends in magnetic field to a line of first order transitions, terminating at a critical endpoint. The critical endpoint is a feature of the H–T phase diagram found in quantum ferromagnets, both experimentally and theoretically, supporting the belief that these systems can never be truly quantum critical. The appearance of a critical endpoint in the phase diagram of a quantum antiferromagnet is unique, and suggests that here quantum critical fluctuations are just strong enough to drive the phase transition second order, once the transition temperature reaches a low enough value. Accordingly, we have found no signs in Yb_3Pt_4 of the non-Fermi liquid behavior which is ubiquitous in other quantum critical antiferromagnets. Unlike these systems, the quantum critical phenomena in Yb_3Pt_4 are not sufficiently strong to overcome the noncritical parts of the specific heat and magnetization, except at the very lowest temperatures. The divergence of the extrapolated $T = 0$ susceptibility χ_0 clearly indicates that quantum criticality ultimately dominates.

The zero temperature phase transition at the quantum critical field appears to be driven by a Fermi surface instability, with no overt role for moment localization. At the highest fields, values for the Fermi liquid parameters A, χ_0, and γ, are all consistent with those found in other heavy electron compounds, suggesting a moderate quasiparticle mass enhancement. As the field is reduced towards the quantum critical value, there is only a slow evolution of γ, with T_{FL} bounded from below by the temperature of the critical end point. There is no divergence of the quasiparticle mass near the quantum critical point in Yb_3Pt_4, although both χ_0 and A are hugely enhanced relative to other heavy electron compounds. This suggests that the formation of local moments is not central to the physics of the quantum critical point in Yb_3Pt_4. Instead, the divergence of the magnetic susceptibility indicates that these quasiparticles interact in a medium which is increasingly polarizable as the quantum critical field is approached, and the range and strength of their interactions expand sufficiently to drive a Fermi surface instability, not unlike the physics of a Stoner instability in a ferromagnet.

The relationship between moment formation and magnetic order, especially at the quantum critical point, is very different in Yb_3Pt_4 and in $YbRh_2Si_2$. These differences are captured in the two schematic phase diagrams presented in Fig. 15. Figure 15a is the phase diagram found in $YbRh_2Si_2$ [44], where moment formation occurs along the line $T^*(H)$, which terminates at the field driven quantum critical point $\Gamma_C = H_{QCP}$. Here, antiferromagnetic order can be considered local moment-like, in the sense that these moments are preserved both above and below the Neel temperature, and are included in the Fermi surface only for $H \geq H_{QCP}$ and $T \leq T^*(H)$. Our experiments suggest a very different scenario for Yb_3Pt_4, where the local moments appear to vanish with the onset of antiferromagnetic order in zero field, which

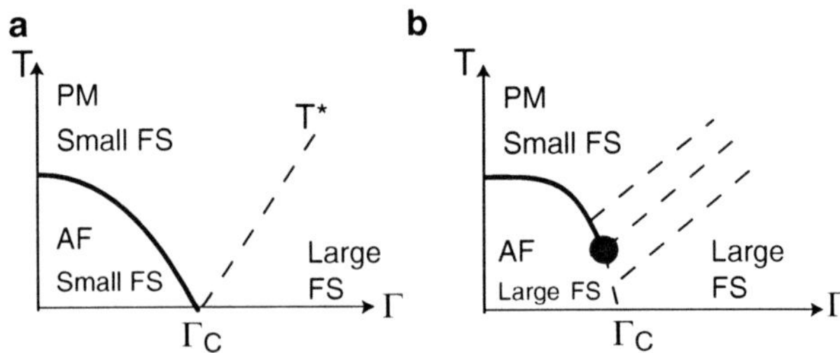

Fig. 15 Schematic heavy electron phase diagrams as functions of temperature and tuning parameter Γ, representing pressure, field, or composition. (**a**) Moment delocalization occurs on *dashed line* T^* in $YbRh_2Si_2$, where magnetic field provides the tuning Γ [44]. (**b**) Phase diagram proposed for Yb_3Pt_4, where moment delocalization occurs along the *dashed line*, terminating on the first order phase boundary (1), perhaps at the critical endpoint (2) or on the second order phase line (3), but does not extend to the QCP at $\Gamma_C = H_{QCP}$.

can be considered a transition between a small Fermi surface at high temperatures, and a large Fermi surface which includes the previously localized f-electron in the ordered state. Conversely, the zero temperature instability at the quantum critical field appears to be between two states *without* localized moments, i.e. where the f-electrons have been absorbed into the Fermi surface. This observation implies that moment delocalization occurs in Yb_3Pt_4 along one of the dashed lines (1)–(3) in Fig. 15b, suggesting that it is always a finite temperature phenomenon in Yb_3Pt_4.

3 Conclusions

Coming full circle, we conclude that the richness of the charge and spin degrees of freedom intrinsic to the heavy electron compounds leads to their diversity of instabilities which emerge from the universally paramagnetic state when the temperature is progressively lowered. It is axiomatic that the underlying electronic structure which drives these instabilities must have a similar richness, making it virtually inevitable that different systems will display different routes to quantum criticality. This work on Yb_3Pt_4 perhaps serves as the first evidence that there are a multiplicity of ways in which magnetic order and moment localization can be achieved at $T = 0$ within the heavy electron compounds.

Acknowledgements The authors acknowledge useful conversations with C. Varma, Q. Si, P. Coleman, P. Chandra, E. Abrahams, G. Zwicknagl, and V. Zlatic. Work at Stony Brook University was supported by the National Science Foundation under grant NSF-DMR-0405961.

References

1. N. D. Mathur, et al., Nature **394**, 39 (1998).
2. S. Saxena, et al., Nature **406**, 587 (2000).
3. R. A. Borzi, et al., Science **315**, 214 (2007).
4. P. Coleman and A. J. Schofield, Nature **433**, 226 (2005).
5. R. B. Laughlin, Adv. Phys. **47**, 943 (1998).
6. C. M. Varma, et al., Phys. Rev. Lett. **63**, 1996 (1999).
7. G. R. Stewart, Rev. Mod. Phys. **73**, 797 (2001).
8. H. v. Lohneysen, et al., Rev. Mod. Phys. **79**, 1015 (2007).
9. P. Gegenwart, et al., Nature Physics **4**, 186 (2008).
10. D. Jaccard, et al., J. Phys.: Cond. Matter **13**, L89 (2001).
11. S. Doniach, in *Valence Instabilities and Related Narrow Band Phenomena*, edited by R. D. Parks (Plenum, New York, 1977), p. 169.
12. P. Gegenwart, J. Custers, C. Geibel , K. Neumaier, T. Tayama, K. Tenya, O. Trovarelli, and F. Steglich, Phys. Rev. Lett. **89**, 056402 (2002).
13. P. Gegenwart, J. Custers, T. Tayama, K. Tenya, C. Geibel, G. Sparn, N. Harrison, P. Kerschl, D. Eckert, K. -H. Muller, and F. Steglich, J. Low. Temp. Phys. **133**, 3 (2003).
14. H. Shishido, R. Settai, H. Harima, and Y. Onuki, J. Phys. Soc. Japan **74**, 1103 (2005).
15. M. C. Aronson, R. Osborn. R. A. Robinson, J. W. Lynn, R. Chau, C. L. Seaman, and M. B. Maple, Phys. Rev. Lett. **75**, 725 (1995).

16. M. C. Aronson, R. Osborn, R. Chau, M. B. Maple, B. D. Rainford, and A. P. Murani, Phys. Rev. Lett. **87**, 197205 (2001).
17. A. Schroder, G. Aeppli, R. Coldea, M. Adams, O. Stockert, H. v. Lohneysen, E. Bucher, R. Ramazashvili, and P. Coleman, Nature **407**, 351 (2000).
18. P. Coleman, C. Pepin, Q. Si, and R. Ramazashvili, J. Phys.: Cond. Matter **13**, R723 (2001).
19. Q. Si, S. Rabello, K. Ingersent, and J. L. Smith, Nature **413**, 804 (2001).
20. J. Custers, P. Gegenwart, H. Wilhelm, K. Neumaier, Y. Tokiwa, O. Trovarelli, C. Geibel, F. Steglich, C. Pepin, and P. Coleman, Nature **424**, 524 (2003).
21. T. Senthil, S. Sachdev, and M. Vojta, Phys. Rev. Lett. **90**, 216403 (2003).
22. T. Senthil, M. Vojta, and S. Sachdev, Phys. Rev. B **69**, 035111 (2004).
23. S. Paschen, T. Luhmann, S. Wirth, P. Gegenwart, O. Trovarelli, C. Geibel, F. Steglich, P. Coleman, and Q Si, Nature **432**, 881 (2004).
24. R. Daou, C. Bergemann, and S. R. Julian, Phys. Rev. Lett. **96**, 026401 (2006).
25. M. C. Bennett, P. Khalifah, D. A. Sokolov, W. J. Gannon, Y. Yiu, M. S. Kim, C. Henderson, and M. C. Aronson, arXiv:0807.3586.
26. Y. Janssen, M. S. Kim, K. S. Park, L. Wu, C. Marques, M. C. Bennett, M. C. Aronson, Y. Chen, J. Li, Q. Huang, and J. W. Lynn (unpublished).
27. Y. Janssen, M. S. Kim, K. Stone, P. Stephens, P. G. Khalifah, Y. Chen, J. Li, Q. Huang, J. W. Lynn, and M. C. Aronson (unpublished).
28. M. C. Bennett, D. A. Sokolov, M. S. Kim, Y. Janssen, Yuen Yiu, W. J. Gannon, and M. C. Aronson, arXiv:0812.1082.
29. A. J. Millis, Phys. Rev. B **48**, 7183 (1993).
30. H. von Lohneysen, J. Phys.: Condensed Matter **8**, 9689 (1996).
31. S. R. Julian, et al., J. Phys.: Condensed Matter **8**, 9675 (1996).
32. S. L. Bud'ko, et al., Phys. Rev. B **69**, 014415 (2004).
33. C. Pfleiderer, J. Phys.: Cond. Matter **17**, S987 (2005).
34. M. Uhlarz, C. Pfleiderer, and S. M. Hayden, Phys. Rev. Lett. **93**, 256404 (2004).
35. D. Belitz, et al., Rev. Mod. Phys. **77**, 579 (2005).
36. Y. J. Uemura, et al., Nature Physics **3**, 29 (2007).
37. Z. Fisk, et al., Japanese Journal of Applied Physics **26** Suppl. 3, 1882 (1987).
38. M. E. Fisher and J. S. Langer, Phys. Rev. Lett. **20**, 665 (1968).
39. N. Tsujii, et al., J. Phys.: Cond. Matt. **15**, 1993 (2003).
40. K. S. Dy and C. J. Pethick, Phys. Rev. **185**, 373 (1969).
41. G. Zwicknagl, Adv. Phys. **41**, 203 (1992).
42. C. D. Immer, J. L. Sarrao, Z. Fisk, A. Lacerda, C. Mielke, and J. D. Thompson, Phys. Rev. B **56**, 71 (1997).
43. J. L. Sarrao, A. P. Ramirez, T. W. Darling, F. Freibert, A. Migliori, C. D. Immer, Z. Fisk, and Y. Uwatoko, Phys. Rev. B **58**, 409 (1998).
44. P. Gegenwart, T. Westerkamp, C. Krellner, Y. Tokiwa, S. Paschen, C. Geible, F. Steglich, E. Abrahams, and Q. Si, Science **315**, 969 (2007).

Resistivity of $Mn_{1-x}Fe_xSi$ Single Crystals: Evidence for Quantum Critical Behavior

C. Meingast, Q. Zhang,* T. Wolf, F. Hardy, K. Grube, W. Knafo,† P. Adelmann, P. Schweiss, and H. v. Löhneysen

Abstract Resistivity measurements have been made on $Mn_{1-x}Fe_xSi$ single crystals between 2 and 300 K for $x = 0$, 0.05, 0.08, 0.12 and 0.15. Fe doping is found to depress the magnetic ordering temperature from 30 K for $x = 0$ to below 2 K for $x = 0.15$. Although Fe doping results in a large increase of the low-temperature residual resistivity, the temperature dependence of the resistivity above the magnetic transition remains practically unaffected by increasing Fe content. An analysis of the temperature derivative of the resistivity provides strong evidence for the existence of a non-Fermi-liquid ground state near $x = 0.15$ and thus for a quantum critical point tuned by Fe content.

1 Introduction

MnSi is thought to be a good example of a weak itinerant ferromagnet [1, 2]. Actually, due to the lack of space inversion symmetry of the B20 crystal structure and the resulting Dzyaloshinski–Moriya spin-orbit interaction, MnSi possesses a long-wavelength (180 Å) helical rotation of the magnetization in zero magnetic field [3]. The magnetic ground state of MnSi is thus better described as incommensurate antiferromagnetic. However, above a field of roughly 0.6 T, the helical order

* Present address: Jinan city, Beijing, China

† Present address: Laboratoire National des Champs Magnétiques Pulsés, UMR CNRS-UPS-INSA 5147, 143 avenue de Rangueil, 31400 Toulouse, France

C. Meingast, Q. Zhang, T. Wolf, F. Hardy, K. Grube, W. Knafo, P. Adelmann, P. Schweiss, and H. v. Löhneysen
Forschungszentrum Karlsruhe, Institut für Festkörperphysik, PO Box 3640, 76021 Karlsruhe
e-mail: Christoph.Meingast@ifp.fzk.de, qinzhang788@yahoo.com.cn, Thomas.Wolf@ifp.fzk.de, frederic.hardy@ifp.fzk.de, Kai.Grube@ifp.fzk.de, knafo@lncmp.org, peter.Adelmann@ifp.fzk.de, peter.schweiss@ifp.fzk.de, h.vL@phys.uni-karlsruhe.de

W. Knafo and H. v. Löhneysen
Physikalisches Institut, Universität Karlsruhe, 76128 Karlsruhe

V. Zlatić and A. C. Hewson (eds.), *Properties and Applications of Thermoelectric Materials*, 261
NATO Science for Peace and Security Series B: Physics and Biophysics,

is quenched and the systems then behaves like a typical weak itinerant ferromagnet [3]. The smallest energy scale in MnSi is the crystal–field interaction, which pins the helices along one of the $\langle 111 \rangle$ crystallographic directions. The interest in MnSi has recently been renewed by the discovery of a large region of non-Fermi liquid behavior in the resistivity with a $T^{3/2}$ power law above the critical pressure (~15 kbar) where the long-range magnetic order disappears [4]. Usually non-Fermi-liquid behavior is limited to a small region around a quantum critical point [5,6], and the reason for this large region of non-Fermi liquid behavior remains unknown [7].

Here we study the resistivity of $Mn_{1-x}Fe_xSi$. Previous work has shown that the magnetic order is suppressed with increasing Fe content [8,9], and it is thus of strong interest to see if a concentration-tuned quantum critical point occurs in $Mn_{1-x}Fe_xSi$. Resistivity measurements provide a simple, but quite powerful, means to investigate such transitions. In particular, there appears to be a very strong correlation of the temperature derivative of the resistivity with the magnetic/electronic heat capacity in strongly correlated systems [10, 11]. A more detailed report of these data, as well as magnetization, specific heat and thermal expansion will be published elsewhere [12].

2 Experimental

Single crystals of $Mn_{1-x}Fe_xSi$ have been grown by the vertical Bridgman method in conical Al_2O_3 crucibles. Stoichiometric mixtures of Mn (Cerac, 4N), Si (Alfa, 6N), and Fe (Fluka, 2N5) were heated to 1460°C at a rate of 250°C/h, and after a holding time of 1 h, cooled down to 1150°C at rates between 1–2°C/h. At the beginning the growth chamber was evacuated overnight to pressures $<10^{-5}$ mbar and then, at temperatures >1000°C, backfilled to 1100 mbar with 6N argon. To further reduce the oxygen content in the growth atmosphere the Al_2O_3 crucible was packed into Zr powders as a getter material. Crystals with typical dimensions of $10 \times 1 \times 0.5\,mm^3$ were cut with a wire saw and mechanically polished to the final dimensions. Some of the crystals had small pores, which made an accurate determination of the absolute resistivity problematic (see below). The resistivity was measured using a four-points ac measurement technique in a PPMS (Physical Property Measurement System) from Quantum Design. The contacts were made using silver paint and gold (or platinum) wires.

3 Results and Discussion

Figure 1 shows the raw resistivity data of $Mn_{1-x}Fe_xSi$ between 2 and 300 K for $x = 0, 0.05, 0.08, 0.12$ and 0.15. The resistivity of pure MnSi is very similar to previous measurements [13–15]. The present residual resistivity ratio (RRR), $\rho(300\,K)/\rho(2\,K) = 21.5$, is smaller than the highest RRR (up to 100–200) reported in the liter-

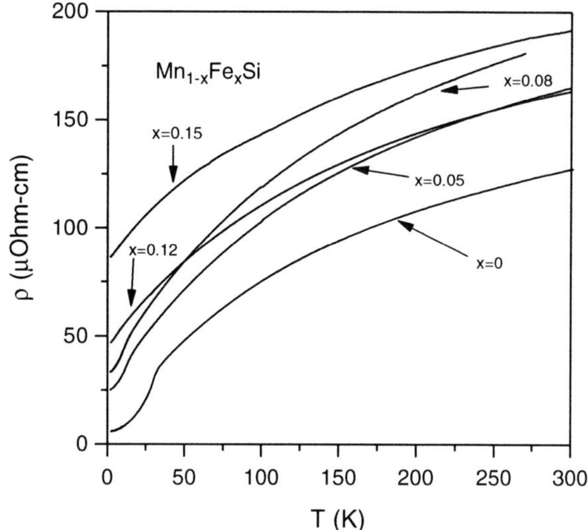

Fig. 1 Resistivity versus temperature of $Mn_{1-x}Fe_xSi$.

ature [13–15]. Interestingly, we also observed higher RRRs for some crystals, which however had broader magnetic transitions. We do not have an explanation for this behavior, and here we present only the data of the crystal with a sharp magnetic phase transitions and a lower RRR. The residual resistivity increases strongly with x. However, the general shape of the resistivity curves above T_N is hardly affected by an increase of the Fe content. The curves for $x = 0.05$ and 0.08 have somewhat steeper slopes, however this may be due to an uncertainty in the absolute values due to the previously mentioned pores in some of the samples. In order to have a better comparison of the shape of the resistivity curves, we have normalized the resistivity curves at 300 K (see Fig. 2). Now it is clearly seen that, besides the increase of the residual resistivity, the curves all have very similar shapes above T_N.

In order to study the magnetic phase transition at T_N in more detail, we calculated the temperature derivative of the resistivity, $d\rho(T)/dT$, which is plotted versus T in Fig. 3. The data have been normalized at 100 K. The phase transition is now clearly seen and strongly resembles the behavior of the specific heat and thermal expansion [12, 15]. We note that the transition takes place in two stages; coming from low temperatures, $d\rho(T)/dT$ first has a very sharp peak, followed by a broadened hump just above the peak. We believe that the sharp peak signals the real phase transition, i.e. the point where helical order is established with long-range phase coherence throughout the crystal. This transition is weakly first-order. The broad hump above this peak, on the other hand, is probably due to the establishment of ferromagnetic-like correlations. This is because, with the application of an applied magnetic field, the sharp peak decreases in temperature, whereas the broad hump becomes broader and moves to higher temperatures, as would be expected for a simple ferromagnetic transition [12, 15]. With increasing Fe content, the transitions remain remarkably sharp for $x = 0.05$ and 0.08. For $x = 0.12$, the transition is somewhat broadened, and the sharp first-order transition is no longer observed. For $x = 0.15$, we see no clear

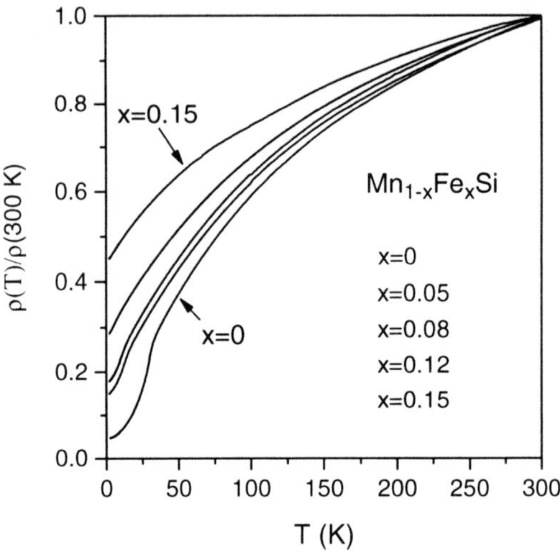

Fig. 2 Normalized resistivity, $\rho(\mathrm{T})/\rho(300\,\mathrm{K})$, versus temperature of $\mathrm{Mn}_{1-x}\mathrm{Fe}_x\mathrm{Si}$.

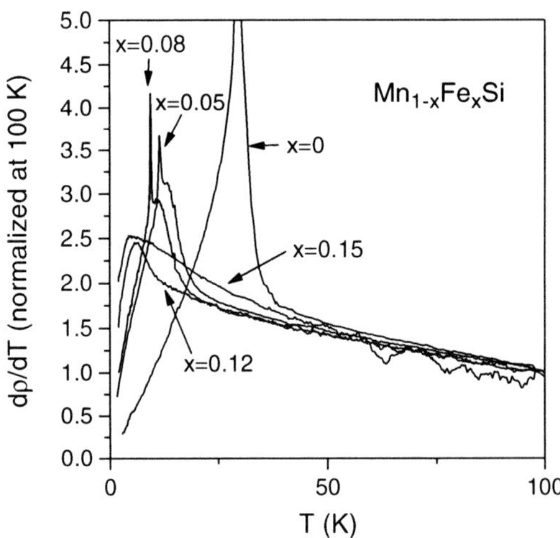

Fig. 3 Normalized (at 100 K) temperature derivative of the resistivity $\rho(\mathrm{T})/d\mathrm{T}$ of $\mathrm{Mn}_{1-x}\mathrm{Fe}_x\mathrm{Si}$. The magnetic phase transition is clearly seen for all Fe concentrations except for $x = 0.15$.

sign of a transition; $d\rho(T)/dT$ first increases with increasing temperature and then goes over a broad maximum at roughly 4 K.

In order to further analyze these resistivity data, it is very useful to look at $d\rho(T)/dT$ divided by temperature (see Fig. 4). $(1/T)\ d\rho(T)/dT$ is analogous to the magnetic/electronic specific heat divided by T, i.e. C_p/T [10, 11, 15]. In Fig. 4 we see that $(1/T)\ d\rho(T)/dT$ becomes constant at low T for pure MnSi, as would be expected for a Fermi-liquid like state with a constant C_p/T. For increasing Fe content x, $(1/T)\ d\rho(T)/dT$ is no longer constant at low temperature, but rather approaches a nearly logarithmic temperature dependence, indicating the approach to a non Fermi-liquid state. For $x = 0.15$, $(1/T)\ d\rho(T)/dT$ increases smoothly to the lowest measured temperature (2 K) and shows no sign of a finite-temperature

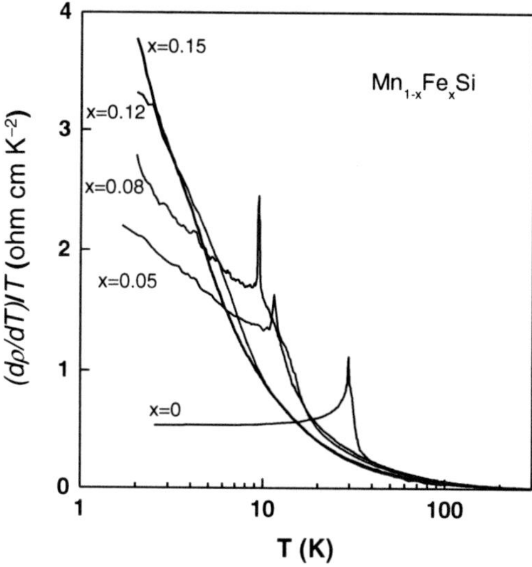

Fig. 4 $(1/T)d\rho(T)/dT$ versus temperature on a log(T) scale for $Mn_{1-x}Fe_xSi$. $(1/T)d\rho(T)/dT$ is constant at low temperature for pure MnSi indicative of a Fermi-liquid state. For $x = 0.16$, on the other hand, $(1/T)d\rho(T)/dT$ diverges nearly logarithmically implying a non-Fermi-liquid ground state.

phase transition. The present data thus provide strong evidence that $Mn_{1-x}Fe_xSi$ approaches a magnetic quantum critical point for x close to 0.15. This nearly logarithmic divergence is in contrast to the $T^{-0.5}$ divergence expected from the $\rho(\mathrm{T}) \sim T^{3/2}$ power law found for pure MnSi in the region above the critical pressure [4]. Here, it is worth recalling that the transition under pressure to a non-magnetic state in pure MnSi is of first order above a pressure p^* close to the critical pressure, whereas the present data point more to a second-order transition with Fe substitution. The difference in critical behavior may, thus, be due to the different tuning parameters (pressure versus composition) employed to approach the quantum critical point.

4 Conclusions

In summary, resistivity measurements of $Mn_{1-x}Fe_xSi$ single crystals between 2 and 300 K nicely demonstrate that Fe doping depresses the magnetic ordering temperature from 30 K for $x = 0$ to below 2 K for $x = 0.15$. A detailed analysis of the temperature derivative of the resistivity provides strong evidence for the existence of a non-Fermi-liquid ground state near $x = 0.15$. This makes $Mn_{1-x}Fe_xSi$ a very interesting system for further studies. Of special interest is the comparison with the behavior of pressure-tuned pure MnSi. Preliminary thermal expansion and specific heat data of $Mn_{1-x}Fe_xSi$ are in accord with the present resistivity data and also provide evidence for a quantum critical point near x = 0.15 [12]. Together these data show that the Grüneisen parameter, which is just the ratio of the thermal expansion coefficient and the specific heat, exhibits a divergence and a sign change, as is typically expected close to a pressure-tuned quantum phase transition [16].

Acknowledgments We would like to acknowledge useful discussion with Daniel Lamago and Dmitry Reznik and to thank Severin Adandogou and Doris Ernst for technical assistance.

References

1. T. Moriya, Spin Fluctuations in Itinerant Electron Magnetism, vol. 56 of Solid-State Sciences (Springer, Berlin, 1985).
2. G. G. Lonzarich and L. Taillefer, J. Phys. **C 18**, 4339 (1985).
3. Y. Ishikawa et al., Phys. Rev. **B 16**, 4956 (1977).
4. N. Doiron-Leyraud et al., Nature **425**, 595 (2003).
5. G. R. Stewart, Rev. Mod. Phys. **73**, 797 (2001).
6. H. von Löhneysen et al., Rev. Mod. Phys. **79**, 1015 (2007).
7. C. Pfleiderer et al., Science **316**, 1871 (2007).
8. N. Manyala et al., Nature **404**, 581–584 (2000).
9. D. Shinoda, Phys. Stat. Sol. (A) **11**, 129 (1972).
10. M. E. Fisher and J. S. Langer, Phys. Rev. Lett. **20**, 665 (1968).
11. N. Sakamoto et al., Phys. Rev **B 69**, 092401 (2004).
12. Q. Zhang et al., (unpublished).
13. C. Pfleiderer et al., Phys. Rev. **B55**, 8330 (1997).
14. F. P. Mena et al., Phys. Rev. **B 67**, 241101 (2003).
15. S. M. Stishov et al., Phys. Rev. **B 76**, 052405 (2007).
16. M. Garst and A. Rosch, Phys. Rev. **B 72**, 205129 (2005).

Electron Spectroscopy of Correlated Transition Metal Oxides

K. Maiti

Abstract It is now realized that electron–electron Coulomb repulsion (electron correlation) strength, U plays the key role in determining the electronic properties of various condensed matter systems. Employing photoelectron spectroscopy, it is observed that the surface and bulk electronic structures can be significantly different in transition metal oxides. One can extract the surface and bulk spectra using variation of surface sensitivity of the technique with the incident photon energies. The bulk spectra in various transition metal oxides show that the magnitude of U reduces with the increase in radial extension of the associated orbitals as expected due to their inverse relation $[U(3d) \geq U(4d) \geq U(5d)]$. However, U is also found to depend on associated crystallographic structures. Particle-hole asymmetry, disorder in these correlated systems leads to plethora of novel properties. In higher d systems, electron–lattice, electron–magnon couplings become more important, which could be probed using high energy resolution photoemission technique. Interestingly, the insulating behavior observed in $5d$ transition metal oxides often does not correspond to a gapped state.

1 Introduction

The investigation of electronic structures of transition metal oxides (TMO) has attracted a great deal of attention both experimentally and theoretically due to many interesting properties exhibited by these systems. Some of the notable examples are unusual metal-insulator (MI) transitions [25, 65], high temperature superconductivity [7], colossal magnetoresistance [28], etc. The metal-insulator transition has been of great interest for nearly more than six decades [26, 30] and yet not been understood. While the observation of an insulating behavior in partially filled systems is interesting in itself, the insulator to metal transition driven by temperature [21],

---

K. Maiti
Department of Condensed Matter Physics and Materials Science, Tata Institute of Fundamental Research, Homi Bhabha Road, Colaba, Mumbai - 400005, India
e-mail: kbmaiti@tifr.res.in

V. Zlatić and A. C. Hewson (eds.), *Properties and Applications of Thermoelectric Materials*, 267
NATO Science for Peace and Security Series B: Physics and Biophysics,

pressure [16, 32, 61–63], magnetic field [28], isotope effect [6], exposure to x-rays [38], etc. has provided an impetus for further research in this direction. Such a wide spectrum of novel electronic and magnetic properties arises primarily due to a competition between on-site Coulomb repulsion strength, U among transition metal d electrons and a large hopping interaction strength between the d and ligand states hybridized to the d states. This enables the transition metal d states to support a local magnetic moment with a concomitant tendency towards delocalized band formation. In this article, we describe, how the electron correlation influences the electronic structure of various transition metal oxides that determines their electronic properties and that can be visualized directly using photoemission spectroscopy.

1.1 Band Picture

The distinction between metals and non-metals (insulators) was first established by Bloch [9] and Wilson [93] in the early years of twentieth century within a quantum mechanical framework. Crystalline materials have a band-like energy spectrum due to finite overlap of the electronic states corresponding to different sites in a crystal lattice. An insulator is a material which has all the bands fully filled (the highest occupied band is termed as valence band, VB) or entirely empty (the lowest unoccupied band is called conduction band, CB) as shown in Fig. 1.

The energy gap between the top of the valence band and the bottom of the conduction band is called band gap, E_g. For a N particle system this can be defined as,

$$\begin{aligned} E_g &= [E(N+1) - E(N)] + [E(N-1) - E(N)] \\ &= E(N+1) + E(N-1) - 2E(N) \end{aligned} \quad (1)$$

where $E(N)$, $E(N+1)$ and $E(N-1)$ are the energies corresponding to the system having N, $(N+1)$ and $(N-1)$ particles, respectively. Here, E_g is essentially the sum

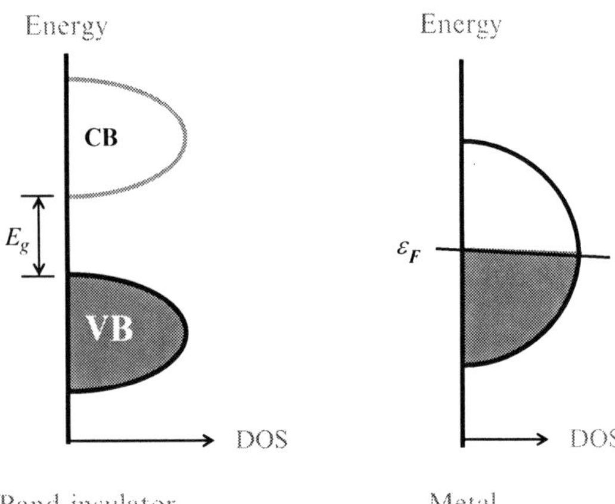

Fig. 1 Schematic diagram of the band picture of insulator and metal within independent electron approximations.

of electron affinity, $[E(N+1)-E(N)]$ and a one-electron ionization energy, $[E(N-1)-E(N)]$. Thus, the minimum energy required to transport an electron from one site to another spatially uncorrelated site is E_g and the electronic transport is activated. These insulators, that are describable within the Bloch–Wilson theories are termed as *band insulators*.

If the energy band is partially filled, the highest occupied and lowest unoccupied energy level are the same energy termed as Fermi energy, ε_F. Therefore, transport of electrons from one site to the other does not require excitation energies. These materials are known as metals, where the resistivity appears due to impurity scattering, thermal disorder etc.

Interestingly, this scenario can be described with great success using ab initio band structure calculations within the local density approximations. Since the two-particle term involving electron–electron Coulomb repulsion is treated in a mean field way, thereby, highly underestimated in the calculations within the local density approximation (LDA), all the LDA results are considered as representation of the band description (effective single particle descriptions).

1.2 Electron Correlation

Although the band picture, in describing metals and insulators has been successful in many cases, such as alkali and alkaline earth metals, the observation of insulating behavior in NiO by de Boer and Verwey [10] cannot be accounted for within this framework. Nickel in NiO is in 2+ oxidation state with a $3d^8$ electronic configuration. In an octahedral crystal field, the $3d$ levels are split into a triply degenerate t_{2g} band and a doubly degenerate e_g band. Thus, 8 electrons in the $3d$ levels will give rise to a fully filled t_{2g} band and half filled e_g band. According to the Bloch–Wilson theory, this system should be a metal in sharp contrast to the highly insulating behavior of NiO. Peierls [74] pointed out the possible importance of electron–electron interactions in this case. More than a decade later, Mott [64] presented a theory suggesting the electron–electron Coulomb interactions to be the driving force for insulating behavior in NiO. Following Mott's description, it was pointed out that if all the sites in the lattice have n electrons in the d levels, the transfer of an electron from one site to another will generate d^{n-1} and d^{n+1} sites as shown in Fig. 2. Due to the presence of electron–electron interactions, the energy required for such an electronic transport is given by,

$$E(d^n d^n \rightarrow d^{n-1} d^{n+1}) = E(d^{n+1}) + E(d^{n-1}) - 2E(d^n) = U \tag{2}$$

where $E(d^n)$ denotes the total energy of a system having n electrons in the d levels. U is the electron–electron Coulomb repulsion strength. Thus, the unoccupied band, $E(d^{n+1})$ (empty band in the figure) will be above the occupied band, $E(d^{n-1})$ (shaded band in the figure) by energy U, which represents the energy cost to transfer an electron from one site to another. On the other hand, finite hopping between

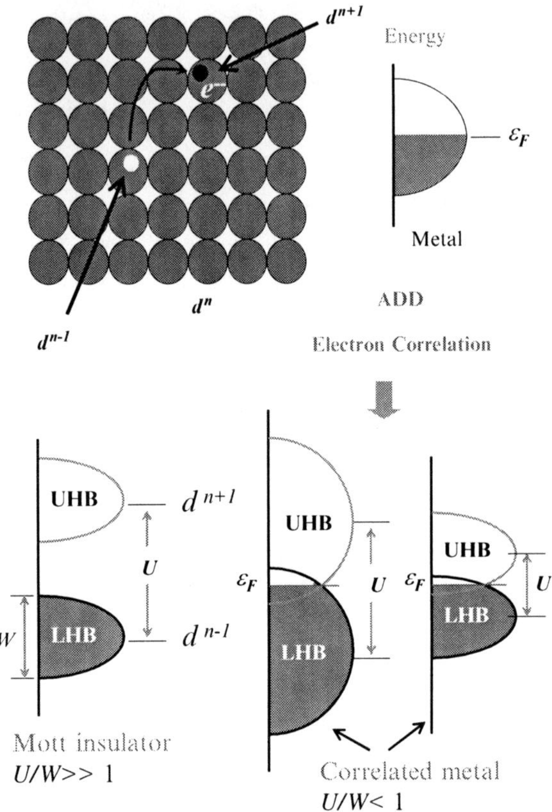

Fig. 2 Schematic diagram demonstrating the insulating phase induced by electron–electron interaction.

different sites in the solid leads to lowering of energy via charge delocalization; it competes against the tendency towards localization due to the electron correlation effect. The hopping interaction strength is directly related to the bandwidth of the system. When U is larger than the band width of d band, W the system will exhibit an insulating behavior, in contrast to the metallic behavior expected from the independent electron models. Otherwise, the system remains metallic despite electron correlation. Clearly, the electronic properties can be described by a single parameter, U/W ($U/W > 1 \Rightarrow$insulator, $U/W < 1 \Rightarrow$metal).

The above scenario has been summarized in a simple model, which is widely known as the Hubbard model as expressed below.

$$H = \sum [t_{ij} c^{\dagger}_{j\sigma} c_{i\sigma} + h.c.] + \sum_i U n_{i\uparrow} n_{i\downarrow} \tag{3}$$

where t_{ij} is the hopping interaction strength for an electron from site i to site j. Thus, the first term in the model Hamiltonian represents the inter-site hopping energy, t ($\propto W =$ bandwidth) and the second term represents on-site Coulomb repulsion energy, U. Correlation effects in 3d transition metal oxides (TMO) have extensively been studied using this model during the past few decades following the discovery of many exotic properties in these systems [26,30]. Despite simplicity of the model, the major difficulty arises from large U leading to non-perturbative nature of the problem, and the presence of various competing effects such as different types of long-range order, interplay between localization and lattice coherence etc. Therefore, various approximation methods have been used to understand the influence of electron correlation in the electronic properties.

In contrast to the predictions of the above model, recent theoretical studies reveal a correlation induced metallic ground state in a band insulator using ionic Hubbard model [22, 24, 33, 72]. Such an unusual transition has been observed in two dimensional bipartite lattice by tuning U/W and the local potential, Δ. The model Hamiltonian in this case can be expressed as,

$$H = \sum_{i\in A,j\in B} [t_{ij}c^{\dagger}_{j\sigma}c_{i\sigma} + h.c.] + \sum_i U n_{i\uparrow}n_{i\downarrow} + \Delta\left(\sum_{i\in A} n_i - \sum_{i\in B} n_i\right) + \sum_i \mu.n_i \tag{4}$$

where, μ is the chemical potential. A and B are the two non-equivalent sites in the bipartite lattice. Clearly, a half filled valence band corresponding to the pristine lattice would lead to an insulating ground state due to the doubling of the lattice translation in the bipartite lattice. Here the electron density will be dependent on the local potential at each site. Introduction of electron correlation resists delocalization of the electrons and hence a metallic phase would maintain until a Mott insulating phase emerges due to strong electron correlations. All these are theoretical predictions and need experimental realization.

1.3 Photoelectron Spectroscopy

Photoemission spectroscopy is a powerful and arguably the only tool to probe the entire electronic structure directly. The basic process involves the photo-electric effect, in which the sample under investigation is irradiated with a photon beam and the energy of the incident photon, $h\nu$ is transferred to an electron of the sample. If $h\nu$ is large enough for the electron to overcome its binding energy and the work function of the solid, the electron is ejected from the system and can be detected by an electron spectrometer. This is a quantum mechanical many body process, where the electron in the ground state (Bloch state in the crystal) is excited to vacuum by the incident photon beam and emits through the sample surface. Thus, the vacuum electron has a damped final state which carries all the informations of intrinsic effects due to photo-excitation from the ground state and various extrinsic effects occur during the transmission through the solid and emission through the sample surface.

The photoemission intensity can be expressed as

$$I(k,\varepsilon) = S(k,\varepsilon).W(\varepsilon).F(\varepsilon,T)\otimes G(\varepsilon,\alpha)\otimes L(\varepsilon,\Gamma_h)\otimes L(\varepsilon,\Gamma_e) \tag{5}$$

where, $S(k,\varepsilon)$ is the spectral function, which is the imaginary part of the Green's function and can be expressed as

$$S(k,\varepsilon) = Im\left[\frac{1}{\varepsilon - \varepsilon_k - \Sigma(k,\varepsilon)}\right] \tag{6}$$

$W(\varepsilon)$ is the transition probability, which can be written as

$$W(\varepsilon) = \langle \psi_f |A.p| \psi_i \rangle .\delta(\varepsilon_f - \varepsilon_i - h\nu). \tag{7}$$

$F(\varepsilon,T)$ is the Fermi–Dirac distribution function, $G(\varepsilon,\alpha)$ is a Gaussian representing the resolution broadening, $L(\varepsilon,\Gamma_h)$ is a Lorentzian representing the photo-hole lifetime broadening and $L(\varepsilon,\Gamma_e)$ is a Lorentzian representing the photoelectron lifetime broadening.

The electronic transition follows all the conservation rules. The energy conservation rule states that

$$h\nu = E_f - E_i = E_{bin} + W_\phi + E_{kin}, \tag{8}$$

where E_f and E_i are the final and initial state energies, respectively. W_ϕ is the work function. This rule helps to determine the binding energy of the photoelectrons via measuring their kinetic energy using a suitable electron detector. The momentum conservation rule can be expressed as,

$$k_f = k_i \pm G, \tag{9}$$

where G is the reciprocal lattice vector. Along with the energy conservation rule, the momentum conservation helps to determine the momentum of the photoelectrons via measuring the emission angle. Thus, one can map the entire Brillouin zone experimentally using this technique and is widely known as ARPES (angle resolved photoemission spectroscopy).

It is clear that photoemission spectroscopy directly probes the occupied part of the electronic structure. If the valence band consists of n d electrons (d^n electronic configuration), the photoemission would lead to d^{n-1} electronic configuration (lower Hubbard band in a correlated system). Thus, the signature of correlation induced features in the electronic structure will be manifested directly in the photoemission spectral functions.

In addition, while ejecting through the sample, the photoelectrons undergo inelastic scattering leading to various excitations such as plasmons, phonons, etc. Thus, the photoelectrons have finite lifetime. If the photo-excitations occur at a depth x from the sample surface as shown in Fig. 3, the intensity of the electrons coming out of the surface without such inelastic loss can be expressed as,

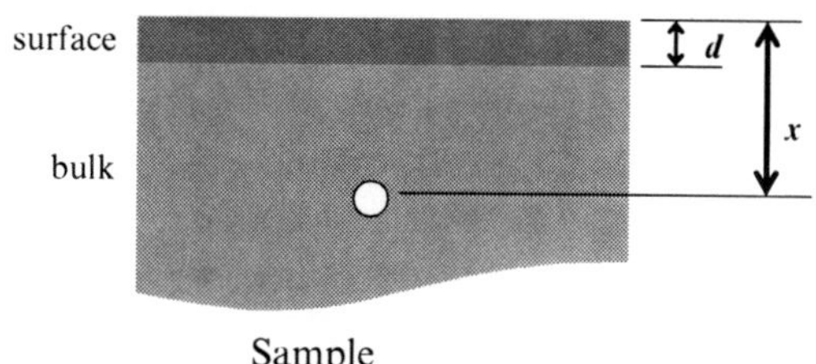

Fig. 3 Photoemission at a depth of x from the sample surface.

$$
\begin{aligned}
I(\varepsilon) &= \int_0^d I^s(\varepsilon)e^{-x/\lambda}dx + \int_d^\infty I^b(\varepsilon)e^{-x/\lambda}dx \\
&= [1-e^{-d/\lambda}]I^s(\varepsilon) + e^{-d/\lambda}I^b(\varepsilon)
\end{aligned}
\tag{10}
$$

where d is the effective surface depth and λ is the mean escape depth. $I^s(\varepsilon)$ and $I^b(\varepsilon)$ represent the surface and bulk spectra, respectively. λ varies with the kinetic energy of the photoelectrons; faster electrons will have lesser impact time with the scattering centers and hence, will have larger escape depth. At very low kinetic energies, the photoelectrons have again large λ as electron kinetic energy is not sufficient to excite other modes. The minima in escape depth strongly depend on the sample (compactness, metallicity, defects etc). It is observed that in most cases, the minima lie in the energy range of 15–100 eV [47–49, 51, 55, 56, 83].

It is evident from Eq. (10) that if one knows d/λ corresponding to two photon energies, it becomes a two unknown equation. Thus, one can delineate the experimental surface and bulk spectral functions analytically. This is an unique feature of photoemission spectroscopy; by changing the surface sensitivity of the technique, one can extract the surface and bulk electronic structures directly. However, due to the surface sensitivity of this technique, one needs an extremely clean surface to probe the electronic structure. Thus, all the photoemission measurements are carried out on freshly in situ prepared sample surface in an ultra high vacuum (UHV) chamber ($P \leq 10^{-10}$ torr).

2 Electron Correlation in 4*d* Transition Metal Oxides

Transition metal oxides have been investigated extensively during past few decades followed by the observation of many interesting properties such as high temperature superconductivity, giant magnetoresistance etc. However, large electron correlation strength, U in these systems, forbids any realistic description of these systems based on ab initio approaches. 4*d* orbitals in 4*d* transition metal oxides (TMOs) are more extended than 3*d* orbitals in 3*d* TMOs. Thus, the correlation effect is expected to be less important in these systems and provides a suitable testing ground for the applicability of various ab initio approaches. Here, we show some of the results manifesting how the electron correlation strength depends on orbital extension in 4*d* transition metal oxides, $SrRuO_3$ and $CaRuO_3$, and hence influence their electronic properties.

2.1 Electron Correlation in $SrRuO_3$ and $CaRuO_3$

Research involving ruthenates has seen an explosive growth in the recent time due to many interesting properties such as superconductivity [46], non-Fermi liquid behavior [35, 39, 41], unusual magnetic ground states [13, 35, 39, 41, 80], etc. observed in these materials. $SrRuO_3$, a perovskite compound exhibits ferromagnetic long range order below the Curie temperature of about 160 K with a large magnetic moment (1.4 μ_B) despite highly extended $4d$ character of the valence electrons [13, 80]. Interestingly, $CaRuO_3$, an isostructural compound possess similar magnetic moment at high temperatures as that observed in $SrRuO_3$ but the magnetic ground state is controversial. While some studies predicted an antiferromagnetic ordering in $CaRuO_3$ [12,44,86,92], various other studies suggest absence of long-range order down to the lowest temperature studied [13,35,39,41,59]. These later investigations predict that the behavior in this compound is in the proximity of quantum critical point, which is manifested in the transport measurements exhibiting non-Fermi liquid behavior [35, 39, 41].

Both $SrRuO_3$ and $CaRuO_3$ form in an orthorhombic perovskite structure (ABO_3-type) as shown in Fig. 4. The space group for $SrRuO_3$ is conventionally defined as *Pbnm* and that for $CaRuO_3$ is *Pnma*, which are essentially the same structure type with a difference in the definition of axis system [68, 76]. The A cation (Sr/Ca) sites help to form the typical building block of this structure. The RuO_6 octahedra in these compounds are connected by corner sharing. The conduction electrons move via this RuO_6 network and hence determine various electronic properties. A recent study, however, suggests significant influence of A-site cations in determining the structural distortions [54, 81]. The tilting and buckling of the RuO_6 octahedra as evident in the figure essentially leads to a $GdFeO_3$ type distortion resulting in an orthorhombic structure. While the structure type is same in both the com-

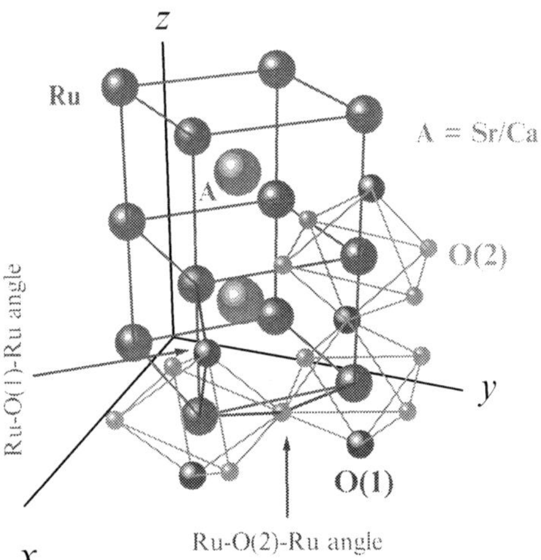

Fig. 4 Crystal structure of $SrRuO_3$ and $CaRuO_3$.

pounds, the extent of distortion is somewhat different resulting in a slightly different Ru-O–Ru bond angle in these compounds. If O(1) represents the apical oxygen in the RuO_6 octahedra along z-axis in the figure and O(2) represents the oxygens in the basal (xy) plane, there are two O(2) and one O(1) atoms in one formula unit. The Ru–O–Ru angles in $SrRuO_3$ are: Ru–O(1)–Ru = 167.6°, Ru–O(2)–Ru = 159.7° and those in $CaRuO_3$ are: Ru–O(1)–Ru = 149.6°, Ru–O(2)–Ru = 149.8°. Thus, the widely different properties of $SrRuO_3$ and $CaRuO_3$ were attributed to the different Ru–O–Ru angle leading to different valence band width and hence change in U/W.

2.1.1 Experimental Details

In order to understand the correlation induced effects in these compounds, detailed photoemission studies have been carried out. For these studies, high quality poly-crystalline samples (large grain size achieved by long sintering at the preparation temperature) were prepared by solid state reaction method using ultra-high purity ingredients and characterized by x-ray diffraction (XRD) patterns and magnetic measurements as described elsewhere [13, 70, 80]. Sharp XRD patterns revealed single phase for all the samples. Magnetic susceptibility exhibit a ferromagnetic transition in $SrRuO_3$ at 165 K and a signature of antiferromagnetic interactions in $CaRuO_3$ at 180 K. The magnetic moment of 2.7 μ_B in $SrRuO_3$ and 3 μ_B in $CaRuO_3$ in the paramagnetic phase is close to their spin-only value of 2.83 μ_B for $t^3_{2g\uparrow}t^1_{2g\downarrow}$ configurations at Ru sites.

Photoemission measurements were performed on in situ (4×10^{-11} torr) scraped samples [96, 97] using SES2002 Scienta analyzer. The experimental resolution was 1.4 meV, 2 meV and 0.3 eV for measurements with monochromatic He I, He II, and Al $K\alpha$ lines, respectively.

2.1.2 Electron Correlation Effect Manifesting in Valence Band

Valence band spectra in the vicinity of the Fermi level, ε_F obtained at He II excitation energies are shown in Fig. 5. This is essentially contributed by the Ru $4d$ density of states. O $2p$ states appear at higher binding energies [51, 54]. The spectral intensity at ε_F is significant suggesting a metallic phase. In order to obtain the band picture of this system,the single-particle electronic density of states (DOS) can be calculated using the full potential linearized augmented plane wave method (FLAPW) [8] with the local spin density approximations (LSDA). The calculational details can be found elsewhere [54]. The calculated results are also shown in the figure. Clearly, the intensity down to about 0.8 eV binding energies corresponds well to the experimental spectra. This feature is, thus, commonly known as *coherent feature* and represents the delocalized density of states of the electronic structure.

The maximum intensity in the experimental spectra appears around 1.2 eV as also observed in earlier studies [70,73]. On the other hand, the theoretical intensity peaks at about 0.5 eV with negligible contributions beyond 1 eV. Thus, this feature is often

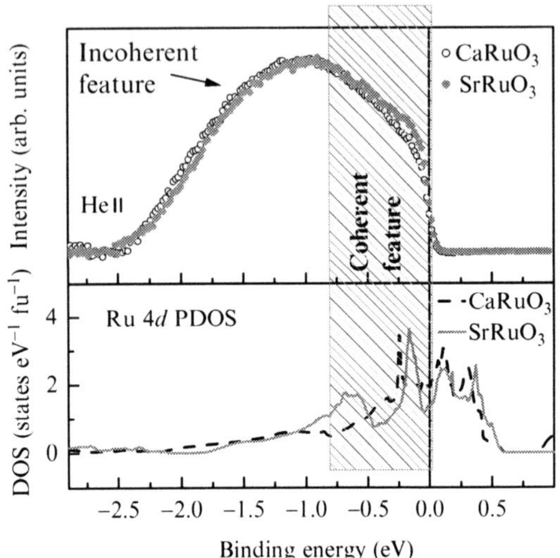

Fig. 5 (**a**) Valence band spectra of $SrRuO_3$ and $CaRuO_3$ collected using He II photons at room temperature. (**b**) Ru 4d partial density of states (PDOS). The feature below 1 eV binding energy was attributed to the lower Hubbard band (incoherent feature) in earlier studies [70, 73].

attributed to the signature of the electronic states essentially localized due to electron correlations (lower Hubbard band) and is termed as *incoherent feature*. Since, the relative intensity of the coherent feature compared to that for the incoherent feature is somewhat larger in $SrRuO_3$ than that in $CaRuO_3$, $CaRuO_3$ was predicted to be more correlated than $SrRuO_3$. While large intensity at E_F indicates metallic character, the complete dominance of this apparent incoherent feature was taken as evidence of the presence of strong correlation effects in previous studies [2,70,73]. This is surprising, considering the fact that Ru 4d orbitals are highly extended.

Valence band spectra collected using Al $K\alpha$ photons are shown in the upper panel of Fig. 6. Interestingly, the XP spectra exhibit highest intensity at 0.5 eV. The spectral lineshape is much closer to the band structure results than the low energy spectra. In order to compare them with the UP (He II) spectra, we have broadened He II spectra upto 0.3 eV, which is similar to the XP measurements. The lineshape of the valence band collected at two different photon energies are significantly different. Such a difference is curious as this energy range is essentially contributed by Ru 4d electronic states (O 2p contributions are very small).

The difference, here, appears to be the change in escape depth of the photoelectrons at these widely different photon energies. The x-ray photoelectrons has significantly larger escape depth ∼20 Å compared to the ultra-violet photoelectrons (≤10 Å), which leads to a significantly larger bulk sensitivity in XPS compared to that in UPS. Thus, the spectra in Fig. 6a suggest that the bulk and surface electronic structures are significantly different in these systems. *Notably, the lineshape of the valence band in the XP spectra of $CaRuO_3$ and $SrRuO_3$ are very similar as expected from the band structure results described above* [51,82].

The photoemission intensity shown in Eq. (10) is a function of escape depth. It is clear that if one has two experimental spectra and corresponding d/λ values, the

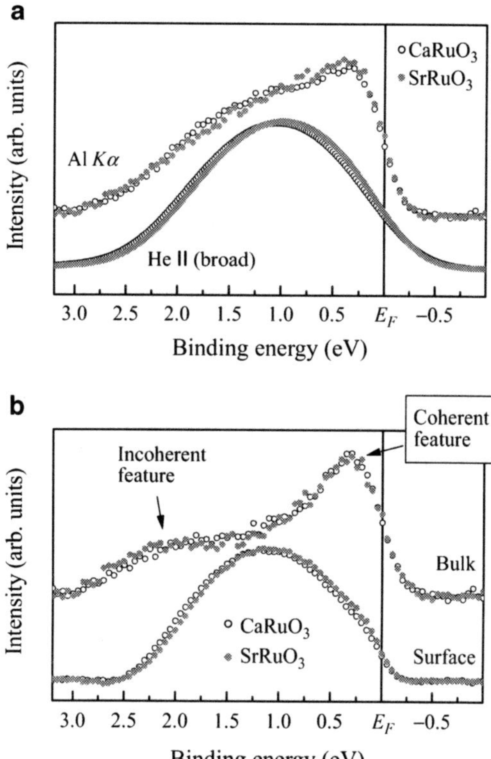

Fig. 6 (**a**) Valence band spectra collected using x-ray photons. The He II spectra broadened upto XP resolution of 0.3 eV are also plotted here to compare with the lineshape of the XP spectra. (**b**) Extracted surface and bulk spectral functions representing the surface and bulk electronic structure.

surface and bulk spectra can be analytically calculated. It is expected that the values of d/λ is very close in various transition metal oxides. Thus, we have extracted the surface and bulk spectral functions using the d/λ values similar to $CaSrVO_3$ [49]. We observe that a change in d/λ values by more than 10% leads to unphysical intensities providing confidence in this procedure. The extracted surface and bulk spectra are shown in Fig. 6b. $I^s(\varepsilon)$ in both the cases is dominated by the intensity centered at 1 eV. $I^b(\varepsilon)$ exhibit large coherent feature with the peak at about 0.5 eV as observed in the ab initio calculations. The feature around 2 eV indicates the presence of some degree of correlation effects, which is estimated below.

Since, U appears to be small, the self-energy, $\Sigma_k(\varepsilon)$, can be calculated using perturbation approach up to the second order term [77, 88, 89] within the local approximations. We have used Ru t_{2g} PDOS for these calculations. The calculated spectral functions are shown by solid lines in Fig. 7. The fit to the experimental spectra is remarkable. Most interestingly, U/W is found to be significantly small (0.24 for $CaRuO_3$ and 0.21 for $SrRuO_3$; $U = 0.6 \pm 0.05$ eV in both the cases) in sharp contrast to all previous predictions [2, 70, 73]. A small variation in U/W leads to significantly large spectral weight transfer as shown in the figure. We calculated the mass enhancement factor following the relation,

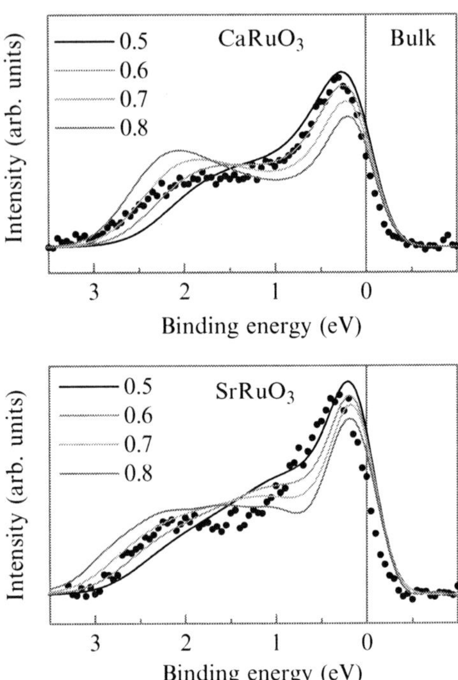

Fig. 7 Simulated bulk spectra of $CaRuO_3$ and $SrRuO_3$ for different values of U. The experimental spectra are shown by symbols.

$$\frac{m^\star}{m_b} = 1 - \frac{\partial}{\partial \varepsilon}(Re\Sigma(\varepsilon))|_{\varepsilon=E_F}.$$

Interestingly, m^*/m_b for $SrRuO_3$ is almost the same (= 2.9) as that obtained (= 3.0) from specific heat measurements. While U/W in $CaRuO_3$ is little larger than that in $SrRuO_3$, m^*/m_b in $CaRuO_3$ is found to be close to $SrRuO_3$ and somewhat smaller than its experimental value.

Differences in the surface and bulk electronic structures in $3d$ TMOs as well as in rare earths were attributed to the enhancement of U/W due to the band narrowing at the surface [42, 49]. However, this seems unreasonable as the surface peak appears at lower binding energies than the incoherent peak in the bulk spectra. It is observed that small distortion of RuO_6 octahedra in the orthorhombic structure lifts the degeneracy of the bulk t_{2g} band. Whether the surface layer consists of Sr/Ca–O or Ru–O layer, the absence of periodicity along the surface normal will enhance this distortion leading to a different crystal field symmetry presumably close to D_{4h} symmetry ($t_{2g} \Rightarrow e_g + b_{2g}$); where d_{xz} and d_{yz} bands exhibit e_g-symmetry and d_{xy} band has b_{2g}-symmetry [50]. Thus, the peak around 1 eV in the surface spectra may be attributed to an essentially filled e_g band with b_{2g} band appearing above E_F.

2.1.3 Non-Fermi Liquid Behavior in Three Dimension: Particle-Hole Asymmetry

All the above results shows that the surface and bulk electronic structures are different in ruthenates and that the electron correlation strength is rather weak as expected for extended $4d$ orbitals. The electron correlation strength is very similar in $SrRuO_3$ and $CaRuO_3$. These results, thus, makes the understanding of non-Fermi liquid behavior in $CaRuO_3$ more uncertain. In order to probe this further, we note here that the surface spectra shown in Fig. 6b for both $CaRuO_3$ and $SrRuO_3$ exhibit very small intensity at ε_F despite the fact that the resolution broadening of 300 meV of the intense higher binding energy features in the surface spectra is expected to enhance the intensity at ε_F. Thus, the spectral density at ε_F in the photoemission spectra is essentially dominated by the bulk spectra and the surface contribution is negligible. The dominance of the bulk contribution in the He I spectra has indeed been demonstrated in a similar system, $SrVO_3$ [95], where the dispersion and mass enhancement of the electronic states at ε_F in the ultraviolet spectra are identical to the bulk of the system.

In order to investigate the anomalous behavior in $CaRuO_3$ further, we probe the spectral evolution close to ε_F. The energy resolution and the Fermi level at each temperature are determined by the experiments on high purity Ag sample. The energy resolution used for these measurements was 1.5 meV. Since transport occurs in the low energy scale (~meV), *it is necessary to achieve such high resolution to investigate critically these properties*. Using this *state-of-the-art* energy resolution achieved in this instrument, we investigate the spectral changes in $SrRuO_3$ and $CaRuO_3$ as shown in Fig. 8a and b, respectively. All the spectra, normalized at the binding energy 100 meV ($\gg k_BT$), appear to cross each other at ε_F as expected in a system of fermions following Fermi-Dirac distribution function.

The electrical conduction, σ, of a Fermi-liquid scales as T^{-2} and can be expressed as $\sigma = \frac{n(\varepsilon)e^2\tau}{m^*}$ ($n(\varepsilon)$ = carrier density in the vicinity of ε_F, τ = scattering rate and $m^\star$ = effective mass). Thus, in addition to the temperature dependence of τ, the shape of $n(\varepsilon)$ and its evolution with temperature play a significant role in determining the temperature dependence of electronic conduction. Various recent studies [20] suggest that a simple power law dependence of $n(\varepsilon)$ captures most of the physical properties associated to the electronic states close to ε_F. Thus, we plot the spectral density of states (SDOS) in Fig. 8c and d. Here, the SDOS were obtained via dividing the experimental spectra by the corresponding Fermi-Dirac distribution function. All the SDOS shown reveal a dip at ε_F, which gradually becomes sharper with the decrease in temperature. The line superimposed on the experimental SDOS represents the $|\varepsilon - \varepsilon_F|^{1/2}$-dependence. Clearly, the spectral lineshape follow $|\varepsilon - \varepsilon_F|^{1/2}$-dependence for $BE > 2k_BT$ suggesting an influence of disorder in the electronic structure as also observed in other oxides [3, 78].

In addition, the spectral weight transfer appears to be different in the two systems; e.g. the spectral weight transfer from 0 to 20 meV binding energy range to the energy region above ε_F for a change in temperature from 22 to 42 K in $SrRuO_3$ is significantly larger compared to that observed for similar temperature change (16–

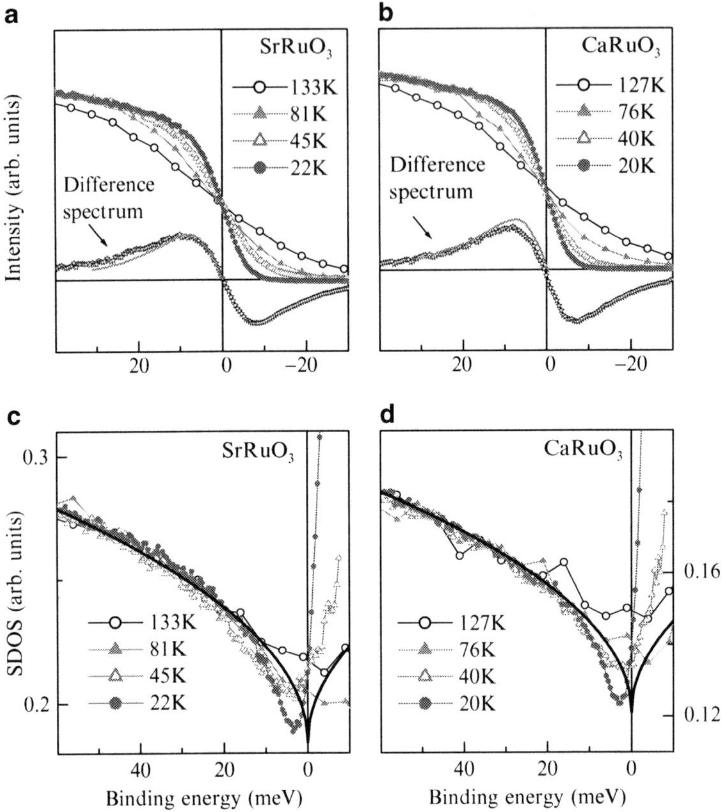

Fig. 8 Temperature evolution of the very high resolution spectra in the vicinity of the Fermi level in (**a**) $SrRuO_3$ and (**b**) $CaRuO_3$. The difference between highest and lowest temperature spectra are shown by *stars*. The *lines* represent the part of the difference spectra above ε_F plotted after inversion. The spectral density of states obtained via division by the Fermi Dirac function at each temperature are shows for (**c**) $SrRuO_3$ and (**d**) $CaRuO_3$.

40 K) in $CaRuO_3$. In order to bring out this point, we subtract the spectral functions at highest temperature from the lowest temperature spectra shown in the figure. The difference is shown by open circles in Fig. 8a and b. The lines represent the spectral differences above ε_F, superimposed after inversion onto the difference spectra below ε_F. In the case of $SrRuO_3$, the difference spectra are almost symmetric with respect to ε_F for $(|\ \varepsilon - \varepsilon_F\ | < k_BT)$ as expected for a Fermi-liquid system. The differences observed around 10–50 meV binding energies may be attributed to the temperature induced gradual population of the coherent feature [26]. In sharp contrast, the spectra in $CaRuO_3$ reveal an unusual evolution with temperature; the spectral weight transferred above ε_F is larger than the reduction in intensity below ε_F. This is clearly in contrast to the expected trend based on band structure results, which shows higher intensity above ε_F in $SrRuO_3$ suggesting a larger spectral weight transfer than that in $CaRuO_3$ [54]. *Increase in temperature populates the hole states, which could be probed*

efficiently using high energy resolution. Thus, the anomaly observed in Fig. 8b clearly reveals the asymmetric excitations between electrons and holes (particle-hole asymmetry) in $CaRuO_3$ [5,57].

NFL behavior is often found experimentally in a phase in the proximity of quantum critical point, where the NFL ground state is related to magnetic instability [85]. Evidence of the proximity of such quantum critical behavior has indeed been observed in the recent studies in high-temperature superconductors [90]. Various studies suggest that existence of low-dimensionality in these systems leads to charge fractionalization and hence NFL behavior manifests as has been shown in one-dimensional systems possessing decoupled charge and spin excitations [71,91]. The signature of particle-hole asymmetry in the spectral functions of $CaRuO_3$ observed in this study provide evidence to consider additional parameters in the study of NFL behavior [57].

2.2 *Disorder in a Weakly Correlated System: $SrRu_{1-x}Ti_xO_3$*

Ti substitution at Ru sites in $SrRuO_3$ introduces a local potential different from Ru sites in the Ru–O sublattice. At 50% substitution, every alternative Ru site is replaced by a Ti site (unit cell is doubled) similar to a bipartite lattice considered in the ionic Hubbard model as described in Eq. (4). In addition, Ti^{4+}-substitution [1,36,78] dilutes Ru–O–Ru connectivity leading to a reduction in Ru 4d bandwidth, W and hence, U/W will increase. Transport measurements [37] in this system exhibit plethora of novel phases such as correlated metal ($x \sim 0.0$), disordered metal ($x \sim 0.3$), Anderson insulator ($x \sim 0.5$), soft Coulomb gap insulator ($x \sim 0.6$), disordered correlated insulator ($x \sim 0.8$), and band insulator ($x = 1.0$). Thus, this system is a good candidate to study the role of electron correlation, disorder in exhibiting varieties of electronic properties in addition to the test case for predictions in the ionic Hubbard model.

2.2.1 Experimental Details

These samples were prepared by solid state reaction route and found to be in single phase. Magnetic measurements on $SrRu_{1-x}Ti_xO_3$ exhibit distinct ferromagnetic transition at each x up to $x = 0.6$ studied, as also evidenced by the Curie–Weiss fits in the paramagnetic region. The fits provide an estimation of effective magnetic moment ($\mu = 2.8$, 2.54, 2.45, 2.18, 2.19, 1.95 and 1.93 μ_B) and Curie temperature ($\theta_P = 164$, 156.6, 150.6, 145.3, 139, 138.6 and 100 K) for $x = 0.0$, 0.15, 0.2, 0.3, 0.4, 0.5 and 0.6, respectively. The photoemission measurements were carried out using SES2002 electron analyzer as described above. All the experimental parameters were set to similar values as described above.

2.2.2 Valence Band: Influence of Correlation and Disorder

In Fig. 9, we show the valence band spectra in the vicinity of ε_F obtained using Al $K\alpha$ and He II lines. The peaks A and B appear due to the electronic states having primarily Ru 4*d* character. Since Ti is in 4+ valence state [56], there is no electron in the Ti 3*d* band. Hence, the valence band is essentially contributed by the Ru 4*d* electronic states. Thus, all the spectra are normalized by the integrated area under the curve and corresponds to Ru 4*d* content. Clearly, in addition to the decrease in intensity due to the decrease in Ru content, the spectral lineshape changes significantly. The coherent feature (feature A in the figure) corresponds to the delocalized DOS observed in ab initio results. The feature B is the incoherent feature, absent in the ab initio results [58], is the signature of correlation induced localized electronic

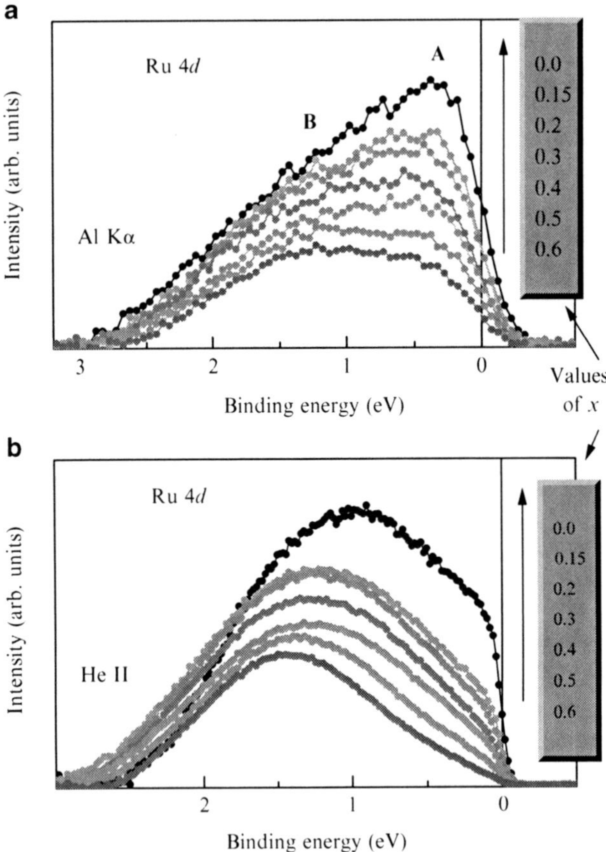

Fig. 9 Valence band spectra obtained by **(a)** Al $K\alpha$ and **(b)** He II photon excitations are shown for different values of x in $SrRu_{1-x}Ti_xO_3$. The spectral intensity is essentially contributed by Ru 4*d* density of states.

states forming the lower Hubbard band. The increase in x leads to a decrease in intensity of A and subsequently, the intensity of B grows gradually. Since the bulk sensitivity of valence electrons at 1486.6 eV photon energy is high (∼60%), the spectral evolution in Fig. 9a manifests primarily the changes in the bulk electronic structure. Ru 4d contributions extracted from the He II spectra are shown in Fig. 9b, where the surface sensitivity is about 80%. The coherent feature intensity in Fig. 9b reduces drastically with the increase in x and becomes almost negligible at $x = 0.6$.

2.2.3 Surface and Bulk Spectra Showing Dimensionality Dependence

We have extracted the surface and bulk spectra using the same parameters used in $CaRuO_3$ and $SrRuO_3$. The bulk spectra are shown in Fig. 10a. The enhancement of U/W due to Ti substitution is expected to reduce the width of the coherent features along with an increase in the incoherent feature intensity [26]. In sharp contrast, the

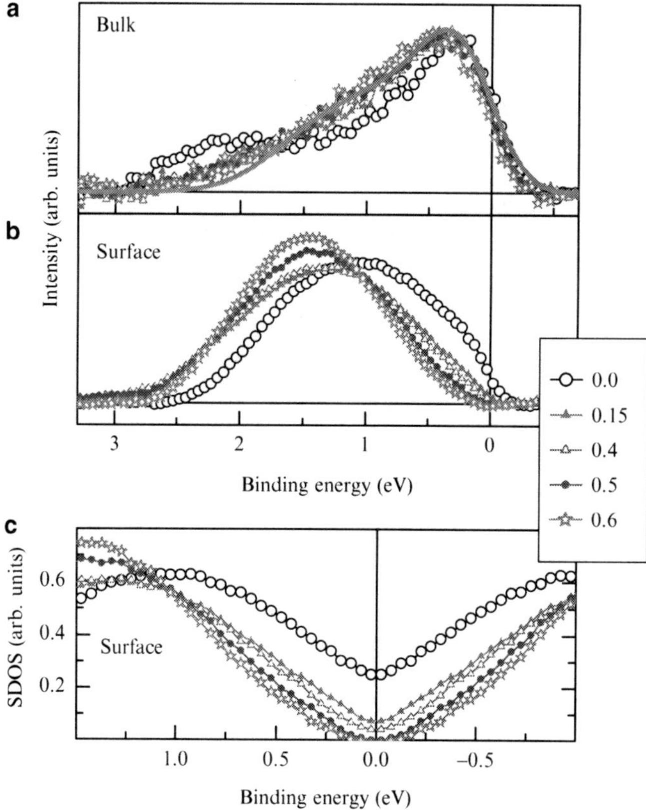

Fig. 10 (**a**) Bulk spectra, (**b**) surface spectra and (**c**) symmetrized surface spectra of $SrRu_{1-x}Ti_xO_3$ for different values of x. All the spectra are normalized by integrated intensity.

distinct signature of the incoherent feature in $SrRuO_3$ almost disappears making a broad valence band in the vicinity of ε_F. As if the coherent feature becomes significantly broad at higher x values and the lineshape is not sensitive to the value of x.

Since, U is weak in these highly extended $4d$ systems [51, 87], a perturbative approach may be useful to understand the role of electron correlation in the spectral lineshape. We have calculated the bare density of states (DOS) for $SrRu_{0.5}Ti_{0.5}O_3$ using state-of-the-art full potential linearized augmented plane wave (FLAPW) method [8, 58]. The self energy and spectral functions were calculated using this t_{2g} partial DOS as done before [77, 88, 89]. The calculated spectral functions are convoluted by Fermi–Dirac distribution function and the Gaussian representing the resolution broadening of 300 meV. The solid line in Fig. 10a represents the spectral function corresponding to $U = 0.5 \pm 0.1$ (similar to that in $SrRuO_3$) and exhibits remarkable representation of the experimental bulk spectrum. These results clearly establish that U remains almost the same across the series. The overall narrowing of the valence band observed in the substituted compounds are essentially a single particle effect and can be attributed to the reduced degree of Ru–O–Ru connectivity in these systems.

The surface spectra shown in Fig. 10b exhibit a gradual decrease in coherent feature intensity with the increase in x and subsequently, the feature around 1.5 eV becomes intense, narrower and slightly shifted towards higher binding energies. The decrease in intensity at ε_F is clearly visible in the symmetrized spectra, $S(\varepsilon) = I(\varepsilon) + I(-\varepsilon)$ shown in Fig. 10c (ε represents binding energy). Intensity at ε_F in $S(\varepsilon)$ of $x = 0.5$ sample becomes zero exhibiting a soft gap. A hard gap appears in $S(\varepsilon)$ corresponding to higher x. This spectral evolution is remarkably consistent with the transport properties [37]. These results corresponding to two-dimensional surface states presumably have strong implication in realizing recent theoretical predictions [22, 24, 33, 72] and the bulk properties of this system.

The effect of resolution broadening of 4 meV in the He II spectra is not significant in the energy scale shown in the figure. The electron and hole lifetime broadening is also negligible in the vicinity of ε_F. Thus, $S(\varepsilon)$ obtained from the He II spectra provides a good testing ground to investigate evolution of the spectral lineshape in the vicinity of ε_F. The lineshape of $S(\varepsilon)$ in Fig. 11a exhibits significant modification with the increase in x. We, thus, replot $S(\varepsilon)$ as a function of $|\varepsilon - \varepsilon_F|^{\alpha}$ for various values of α. Two extremal cases representing $\alpha = 0.5$ and 1.25 are shown in Fig. 11b and c, respectively. It is evident that $S(\varepsilon)$ of $SrRuO_3$ exhibits a straight line behavior in Fig. 11b suggesting a significant role of disorder in the electronic structure as shown earlier [3, 78].

The lineshape modifies significantly with the increase in x and becomes 1.25 in the 60% Ti substituted sample. Systems consisting of localized electronic states in the vicinity of ε_F, a soft Coulomb gap opens up due to electron–electron Coulomb repulsion; in such a situation, the ground state is stable with respect to single-particle excitations, when SDOS is characterized by $(\varepsilon - \varepsilon_F)^2$-dependence [18, 60]. Here, Ti substitution introduces defects in the Ru–O network, where Ti^{4+} having

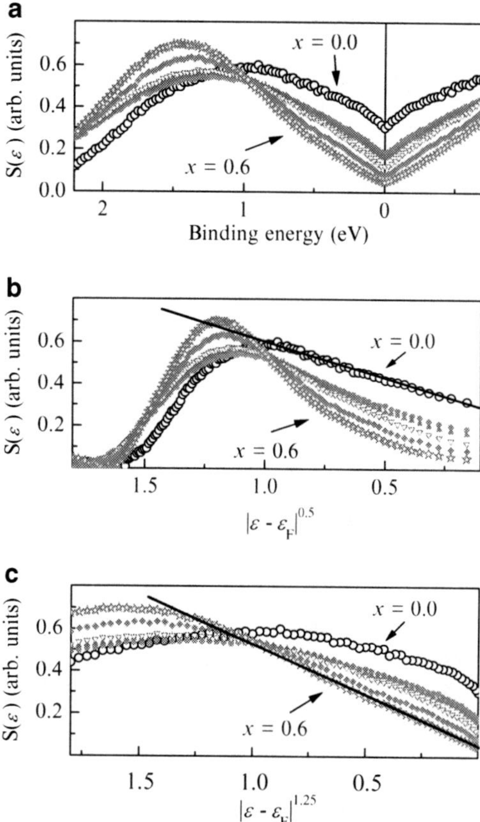

Fig. 11 (**a**) Symmetrized He II spectra, $S(\varepsilon)$ in $SrRu_{1-x}Ti_xO_3$. $S(\varepsilon)$ plotted as a function of (**b**) $|\varepsilon - \varepsilon_F|^{0.5}$ and (**c**) $|\varepsilon - \varepsilon_F|^{1.25}$.

no d-electron, does not contribute in the valence band. In addition, the reduced degree of Ru–O–Ru connectivity leads to a decrease in bandwidth, W and hence U/W enhances. Thus, gradual increase in α with the increase in x in the intermediate compositions is curious and indicates strong interplay between correlation effect and disorder in this system.

In summary, introduction of Ti^{4+} sublattice within the Ru^{4+} sublattice provides a paradigmatic example, where the charge density near Ti^{4+} sites is close to zero and each Ru^{4+} site contributes four electrons in the valence band. Such large charge fluctuation leads to a significant change in spectral lineshape and a dip appears at ε_F (*pseudogap*). The effects are much stronger in the (two dimensional) surface electronic structure leading to a soft gap at 50% substitution and eventually a hard gap appears. Bulk electronic structure (three-dimensional), however, remains less influenced. A theoretical understanding of these effects needs consideration of strong disorder in addition to the electron correlation effects.

3 Electron Correlation in 5*d* Transition Metal Oxides

The radial extension of the *d* orbitals is further extended in 5*d* transition metal oxides. Thus, the correlation strength is expected to reduce further in these compounds. Concomitantly, the *d* orbitals will be more exposed to the crystal field leading to stronger lattice coupling. Interestingly, 5*d* transition metal oxides often exhibit anomalous properties in contrast to the general belief. These systems are not studied so well like 3*d* and 4*d* transition metal oxides. Here we present the electronic structure study of two 5*d* transition metal oxides exhibiting unusual electronic properties.

3.1 Collective Excitation Induced Giant Non-linear Conductivity: $BaIrO_3$

$BaIrO_3$ exhibits many unusual crystallographic and magnetic properties [14, 15, 17, 43, 69, 75]. The crystal structure consists of Ir_3O_{12} trimers, where the IrO_6 octahedra are face shared. Inter-trimer link occurs by corner-sharing to form columns parallel to *c*-axis. The trimers are tilted by about 12° leading to monoclinic structure (space group C2/m) [43]. This quasi-one-dimensional structure is manifested in the anisotropic transport behavior [14]. The complexity of the structure results in twisting and buckling of the trimers, and in multiplicity of Ir–O and Ba–O bond distances, thereby creating four types of Ir and three types of Ba [75, 79].

$BaIrO_3$ is found to be insulating in the whole temperature range studied. Interestingly, various bulk measurements suggest a charge density wave (CDW) transition around 175 K [14]. CDW state usually appears in low-dimensional metallic systems due to the spatial modulation of the conduction electron density; the periodicity of the charge modulation is different from that of the unit-cell. As a consequence, a gap opens up in the single particle excitation spectrum below the transition temperature. Thus, the observation of CDW state in an insulating compound, $BaIrO_3$ is unusual.

Subsequently, several studies have been performed on this system to understand this unusual electronic transition. Recently, it is proposed [66] that the electronic phase transition occurring at about 180 K in $BaIrO_3$ may not be attributed to CDW transition as the abrupt jump in resistivity typical of a CDW transition is not observed. In this study, the negative differential resistance below the transition temperature was described as a giant non-linear conduction which presumably corresponds to a different class of collective excitations. Despite controversy over the identification of the ground state, it is clear that the presence of a collective excitation is necessary to explain the bulk properties, which is unusual in an insulating material.

A finding unique to this compound is that ferromagnetism sets in at the same temperature of about 180 K. Thus, the ground state also corresponds to a magnetically ordered state indicating intimate relationship between magnetism, lattice and electronic properties. In addition, it exhibits low magnetic moment, additional transitions around 80 and 26 K, non-metallicity despite very short Ir–Ir distances etc. Here, we present the temperature evolution of the electronic structure that reveals the microscopic origin of such unusual properties.

3.1.1 Experimental Details

The sample was prepared by a conventional solid state reaction method in the polycrystalline form using ultra-high pure ingredients ($BaCO_3$ and Ir metal powders). In order to achieve large grain size and stoichiometry, the final product was pelletized and sintered at a rather high temperature (1,000°C) for more than 2 days and furnace-cooled (rather than quenching) which ensures the oxygen stoichiometry close to 3 [17]. The x-ray diffraction pattern exhibits single phase without any signature of impurity and the lattice constants ($a = 10.015$ Å, $b = 5.748$ Å, $c = 15.157$ Å and $\beta = 103.27°$) are found to be in excellent agreement with those reported for single crystals [14]. In addition, we performed electrical resistivity measurements by a conventional four-probe method and establish the presence of a CDW transition at 183 K from an insulating phase; the onset of a ferromagnetic transition at the same temperature is observed by dc magnetic susceptibility measurements taken in the presence of a magnetic field of 5 kOe. Somewhat higher transition temperature observed in this compound compared to the single crystals [14] ensures good quality of the sample.

Photoemission measurements were performed using a Gammadata Scienta analyzer, SES2002 at a base pressure of 4×10^{-11} torr. The experimental resolution was 0.8 eV for Al $K\alpha$ (1486.6 eV), 4.5 meV for He II (40.8 eV) and 1.4 meV for He I (21.2 eV) measurements. The sample surface was cleaned by in situ scraping and the cleanliness was ascertained by tracking the sharpness of O 1s feature and absence of C 1s peaks. The low temperature photoemission measurements were performed using an open cycle He cryostat and the sample temperature was measured by thermocouple mounted close to the samples. At each temperature the Fermi edge was determined by the Fermi edge of in-situ scraped polycrystalline silver sample mounted on the same sample holder in electrical contact with the other samples and ground.

3.1.2 Electron Correlation Among Ir 5*d* Electrons in $BaIrO_3$

We show the valence band spectra collected using Al $K\alpha$ and He II radiations in Fig. 12. Interestingly, the spectral lineshape of both He II and Al $K\alpha$-spectra are identical despite their large difference in probing depth; this is demonstrated by superposing the resolution broadened He II spectrum (solid line in the figure) over the Al $K\alpha$-spectrum. This establishes that the surface and the bulk electronic structures are essentially identical in contrast to the observations in 3d and 4d transition metal oxides [47–49, 51, 56].

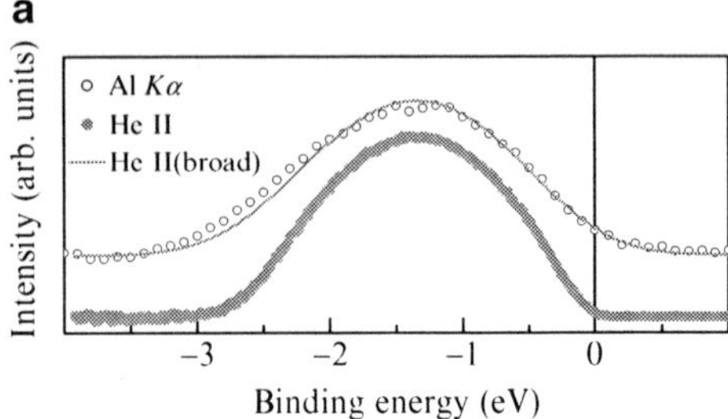

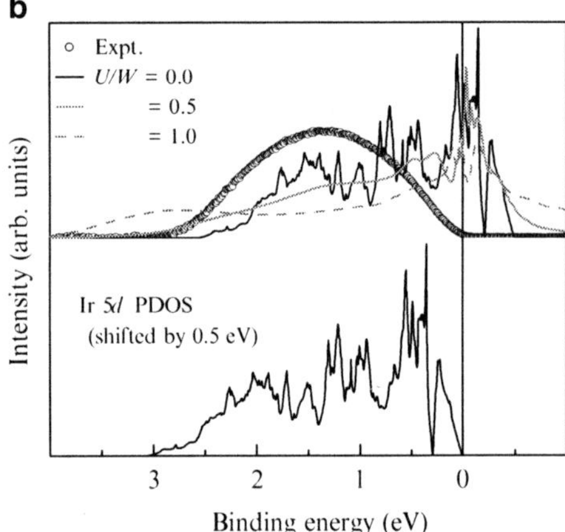

Fig. 12 (**a**) Valence band spectra at Al $K\alpha$ and He II photon energies. The *solid line* represent the resolution broadened He II spectrum exhibiting essentially identical surface and bulk electronic structure. (**b**) He II spectrum is compared with the band structure results and the spectral functions obtained introducing U via perturbation method.

Now, we compare the experimental photoemission spectra with the band structure results to investigate the influence of electron correlations. The calculated t_{2g} band of Ir 5d PDOS are shown by thin solid line in the figure exhibiting large intensity at ε_F suggesting highly metallic phase of the system; the experimental Ir 5d feature at room temperature exhibit only very small intensity. The total width of the occupied part in the calculated PDOS is also substantially smaller compared to the Ir 5d signal in the experimental spectra. Such observations of significant reduction of spectral weight at the Fermi level and the population of the higher binding energy region often attributed to the underestimation of electron–electron Coulomb repulsion effect in the ab initio approaches. However, such a strong correlation effect in these highly extended 5d electrons is unexpected.

It is to note here that the difference between experimental bulk spectral functions and the ab initio results could well be explained in the 4d transition metal oxides [51] using a perturbative approach to introduce the correlation effect in the self energy of the system. The effective correlation strength U/W was found to be small ($U/W \sim 0.2$) in 4d systems. We thus calculate the spectral function for $BaIrO_3$ in the same way, using the second order perturbation method employed by Treglia et al. [77,88,89]. The calculated spectral functions for $U/W = 0.5$ and 1.0 are shown in Fig. 12b. The increase in U/W leads to a significant spectral weight transfer to higher energies. However, the intensity at ε_F remains almost the same for U/W

as large as 1.0. The spectral function obtained in this process could not generate the spectral lineshape observed in the experimental spectrum for any value of U/W. This suggests that this procedure may not be adequate to determine the experimental results.

Interestingly, a rigid shift of the calculated t_{2g} band by 0.5 eV towards higher binding energies provides a significantly good description of the experimental spectrum. The width and spectral distribution in the shifted spectral function is remarkably similar to that observed in the He II spectrum. Thus, the Fermi level may be shifted to the upper edge of the t_{2g} band due to the electron doping via oxygen non-stoichiometry. Such a shift would stabilize the system due to the closeness to the fully filled electronic configuration resulting to lowering in energy. Interestingly, various experimental results indeed indicate that the oxygen content is often found to be less than 3.0 [17, 75], which effectively leads to electron doping into the Ir 5*d* band.

All the above observations indicate that the electron correlation may not be the primary origin for the differences between experimental and theoretical results. The electron doping due to the oxygen non-stoichiometry intrinsic to $BaIrO_3$ possibly plays significant role in determining the electronic structure and subsequently various physical properties in this system. While a direct verification of this effect is difficult due to the limitations of the computing power to handle a large supercell corresponding to a composition having small oxygen non-stoichiometry, further photoemission studies on samples having different oxygen non-stoichiometry may ensure such an effect experimentally.

3.1.3 Low Temperature Phase: Origin of Non-linear Effects

We now focus on the evolution of the valence band spectral intensities as a function of temperature in order to investigate the origin of unusual ground state. High resolution spectra are shown in Fig. 13. No spectral modification is observed down to 183 K. As the temperature is lowered across 183 K, the intensity at ε_F decreases and the leading edge shifts towards higher binding energies. A further decrease in temperature leads to an opening of a band gap of the order of 50 meV below ε_F.

The photoemission response function in Eq. (5) represent a Gaussian convolution of the spectral DOS multiplied by the Fermi-distribution function. Since the resolution broadening represented by the Gaussian (full width at half maximum = 1.4 meV) is negligible compared to the energy scale of investigation, $I(\varepsilon)/F(\varepsilon,T)$ provide a good representation of the spectral density of states, SDOS.

The SDOS thus obtained are shown by solid lines superimposed on the experimental spectra in Fig. 12a. It is important to note here that such an estimate of SDOS is sensitive to the precise location of ε_F. Therefore, we have carefully determined ε_F at each temperature by the Fermi cut off observed for silver mounted on the sample holder together with the sample. The representative spectra at 20 and 300 K are shown in the figure. In order to verify the reliability of our analysis, we carry out the same exercise for silver and found the flat density of states of silver in a wide energy range ($>3.5\,k_BT$).

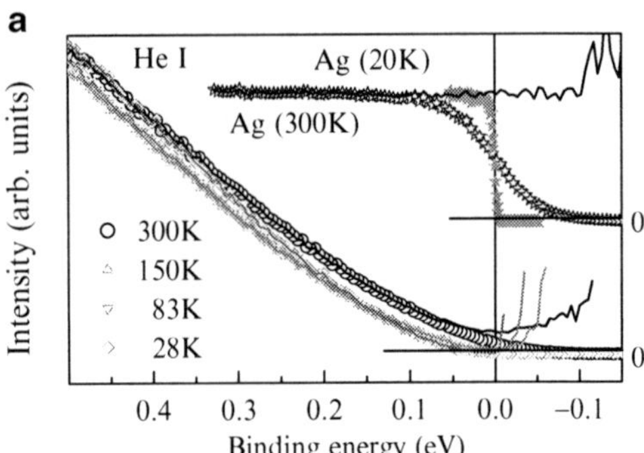

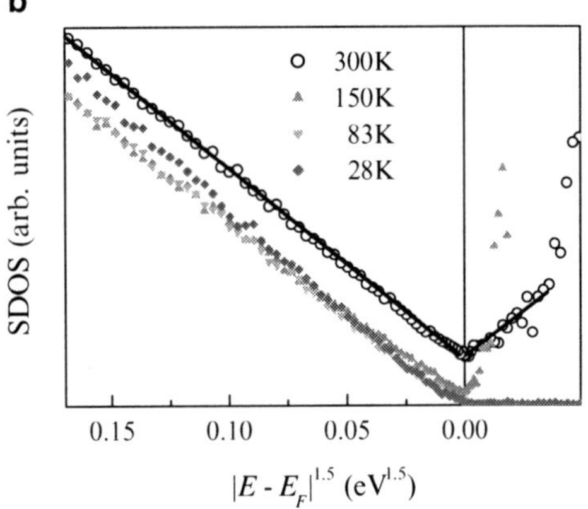

Fig. 13 (**a**) High resolution He I spectra at different temperatures. The spectral density of states (SDOS) are shown by *solid lines* superimposed on experimental spectra. Fermi level determination is also demonstrated using Ag spectra. (**b**) Experimental SDOS are plotted as a function of $|\varepsilon - \varepsilon_F|^{1.5}$.

Interestingly, the spectral DOS at room temperature exhibits a finite intensity and a distinct dip at ε_F suggesting the signature of a '*pseudogap*' [65]. This observation in conjunction with the insulating transport behavior [14] suggests that all these electronic states are essentially localized. This is not surprising for such a quasi-one dimensional system where a small lattice distortion and/or impurity leads to localization of the electronic states. Existence of such localized states forming a *pseudogap* at E_F has been predicted long before [4, 65]. High resolution employed for these measurements makes it possible to directly probe such electronic states experimentally. A notable finding is that the intensity at ε_F becomes close to zero below the phase transition, thus forming a *soft gap* at ε_F. Eventually a gap of the order of 50 meV below ε_F opens up around 83 K.

It is well known that the disorder leads to a $|\varepsilon - \varepsilon_F|^{1/2}$ (ε is energy) cusp at ε_F in metals [3, 78]. In an insulator consisting of localized electronic states at ε_F, a soft Coulomb gap opens up due to electron–electron Coulomb repulsion; in such a situation, the ground state is stable with respect to a single-particle excitation only if SDOS can be characterized by $(\varepsilon - \varepsilon_F)^2$-dependence [18, 60]. In contrast, in the present case, the SDOS exhibits a $|\varepsilon - \varepsilon_F|^{3/2}$-dependence spanning a large energy range close to ε_F (BE $\leq$ 300 meV) as shown in Fig. 12b. Most interestingly, this dependence remain unchanged down to the lowest temperature studied. Thus, the opening of a *soft gap* observed here has an origin different from the electron–electron Coulomb repulsion observed in other systems [60]. The exponent of $3/2$ in SDOS suggests strong influence of electron–magnon coupling on the electronic structure [31] presumably responsible for the ferromagnetic ground state.

3.2 Strong Correlation in 5d Transition Metal Oxide: $Y_2Ir_2O_7$

It is now clear that the electron correlation strength is weak in 4*d* transition metal oxides. Signature of electron correlation in 5*d* transition metal compounds has not been observed so far [19, 29, 53]. Here, we show a clear evidence of strong electron correlation among 5*d* electrons in a pyrochlore, $Y_2Ir_2O_7$.

Recently, pyrochlores have drawn significant attention due to the possibility of geometrical frustration leading to a varieties of novel phenomena e.g., spin ice behavior [11], superconductivity [27], correlation induced metal insulator transitions [34] etc. In particular, a Ir based pyrochlore, $Pr_2Ir_2O_7$ has been recently identified to show spin-liquid behavior [67], anomalous Hall effect [45] etc. While $Pr_2Ir_2O_7$ is a metal, the Y-analogue, $Y_2Ir_2O_7$ is described to be a *Mott insulator* [23, 84, 94]. In addition, $Y_2Ir_2O_7$ has been proposed to exhibit a weak ferromagnetic transition at around 150 K [23, 84]. Therefore, this is an ideal system to probe 5*d* electron correlation.

3.2.1 Experimental Details

$Y_2Ir_2O_7$ was prepared by solid state reaction route; high-purity (>99.9%) Y_2O_3 and Ir powder were ground together and heated at 900°C for a day, 1150°C for 6 days. To achieve large grain size and good inter-grain binding, the sample was finally sintered in pellet form at 1150°C for 10 days. X-ray diffraction (XRD) pattern does not exhibit any trace of impurity. Rietveld refinement reveals a single cubic phase ($a = 10.20$ Å; space group $Fd\bar{3}m$). Energy dispersive x-ray analysis using scanning electron microscopy reveals Y:Ir ratio to be uniform (1:1) throughout the specimen.

The energy resolutions in the photoemission measurements using SES2002 Gammadata Scienta analyzer and LT-3M cryostat from Advanced Research Systems were set to 300, 4 and 1.4 meV for the measurements with x-ray (XP) (1486.6 eV), He II (40.8 eV) and He I (21.2 eV) photons, respectively. The sample surface was cleaned by in situ scraping (base pressure $= 3 \times 10^{-11}$ torr) and cleanliness of the sample surface was ensured by negligible (<2%) impurity contributions in the O 1*s* spectral region and the absence of C 1*s* peak. The electronic band structure calculations were carried out using full potential linearized augmented plane wave method (WIEN2k software) [8] within the local spin density approximations, LSDA. Structural parameters used for these calculations were estimated via Rietveld refinement of the XRD pattern.

3.2.2 Valence Band: Surface-bulk Issue and Electron Correlation

Like other transition metal oxides presented here, the valence band close to ε_F is essentially contributed by the Ir 5*d* density of states. All other contributions in this energy range are negligible. The valence spectra collected using He I and Al $K\alpha$ (XP) radiations are shown in Fig. 14a. Both the spectra exhibit a broad feature but the lineshape is significantly different as well as the energy position of the peak. This is demonstrated by overlapping the resolution broadened He I spectrum (solid line) over the XP spectrum. The x-ray photoelectrons have larger escape depth than the ultraviolet (UV) photoelectrons. Thus, the difference in the XP and UV spectra is attributed to the different bulk and surface electronic structures. We have extracted the surface and bulk spectra following the procedure and parameters experimentally found in earlier studies on similar systems [51, 55]. The extracted surface and bulk spectral functions are plotted in Fig. 14c. The surface spectrum exhibits a peak around 1.8 eV with no intensity at ε_F suggesting insulating character of the surface electronic structure. The bulk spectrum, on the other hand, exhibits a substantial intensity at ε_F suggesting metallic phase in addition to an intense peak at around 1 eV. This is in sharp contrast to the prediction of Mott insulating phase in earlier studies.

In order to compare with the LSDA results, the calculated density of states (DOS) are shown in Fig. 14b. The total DOS reveal metallic ground state with DOS spreading down to 1 eV binding energy. This is significantly different from

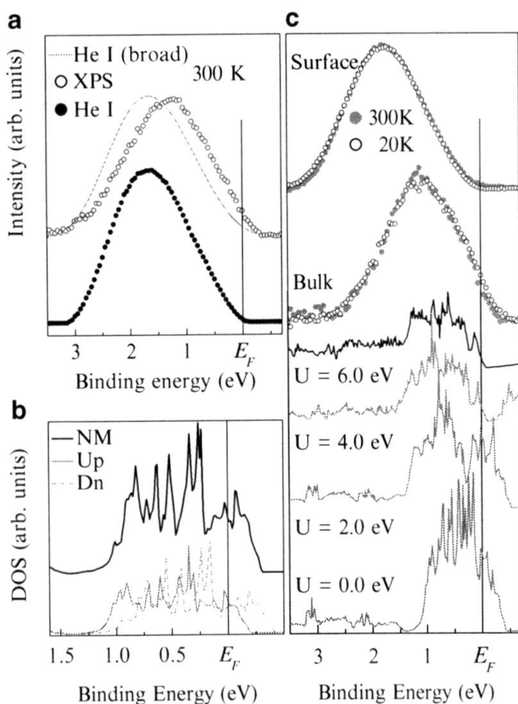

Fig. 14 (**a**) Al $K\alpha$ and He I spectra at room temperature. The solid line represents the resolution broadened He I spectrum. (**b**) Calculated density of states for the nonmagnetic (NM) and ferromagnetic ground states. Up and Dn represent the up spin and down spin density of states. (**c**) Surface and bulk spectra at 300 and 20 K. The bulk spectra are compared with the calculated spectral density of states obtained via LSDA + U calculations for different values of U.

the experimental bulk spectra exhibiting a peak at about 1 eV. Consideration of spin polarization in the calculations helps to capture the magnetic phase with a small finite magnetic moment centered at Ir sites. However, the spectral lineshape does not become better.

Since, the electron correlation is significantly underestimated in the band structure calculations within LSDA, the difference between the calculated and experimental spectra can be attributed to the electron correlation effects [30]. Thus, the feature at about 1.1 eV in the bulk spectrum can be assigned to the contributions from the lower Hubbard band (incoherent feature). The large intensities near ε_F represent the signature of delocalized electronic states (coherent feature). The large intensity of the incoherent feature compared to the coherent feature intensity indicates strong electron correlation effects, in sharp contrast to the observations in other Ir compounds [53]. The deviation from LSDA results is most evident in the surface spectrum. Only the incoherent feature is observed along with a large gap at ε_F suggesting a Mott insulating phase corresponding to the two dimensional electronic structure at the surface.

In order to estimate the electron correlation strength among 5*d* electrons, we compare the experimental spectra with the ab initio results, where the calculations were carried out for different values of U using LSDA + U method. The calculated Ir 5*d* PDOS are shown in Fig. 14c. The PDOS for $U = 0\,\text{eV}$ exhibits a peak at around 0.5 eV, spreading down to about 1 eV binding energy, which is significantly narrower than the width of the experimental bulk spectra. The increase in U leads to a spectral weight transfer away from ε_F. It is evident that the results corresponding to $U \geq 4\,\text{eV}$ provide the best description of the experimental spectrum. This value of U is significantly large ($U/W \gg 2$) and comparable to that found in 3*d* transition metal oxides. Such a large U corresponding to highly extended Ir 5*d* electrons is unusual. This study, thus, establishes that in addition to radial extension, one needs to consider the associated crystal structure to understand the correlation effects in various transition metal oxides.

3.2.3 Issue of Insulating Behavior and Magnetic Ground State

We now focus on the behavior of correlated 5*d* electrons in the magnetically ordered phase. For this purpose, we first discuss the surface and bulk spectra at 20 K shown by open symbols in Fig. 14c. The surface spectra remain unaffected with the change in temperature. No hard gap is observed in the bulk spectra down to the lowest temperature studied, which evidently rules out the possibility of Mott insulating phase in the bulk even at low temperatures. This suggests that the insulating transport observed in this system is unusual. Clearly, other localization effects are operative here. This is an important issue that came out of this study and hope these results will help to initiate future studies in this direction. The lineshape of the 20 K bulk spectrum is very similar to that at 300 K indicating that the magnetic phase transition has insignificant influence if viewed in the energy scale of the figure.

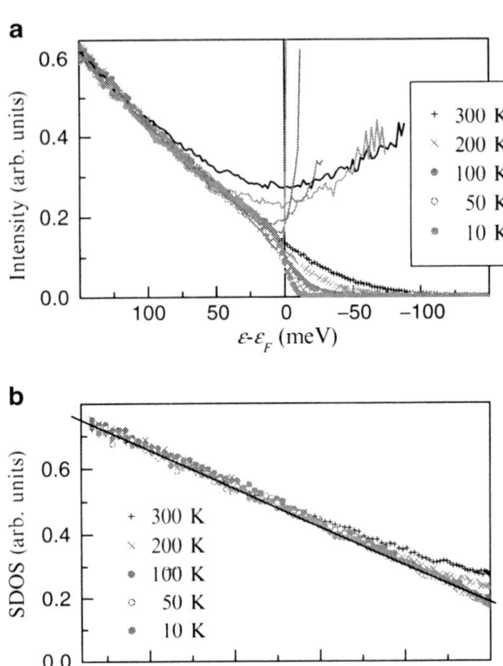

Fig. 15 (**a**) High resolution spectra at different temperatures in the vicinity of ε_F. The spectral density of states obtained via dividing the spectra by the Fermi–Dirac functions are shown by lines overlapped onto the raw data. (**b**) The spectral density of states are plotted as a functions of $|\varepsilon - \varepsilon_F|^{1.5}$.

Although the electron correlation effects are manifested in the large energy scale as described above, various thermodynamic properties are essentially determined by the electronic states near ε_F ($|\varepsilon - \varepsilon_F| \approx k_B T$). High energy resolution employed in the present investigation enables one to address this issue. We investigate the evolution of the He I spectra near ε_F as a function of temperature in Fig. 15a, which represent the bulk features as the surface spectra exhibit a large gap. Normalization of all the spectra at around 200 meV binding energy shows similar line shape down to about 50 meV binding energy at all the temperatures. The spectra in the energy range closer to ε_F reveal interesting evolution along with the appearance of a sharp Fermi cut off at low temperatures.

Since the energy resolution is high and various lifetime broadenings are insignificant in the vicinity of ε_F, one can extract the spectral density of states (SDOS) directly from the raw data, dividing them by the Fermi Dirac distribution function. There is a dip in SDOS at ε_F, which gradually increases with the decrease in temperature. It is often observed that the intensity at ε_F decreases with the decrease in temperature due to the disorder induced localization of the electronic states at ε_F [40, 78]. In such a case, the DOS at ε_F follows ($a + b\sqrt{T}$; T is temperature) behavior [3, 40], which is very different from the behavior in the present case.

In order to investigate the energy dependence of the spectral lineshape, we analyzed SDOS as a function of $|\varepsilon - \varepsilon_F|^{\alpha}$ for different values of α. All the spectra could not be simulated for one value of α. We show an extreme case in Fig. 15b ($\alpha = 1.5$). Clearly, disorder does not play a major role as corresponding behavior of $\alpha = 0.5$ is not observed [3, 40, 78]. The spectra in the magnetically ordered phase ($T < 150$ K) exhibit a deviation from $\alpha = 2$ behavior and are better represented by $\alpha = 1.5$. High resolution spectra of another 5*d* compound, $BaIrO_3$, exhibiting ferromagnetic ground state below 183 K is also characterized by $\alpha = 1.5$ [52]. Thus, it is clear that electron–magnon coupling [31] plays a significant role to determine the magnetically ordered states in these highly extended 5*d* systems.

The SDOS at non-magnetic phase ($T > 150$ K) exhibit deviation from $\alpha = 1.5$ behavior. It appears that 300 K is better represented by $\alpha = 2$ indicating strong influence of electron correlation effect in the electronic structure. These findings may serve as an experimental demonstration of various theoretical predictions of correlation induced effects, disorder, collective excitations etc. in a Fermi liquid [18, 60].

4 Conclusions

In summary, we have addressed the issue of electron correlation among 4*d* and 5*d* electrons in various crystallographic structures. Photoemission spectroscopy is arguably the best technique to probe the electronic structure directly. However, one needs to be careful while comparing the experimental spectra directly to various bulk properties and theoretical predictions as the surface and bulk electronic structure can be significantly different. One can use the variation of probing depth in photoemission to verify this and extract the surface and bulk spectral functions. Probing exotic bulk properties requires high energy resolution that reveals subtle changes in the spectra in the vicinity of the Fermi level.

Experimental results show that the electron correlation strength is a function of radial extension of the associated orbitals. The correlation strength reduces with the increase in radial extensions. However, the correlation strength can be strong even in highly extended 5*d* transition metal oxides depending on the crystallographic structure. We see that the bandwidth is significantly small in $Y_2Ir_2O_7$, which leads to an effective local character in these electronic states. Disorder plays a significant role in determining the electronic properties. While disorder induced effects are often described via $|\varepsilon - \varepsilon_F|^{0.5}$ dependence, other exponents are also manifested in the experimental results depending on electron correlation, collective excitations etc.

The deviation from the Fermi liquid behavior is usually observed in low dimensional systems. Recently a three dimensional system, $CaRuO_3$ exhibits such behavior in various bulk measurements. The high resolution photoemission study here reveals a signature of particle-hole asymmetry in this system paving a new way to describe non-Fermi liquids.

Although 5*d* orbitals have large radial extensions, most of the oxides involving 5*d* transition metals exhibit insulating behavior. The spectral density in such cases is finite at the Fermi level both in experiment and band structure calculations. This is surprising and a new mechanism is necessary to explain the insulating ground state in these systems. It is observed that often these compounds exhibit ferromagnetism at low temperatures. Subsequently, the spectral functions show a $|\varepsilon - \varepsilon_F|^{1.5}$ dependence suggesting an important role of electron-magnon coupling in their properties. This is natural as a large radial extension of the $5d$ orbitals would favor collective excitation modes. This is also manifested by the giant-non linear conductivity in a quasi one dimensional system, $BaIrO_3$.

References

1. Abbate, M., J.A. Guevara, S.L. Cuffini, Y.P. Mascarenhas, and E. Morikawa, Eur. Phys. J. B **25**, 203 (2002).
2. Ahn, J.S. et al., Phys. Rev. Lett. **82**, 5321 (1999).
3. Altshuler, B.L. and A.G. Aronov, Solid State Commun. **30**, 115 (1979).
4. Anderson, P.W., Phys. Rev **109**, 1492 (1958).
5. Anderson, P.W., Nature Phys. **2**, 626 (2006).
6. Babushkina, N.A. et al., Nature **391**, 159 (1998).
7. Bednorz, G. and K.A. Muller, Z. Phys. B **64**, 189 (1986).
8. Blaha, P., K. Schwarz, G.K.H. Madsen, D. Kvasnicka, and J. Luitz, WIEN2k, An Augmented Plane Wave + Local Orbitals Program for Calculating Crystal Properties (Karlheinz Schwarz, Techn. Universität Wien, Austria), 2001. ISBN 3-9501031-1-2.
9. Bloch, F., Z. Phys. **57**, 545 (1929).
10. Boer, J.H. de and E.J.W. Verwey, Proc. Phys. Soc. A **49**, 59 (1937).
11. Bramwell, S.T. and M.J.P. Gingras, Science **294**, 1495 (2001).
12. Callaghan, A., C.W. Moeller, and R. Ward, Inorg. Chem. **5**, 1572 (1966).
13. Cao, G., S. McCall, M. Shepard, J.E. Crow, and R.P. Guertin, Phys. Rev. B **56**, 321 (1997).
14. Cao, G. et al., Solid State Commun. **113**, 657 (2000).
15. Cao, G. et al., Phys. Rev. B **69**, 174418 (2004).
16. Carter, S.A. et al., Phys. Rev. B **48**, 16841 (1993).
17. Chamberland, B.L., J. Less-Common Met. **171**, 377 (1991).
18. Efros, A.L. and B.I. Shklovskii, J. Phys. C: Solid State Phys. **8**, L49 (1975).
19. Eguchi, R. et al., Phys. Rev. B **66**, 012516 (2002).
20. Florens, S. and M. Vojta, Phys. Rev. B **72**, 115117 (2005); G.-H. Gweon et al., Phys. Rev. Lett. **85**, 3985 (2000).
21. Föex, M., C.R. Acad. Sci. **223**, 1126 (1946); **229**, 880 (1949); F.J. Morin, Phys. Rev. Lett. **3**, 34.
22. Fuhrmann, A., D. Heilmann, and H. Monien, Phys. Rev. B **73**, 245118 (2006).
23. Fukazawa, H. and Y. Maeno, J. Phys. Soc. Jpn. **71**, 2578 (2002).
24. Garg, A., H.R. Krishnamurthy, and Mohit Randeria, Phys. Rev. Lett. **97**, 046403 (2006).
25. Gebhard, F., *The Mott Metal-Insulator Transition*, (Springer-Verlag, Berlin, 1997).
26. Georges, A., G. Kotliar, W. Krauth, and M.J. Rozenberg, Rev. Mod. Phys. **68**, 13 (1996).
27. Hanawa, M. et al., Phys. Rev. Lett. **87**, 187001 (2001).
28. Helmolt, R. von et al., Phys. Rev. Lett. **71**, 2331 (1993); S. Jin et al., Science **264**, 413 (1994).
29. Hsu, L.S., G.Y. Guo, J.D. Denlinger, and J.W. Allen, Phys. Rev. B **63**, 155105 (2001).
30. Imada, M., A. Fujimori, and Y. Tokura, Rev. Mod. Phys. **70**, 1039 (1998).
31. Irkhin, V. Yu et al., J. Phys.: Condens. Matter **5**, 8763 (1993); V. Yu Irkhin and M.I. Katsnelson, J. Phys.: Condens. Matter **2**, 7151 (1990).
32. Jayaraman, A., D.B. Mc Whan, J.P. Remeika, and P.D. Dernier, Phys. Rev. B **2**, 3751 (1970).
33. Kancharla, S.S. and E. Dagotto, Phys. Rev. Lett. **98**, 016402 (2007).
34. Kézsmárki, I. et al., Phys. Rev. Lett. **93**, 266401 (2004).

35. Khalifah, P., I. Ohkubo, H. Christen, and D. Mandrus, Phys. Rev. B **70**, 134426 (2004).
36. Kim, J., J.-Y. Kim, B.-G. Park, and S.-J. Oh, Phys. Rev. B **73**, 235109 (2006).
37. Kim, K.W., J.S. Lee, T.W. Noh, S.R. Lee, and K. Char, Phys. Rev. B **71**, 125104 (2005).
38. Kiryukhin, V. et al., Nature **386**, 813 (1997).
39. Klein, L., L. Antognazza, T.H. Geballe, M.R. Beasley, and A. Kapitulnik, Phys. Rev. B **60**, 1448 (1999).
40. Kobayashi, H., M. Nagata, R. Kanno, and Y. Kawamoto, Mater. Res. Bull. **29**, 1271 (1994).
41. Lee, Y.S., Jaejun Yu, J.S. Lee, T.W. Noh, T.-H. Gimm, Han-Yong Choi, and C.B. Eom, Phys. Rev. B **66**, 041104(R) (2002).
42. Liebsch, A., Phys. Rev. Lett. **90**, 096401 (2003); E. Pavarini et al., Phys. Rev. Lett. **92**, 176403 (2004).
43. Lindsay, R. et al., Solid State Commun. **86**, 759 (1993).
44. Longo, J.M., P.M. Raccah, and J.B. Goodenough, J. Appl. Phys. **39**, 1327 (1968).
45. Machida, Y. et al., Phys. Rev. Lett. **98**, 057203 (2007).
46. Maeno, Y. et al., Nature **372**, 532 (1994).
47. Maiti, K., P. Mahadevan, and D.D. Sarma, Phys. Rev. Lett. **80**, 2885 (1998).
48. Maiti, K. and D.D. Sarma, Phys. Rev. B **61**, 2525 (2000).
49. Maiti, K. et al., Europhys. Lett. **55**, 246 (2001).
50. Maiti, K. et al., Phys. Rev B **70**, 195112 (2004).
51. Maiti, K. and R.S. Singh, Phys. Rev. B **71**, 161102(R) (2005).
52. Maiti, K., R.S. Singh, V.R.R. Medicherla, S. Rayaprol, and E.V. Sampathkumaran, Phys. Rev. Lett. **95**, 016404 (2005).
53. Maiti, K. Phys. Rev. B **73**, 115119 (2006).
54. Maiti, K., Phys. Rev. B **73**, 235110 (2006).
55. Maiti, K., U. Manju, S. Ray, P. Mahadevan, I.H. Inoue, C. Carbone, and D.D. Sarma, Phys. Rev. B **73**, 052508 (2006).
56. Maiti, K., R.S. Singh, and V.R.R. Medicherla, Phys. Rev. B **76**, 165128 (2007).
57. Maiti, K., R.S. Singh, and V.R.R. Medicherla, Europhys. Lett. **78**, 17002 (2007).
58. Maiti, K., Phys. Rev. B **77**, 212407 (2008).
59. Martinez, J.L., C. Prieto, J. Rodriguez-Carvajal, A. de Andrés, M. Valet-Regi, and J.M. Gonzaler-Calbet, J. Magn. Magn. Mater. **140–144**, 179 (1995).
60. Massey, J.G. and M. Lee, Phys. Rev. Lett. **75**, 4266 (1995).
61. Mc Whan, D.B., T.M. Rice, and J.P. Remeika, Phys. Rev. Lett. **23**, 1384 (1969).
62. Mc Whan, D.B. and J.P. Remeika, Phys. Rev. B **2**, 3734 (1970).
63. Mc Whan, D.B., A. Menth, J.P. Remeika, W.F. Brinkman, and T.M. Rice, Phys. Rev. B **7**, 1920 (1973).
64. Mott, N.F., Proc. Roy. Soc. A **62**, 416 (1949).
65. Mott, N.F., *Metal-Insulator Transitions*, 2nd ed. (Taylor & Francis, London, 1990).
66. Nakano, T. and I. Terasaki, Phys. Rev. B **73** 195106 (2006).
67. Nakatsuji, S. et al., Phys. Rev. Lett. **96**, 087204 (2006).
68. Nakatsugawa, H., E. Iguchi, and Y. Oohara, J. Phys.: Condens. Mater **14**, 415 (2002).
69. A. Ohnishi et al., Physica B **329**, 930 (2003).
70. Okamoto, J. et al., Phys. Rev. B **60**, 2281 (1999).
71. Orgad, D. et al., Phys. Rev. Lett. **86**, 4362 (2001).
72. Paris, N., K. Bouadim, F. Hebert, G.G. Batrouni, and R.T. Scalettar, Phys. Rev. Lett. **98**, 046403 (2007).
73. Park, J. et al., Phys. Rev. B **69**, 085108 (2004).
74. Peierls, R., Proc. Phys. Soc. A **49** (1937).
75. Powell, A.V. and P.D. Battle, J. Alloys and Compounds **232**, 147 (1996).
76. Rama Rao, M.V., V.G. Sathe, D. Sornadurai, B. Panigrahi, and T. Shripathi, J. Phys. Chem. Solids **62**, 797 (2001).
77. Sarma D.D., et al., Phys. Rev. Lett. **57**, 2215 (1986).
78. Sarma, D.D., et al., Phys. Rev. Lett. **80**, 4004 (1998).
79. Siegrist, T. and B.L. Chamberland, J. Less-Common Met. **170**, 93 (1991).
80. Singh, R.S. and K. Maiti, Solid State Commun **140**, 188 (2006).

81. Singh, R.S. and K. Maiti, Phys. Rev. B **76**, 085102 (2007).
82. Singh, R.S., V.R.R. Mdeicherla, and K. Maiti, Appl. Phys. Lett. **91**, 132503 (2007).
83. Singh, R.S., V.R.R. Medicherla, K. Maiti, and E.V. Sampathkumaran, Phys. Rev. B **77**, 201102(R) (2008).
84. Soda, M. et al., Physica B **329–333**, 1071 (2003).
85. Stewart, G.R., Rev. Mod. Phys. **73**, 797 (2001).
86. Sugiyama, T. and N. Tsuda, J. Phys. Soc. Jpn. **68**, 3980 (1999).
87. Takizawa, M. et al., Phys. Rev. B **72**, 060404(R) (2005).
88. Treglia, G. et al., J. Physique **41**, 281 (1980).
89. Treglia, G. et al., Phys. Rev. B **21**, 3729 (1980).
90. Valla, T. et al., Science **285**, 2110 (1999).
91. Valla, T. et al., Nature **417**, 627 (2002).
92. Vidya, R., P. Ravindran, A. Kjekshus, H. Fjellvag, and B.C. Hauback, J. Solid State Chem. **177**, 146 (2004).
93. Wilson, A.H., Proc. Roy. Soc. A **133**, 458 (1931).
94. Yanagishima, D. and Y. Maeno, J. Phys. Soc. Jpn. **70**, 2880 (2001).
95. Yoshida, T. et al., Phys. Rev. Lett. **95**, 146404 (2005).
96. Since, perovskite structure has no cleavage plane, scraping has sufficiently high probability to expose intrinsic bulk material through intragrain fracture [Susaki et al. Phys. Rev. Letts. **78**, 1832 (1997)]. We have verified this on a similar system, $SrVO_3$. We find that the spectra collected from scraped and cleaved single crystals are identical [55].
97. A small impurity contribution could not be avoided as that also observed in the high quality films studied recently [73]. Reproducibility of the spectra was confirmed after every scraping cycle.

Orbital Fluctuations in the RVO_3 Perovskites

A. M. Oleś and P. Horsch

Abstract The properties of Mott insulators with orbital degrees of freedom are described by spin-orbital superexchange models, which provide a theoretical framework for understanding their magnetic and optical properties. We introduce such a model derived for $(xy)^1(yz/zx)^1$ configuration of V^{3+} ions in the RVO_3 perovskites, $R = \text{Lu,Yb,}\cdots\text{,La}$, and demonstrate that $\{yz,zx\}$ orbital fluctuations along the c axis are responsible for the huge magnetic and optical anisotropies observed in the almost perfectly cubic compound $LaVO_3$. We argue that the $GdFeO_3$ distortion and the large difference in entropy of C-AF and G-AF phases is responsible for the second magnetic transition observed at T_{N2} in YVO_3. Next we address the variation of orbital and magnetic transition temperature, T_{OO} and T_{N1}, in the RVO_3 perovskites, after extending the spin-orbital model by the crystal-field and the orbital interactions which arise from the $GdFeO_3$ and Jahn–Teller distortions of the VO_6 octahedra. We further find that the orthorhombic distortion which increases from $LaVO_3$ to $LuVO_3$ plays a crucial role by controlling the orbital fluctuations, and via the modified orbital correlations influences the onset of both magnetic and orbital order.

1 Orbital Degrees of Freedom in Strongly Correlated Systems

Orbital degrees of freedom play a key role for many intriguing phenomena in strongly correlated transition metal oxides, such as the colossal magnetoresistance in the manganites or the effective reduction of dimensionality in $KCuF_3$ [1]. Before addressing complex phenomena in doped Mott insulators, it is necessary to describe first the undoped materials, such as $LaMnO_3$ or $LaVO_3$. These two systems

A. M. Oleś
M. Smoluchowski Institute of Physics, Jagellonian University, Reymonta 4, 30059 Kraków, Poland
Max-Planck-Institut für Festkörperforschung, Heisenbergstrasse 1, 70569 Stuttgart, Germany
e-mail: A.M.Oles@fkf.mpg.de

P. Horsch
Max-Planck-Institut für Festkörperforschung, Heisenbergstrasse 1, 70569 Stuttgart, Germany
e-mail: P.Horsch@fkf.mpg.de

V. Zlatić and A. C. Hewson (eds.), *Properties and Applications of Thermoelectric Materials*, 299
NATO Science for Peace and Security Series B: Physics and Biophysics,

are canonical examples of correlated insulators with coexisting magnetic and orbital order [1]. In both cases large local Coulomb interaction U suppresses charge fluctuations, leading to low-energy effective Hamiltonians with superexchange interactions which stabilize antiferromagnetic (AF) spin order at low temperature [2, 3]. However, the AF order is different in both cases: ferromagnetic (FM) planes are coupled by AF interactions in the A-type AF phase of $LaMnO_3$, while FM chains along the c cubic axis are coupled by AF interactions in the ab planes in the C-type AF (C-AF) phase of $LaVO_3$. The superexchange Hamiltonians which describe both systems are just examples for the spin-orbital physics [4], where orbital (pseudospin) operators contribute explicitly to the structure of the superexchange interactions – their actual form depends on the number of $3d$ electrons (holes) at transition metal ions which determines the value of spin S, and on the type of active orbital degrees of freedom, e_g or t_{2g}. In simple terms, the magnetic structure is determined by the pattern of occupied and empty orbitals, and the associated rules are known as Goodenough–Kanamori rules (GKR). The central focus of this overview are t_{2g} orbital degenerate systems, where quantum fluctuations of orbitals play a central role for the electronic properties [3, 5, 6] and modify the predictions of the GKR.

In the last two decades several new concepts were developed in the field of orbital physics [1]. The best known spin-orbital superexchange Hamiltonian is the Kugel-Khomskii model [7], which describes the e_g orbital $\{x^2-y^2, 3z^2-r^2\}$ degrees of freedom coupled to $S=1/2$ spins at Cu^{2+} (d^9) ions in $KCuF_3$. The spins interact by either FM and AF exchange interactions, depending on the type of occupied and empty orbitals on two neighboring ions. It has been found that enhanced quantum fluctuations due to orbital degrees of freedom, which contribute to joint spin-orbital dynamics, may destabilize long-range magnetic order near the quantum critical point of the Kugel–Khomskii model [8]. The orbital part of the superexchange is thereby intrinsically frustrated even on geometrically non-frustrated lattices, as in the perovskite lattice [8, 9], which is a second important concept in the field of orbital physics. Finally, although spin and orbital operators commute, there are situations where joint spin-orbital dynamics plays a crucial role, and spin and orbital operators cannot be separated from each other. This situation is called spin-orbital entanglement [10], and its best example are the entangled SU(4) singlets in the one-dimensional (1D) SU(4) model [11]. There is no doubt that these recent developments in the orbital physics provide many challenges both for the experimental studies and for the theoretical understanding of the experimental consequences of the spin-orbital superexchange.

Let us consider first the orbital part of the superexchange. Its intrinsic frustration results from the directional nature of orbital pseudospin interactions [8, 9] – this implies that the pair of orbitals which would minimize the energy depends on the direction of a bond $\langle ij\rangle$ in a cubic (perovskite) lattice. In case of e_g orbitals the superexchange interactions are Ising-like as only one orbital flavor allows for electron hopping t and the electron exchange process does not occur. This favors occupation of a pair of orthogonal orbitals on both sites of the considered bond [12], for instance $|z\rangle \sim (3z^2-r^2)/\sqrt{6}$ and $|x\rangle \sim x^2-y^2$ orbital for a bond along the c axis. When the two above orbital states are represented as components of $\tau=1/2$ pseudospin, this configuration gives the energy of $-\frac{1}{4}J$, where J is the superexchange constant. Unlike in the 1D model [13], such an optimal orbital configuration cannot be realized simultaneously on all the bonds in a two-dimensional (2D) or three-dimensional (3D) system. Thus, in contrast to spin systems, the tendency towards an orbital disordered state (orbital liquid) is *enhanced* with increasing system dimension [14, 15].

The essence of orbital frustration is captured by the 2D compass model, originally developed as a model for Mott insulators [7]. Intersite interactions in the compass model are described by products $\tau_i^\alpha \tau_j^\alpha$ of single pseudospin components,

$$\tau_i^x = \frac{1}{2}\sigma_i^x, \qquad \tau_i^y = \frac{1}{2}\sigma_i^y, \qquad \tau_i^z = \frac{1}{2}\sigma_i^z. \tag{1}$$

for a bond $\langle ij \rangle \parallel \gamma$, where $\alpha = x, y, z$, rather than by a pseudospin scalar product $\tau_i \cdot \tau_j$. For instance, in the 2D case of a single ab plane, the compass model [16],

$$H_{2D} = J_x \sum_{\langle ij \rangle \parallel a} \tau_i^x \tau_j^x + J_z \sum_{\langle ij \rangle \parallel b} \tau_i^z \tau_j^z , \tag{2}$$

describes the competition between $\tau_i^x \tau_j^x$ and $\tau_i^z \tau_j^z$ interactions for the bonds along a and b axis, respectively. This competition of pseudospin interactions along different directions results in intersite correlations similar to those in the anisotropic XY model, and generates a quantum critical point at $J_x = J_z$, with high degeneracy of the ground state [17]. So, despite certain similarities of the compass model to ordinary models used in quantum magnetism, an ordered phase with finite magnetization is absent. It is interesting to note that a similar quantum phase transition exists also in the 1D chain compass model [18] ($N' = N/2$ is the number of unit cells):

$$H_{1D} = \sum_{i=1}^{N'} \left\{ J_x \tau_{2i-1}^x \tau_{2i}^x + J_z \tau_{2i}^z \tau_{2i+1}^z \right\} . \tag{3}$$

Recently this 1D compass model was solved exactly in the whole range of $\{J_x, J_z\}$ parameters [18] by mapping to the exactly solvable (quantum) Ising model in a transverse field. It provides a beautiful example of a first order quantum phase transition between two phases with large $\langle \tau_{2i-1}^x \tau_{2i}^x \rangle$ or $\langle \tau_{2i}^z \tau_{2i+1}^z \rangle$ correlations, and a discontinuous change at the transition point of intersite correlation functions.

In realistic spin-orbital superexchange models transitions between different ordered or disordered orbital states are accompanied by magnetic transitions. This field is very rich, and several problems remain unsolved as simple mean-field (MF) approaches do not suffice in general, even for the systems with perovskite lattices [4]. In this chapter we shall address the physical properties of the RVO_3 perovskites (R = Lu,Yb,$\cdots$,La), where not only the above intrinsic frustration of the orbital superexchange, but also the structure of the spin-orbital superexchange arising from multiplet splittings due to Hund's exchange plays a role and determines the observed physical properties at finite temperature. Moreover, we shall see that the coupling of the orbitals to the lattice, i.e., via Jahn–Teller (JT) coupling, $GdFeO_3$-like and orthorhombic distortion, are important control parameters. First we analyze the structure of the spin-orbital superexchange in Section 2 and show its consequences for the magnetic and optical properties of strongly correlated transition metal compounds. Here we also address the entanglement of spin and orbital variables which is ignored in the MF decoupling, and we point out that it fails in certain situations.

The coupling between the orbital and spin variables is capable of generating qualitatively new phenomena which do not occur in the absence of orbital interactions, such as anisotropic magnetic interactions, and novel quantum phenomena at finite temperature, discussed for the example of $LaVO_3$ in Section 3. One of such novel and puzzling phenomena is the magnetic phase transition between two different types of AF order observed in YVO_3 – this compound has G-type AF order (staggered in all three directions, called below G-AF phase) at low temperature $T < T_{N2}$, while the magnetic order changes in the first order magnetic transition at $T_{N2} = 77$ K to C-AF phase which remains stable up to $T_{N1} \simeq 116$ K. The latter C-AF phase has rather exotic magnetic properties, and the magnon spectra show dimerization of the FM interactions along the c axis [19], see Fig. 1. In fact, the G-AF phase occurs in systems with large $GdFeO_3$-like distortion [20]. In Ref. [3] an orbital interaction favoring C-type alternating orbital (C-AO) order was invoked to explain the G-AF phase. We also address this problem in Section 3 and present arguments that at higher $T > T_{N2}$ C-AF phase reappears is due to its higher entropy [3].

In Section 4 we address the experimental phase diagram of the RVO$_3$ perovskites. It is quite different from the (also puzzling) phase diagram of the RMnO$_3$ perovskites, where the orbital order (OO) appears first at T_{OO} upon lowering the temperature, and spin order follows [21] at the Néel temperature, $T_{N1} \ll T_{OO}$. In contrast, in the RVO$_3$ perovskites the two transitions appear at similar temperatures [22]. For instance, in $LaVO_3$ they occur even almost simultaneously i.e., $T_{N1} \simeq T_{OO}$. However, they become separated from each other in the RVO$_3$ systems with smaller ionic radii of R ions – whereas T_{N1} gets reduced for decreasing r_R, T_{OO} exhibits a *nonmonotonic* dependence on r_R [20]. A short summary is presented in Section 5. We also point out a few unsolved problems of current interest in the field of orbital physics.

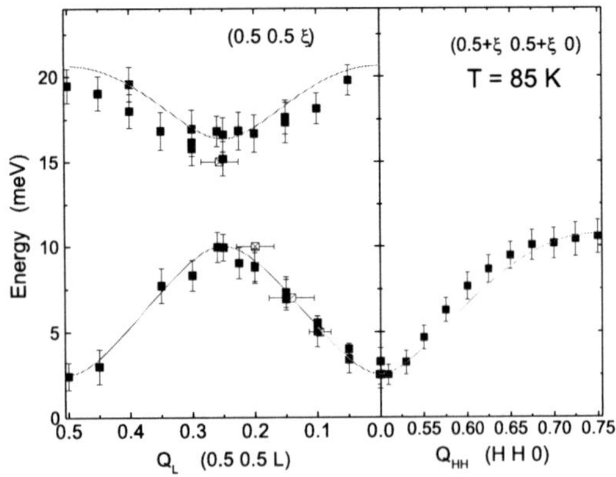

Fig. 1 Magnon dispersion relation obtained by neutron scattering for the C-AF phase of YVO_3 at $T = 85$ K. The lines are an interpolation between the experimental points (*squares* with error bars) along two high symmetry directions in the Brillouin zone. (Image courtesy of Clemens Ulrich.)

2 Spin-orbital Superexchange and Entanglement

Spin-orbital models derived for real systems are rather complex and the orbital part of the superexchange is typically much more complicated than the compass model (2) of Section 1 [4]. The main feature of these superexchange models is that not only the orbital interactions are directional and frustrated, but also spin correlations may influence orbital interactions and vice versa. This is best seen in the Kugel–Khomskii (d^9) model [7], where the G-AF and A-AF order compete with each other, and the long-range order is destabilized by quantum fluctuations in the vicinity of the quantum critical point $(J_H, E_z) = (0,0)$ [23]. Here J_H is the local exchange (see below), and E_z is the splitting of two e_g orbitals. Although this model is a possible realization of disordered spin-orbital liquid, its phase diagram remains unexplored beyond the MF approach and simple valence-bond wave functions of Ref. [8] – it remains one of the challenges in this field.

In this chapter we consider the superexchange derived for an (idealized) perovskite structure of RVO_3, with V^{3+} ions occupying the cubic lattice. The kinetic energy is given by:

$$H_t = -t \sum_{\langle ij \rangle \parallel \gamma} \sum_{\alpha(\gamma),\sigma} \left(d^\dagger_{i\alpha\sigma} d_{j\alpha\sigma} + d^\dagger_{j\alpha\sigma} d_{i\alpha\sigma} \right), \tag{4}$$

where $d^\dagger_{i\alpha\sigma}$ is electron creation operator for an electron with spin $\sigma = \uparrow, \downarrow$ in orbital α at site i. The summation runs over the bonds $\langle ij \rangle \parallel \gamma$ along three cubic axes, $\gamma = a, b, c$, with the hopping elements t between active t_{2g} orbitals. They originate from two subsequent hopping processes via the intermediate $2p_\pi$ oxygen orbital along each V–O–V bond. Its value can in principle be derived from the charge-transfer model [24], and one expects $t = t_{pd}^2/\Delta \sim 0.2$ eV [3]. Only two out of three t_{2g} orbitals, labelled by $\alpha(\gamma)$, are active along each bond $\langle ij \rangle$ and contribute to the kinetic energy (4), while the third orbital lies in the plane perpendicular to the γ axis and the hopping via the intermediate oxygen $2p_\pi$ oxygen is forbidden by symmetry. This motivates a convenient notation used below,

$$|a\rangle \equiv |yz\rangle, \qquad |b\rangle \equiv |xz\rangle, \qquad |c\rangle \equiv |xy\rangle, \tag{5}$$

where the orbital inactive along a cubic direction γ is labelled by its index as $|\gamma\rangle$.

The superexchange model for the RVO_3 perovskites arises from virtual charge excitations between V^{3+} ions in the high-spin $S = 1$ state. The number of d electrons is 2 at each V^{3+} ion (d^2 configuration), and the superexchange is derived from all possible virtual $d_i^2 d_j^2 \rightleftharpoons d_i^3 d_j^1$ excitation processes (for more details see Ref. [25]). It is parametrized by the superexchange constant J and Hund's parameter η,

$$J = \frac{4t^2}{U}, \qquad \eta = \frac{J_H}{U}, \tag{6}$$

where U is the intraorbital Coulomb interaction and J_H is Hund's exchange between t_{2g} electrons. Here we use the usual convention and write the local Coulomb interactions between $3d$ electrons at V^{3+} ions by limiting ourselves to intraorbital and two-orbital interaction elements [4]:

$$H_{\rm int} = U\sum_{i\alpha} n_{i\alpha\uparrow}n_{i\alpha\downarrow} + \left(U-\frac{5}{2}J_H\right)\sum_{i,\alpha<\beta} n_{i\alpha}n_{i\beta} - 2J_H\sum_{i,\alpha<\beta}\mathbf{S}_{i\alpha}\cdot\mathbf{S}_{i\beta}$$
$$+J_H\sum_{i,\alpha<\beta}\left(d^\dagger_{i\alpha\uparrow}d^\dagger_{i\alpha\downarrow}d_{i\beta\downarrow}d_{i\beta\uparrow} + d^\dagger_{i\beta\uparrow}d^\dagger_{i\beta\downarrow}d_{i\alpha\downarrow}d_{i\alpha\uparrow}\right). \quad (7)$$

When only one type of orbital is partly occupied (as in the present case of the RVO$_3$ perovskites or in KCuF$_3$), the two parameters $\{U, J_H\}$ are sufficient to describe these interactions in Eq. (7): (*i*) the intraorbital Coulomb element U and (*ii*) the interorbital (Hund's) exchange element J_H, where $\{A,B,C\}$ are the Racah parameters [26]. In such cases the above expression is exact; in other cases when both e_g and t_{2g} electrons contribute to charge excitations (as for instance in the RMnO$_3$ perovskites), Eq. (7) is only an approximation – the anisotropy on the interorbital interaction elements has to be then included to reproduce accurately the multiplet spectra of the transition metal ions [26]. The intraorbital interaction is $U = A+4B+3C$, while J_H depends on the orbital type – for t_{2g} electrons one finds [4, 26] $J_H = 3B + C$.

The perturbative treatment of intersite charge excitations $d_i^2d_j^2 \rightleftharpoons d_i^3d_j^1$ in the regime of $t \ll U$ leads for the RVO$_3$ perovskites (and in each similar case [4]) to the spin-orbital superexchange model:

$$\mathscr{H}_J = \sum_{\langle ij\rangle\parallel\gamma} H^{(\gamma)}(ij) = J\sum_{\langle ij\rangle\parallel\gamma}\left\{\left(\mathbf{S}_i\cdot\mathbf{S}_j + S^2\right)\hat{J}^{(\gamma)}_{ij} + \hat{K}^{(\gamma)}_{ij}\right\}. \quad (8)$$

The spin interactions $\propto \mathbf{S}_i\cdot\mathbf{S}_j$ obey the SU(2) symmetry. In contrast, the orbital interaction operators $\hat{J}^{(\gamma)}_{ij}$ and $\hat{K}^{(\gamma)}_{ij}$ involve directional (here t_{2g}) orbitals on each individual bond $\langle ij\rangle \parallel \gamma$, so they have a lower (cubic) symmetry. The above form of the spin-orbital interactions is general and the spin value S depends on the electronic configuration d^n of the transition metal ions involved (here $n = 2$ and $S = 1$). For convenience, we introduced also a constant S^2 in the spin part, so for the classical Néel order the first term $\propto \hat{J}^{(\gamma)}_{ij}$ vanishes.

In the RVO$_3$ perovskites one finds the orbital operators [25]:

$$\hat{J}^{(\gamma)}_{ij} = \frac{1}{2}\left\{(1+2\eta r_1)\left(\tau_i\cdot\tau_j + \frac{1}{4}n_in_j\right)\right.$$
$$\left. - \eta r_3\left(\tau_i\times\tau_j + \frac{1}{4}n_in_j\right) - \frac{1}{2}\eta r_1(n_i+n_j)\right\}^{(\gamma)}, \quad (9)$$

$$\hat{K}^{(\gamma)}_{ij} = \left\{\eta r_1\left(\tau_i\cdot\tau_j + \frac{1}{4}n_in_j\right) + \eta r_3\left(\tau_i\times\tau_j + \frac{1}{4}n_in_j\right)\right.$$
$$\left. - \frac{1}{4}(1+\eta r_1)(n_i+n_j)\right\}^{(\gamma)}, \quad (10)$$

where the scalar product $(\tau_i\cdot\tau_j)^{(\gamma)}$ and the cross-product,

$$(\tau_i \times \tau_j)^{(\gamma)} = \frac{1}{2}\left(\tau_i^+\tau_j^+ + \tau_i^-\tau_j^-\right) + \tau_i^z\tau_j^z, \tag{11}$$

involve orbital (pseudospin) operators corresponding to two active t_{2g} orbitals along the γ axis, with $\tau_i = \{\tau_i^+, \tau_i^- \tau_i^z\}$, and

$$\tau_i^z = \tfrac{1}{2}(n_{i,yz} - n_{i,zx})\,. \tag{12}$$

They follow from the structure of local Coulomb interaction (7). The term (11) leads to the nonconservation of total pseudospin quantum number. Density operators $n_i^{(\gamma)}$ in Eqs. (9) and (10) stand for the number of d electrons in active orbitals for the considered bond $\langle ij\rangle$, e.g. $n_i^{(c)} = n_{ia} + n_{ib}$. The coefficients,

$$r_1 = \frac{1}{1-3\eta}, \qquad r_3 = \frac{1}{1+2\eta}, \tag{13}$$

follow from the energies of $d_i^2 d_j^2 \rightleftharpoons d_i^3 d_j^1$ excitations in the units of U: (i) r_1 represents the high-spin $^4A_2$ excitation of energy $(U - 3J_H)$, while the low-spin excitations are given by (ii) $r_2 = 1$ originating from the low-spin $^2T_1$ and ^2E excitations of energy U, and (iii) r_3 represents the low-spin $^2T_2$ states of energy $(U + 2J_H)$.

Magnetic order observed in Mott insulators is usually understood in terms of the GKR which are based on the MF picture and ignore entangled quantum states. These rules state that the pattern of occupied orbitals determines the spin structure. For example, for 180° bonds (e.g. Mn–O–Mn bonds in $LaMnO_3$) there are two key rules: (i) if two partially occupied $3d$ orbitals point towards each other, the interaction is AF, however, (ii) if an occupied orbital on one site has a large overlap with an empty orbital on the other site of a bond $\langle ij\rangle$, the interaction is weak and FM due to finite Hund's exchange. This means that spin order and orbital order are complementary – ferro-like (uniform) orbital (FO) order supports AF spin order, while AO order supports FM spin order. Indeed, these celebrated rules are well followed in $LaMnO_3$ [27] and in $KCuF_3$ [28], where strong JT effect stabilizes the orbital order and suppresses the orbital fluctuations. The AO order is here robust in the FM *ab* planes, while the orbitals obey the FO order along the c axis, supporting the AF coupling and leading to the A-AF phase for both systems. In such cases the GKR directly apply. Therefore, one may disentangle the spin and orbital operators, and it has been shown that this procedure is sufficient to explain both the magnetic [2] and optical [4] properties of $LaMnO_3$.

As another prominent example of the Goodenough–Kanamori complementarity we would like to mention the AF phases realized in YVO_3 [19], which have been the subject of intense research in recent years. A priori, the orbital interactions between V^{3+} ions in d^2 configuration obey the cubic symmetry, if the t_{2g} orbitals are randomly occupied. However, the symmetry breaking at the structural transition where the symmetry is reduced from cubic to orthorhombic, which persists in the magnetic phases, suggests that the electronic configuration is different. Indeed, the $GdFeO_3$ distortions in the RVO_3 structure break the symmetry in the orbital space, and both the electronic structure calculations [29] and the analysis using the point charge model [30] indicate that the electronic configuration $(xy)^1(yz,zx)^1$ is induced at every site, i.e.,

$$n_{ic} = 1\,, \qquad n_{ia} + n_{ib} = 1\,. \tag{14}$$

The partly filled $\{a,b\}$ orbitals are both active along the c axis, and may lead either to FO or to AO order. Indeed, depending on this orbital pattern, the magnetic correlations are there either AF or FM, explaining the origin of the two observed types of AF order: (i) the C-AF phase, and (ii) the G-AF phase. However, the situation is more subtle as both orbitals in the orbital doublet $\{|yz\rangle,|xz\rangle\} \equiv \{|a\rangle,|b\rangle\}$ at each site i are active on the bonds along the c axis. This demonstrates an important difference between the e_g (with one electron or one hole in active e_g orbital at each site [13]) and a t_{2g} system, such as RVO_3 perovskite vanadates, where electrons occupying two active t_{2g} orbitals may fluctuate and form an *orbital singlet* [3]. The cubic symmetry is thus broken as both orbital flavors are active only along the c axis, and the bonds in the ab planes and along the c axis are nonequivalent. Consequently, superexchange orbital operators (9) and (10) take different forms along these two distinct directions,

$$\hat{J}_{ij}^{(c)} = \frac{1}{2}\left\{(1+2\eta r_1)\left(\tau_i\cdot\tau_j+\frac{1}{4}\right) - \eta r_3\left(\tau_i\times\tau_j+\frac{1}{4}\right) - \eta r_1\right\}, \tag{15}$$

$$\hat{K}_{ij}^{(c)} = \left\{\eta r_1\left(\tau_i\cdot\tau_j+\frac{1}{4}\right) + \eta r_3\left(\tau_i\times\tau_j+\frac{1}{4}\right) - \frac{1}{2}(1+\eta r_1)\right\}, \tag{16}$$

$$\hat{J}_{ij}^{(a)} = \frac{1}{4}\left\{(1-\eta r_3)(1+n_{ib}n_{jb}) - r_1(n_{ib}-n_{jb})^2\right\}, \tag{17}$$

$$\hat{K}_{ij}^{(a)} = \frac{1}{2}\eta(\eta r_1 + r_3)(1+n_{ib}n_{jb})\,. \tag{18}$$

The general form of spin-orbital superexchange model (8) suggests that the above symmetry breaking leads indeed to an effective spin model with broken symmetry between magnetic interactions along different cubic axes. By averaging over the orbital operators one finds indeed different effective magnetic exchange interactions, J_c along the c axis and J_{ab} within the ab planes:

$$J_c = \left\langle \hat{J}_{ij}^{(c)} \right\rangle, \qquad J_{ab} = \left\langle \hat{J}_{ij}^{(a)} \right\rangle. \tag{19}$$

The interactions in the ab planes could in principle still take two different values in case of finite lattice strain discussed below, making both $\{a,b\}$ axes inequivalent, but here we want just to point out the symmetry breaking between the c axis and the ab planes, which follows from the density distribution (14) and explains the nonequivalence of spin interactions in the C-AF phase of the RVO_3 perovskites [3].

Apart from the superexchange there are in general also interactions due to the couplings to the lattice that control the orbitals. In the cubic vanadates these interactions are expected to be weak, but nevertheless they influence significantly the spin-orbital fluctuations and decide about the observed properties in the RVO_3 family. We write the orbital interactions, $\propto \tau_i^z\tau_j^z$, induced by the $GdFeO_3$ distortions and by the JT distortions of the lattice using two parameters, V_{ab} and V_c,

$$\mathscr{H}_V = V_{ab} \sum_{\langle ij \rangle \| c} \tau_i^z \tau_j^z - V_c \sum_{\langle ij \rangle \| c} \tau_i^z \tau_j^z . \tag{20}$$

The orbital interaction along the c axis V_c plays here a crucial role and allows one to switch between the two types of magnetic order, C-AF and G-AF phase [31], stabilizing simultaneously a complementary OO, either G-AO or C-AO order.

However, the description in terms of the GKR does not suffice and the ground state of the spin-orbital model for the RVO_3 perovskites, which consists of the superexchange and the effective orbital interactions,

$$\mathscr{H}_{S\tau} = \mathscr{H}_J + \mathscr{H}_V . \tag{21}$$

may also be entangled due to the quantum coupling between spin $S = 1$ and orbital $\tau = 1/2$ operators along the c axis, see Eq. (15). In contrast, the orbital fluctuations in the ab planes are quenched due to the occupied c orbitals at each site (14), so spins and orbitals disentangle. Possible entanglement between spin $(\mathbf{S}_i \cdot \mathbf{S}_j)$ and orbital $(\tau_i \cdot \tau_j)$ operators along the bonds $\langle ij \rangle \| c$ in the RVO_3 perovskites, and the applicability of the GKR to these systems, may be investigated by evaluating intersite spin and orbital correlations (to make these two functions comparable, we renormalized the spin correlations by the factor 1/4),

$$S_{ij} = \frac{1}{4} \langle \mathbf{S}_i \cdot \mathbf{S}_j \rangle , \tag{22}$$

$$T_{ij} = \langle \tau_i \cdot \tau_j \rangle , \tag{23}$$

and comparing them with each other. A key quantity that measures spin-orbital entanglement is the composite correlation function [10],

$$C_{ij} = \frac{1}{4} \left\{ \langle (\mathbf{S}_i \cdot \mathbf{S}_j)(\tau_i \cdot \tau_j) \rangle - \langle \mathbf{S}_i \cdot \mathbf{S}_j \rangle \langle \tau_i \cdot \tau_j \rangle \right\} . \tag{24}$$

When $C_{ij} = 0$, the spin and orbital operators are disentangled and their MF decoupling is exact, while if $C_{ij} < 0$ – spin and orbital operators are entangled and the MF decoupling not justified.

The numerical results for a 1D chain along the c axis described by vanadate spin-orbital model (21) are shown in Fig. 2. One finds entangled spin-orbital states with all three S_{ij}, T_{ij} and C_{ij} correlations being negative in the spin-singlet ($S = 0$) regime of fluctuating yz and zx orbitals, obtained for $\eta < 0.07$ (Fig. 2a). Therefore, the complementary behavior of spin (22) and orbital (23) correlations is absent in this regime of parameters and the GKR are violated. In addition, composite spin-orbital correlations (24) are here finite ($C_{ij} < 0$), so spin and orbital variables are entangled, and the MF factorization of spin-orbital operators fails. In a similar d^1 model for the perovskite titanates (with $S = 1/2$) one finds even somewhat stronger spin-orbital entanglement and the regime of η with $C_{ij} < 0$ is broader (i.e., $\eta <$

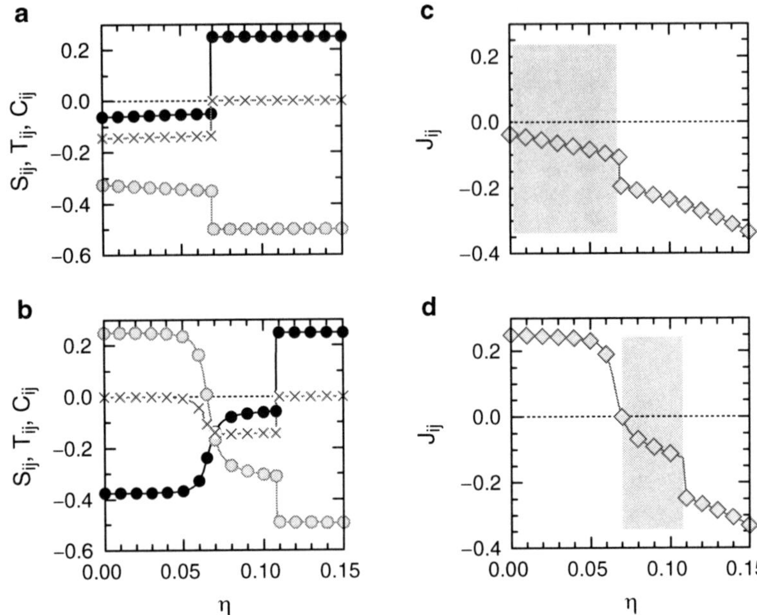

Fig. 2 Evolution of intersite correlations and exchange constants along the c axis obtained by exact diagonalizaton of spin-orbital model on a chain of $N = 4$ sites with periodic boundary conditions, with $\hat{J}_{ij}^{(c)}$ and $\hat{K}_{ij}^{(c)}$ given by Eqs. (15) and (16), for increasing Hund's exchange η: (**a**),(**b**) intersite spin S_{ij} (22) (*filled circles*), orbital (23) (*empty circles*), and spin-orbital C_{ij} (24) ($\times$) correlations; **(c),(d)** the corresponding spin exchange constants J_{ij} (19). In the *shaded areas* of (**c**) and (**d**) the spin correlations $S_{ij} < 0$ do not follow the sign of the exchange constant $J_{ij} < 0$, and the classical GKR are violated. Parameters: (**a**),(**c**) $V_c = 0$, and (**b**),(**d**) $V_c = J$.

0.21) [10]. At the point $\eta = 0$ one recovers then the SU(4) model with $S_{ij} = T_{ij} = C_{ij} = -0.25$, and the ground state is an entangled SU(4) singlet, involving a linear combination of (spin singlet/orbital triplet) and (spin triplet/orbital singlet) states.

To provide further evidence that the GKR do not apply to spin-orbital model (21) in the regime of small η, we compare spin exchange constants J_{ij} (19) shown in Fig. 2c with spin correlations S_{ij} (22), see Fig. 2a. One finds that exchange interaction is formally FM ($J_{ij} < 0$) in the orbital-disordered phase in the regime of $\eta < 0.07$, but it is accompanied by AF spin correlations ($S_{ij} < 0$). Therefore $J_{ij}S_{ij} > 0$ and the ground state energy would be enhanced in an ordered state, when calculated in the MF decoupling of spin-orbital operators [10]. This at first instance somewhat surprising result is a consequence of *'dynamical'* nature of exchange constants $\hat{J}_{ij}^{(c)}$ which exhibit large fluctuations [10], measured by the second moment, $\delta J = \{\langle(\hat{J}_{ij}^{(\gamma)})^2\rangle - J_{ij}^2\}^{1/2}$. For instance, in d^2 model (21) the orbital bond correlations change dynamically from singlet to triplet, resulting in large $\delta J = \frac{1}{4}\{1-(2T_{ij}+\frac{1}{2})^2\}^{1/2} \simeq 0.247$, i.e., $\delta J > |J_{ij}|$.

Remarkably, finite spin-orbital correlations $C_{ij} < 0$ and similar violation of the GKR are found also at finite orbital interaction (20) induced by the lattice, $V_c > 0$.

Representative results obtained for $V_c = J$ are shown in Figs. 2b and d. At small η FO order is induced, and in this regime the GKR are followed by the AF/FO phase (similar to the FM/AO phase at large η which also follows the GKR). However, for intermediate Hund's exchange $\eta \sim 0.07$ FO order is destabilized and the entangled AF/AO phase appears, with similar spin, orbital and composite spin-orbital correlations as found before at $V_c = 0$ and $\eta = 0$ (Fig. 2a). Also in this case FM exchange ($J_{ij} < 0$) coexists with AF spin correlations ($S_{ij} < 0$). Thus we conclude that orbital interactions induced by the lattice modify the regime of entangled spin-orbital states in the intermediate AF/AO phase which may be moved to more realistic values of $\eta \sim 0.1$, and cannot eliminate it completely. In addition, the transition between the FO/AF and AO/AF phase is *continuous* [10] due to the structure of orbital superexchange which contains terms (11) responsible for non-conservation of orbital quantum numbers.

3 Experimental Evidence of Orbital Fluctuations in $LaVO_3/YVO_3$

Before discussing the exotic magnetic properties and the phase diagram of the RVO_3 perovskites, we will consider the influence of magnetism on the optical spectra of $LaVO_3$, starting with a general formulation of the theory. While exchange constants may be extracted from the spin-orbital superexchange model (19), it is frequently not realized that virtual charge excitations that contribute to the superexchange are responsible as well for the optical absorption, thus the superexchange and the optical absorption are intimately related to each other via the optical sum rule [33]. This is not so surprising as when electrons are almost localized in a Mott insulator, the only kinetic energy which is left and decides about the optical spectral weight is associated with virtual excitations contributing to superexchange. Therefore, in Mott insulators the thermal evolution of optical spectral weight can be deduced from the superexchange [34]. In a system with orbital degeneracy the optical spectra consist of several multiplet transitions, and the kinetic energy $K_n^{(\gamma)}$ (due to $d-d$ excitations) associated with each of them can be determined from the superexchange (8) using the Hellman–Feynman theorem [32],

$$K_n^{(\gamma)} = 2\left\langle H_n^{(\gamma)}(ij)\right\rangle. \quad (25)$$

Note that $K_n^{(\gamma)}$ is negative and corresponds to the nth multiplet state of the transition metal ion, created by a charge excitation along a bond $\langle ij\rangle \parallel \gamma$. It is obvious that the thermal excitation values $\langle\cdots\rangle$ depend sensitively on the magnetic structure, i.e., whether spin correlations on a bond $\langle ij\rangle$ are FM or AF.

Thus it is natural to decompose the optical sum rule which is usually formulated in terms of the *total* kinetic energy for polarization γ,

$$K^{(\gamma)} = 2J\sum_n \langle H_n^{(\gamma)}(ij)\rangle, \tag{26}$$

into *partial optical sum rules* for individual Hubbard subbands [32],

$$\frac{a_0\hbar^2}{e^2}\int_0^\infty \sigma_n^{(\gamma)}(\omega)d\omega = -\frac{\pi}{2}K_n^{(\gamma)} = -\pi\left\langle H_n^{(\gamma)}(ij)\right\rangle, \tag{27}$$

where $\sigma_n^{(\gamma)}(\omega)$ is the contribution of band n to the optical conductivity for polarization along the γ axis, a_0 is the distance between transition metal ions, and the tight-binding model with nearest neighbor hopping is assumed. Equation (27) provides a practical way of calculating the optical spectral weights from spin-orbital superexchange models, such as the one derived for the RVO$_3$ perovskites (21). Note that the total optical intensity (26) is of less interest here as it has a much weaker temperature dependence and does not allow one a direct insight into the nature of the electronic structure. In addition, it might be also more difficult to resolve from experiment.

In order to apply the above theory to the RVO$_3$ perovskites, we write the superexchange operator $H^{(\gamma)}(ij)$ for a bond $\langle ij\rangle \parallel \gamma$, contributing to operator $\mathscr{H}_J$ (8), as a superposition of $d_i^2d_j^2 \rightleftharpoons d_i^3d_j^1$ charge excitations to different upper Hubbard subbands labelled by n [32],

$$H^{(\gamma)}(ij) = \sum_n H_n^{(\gamma)}(ij). \tag{28}$$

One finds the superexchange terms $H_n^{(c)}(ij)$ for a bond $\langle ij\rangle$ along the c axis [32],

$$H_1^{(c)}(ij) = -\frac{1}{3}Jr_1(2+\mathbf{S}_i\cdot\mathbf{S}_j)\left(\frac{1}{4}-\tau_i\cdot\tau_j\right), \tag{29}$$

$$H_2^{(c)}(ij) = -\frac{1}{12}J(1-\mathbf{S}_i\cdot\mathbf{S}_j)\left(\frac{7}{4}-\tau_i^z\tau_j^z-\tau_i^x\tau_j^x+5\tau_i^y\tau_j^y\right), \tag{30}$$

$$H_3^{(c)}(ij) = -\frac{1}{4}Jr_3(1-\mathbf{S}_i\cdot\mathbf{S}_j)\left(\frac{1}{4}+\tau_i^z\tau_j^z+\tau_i^x\tau_j^x-\tau_i^y\tau_j^y\right), \tag{31}$$

and $H_n^{(ab)}(ij)$ for a bond in the ab plane,

$$H_1^{(ab)}(ij) = -\frac{1}{6}Jr_1(2+\mathbf{S}_i\cdot\mathbf{S}_j)\left(\frac{1}{4}-\tau_i^z\tau_j^z\right), \tag{32}$$

$$H_2^{(ab)}(ij) = -\frac{1}{8}J(1-\mathbf{S}_i\cdot\mathbf{S}_j)\left(\frac{19}{12}\mp\frac{1}{2}\tau_i^z\mp\frac{1}{2}\tau_j^z-\frac{1}{3}\tau_i^z\tau_j^z\right), \tag{33}$$

$$H_3^{(ab)}(ij) = -\frac{1}{8}Jr_3(1-\mathbf{S}_i\cdot\mathbf{S}_j)\left(\frac{5}{4}\mp\frac{1}{2}\tau_i^z\mp\frac{1}{2}\tau_j^z+\tau_i^z\tau_j^z\right). \tag{34}$$

These expressions show that the spin correlations along the c axis and within the ab planes,

$$s_c = \langle \mathbf{S}_i \cdot \mathbf{S}_j \rangle_c\,, \qquad s_{ab} = \langle \mathbf{S}_i \cdot \mathbf{S}_j \rangle_{ab}\,, \tag{35}$$

as well as the orbital correlations, play an important role in the intensity distribution in optical spectroscopy. From the form of the above superexchange contributions one sees that high-spin excitations $H_1^{(\gamma)}(ij)$ support the FM coupling while the low-spin ones, $H_2^{(\gamma)}(ij)$ and $H_3^{(\gamma)}(ij)$, contribute with AF couplings.

We have determined the exchange constants in LaVO$_3$ by averaging over the orbital operators, see Eqs. (19). The case of the ab planes is straightforward as only the average densities $\langle n_{ia} \rangle$ and $\langle n_{ib} \rangle$ are needed to determine J_{ab}, and at large η they follow from the G-AO order in these planes. At $\eta = 0$ the orbital correlations along the c axis result from orbital fluctuations in the 1D orbital chain. In this limit the orbital correlations are the same as for the AF Heisenberg chain, i.e., $\langle \boldsymbol{\tau}_i \cdot \boldsymbol{\tau}_j \rangle = -0.4431$ and the ground state is disordered, with $\langle \tau_i^z \rangle = 0$. Nevertheless, for this disordered state the result for J_{ab} is similar as for the G-AO phase [25].

For the disordered (fluctuating) $\{a,b\}$ orbital state at $\eta = 0$, the AF exchange interactions in ab planes (see Fig. 3) result solely from singly occupied c orbitals (14), which are active in these planes and contribute by their double occupancies in an excited state with AF superexchange. One expects that the exchange constants along the c axis in the C-AF phase could be deduced from Eqs. (19), as spin and orbital order are complementary [22]. It is quite remarkable that at the same time finite FM interactions $-J_c \simeq 3$ meV are obtained at $\eta = 0$ (Fig. 3). They follow from the orbital fluctuations which dominate at low values of η. This mechanism of FM exchange adds to the one known in systems with real orbital order at finite η – the latter mechanism gradually takes over when η increases and the G-AO order develops and reduces the orbital fluctuations. At finite $\eta > 0$ we used the linear orbital-wave theory [12] to determine the intersite orbital correlations $\langle \boldsymbol{\tau}_i \cdot \boldsymbol{\tau}_j \rangle$ and the order parameter $\langle \tau_i^z \rangle$, for more details see Ref. [25]. At $\eta = 0.14$ representative for LaVO$_3$, the FM interactions are stronger than from AF ones, $|J_c| > J_{ab}$. Indeed,

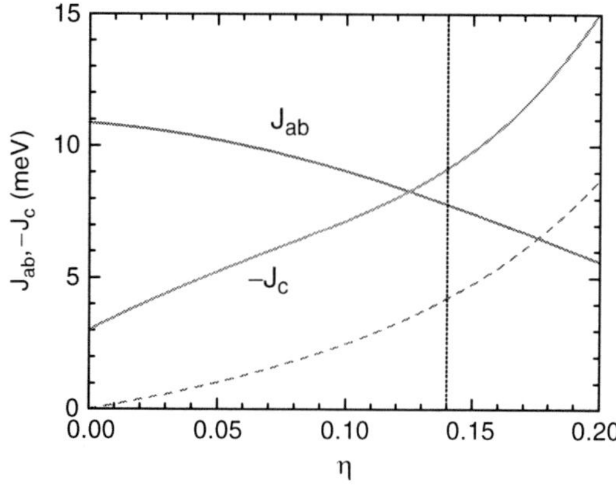

Fig. 3 Exchange constants J_{ab} and $-J_c$ (19) calculated from Eqs. (17) and (15) in the C-AF phase of LaVO$_3$ for increasing η (*solid lines*). *Dashed line* shows the value of $-J_c$ obtained for classical orbital order (36) according to GKR, $\langle \boldsymbol{\tau}_i \cdot \boldsymbol{\tau}_j \rangle = -\frac{1}{4}$. A representative value of $\eta = 0.14$ (for $U = 5.0$ and $J_H = 0.7$ eV) is marked by dotted line. Parameters of the model (21): $J = 35$ meV, $V_c = V_{ab} = 0$.

this early prediction of the theory [3] agrees qualitatively with larger average FM exchange $J_c < 0$ in the C-AF phase of YVO_3 than the AF exchange $J_{ab} > 0$ in the ab planes, see below.

We emphasize that the strong FM exchange along the c axis follows from the orbital fluctuations, and the rigid G-AO order obtained in the limit of strong orbital interactions $\{V_{ab}, V_c\}$ (20) would give a much weaker FM interaction,

$$J_c^{G-\mathrm{AO}} = -\frac{1}{2}\eta r_1 J, \tag{36}$$

see Fig. 3. The FM interaction $J_c^{G\text{-AO}}$ (36) vanishes at $\eta = 0$, is triggered by finite Hund's exchange η and increases in lowest order linearly with η. This behavior follows the conventional mechanism of FM interactions induced by finite Hund's exchange in the states with AO order, as for instance in $KCuF_3$ [8] or in $LaMnO_3$ [2].

A crucial test of the present theory which demonstrates that orbital fluctuations are indeed present in $LaVO_3$, concerns the temperature dependence of the low-energy (high-spin) spectral weight in optical absorption along the c axis $-K_1^{(c)}/2J$. According to experiment [35] it decreases by about 50% between low temperature and $T = 300$ K. In contrast, the result obtained by averaging the high-spin superexchange term $H_1^{(c)}(ij)$ (29) for polarization along the c axis assuming robust G-AO order is,

$$w_{c1}^{G-\mathrm{AO}} = \frac{1}{6} r_1 \left(s_c + 2\right), \tag{37}$$

where the spin correlation function s_c (35) is responsible for the entire temperature dependence of the low-energy spectral weight. Equation (37) predicts decrease of w_{c1} of only about 27%, see Fig. 4, and the maximal possible reduction of $K_1^{(c)}$ reached at $s_c = 0$ in the limit of $T \to \infty$ is by 33%. This result proves that the scenario with frozen G-AO order in $LaVO_3$ is *excluded by experiment* [5].

In contrast, when a cluster method, which allows one to include orbital fluctuations along the c axis, is used to determine the optical spectral weight from the high-spin superexchange term (29) [32], the temperature dependence resulting from the theory follows the experimental data [35]. This may be considered as a remarkable success of the theory based on the spin-orbital superexchange model derived for the RVO_3 perovskites.

However, the experimental situation in the cubic vanadates is more complex and full of puzzles. One is connected with the second magnetic transition in YVO_3, as we already mentioned in Section 2. The magnetic transition at $T_{N2} = 77$ K is particularly surprising as the staggered moments are approximately parallel to the c axis in the G-AF phase, and rotate above T_{N2} to the ab planes in the C-AF phase, with some small alternating G-AF component along the c axis [36]. While the orientation of spins in C-AF and G-AF phase follow in a straightforward manner from the model, i.e., are consistent with the expected anisotropy due to spin-orbit coupling [31], the observed magnetization reversal with the weak FM component remains puzzling. Therefore, in spite of the suggested mechanism based on the entropy increase in the C-AF phase [25], the lower magnetic transition in YVO_3 remains mysterious.

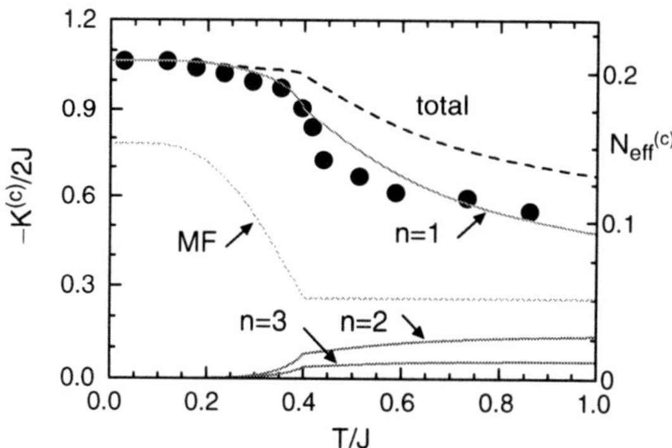

Fig. 4 Kinetic energy $K_n^{(c)}$ (*solid lines*) for the optical subband n and total $K^{(c)}$ (*dashed line*) obtained from the spin-orbital model (21). *Filled circles* show the effective carrier number $N_{eff}^{(c)}$ (in the energy range $\omega < 3$ eV) for $LaVO_3$, presented in Fig. 5 of Ref. [35]. *Dotted line* shows $K_1^{(c)}$ obtained from the MF decoupling (37). Parameters: $\eta = 0.12$, $V_c = 0.9J$, $V_{ab} = 0.2J$.

Secondly, the scale of magnetic excitations is considerably reduced for the C-AF phase (by a factor close to 2) as compared with the exchange constants deduced from magnons measured in the G-AF phase [19]. In addition, the magnetic order parameter in the C-AF phase of $LaVO_3$ is strongly reduced to $\simeq 1.3\mu_B$, which cannot be explained by the quantum fluctuations in the C-AF phase (being only 6% for $S=1$ spins [39]). Finally, the C-AF phase of YVO_3 is dimerized. Until now, only this last feature found a satisfactory explanation in the theory [37, 38], see below.

We remark that the observed dimerization in the magnon dispersions may be seen as a signature of *entanglement in excited states* which becomes active at finite temperature. The microscopic reason for the anisotropy in the exchange constants $\mathscr{J}_{c1} \equiv \mathscr{J}_c(1+\delta_s)$ and $\mathscr{J}_{c2} \equiv \mathscr{J}_c(1-\delta_s)$ is the tendency of the orbital chain to dimerize, activated by thermal fluctuations in the FM spin chain [38] which support a dimerized structure in the orbital sector. As a result one finds alternating stronger $\propto \mathscr{J}_c(1+\delta_s)$ and weaker $\propto \mathscr{J}_c(1-\delta_s)$ FM bonds along the c axis in the dimerized C-AF phase (with $\delta_s > 0$). The observed spin waves may be explained by the following effective spin Hamiltonian for this phase (assuming again that the spin and orbital operators may be disentangled which is strictly valid only at $T=0$):

$$\mathscr{H}_s = \mathscr{J}_c \sum_{\langle i,i+1\rangle \| c} \left[1+(-1)^i\delta_s\right] \mathbf{S}_i \cdot \mathbf{S}_{i+1} + \mathscr{J}_{ab}^C \sum_{\langle ij\rangle \| ab} \mathbf{S}_i \cdot \mathbf{S}_j + K_z \sum_i (S_i^z)^2 . \tag{38}$$

Following the linear spin-wave theory [25], the magnon dispersion is given by

$$\omega_\pm(\mathbf{k}) = 2\left\{\left[2\mathscr{J}_{ab} + |\mathscr{J}_c| + \frac{1}{2}K_z \pm \mathscr{J}_c\eta_{\mathbf{k}}^{1/2}\right]^2 - \left(2\mathscr{J}_{ab}\gamma_{\mathbf{k}}\right)^2\right\}^{1/2}, \tag{39}$$

with

$$\gamma_{\mathbf{k}} = \frac{1}{2}(\cos_k x + \cos k_y) , \tag{40}$$

$$\eta_{\mathbf{k}} = \cos^2 k_z + \delta_s^2 \sin^2 k_z . \tag{41}$$

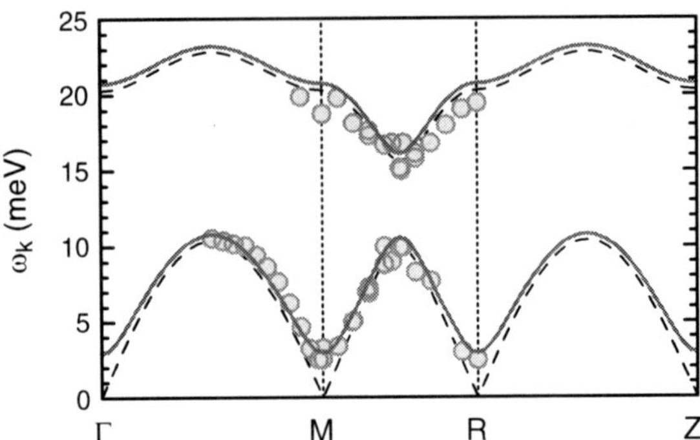

Fig. 5 Spin-wave dispersions $\omega_{\mathbf{k}}$ obtained in the LSW theory (39) for the C-AF phase of YVO_3 (*lines*), and measured by neutron scattering at $T = 85$ K [19] (*circles*). Parameters: $J_{ab} = 2.6$ meV, $J_c = -3.1$ meV, $\delta_s = 0.35$, and $K_z = 0.4$ eV (*full lines*), $K_z = 0$ (*dashed lines*). The high symmetry points are: $\Gamma = (0,0,0)$, $M = (\pi,\pi,0)$, $R = (\pi,\pi,\pi)$, $Z = (0,0,\pi)$.

For the numerical evaluation of Fig. 5 we have used the experimental exchange interactions [19]: $\mathscr{J}_{ab} = 2.6$ meV, $\mathscr{J}_c = -3.1$ meV, $\delta_s = 0.35$. Indeed, a large gap is found between two modes halfway in between the M and R points, and between the Z and Γ points (not shown). Two modes measured by neutron scattering [19] (see also Fig. 1), and obtained from the present theory in the unfolded Brillouin zone, are well reproduced by the dimerized FM exchange couplings in spin Hamiltonian (39). We note that a somewhat different Hamiltonian with more involved interactions was introduced in ref. [19], but the essential features seen in the experiment are reproduced already by the present model H_s with a single ion anisotropy term $\propto K_z$.

As the transition between the two magnetic phases, G-AF and C-AF phase, occurs in YVO_3 at finite temperature, the entropy has to play an important role. As mentioned above, the exchange constants found in the C-AF phase of YVO_3 (Fig. 5) are considerably lower than the corresponding values in the G-AF phase, $J_{ab} = J_c \simeq 5.7$ meV [19]. As a result of weaker exchange interactions, the spin entropy of the C phase will grow faster than that of the G phase, and induce the $G \to C$ transition. However, starting from our model (21) we do not find this strong reduction of energy scale in the C-AF phase. Another mechanism like the fluctuation of n_{xy} occupancy has been invoked to account for this reduction [25]. Here we will simply adopt the experimental values for the exchange constants in the C-AF phase.

Using linear spin-wave and orbital-wave theory, the spin and orbital entropy normalized per one vanadium ion was calculated and compared for both magnetic phases of YVO_3 [25]. Using the experimental parameters [19] one finds that: (i) the entropy $\mathscr{S}_C$ for the C-AF phase is larger that $\mathscr{S}_G$ for the G-AF phase, and (ii) the spin entropy grows significantly faster with temperature than the orbital entropy for each phase. Therefore, we conclude that the spin entropy gives here the leading contribution and is responsible for a fast decrease of the free energy in the C-AF phase which is responsible for the observed magnetic transition at T_{N2} [25], see Fig. 6.

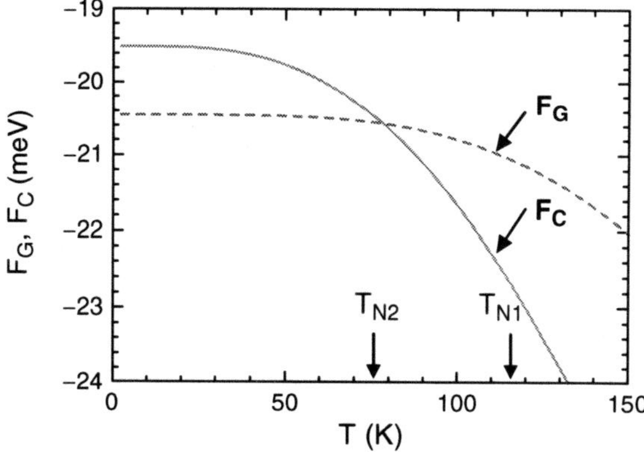

Fig. 6 Free energies $\mathscr{F}_C$ (*C*-AF, *solid line*) and $\mathscr{F}_G$ (*G*-AF, *dashed line*) as obtained for the spin-orbital model (21) using the experimental values of magnetic exchange constants in both phases [19]. The experimental magnetic transition temperatures, $T_{N2} \simeq 77$ K and $T_{N1} \simeq 116$ K, are indicated by *arrows*. Parameters: $J = 40$ meV, $\eta = 0.13$, $V_a = 0.30J$, $V_c = 0.84J$.

4 Orbital and Magnetic Transition in the $R\text{VO}_3$ Perovskites

4.1 Spin-Orbital-Lattice Coupling

Experimental studies have shown that the *C*-AF order is common to the entire family of the $R\text{VO}_3$ vanadates, where $R = \text{Lu}, \cdots, \text{La}$. In general the structural (orbital) transition occurs first. i.e., $T_{N1} < T_{\text{OO}}$, except for LaVO_3 with $T_{N1} \simeq T_{\text{OO}}$ [20, 22]. When the ionic radius r_R decreases, the Néel temperature T_{N1} also decreases, while the orbital transition temperature T_{OO} increases first, passes through a maximum close to YVO_3, and decreases afterwards when LuVO_3 is approached. Knowing that quantum fluctuations and spin-orbital entanglement play so important role in the perovskite vanadates, it is of interest to ask whether the spin-orbital model (21) is able to describe this variation of T_{OO} and T_{N1} with decreasing radius r_R of R ions in $R\text{VO}_3$ [20]. It is clear that the nonmonotonic dependence of T_{OO} on r_R cannot be reproduced just by the superexchange, as a maximum in T_{OO} requires two mechanisms which oppose each other. In fact, the decreasing V–O–V angle (Θ along the c axis) with decreasing ionic radius r_R along the $R\text{VO}_3$ perovskites [40–43] reduces somewhat both the hopping t and superexchange J (6), but we shall ignore this effect here and concentrate ourselves on the leading dependence on orbital correlations which are controlled by lattice distortions.

The model introduced in Ref. [30] to describe the phase diagram of $R\text{VO}_3$ includes a spin-orbital-lattice coupling by the terms: (i) the superexchange H_J (8) between V^{3+} ions in the d^2 configuration with $S = 1$ spins [3], (ii) intersite orbital interactions H_V (20) (which originate from the coupling to the lattice and play an important role in the transition between the C-AF and G-AF phase), (iii) the crystal-field splitting $\propto E_z$ between yz and zx orbitals, and (iv) orbital-lattice term $\propto gu$ which induces orbital polarization when the lattice strain (orthorhombic distortion) u increases. The Hamiltonian consists thus of several terms [30],

$$\mathscr{H} = \mathscr{H}_J + \mathscr{H}_V(\vartheta) + E_z(\vartheta)\sum_i e^{i\mathbf{R}_i\mathbf{Q}}\tau_i^z - gu\sum_i \tau_i^x + \frac{1}{2}NK\{u - u_0(\vartheta)\}^2. \quad (42)$$

Except for the superexchange $\mathscr{H}_J$ (8), all the other terms in Eq. (42) depend on the tilting angle ϑ, which we use to parameterize the RVO$_3$ perovskites below. It is related to the V–O–V angle $\Theta = \pi - 2\vartheta$, which decreases with increasing ionic radius r_R ($\Theta = 180°$ corresponds to an ideal perovskite structure). By analyzing the structural data of the RVO$_3$ perovskites [40–43] one arrives at the following empirical relation between r_R and ϑ:

$$r_R = r_0 - \alpha \sin^2 2\vartheta\,, \quad (43)$$

with $r_0 = 1.5$ Å and $\alpha = 0.95$ Å.

The crystal-field splitting of $\{yz, zx\}$ orbitals ($E_z > 0$) alternates in the ab planes and is uniform along the c axis, with a modulation vector $\mathbf{Q} = (\pi, \pi, 0)$ in cubic notation – it supports the G-AO order, and not the observed (weak) G-AO order. The orbital interactions induced by the distortions of the VO$_6$ octahedra and by GdFeO$_3$ distortions of the lattice, $V_{ab} > 0$ and $V_c > 0$ (see Eg. (20)), also favor C-AO order (like $E_z > 0$). The orbital interaction V_c counteracts the orbital superexchange $\propto J$ (16), and has only rather weak dependence on ϑ, so it suffices to choose a constant $V_c = 0.26J$ to reproduce an almost simultaneous onset of spin and orbital order in LaVO$_3$, with $T_{OO} \simeq T_{N1}$, as observed [20]. One finds $T_{N1}^{\rm exp} = 147$ K taking $J = 200$ K in the present model (42), which reproduces well the experimental value $T_{N1}^{\rm exp} = 143$ K for LaVO$_3$ [20].

The last two terms in Eq. (42) describe the orbital-lattice coupling via the orthorhombic strain $u = (b-a)/a$, where a and b are the lattice parameters of the $Pbnm$ structure, K is the force constant, and N is the number of V^{3+} ions. Unlike E_z, the coupling $gu > 0$ acts as a transverse field in the pseudospin space and favors that one of the two linear combinations $\frac{1}{\sqrt{2}}(|a\rangle_i \pm |b\rangle_i)$ of active t_{2g} orbitals is occupied at site i. By minimizing the energy over u, one finds

$$g_{\rm eff}(\vartheta;T) \equiv gu(\vartheta;T) = gu_0(\vartheta) + \frac{g^2}{K}\langle\tau^x\rangle_T\,, \quad (44)$$

which shows that the global distortion $u(\vartheta;T)$ consists of (i) a pure lattice contribution $u_0(\vartheta)$, and (ii) a contribution due the orbital polarization $\propto \langle\tau^x\rangle$ which is determined self-consistently.

4.2 Dependence on Lattice Distortion

In order to investigate the phase diagram of the RVO_3 perovskites one needs still information on the functional dependence of the parameters $\{E_z, V_{ab}, g_{\text{eff}}\}$ of the microscopic model (42) on the tilting angle ϑ. The $GdFeO_3$-like distortion is parametrized by two angles $\{\vartheta, \varphi\}$ describing rotations around the b and c cubic axes, as explained in Ref. [44]. Here we adopted a representative value of $\varphi = \vartheta/2$, similar as in the perovskite titanates. Therefore, we used only a single rotation angle ϑ in Eq. (42), which is related to the ionic size by Eq. (43). Functional dependence of the crystal-field splitting $E_z \propto \sin^3\vartheta\cos\vartheta$ on the angle ϑ may be derived from the point charge model [30], using the structural data for the RVO_3 perovskites [40–43]. It is expected that the functional dependence of V_{ab} follows the crystal-field term, so we write:

$$E_z(\vartheta) = J v_z \sin^3\vartheta\cos\vartheta\,, \tag{45}$$

$$V_{ab}(\vartheta) = J v_{ab} \sin^3\vartheta\cos\vartheta\,. \tag{46}$$

Qualitatively, increasing E_z and V_{ab} with increasing lattice distortion and tilting angle ϑ do favor the orbital order, so the temperature T_{OO} is expected to increase.

A maximum observed in the dependence of T_{OO} on r_R (or ϑ) may be reproduced within the present model (42) only when a competing orbital polarization interaction $g_{\text{eff}}(\vartheta;T)$ (44) increases faster with ϑ when the ionic radius r_R is reduced than $\{E_z, V_{ab}\}$. Both u_0 and $\langle\tau^x\rangle$ in Eq. (44) are expected to increase with increasing tilting angle ϑ. Below we present the results obtained with a semiempirical relation,

$$g_{\text{eff}}(\vartheta) = J v_g \sin^5\vartheta\cos\vartheta\,, \tag{47}$$

as postulated in Ref. [30]. Altogether, model (21) depends on three parameters: $\{v_z, v_{ab}, v_g\}$ which could be selected [30] to reproduce the observed dependence of orbital and magnetic transition temperature on the ionic radius r_R in the RVO_3 perovskites, see below.

4.3 Evolution of Spin and Orbital Order in RVO_3

Hamiltonian (42) poses a many-body problem which includes an interplay between spin, orbital, and lattice degrees of freedom. A standard approach to investigate the onset of spin and orbital order is to use the mean-field (MF) theory with on-site order parameters $\langle S^z\rangle$ (corresponding to C-AF phase) and

$$\langle\tau^z\rangle_G \equiv \frac{1}{2}\left|\langle\tau_i^z - \tau_j^z\rangle\right|\,, \tag{48}$$

as well as the coupling between them which modifies the MF equations, similar to the situation encountered in the Ashkin–Teller model [45]. This approach was successfully implemented to determine the orbital and magnetic transition temperature, $T_{\rm OO}$ and T_{N1} in $LaMnO_3$ [2]. It was also applied to the RVO_3 perovskites [46] to demonstrate that either spin or orbital order may occur first at decreasing temperature, depending on the amplitude of hopping parameters. However, such a MF approach uses only on-site order parameters and cannot suffice when orbital fluctuations also contribute, e.g. stabilizing the C-AF phase in $LaVO_3$ [3] – then it becomes essential to determine self-consistently the above on-site order parameters together with orbital singlet correlations (23) on the bonds $\langle ij\rangle \parallel c$. The simplest approach which allows us to determine these correlations is a cluster MF theory for a bond coupled to effective spin and orbital symmetry breaking fields which originate from its neighbors in an ordered phase. The respective transition temperatures are obtained when $\langle S^z\rangle > 0$ ($\langle S^z\rangle = 0$) for $T < T_{N1}$ ($T > T_{N1}$), and $\langle \tau^z\rangle_G > 0$ ($\langle \tau^z\rangle_G = 0$) for $T < T_{\rm OO}$ ($T > T_{\rm OO}$).

Making a proper selection of the model parameters $\{v_z, v_{ab}, v_g\}$ one is able to reproduce the experimental phase diagram for the onset of G-AO and C-AF order in the RVO_3 family in the entire range of r_R, see below. We start with presenting an example of the orbital and spin phase transition in $LaVO_3$ and in $SmVO_3$, see Fig. 7. By selecting $V_c = 0.26J$ both G-AO and C-AF order occur simultaneously in $LaVO_3$ below $T_{\rm OO} = T_{N1} \simeq 0.73J$. The crystal field splitting E_z, orbital interaction V_{ab}, and the coupling to the lattice $g_{\rm eff}$ are rather small and do not influence the order in $LaVO_3$. We emphasize that orbital correlations along the c axis are here practically as in the AF Heisenberg chain, $\langle \tau_i \cdot \tau_j\rangle \simeq -0.44$, and the orbital order is considerably reduced, $\langle \tau^z\rangle_G \simeq 0.32$. The orbital polarization in $LaVO_3$ $\langle \tau^x\rangle \simeq 0.03$ is rather weak at T_{N1}, and is further reduced with decreasing $T < T_{\rm OO}$. Note that also spin order parameter is expected to be reduced below $\langle S^z\rangle = 1$, but weak quantum fluctuations in the C-AF phase [39] were neglected here. In contrast, in $SmVO_3$ the phase transitions separate: the orbital transition occurs first at $T_{\rm OO} \simeq 0.86J$, and the magnetic one follows at a lower $T_{N1} \simeq 0.65J$. Already in this case the transverse orbital polarization is considerably increased, with $\langle \tau^x\rangle \simeq 0.20$ at T_{N1} (see Fig. 7),

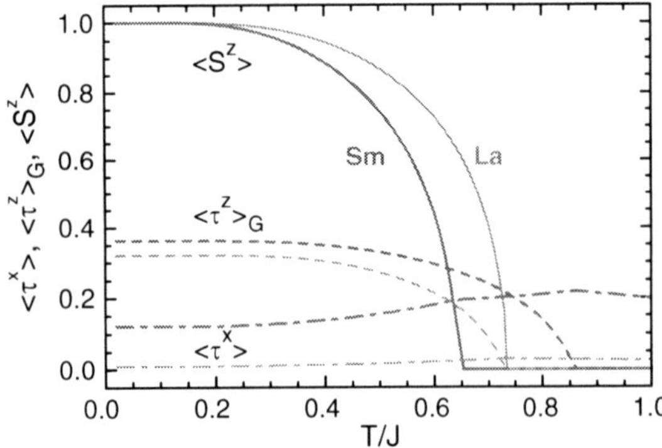

Fig. 7 The orbital polarization $\langle \tau^x\rangle$ (*dashed-dotted lines*), G-type orbital order parameter $\langle \tau^z\rangle_G$ (48) (*dashed lines*), and spin order parameter $\langle S^z\rangle$ (*solid lines*) for $LaVO_3$ and $SmVO_3$ (*thin* and *heavy lines*). Parameters: $V_c = 0.26J$, $v_z = 17$, $v_{ab} = 22$, $v_g = 740$.

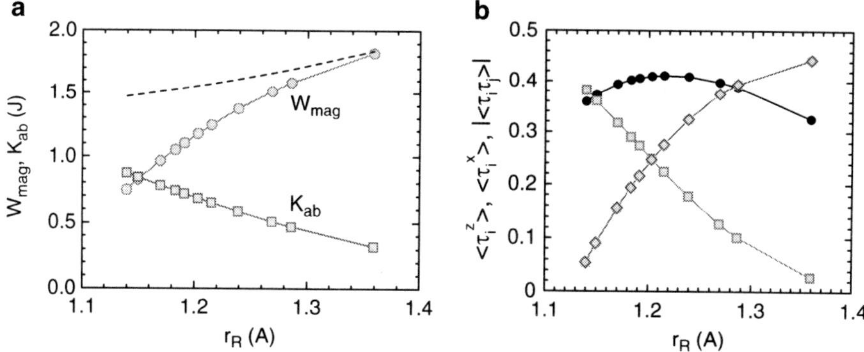

Fig. 8 (**a**) The width of magnon band W_{C-AF} for finite $g_{\rm eff}$ (*circles*) and without orbital-strain coupling ($g_{\rm eff} = 0$, *dashed*), and orbital interactions in *ab* planes K_{ab} (*squares*) in the *C*-AF phase of cubic vanadates (the points correspond to the RVO_3 compounds of Fig. 9). (**b**) Evolution of the orbital order parameter $\langle\tau_i^z\rangle_G$ (*filled circles*), transverse orbital polarization $\langle\tau_i^x\rangle$ (*squares*), and orbital intersite correlations $|\langle\tau_i\cdot\tau_j\rangle|$ (*diamonds*) along *c* axis at $T = 0$. Parameters: $v_z = 17$, $v_{ab} = 22$, $v_g = 740$.

and further increases with decreasing r_R (not shown). Note that the polarization $\langle\tau^x\rangle$ does not change close to $T_{\rm OO}$, and only below T_{N1} gets weakly reduced due to the developing magnetic order, in agreement with experiment [43]. The *G*-OO parameter is here stronger as the singlet orbital fluctuations are not so pronounced when $T \to 0$, being $\langle\tau^z\rangle_G \simeq 0.37$.

The key features of the present spin-orbital system which drive the observed dependence of $T_{\rm OO}$ and T_{N1} on r_R [20] is the evolution of intersite orbital correlations are: (i) the gradual increase of the orbital interactions $K_{ab}\tau_i^z\tau_j^z$ (Fig. 8a), and (ii) the reduction of orbital fluctuations on the bonds along the *c* axis, described by the bond singlet correlations $\langle\tau_i\cdot\tau_j\rangle$ (Fig. 8b). The parameter K_{ab} in Fig. 8a consists of the superexchange contribution $\propto J$ (18) and orbital interaction V_{ab} (20) induced by the lattice distortion. While the superexchange does not change with decreasing r_R, the latter term increases and induces the increase of $T_{\rm OO}$ from $LaVO_3$ to YVO_3. This increase is similar to that observed in the $RMnO_3$ manganites [21]. Thereby the bond angle Θ decreases from 157.4° in $LaVO_3$ to 144.8° in YVO_3.

While the singlet correlations are drastically suppressed from $LaVO_3$ towards $LuVO_3$, the orbital order parameter $\langle\tau^z\rangle_G$ somewhat increases from $LaVO_3$ to $SmVO_3$ (see also Fig. 7). At the same time the orbital polarization $\langle\tau^x\rangle$ increases and soon becomes as important as the orbital order parameter, i.e., $\langle\tau^x\rangle \simeq \langle\tau^z\rangle_G$. Further increase of the orbital polarization towards $LuVO_3$ suppresses the *G*-AO parameter, so $\langle\tau^z\rangle_G$ passes through a maximum and decreases for $r_R < 1.22$ Å.

It is remarkable that the above changes in orbital correlations induced by the lattice suppress gradually the magnetic interactions in the *C*-AF phase, although the value of J remains unchanged. This is well visible in the total width of the magnon band, $W_{C-AF} = 4(J_{ab} + |J_c|)$ (at $T = 0$) [25], shown in Fig. 8a, being reduced from $\sim 1.84J$ in $LaVO_3$ to $\sim 1.05J$ in YVO_3. This large reduction qualitatively agrees with

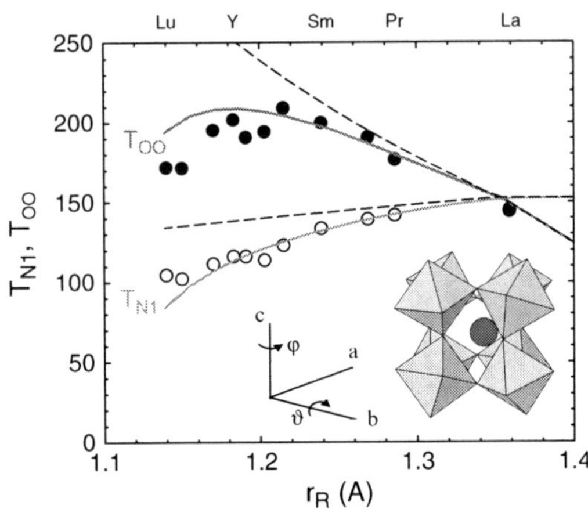

Fig. 9 The orbital T_{OO} and magnetic T_{N1} transition temperature for varying r_R in the RVO_3 perovskites, obtained from model (42) for: $v_g = 740$ (*solid lines*) and $v_g = 0$ (*dashed lines*). *Circles* show the experimental data of Ref. [20]. The inset shows the $GdFeO_3$-type distortion, with the rotation angles ϑ and φ. Other parameters as in Fig. 8. (Reproduced from Ref. [30].)

the rather small values of the exchange constants in the C-AF phase of YVO_3 [19], see also Fig. 5. This reduction is caused by the suppression of the singlet orbital correlations $\langle \tau_i \cdot \tau_j \rangle$ by the increasing coupling to the lattice $g_{\rm eff}(\vartheta)$ when r_R decreases. Note also that this effect would be rather small for $g_{\rm eff} = 0$ – this behavior is excluded, shown by dashed lines in Fig. 9, is excluded by experiment.

Following Ref. [30], we argue that the gradual reduction of the orbital singlet correlations in favor of increasing orbital polarization is responsible for the evolution of the orbital transition temperature T_{OO} in the experimental phase diagram of Fig. 9, which is reproduced by the theory in the entire range of available r_R. The transition temperature T_{OO} changes in a nonmonotonic way, similar to the orbital order parameter $\langle \tau^z \rangle_G$ at $T = 0$ (Fig. 8b). After analyzing the changes in the orbital correlations, we see that the physical reasons of the decrease of T_{OO} for small (large) r_R are quite different. While the orbital fluctuations dominate and largely suppress the orbital order in $LaVO_3$, the orbital polarization connected with orthorhombic distortion takes over near YVO_3 and competes with G-AO order.

While the above fast dependence on the tilting angle ϑ of VO_6 octahedra in the RVO_3 family was introduced in order to reproduce the experimentally observed dependence of T_{OO} on r_R, see Fig. 9, it may be justified *a posteriori*. It turns out that the dependence of $g_{\rm eff}$ on the ionic radius r_R in Eq. (47) follows the actual lattice distortion u in RVO_3 measured at $T = 0$ (u_0), or just above T_{N1} (u_1) [30]. Also the orbital polarization $\langle \tau^x \rangle$ is approximately $\propto \sin^5 \vartheta \cos \vartheta$, and follows the same fast dependence of $g_{\rm eff}(\vartheta)$ for the RVO_3 perovskites (Fig. 10). This result is somewhat unexpected, as information about the actual lattice distortions has not been used in constructing the miscroscopic model (21). These results indicate that the bare coupling parameters $\{g, K\}$ are nearly constant and independent of r_R, which may be treated as a prediction of the theory to be verified by future experiments.

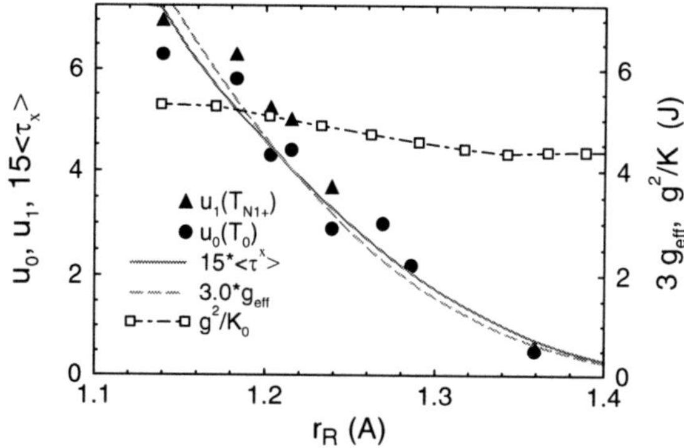

Fig. 10 Experimental distortion (in percent) at $T = 0$ (u_0, *circles*) and above T_{N1} (u_1, *triangles*) for the RVO_3 compounds [40,41,43], compared with the orbital polarization $\langle\tau^x\rangle_{T=0}$ and with $g_{\rm eff}$ (47); $g_{\rm eff}$ and g^2/K are in units of J. *Squares* show the upper bound for g^2/K predicted by the theory. Parameters: $v_z = 17$, $v_{ab} = 22$, $v_g = 740$. (Reproduced from Ref. [30].)

5 Summary and Outlook

Summarizing, spin-orbital superexchange model (42) augmented by orbital-lattice couplings provides an explanation of the observed variation of the orbital $T_{\rm OO}$ and magnetic T_{N1} transition temperatures for the whole class of the RVO_3 perovskites. A more complete theoretical understanding including a description of the second magnetic transition from C-AF to G-AF phase, which occurs at T_{N2} for small ionic radii r_R [47], remains to be addressed by future theory, which should include the spin-orbit relativistic coupling [31].

We conclude by mentioning a few open issues and future directions of research in the field of perovskite vanadates. Rapid progress of the field of orbital physics results mainly from experiment, and is triggered by the synthesis of novel materials. Although experiment is ahead of theory in most cases, there are some exceptions. One of them was a theoretical prediction of the energy and dispersion of orbital excitations [12, 48, 49]. Only recently orbital excitations (orbitons) could be observed by Raman scattering in the Mott insulators $LaTiO_3$ and $YTiO_3$ [50, 51]. They were also identified in the optical absorption spectra of YVO_3 and $HoVO_3$ [52]. The exchange of two orbitals along the c axis in the intermediate C-AF phase was shown to contribute to the optical conductivity $\sigma(\omega)$.

An interesting question which arises in this context is the carrier propagation in a Mott insulator with orbital order. This problem is rather complex as in a spin-orbital polaron, created by doping of a single hole, both spin and orbital excitations contribute to the hole scattering [24], which may even become localized by string excitations as in the t-J^z model [53]. Indeed, the coupling to orbitons increases the effective mass of a moving hole in e_g systems [54]. The orbital part of the superexchange is classical (compass-like) in t_{2g} systems, but nevertheless the hole is not confined as weak quasiparticle dispersion arises from three-site processes [55, 56].

As in the doped manganites, also in doped R_{1-x}(Sr,Ca)$_x$VO$_3$ systems the G-AO order gradually disappears [57]. The C-AF spin order survives, however, in a broad range of doping, in contrast to La$_{1-x}$Sr$_x$MnO$_3$, where FM order replaces the A-AF phase already at $x \sim 0.10$, and is accompanied by the e_g orbital liquid [58] at higher doping. It is quite remarkable that the complementary G-AF/C-AO order is fragile and disappears in Y$_{1-x}$Ca$_x$VO$_3$ already at $x = 0.02$ [57]. The doped holes in C-AF/G-AO phase are localized in polaron-like states [59], so the pure electronic model, such as the one of Ref. [55], is too crude to capture both the evolution of the spin-orbital order in doped vanadates and the gradual decrease of the energy scale for spin-orbital fluctuations. Theoretical studies at finite hole concentration are still nonexistent for 3D models, but one may expect a transition from a phase with AF order to a phase with FM spin polarization at large Hund's coupling, as shown both for e_g [13] and t_{2g} [57] systems.

A few representative problems related to the properties of RVO$_3$ perovskites discussed above demonstrate that the orbital physics is a very rich field, with intrinsically frustrated interactions and rather exotic ordered or disordered phases, with their behavior dominated by quantum fluctuations. While valuable information about the electronic structure is obtained from density functional theory [61], the many-body aspects have to be studied simultaneously using models of correlated electrons. The RVO$_3$ perovskites remain an interesting field of research, as it turned out that electron-lattice coupling is here not strong enough to suppress (quench) the orbital fluctuations [30]. Thus the composite quantum fluctuations described by the spin-orbital model (42) remain active. Nevertheless, there is significant control of the electronic properties due to the electron lattice coupling. Thus, the lattice distortions may also influence the onset of magnetic order in systems with active orbital degrees of freedom. If they are absent and the lattice is frustrated in addition, a very interesting situation arises, with strong tendency towards truly exotic quantum states [15]. Examples of this behavior were considered recently for the triangular lattice, both for e_g orbitals in LiNiO$_2$ [62] and t_{2g} orbitals in NaTiO$_2$ [63]. None of these models could really be solved, but a generic tendency towards dimer correlations with spin singlets on the bonds for particular orbital states has been shown. Yet, the question whether novel types of orbital order, such as e.g. nematic order in spin models [64], could be found in certain situations remains open.

Acknowledgements It is our great pleasure to thank G. Khaliullin and L.F. Feiner for very stimulating collaboration which significantly contributed to our present understanding of the subject. We thank B. Keimer, G.A. Sawatzky, Y. Tokura and particularly C. Ulrich for numerous insightful discussions. A.M. Oleś acknowledges financial support by the Foundation for Polish Science (FNP) and by the Polish Ministry of Science and Education under Project No. N202 068 32/1481.

References

1. Y. Tokura, N. Nagaosa, Science **288**, 462 (2000)
2. L.F. Feiner, A.M. Oleś, Phys. Rev. B **59**, 3295 (1999)

3. G. Khaliullin, P. Horsch, A.M. Oleś, Phys. Rev. Lett. **86**, 3879 (2001)
4. A.M. Oleś, P. Horsch, G. Khaliullin, L.F. Feiner, Phys. Rev. B **72**, 214431 (2005)
5. S. Miyasaka, S. Onoda, Y. Okimoto, J. Fujioka, M. Iwama, N. Nagaosa, Y. Tokura, Phys. Rev. Lett. **94**, 076405 (2005)
6. J.-Q. Yan, J.-S. Zhou, J.B. Goodenough, Y. Ren, J.G. Cheng, S. Chang, J. Zarestky, O. Garlea, A. Liobet, H.D. Zhou, Y. Sui, W.H. Su, R.J. McQueeney, Phys. Rev. Lett. **99**, 197201 (2007)
7. K.I. Kugel, D.I. Khomskii, Sov. Phys. Usp. **25**, 231 (1982)
8. L.F. Feiner, A.M. Oleś, J. Zaanen, Phys. Rev. Lett. **78**, 2799 (1997)
9. J. van den Brink, New J. Phys. **6**, 201 (2004)
10. A.M. Oleś, P. Horsch, L.F. Feiner, G. Khaliullin, Phys. Rev. Lett. **96**, 147205 (2006)
11. B. Frishmuth, F. Mila, M. Troyer, Phys. Rev. Lett. **82**, 835 (1999)
12. J. van den Brink, P. Horsch, F. Mack, A.M. Oleś, Phys. Rev. B **59**, 6795 (1999)
13. M. Daghofer, A.M. Oleś, W. von der Linden, Phys. Rev. B **70**, 184430 (2004)
14. L.F. Feiner, A.M. Oleś, Phys. Rev. B **71**, 144422 (2005)
15. G. Khaliullin, Prog. Theor. Phys. Suppl. **160**, 155 (2005)
16. D.I. Khomskii, M.V. Mostovoy, J. Phys. A **36**, 9197 (2003)
17. J. Dorier, F. Becca, F. Mila, Phys. Rev. B **72**, 024448 (2005)
18. W. Brzezicki, J. Dziarmaga, A.M. Oleś, Phys. Rev. B **75** 134415 (2007)
19. C. Ulrich, G. Khaliullin, J. Sirker, M. Reehuis, M. Ohl, S. Miyasaka, Y. Tokura, B. Keimer, Phys. Rev. Lett. **91**, 257202 (2003)
20. S. Miyasaka, Y. Okimoto, M. Iwama, Y. Tokura, Phys. Rev. B **68**, 100406 (2003)
21. J.-S. Zhou, J.B. Goodenough, Phys. Rev. Lett. **96**, 247202 (2006)
22. S. Miyasaka, J. Fujioka, M. Iwama, Y. Okimoto, Y. Tokura, Phys. Rev. B **73**, 224436 (2006)
23. L.F. Feiner, A.M. Oleś, J. Zaanen, J. Phys.: Condens. Matter **10**, L555 (1998)
24. J. Zaanen, A.M. Oleś, Phys. Rev. B **48**, 7197 (1993)
25. A.M. Oleś, P. Horsch, G. Khaliullin, Phys. Rev. B **75**, 184434 (2007)
26. J.S. Griffith, *The Theory of Transition Metal Ions* (Cambridge University Press, Cambridge, London 1971)
27. A. Weiße, H. Fehske, New J. Phys. **6**, 158 (2004)
28. A.M. Oleś, L.F. Feiner, J. Zaanen, Phys. Rev. B **61**, 6257 (2000)
29. M. De Raychaudhury, E. Pavarini, O.K. Andersen, Phys. Rev. Lett. **99**, 126402 (2007)
30. P. Horsch, A.M. Oleś, L.F. Feiner, G. Khaliullin, Phys. Rev. Lett. **100**, 167205 (2008)
31. P. Horsch, G. Khaliullin, A.M. Oleś, Phys. Rev. Lett. **91**, 257203 (2003)
32. G. Khaliullin, P. Horsch, A.M. Oleś, Phys. Rev. B **70**, 195103 (2004)
33. D. Baeriswyl, J. Carmelo, A. Luther, Phys. Rev. B **33**, 7247 (1986)
34. M. Aichhorn, P. Horsch, W. von der Linden, M. Cuoco, Phys. Rev. B **65**, 201101 (2002)
35. S. Miyasaka, Y. Okimoto, Y. Tokura, J. Phys. Soc. Jpn. **71**, 2086 (2002)
36. Y. Ren, T.T.M. Palstra, D.I. Khomskii, A.A. Nugroho, A.A. Menovsky, G.A. Sawatzky, Phys. Rev. B **62**, 6577 (2000)
37. J. Sirker, G. Khaliullin, Phys. Rev. B **67**, 100408 (2003)
38. J. Sirker, A. Herzog, A.M. Oleś, P. Horsch, Phys. Rev. Lett. **101**, 157204 (2008)
39. M. Raczkowski, A.M. Oleś, Phys. Rev. B **66**, 094431 (2002)
40. Y. Ren, A.A. Nugroho, A.A. Menovsky, J. Strempfer, U. Rütt, F. Iga, T. Takabatake, C.W. Kimball, Phys. Rev. B **67**, 014107 (2003)
41. M. Reehuis, C. Ulrich, P. Pattison, B. Ouladdiaf, M.C. Rheinstädter, M. Ohl, L.P. Regnault, M. Miyasaka, Y. Tokura, B. Keimer, Phys. Rev. B **73**, 094440 (2006)
42. M.H. Sage, G.R. Blake, G.J. Nieuwenhuys, T.T.M. Palstra, Phys. Rev. Lett. **96**, 036401 (2006)
43. M.H. Sage, G.R. Blake, C. Marquina, T.T.M. Palstra, Phys. Rev. B **76**, 195102 (2007)
44. E. Pavarini, A. Yamasaki, J. Nuss, O.K. Andersen, New J. Phys. **7**, 188 (2005)
45. R.V. Ditzian, J.R. Banavar, G.S. Grest, L.P. Kadanoff, Phys. Rev. B **22**, 2542 (1980)
46. T.N. De Silva, A. Joshi, M. Ma, F.C. Zhang, Phys. Rev. B **68**, 184402 (2003)
47. D.A. Mazurenko, A.A. Nugroho, T.T.M. Palstra, P.H.M. van Loosdrecht, Phys. Rev. Lett. **101**, 245702 (2008)
48. S. Ishihara, J. Inoue, S. Maekawa, Phys. Rev. B **55**, 8280 (1997)
49. J. van den Brink, Phys. Rev. Lett. **87**, 217202 (2001)

50. C. Ulrich, A. Gössling, M. Grüninger, M. Guennou, H. Roth, M. Cwik, T. Lorenz, G. Khaliullin, B. Keimer, Phys. Rev. Lett. **97**, 157401 (2006)
51. C. Ulrich, G. Ghiringhelli, A. Piazzalunga, L. Braicovich, N.B. Brookes, H. Roth, T. Lorenz, B. Keimer, Phys. Rev. B **77**, 113102 (2008)
52. E. Benckiser, R. Rückamp, T. Möller, T. Taetz, A. Möller, A.A. Nugroho, T.T.M. Palstra, G.S. Uhrig, M. Grüninger, New J. Phys. **10**, 053027 (2008)
53. G. Martínez, P. Horsch, Phys. Rev. B **44**, 317 (1991)
54. J. van den Brink, P. Horsch, A.M. Oleś, Phys. Rev. Lett. **85**, 5174 (2000)
55. M. Daghofer, K. Wohlfeld, A.M. Oleś, E. Arrigoni, P. Horsch, Phys. Rev. Lett. **100**, 066403 (2008)
56. K. Wohlfeld, M. Daghofer, A.M. Oleś, P. Horsch, Phys. Rev. B **78**, 214423 (2008)
57. J. Fujioka, S. Miyasaka, Y. Tokura, Phys. Rev. B **72**, 024460 (2005)
58. A.M. Oleś, L.F. Feiner, Phys. Rev. B **65**, 052414 (2002)
59. J. Fujioka, S. Miyasaka, Y. Tokura, Phys. Rev. B **77**, 144402 (2008)
60. J. Sirker, J. Damerau, A. Klümper, Phys. Rev. B **78**, 235125 (2008)
61. I.V. Solovyev, J. Phys.: Condens. Matter **20**, 293201 (2008)
62. A.J.W. Reitsma, L.F. Feiner, A.M. Oleś, New J. Phys. **7**, 121 (2005)
63. B. Normand, A.M. Oleś, Phys. Rev. B **78**, 094427 (2008)
64. N. Shannon, T. Momoi, P. Sindzingre, Phys. Rev. Lett. **96**, 027213 (2006)

Low Energy Scales of Kondo Lattices: Mean-field Perspective

S. Burdin

Abstract A review of the low temperature properties of Kondo lattice systems is presented within the mean-field approximation, focusing on the different characteristic energy scales. The Kondo temperature, T_K, and the Fermi liquid coherence energy, T_0, are analyzed as functions of the electronic filling, the shape of the non-interacting density of states, and the concentration of magnetic moments. These two scales can vanish, corresponding to a breakdown of the Kondo effect when an external magnetic field is applied. The Kondo breakdown can also be reached by adding a superexchange term to the Kondo lattice model, which mimics the intersite magnetic correlations neglected at the mean-field level.

1 Introduction

Rare-earth and actinide based compounds exhibit extremely rich phase diagrams, with signatures of heavy fermion behavior, unconventional magnetism, or superconductivity [1, 2]. At high temperature, the main physical properties of these systems are well reproduced by single impurity models, which describe the coupling between conduction electrons and one $4f$ or $5f$ ion. For dense systems, the single impurity models fail to describe the low temperature properties, which are characterized by the formation of a non local coherent state. In this regime, models with a periodic lattice of f ions are more appropriate.

Here, we consider more specifically dense compounds where the f orbital is occupied by one electron (Cerium) or one hole (Ytterbium). In the low temperature regime where the crystal field splitting lifts the degeneracy of the f orbital, these impurities are modeled by effective local spins $S_i = 1/2$. The system is thus described by the Kondo lattice Hamiltonian,

S. Burdin
Institut für Theoretische Physik, Universität zu Köln, Zülpicher Str. 77, 50937 Köln, Germany
and
Max Planck Institut für Physik Komplexer Systeme, Nöthnitzer Str. 38, 01187 Dresden, Germany
e-mail: burdin@thp.uni-koeln.de

V. Zlatić and A. C. Hewson (eds.), *Properties and Applications of Thermoelectric Materials*, 325
NATO Science for Peace and Security Series B: Physics and Biophysics,

$$H = \sum_{\mathbf{k}\sigma} (\varepsilon_{\mathbf{k}} - \mu) c^{\dagger}_{\mathbf{k}\sigma} c_{\mathbf{k}\sigma} + J_K \sum_i \mathbf{s}_i \mathbf{S}_i \ , \tag{1}$$

where $c^{\dagger}_{\mathbf{k}\sigma}$ ($c_{\mathbf{k}\sigma}$) describe creation (annihilation) operators of conduction electrons with spin $\sigma = \uparrow, \downarrow$ and momentum $\mathbf{k}$. The Kondo interaction results from a local antiferromagnetic coupling J_K between the density of spin of conduction electrons at site i, $\mathbf{s}_i$, and the Kondo impurities, $\mathbf{S}_i$. The chemical potential μ fixes the electronic filling to n_c conduction electrons per site.

The Kondo model has been extensively studied throughout the last decades [3]. At high temperature, the Kondo interaction can be considered a small perturbation: conduction electrons and Kondo ions are weakly coupled. The transport properties, which are determined by the conduction band, correspond to those of a normal metal. The magnetic susceptibility, governed by the Kondo free moments, has a Curie–Weiss form. The entropy is large, of the order of $\ln 2$ per site. A crossover occurs at the Kondo temperature, T_K, below which the Kondo interaction cannot be treated by perturbative methods. Experimental signatures of this crossover include, for example, a logarithmic increase of the resistivity when the temperature decreases. Other signatures involve a saturation of the magnetic susceptibility, and a significant decrease of the entropy. At lower temperature, if we neglect magnetic or superconducting instabilities, the physical properties are characteristic of a universal heavy Fermi liquid: the specific heat vanishes linearly with the temperature, $C_V(T) \approx \gamma T$; the local magnetic susceptibility, $\chi(T)$, as well as the resistivity are constant with quadratic (i.e., T^2 like) variations. This low temperature regime is characterized by an energy scale, the coherence temperature, T_0, which can be equivalently determined from the specific heat Sommerfeld coefficient, $T_0 \equiv 1/\gamma$, or the zero temperature magnetic susceptibility, $T_0 \equiv 1/\chi(T=0)$. Figure 1 depicts the schematic phase diagram of the model, as a function of T_0/T_K. The determination of T_0 and T_K from the entropy is illustrated by Fig. 2. The connection between T_0 and the thermal and electric transport properties have been analyzed by Zlatić et al. [5,6].

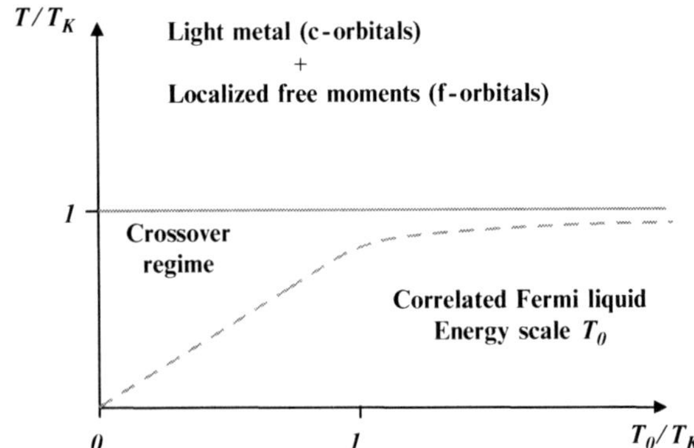

Fig. 1 Schematic phase diagram. The reduced temperature T/T_K is plotted versus T_0/T_K, which is considered a tunable parameter that can vary with the electronic filling, impurity concentration, magnetic field, or the shape of the non-interacting DOS. (From Ref. [4].)

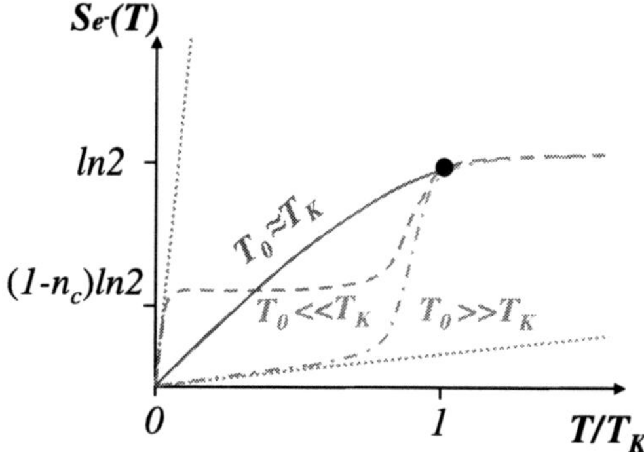

Fig. 2 Schematic plot of the electronic contribution to the entropy S_{e^-} as a function of the normalized temperature T/T_K for three cases: $T_0 \gg T_K$ (*dash dotted line*), $T_0 \approx T_K$ (*solid line*), and $T_0 \ll T_K$ (*dashed line*). The *dotted lines* indicate the linear Fermi liquid regime $S_{e^-}(T) = T/T_0$, with a rescaled slope T_K/T_0. For $T > T_K$ the three curves are identical, reflecting the linear contribution from the conduction band $S_{e^-}(T) = \ln 2 + T/D$. The *black dot* refers to a standard experimental determination of T_K from the electronic entropy: $S_{e^-}(T_K) = p \ln 2$. On this schematic plot, we used $p = 1$, which coincides with the vanishing of the effective hybridization, $r = 0$, defining T_K within the mean-field approach. Experimentally, where T_K is a crossover, one can choose, e.g., $p = 1/2$. (From Ref. [4].)

Note that the authors of Ref. [7] discussed two energy scales: one of them corresponds to our definition of the Kondo lattice temperature, T_K, from the temperature dependence of the entropy [see Fig. 2]. The second is the single impurity Kondo temperature, which, within the mean-field approach, is equal to the lattice Kondo temperature. The coherence temperature T_0 that we analyze here is not considered in Ref. [7].

The physical properties of systems with a small concentration of magnetic ions are universal and characterized by a single energy scale, $T_0 = T_K$. The identity relating the coherence and the Kondo temperatures is consistent with the exact solution of the single impurity model. The situation is different for dense systems for which more than one energy scale can be identified. As an example, the Ruderman–Kittel–Kasuya–Yosida (RKKY) interaction, J_{RKKY}, can compete with T_K, and a magnetic instability can be obtained for a lattice system [8]. For a long time, it has been believed that T_0 and T_K remain equal to each other in the heavy Fermi liquid phase of a Kondo lattice model. In this context and in response to discrepancies between theory and photoemission spectroscopy on Ce-based compounds with respect to the scaling with T_K, alternative scenarios involving phonons have been proposed [9]. On this basis, whether *the Kondo model* is capable of predicting photoemission spectroscopy on $YbAl_3$ has been a subject of controversy [10–12]. The specificity of Kondo lattice systems, with T_0 different from T_K, has been confirmed by experiments including magnetic susceptibility, specific heat, Hall coefficient measurements and x-ray absorption spectroscopy in cerium, $Ce_{1-x}La_xIr_2Ge_2$, $CeIr_{2-x}(Rh,Pt)_xGe_2$, $CeIr_2Ge_{2-x}(Si,Sn)_x$ [13], $CeNiSi_2$ [14], and ytterbium compounds, $YbXCu_4$(X = Ag, Cd, In, Mg, Tl, Zn) [15], $Yb_{1-x}Lu_xAl_3$ [16].

The first suggestion that T_0 could be much smaller than T_K was discussed by Nozières in the framework of the exhaustion problem [17, 18]. This prediction stimulated complementary theoretical works, including approximate methods based on a strong coupling approach [19] or mean-field calculations [20–23], and numerical simulations [24–28] using dynamical mean-field theory (DMFT) [29, 30]. Even if the initial prediction of Nozières turned to be quantitatively wrong [21, 31], all the theoretical calculations converge to the same qualitative conclusion: in a Kondo lattice, T_0 can be different from T_K, and the ratio between these two energy scales can be tuned, for example, by varying the electronic filling of the system. Finally, these theoretical works shed some new light on the formerly controversial photoemission analysis.

The aim of this work is, first, to review how the two energy scales, T_0 and T_K depend on physical parameters: electronic filling, shape of the density of states, concentration of magnetic ions. This is done within the simplest relevant approximation for the Kondo interaction: a mean-field decoupling. Then, we review how, within the mean-field approach, the Kondo phase can be destabilized by a magnetic field or by magnetic inter-ion interactions.

2 Mean-field Formalism

The mean-field approximation for the Kondo lattice was first introduced by Lacroix and Cyrot [32]. It was reformulated by Coleman [33] and by Read et al. [34], as a large-N approximation for the N-fold degenerate Coqblin-Schrieffer model [35]. It was shown that magnetic instabilities require expansions up to the order $1/N$, i.e. fluctuations around the mean-field. Nevertheless, the heavy Fermi liquid phase is well described in the limit $N = \infty$. The analysis of T_0 and T_K can thus be already performed at the mean-field level. Here, we describe the main lines of the mean-field approximation for the Kondo lattice Hamiltonian (1).

The Kondo impurities are represented by local auxiliary fermions as follows: $S_i^z = \frac{1}{2}(f_{i\uparrow}^\dagger f_{i\uparrow} - f_{i\downarrow}^\dagger f_{i\downarrow})$, $S_i^+ = f_{i\uparrow}^\dagger f_{i\downarrow}$, and $S_i^- = f_{i\downarrow}^\dagger f_{i\uparrow}$. The Kondo interaction is thus rewritten as a two-body interaction, $J_K \mathbf{s}_i \mathbf{S}_i \mapsto \frac{J_K}{2} \sum_{\sigma\sigma'} c_{i\sigma}^\dagger c_{i\sigma'} f_{i\sigma'}^\dagger f_{i\sigma}$, which describes the spin-flip processes between conduction electrons and local moments. This mapping is exact as far as the Hilbert space is restricted to the sector of one auxiliary fermion per site, $f_{i\uparrow}^\dagger f_{i\uparrow} + f_{i\downarrow}^\dagger f_{i\downarrow} = 1$. The mean-field solution is obtained within the two following approximations: (i) The local occupation of the auxiliary fermions is equal to one only on average. This corresponds to describing the Kondo spins by an effective local f-level that is half full. This effective filling is controlled by introducing a second chemical potential, λ. (ii) The two-body Kondo interaction is replaced by an effective one-body term, obtained from a mean-field decoupling of the $\uparrow$ and $\downarrow$ components. The mean-field approximation for the Kondo lattice Hamiltonian (1) yields

$$H = \sum_{\mathbf{k}\sigma}(\varepsilon_{\mathbf{k}} - \mu)c^{\dagger}_{\mathbf{k}\sigma}c_{\mathbf{k}\sigma} + r\sum_{i\sigma}[c^{\dagger}_{i\sigma}f_{i\sigma} + f^{\dagger}_{i\sigma}c_{i\sigma}] - \lambda\sum_{i\sigma}f^{\dagger}_{i\sigma}f_{i\sigma}\,, \tag{2}$$

where the effective hybridization is determined by the self-consistent relation

$$r = \frac{J_K}{2\mathcal{N}}\sum_{i\sigma}\langle f^{\dagger}_{i\sigma}c_{i\sigma}\rangle\,. \tag{3}$$

Here $\mathcal{N}$ is the number of lattice sites. The chemical potentials for c-electrons, μ, and f-fermions, λ, are determined by the constraints $n_c = \frac{1}{\mathcal{N}}\sum_{i\sigma}\langle c^{\dagger}_{i\sigma}c_{i\sigma}\rangle$ and $1 = \frac{1}{\mathcal{N}}\sum_{i\sigma}\langle f^{\dagger}_{i\sigma}f_{i\sigma}\rangle$. The mean-field Hamiltonian (2) describes an effective system of conduction electrons hybridized with local f levels. The correlation effects are renormalized into the self-consistent hybridization r. This approximation captures two important features of the Kondo lattice: at high temperature, we find $r = 0$, and the system is described as a paramagnetic light metal (c electrons) decoupled from local free moments (f ions). This picture is oversimplified but it succeeds in qualitatively characterizing the experimental situation, in which, above the Kondo temperature, conduction electrons are weakly coupled to the local moments. Within the mean-field approach, the Kondo temperature T_K is thus defined as the temperature for which a hybridization $r \neq 0$ occurs. At very low temperature, the physical properties correspond to a Fermi liquid and the excitations of the system correspond to the creation of non-interacting, heavy, fermionic quasiparticles. The latter are a linear combination of the light c-electrons and the heavy f-fermions. The Fermi-liquid regime is characterized by an energy scale, T_0, which can be defined identically from different physical properties of the ground state: the quasiparticle density of states, $\rho = 1/T_0$, the Sommerfeld coefficient, $\gamma = lim_{T\to 0}C_V(T)/T = 1/T_0$, or the local impurity spin susceptibility, $\chi_{loc}(T = 0) = 1/T_0$. We have obtained explicit expressions for T_K and T_0 in the limit of small J_K [4, 21]:

$$T_K = F_K[n_c, \rho_0]e^{-1/J_K\rho_0(\mu_0)}, \tag{4}$$

and

$$T_0 = F_0[n_c, \rho_0]e^{-1/J_K\rho_0(\mu_0)}, \tag{5}$$

which depend on the Kondo coupling J_K only within the non-analytic exponential factor $e^{-1/J_K\rho_0(\mu_0)}$. The prefactors F_K and F_0 are functions of the electronic filling, n_c, and the non-interacting density of states (DOS), $\rho_0(\omega) \equiv \sum_{\mathbf{k}}\delta(\omega - \varepsilon_{\mathbf{k}})$. For the sake of clarity, we do not write the explicit expressions of F_K and F_0 here, as they are given in Refs. [4, 21]. The exponential factor in both T_0 and T_K, results in a very sensitive dependence of these energy scales with respect to small changes in the system (e.g., pressure or doping). Since the ratio T_0/T_K does not depend on J_K, it is more likely to be analyzed experimentally as a universal function. The mean-field solution yields [4]

$$\frac{T_0}{T_K} = \left(\frac{D+\mu_0}{D-\mu_0}\right)^{1/2} \frac{F_{shape}}{\alpha\rho_0(\mu_L)\Delta\mu} . \tag{6}$$

Here, D is the half-bandwidth of the non-interacting DOS. μ_0 is the non-interacting chemical potential, corresponding to an electronic filling $n_c = 2\int_{-D}^{\mu_0} \rho_0(\omega)d\omega$. The interacting chemical potential, μ_L, corresponds to a filling $n_c + 1$, and the energy $\Delta\mu \equiv \mu_L - \mu_0$ is related to the enlargement of the Fermi surface. $\alpha = 1.13$ is a number. F_{shape} is an explicit function depending on the shape of ρ_0 as follows:

$$F_{shape} \equiv \exp\left[\left(\int_{-(D+\mu_0)}^{\Delta\mu} - \frac{1}{2}\int_{-(D+\mu_0)}^{D-\mu_0}\right) \frac{\rho_0(\mu_0+\omega) - \rho_0(\mu_0)}{|\omega|\rho_0(\mu_0)} d\omega\right] . \tag{7}$$

3 Tuning T_0 and T_K

3.1 Variation of Electronic Filling

The effect of electronic filling on the temperature scales of the Kondo lattice has been discussed by Nozières who suggested a possible exhaustion problem [17, 18]. Considering N_S Kondo spins coupled to $N_c \leq N_S$ conduction electrons, Nozières started with the following remark: the magnetic entropy which can be released at the temperature $T \approx T_K$ by the formation of incoherent Kondo singlets (i.e., by the single impurity Kondo effect) is $\Delta S = N_c \ln 2$. However, the formation of a coherent Fermi liquid ground state is characterized by a vanishing entropy, $S(T) \approx T/T_0$. The freezing of the $S(T_K) \approx (N_S - N_c)\ln 2$ remaining entropy thus requires a collective mechanism, and can lead to $T_0 \ll T_K$ when $N_c \ll N_S$.

The exhaustion problem remained for several years an open issue, and its first solution was obtained within a mean-field approach [21, 31], which provides a quantitative description of the electronic filling effect. First, from the analytical expressions (4–5), T_K and T_0 have the same J_K dependence, $e^{-1/J_K\rho_0(\mu_0)}$ factor, and the ratio T_K/T_0 does not depend on the Kondo coupling. This is in contradiction with the result of Nozières, who predicted a ratio $T_0/T_K \approx T_K/D \approx e^{-1/J_K\rho_0(\mu_0)}$ [17, 18]. The mean-field result, T_0/T_K independent of J_K, was later confirmed by DMFT calculations [36], and finally accepted by Nozières [31]. Nevertheless, T_0 and T_K define two energy scales with different dependencies with respect to the electronic filling, as depicted by Fig. 3, where $T_0 \ll T_K$ in the limit $n_c \to 0$. The mean-field result obtained here is remarkably similar to the one obtained within DMFT combined with Quantum Monte Carlo simulation (see Fig. 1 in Ref. [26]). The filling effects can also be analyzed from the expression (6). Neglecting the band shape effects (discussed in Section 3.2), we find $T_0/T_K \approx \left(\frac{D+\mu_0}{D-\mu_0}\right)^{1/2}$, which vanishes when μ_0 approaches the band edge $-D$, i.e., when $n_c \to 0$.

The analysis of the electronic occupation, $n_c(\varepsilon_{\mathbf{k}}) \equiv \langle c^\dagger_{\mathbf{k}} c_{\mathbf{k}}\rangle$, provides an important insight for understanding the physical mechanism leading to two different energy

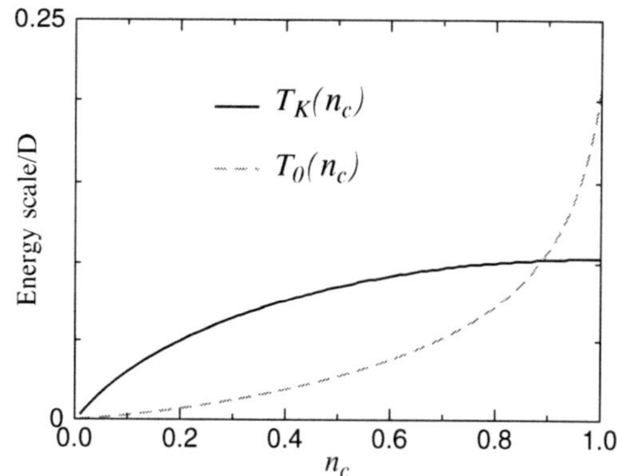

Fig. 3 T_K/D (*solid line*) and T_0/D (*dashed line*) versus electronic filling, for the Kondo lattice. Numerical result obtained within the mean-field approximation, for a semi-elliptic non-interacting DOS and $J_K/D = 0.75$ [21].

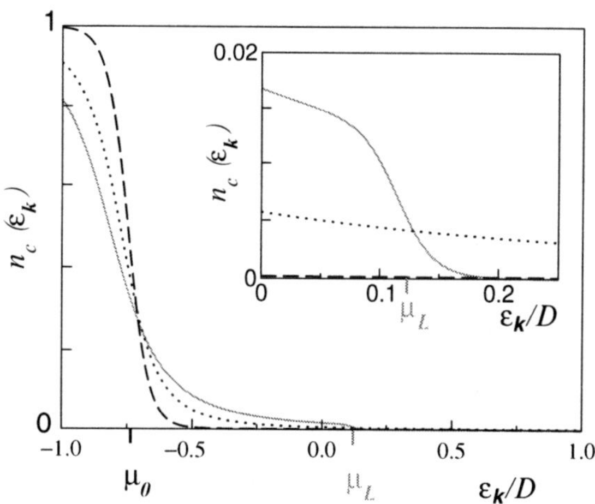

Fig. 4 Electronic occupation $n_c(\varepsilon_{\mathbf{k}})$ for $T/T_K = 1.0$ (*dashed line*), $T/T_K = 0.5$ (*dotted line*), and $T/T_K = 0.005$ (*solid line*). Numerical result obtained within the mean-field approximation for a semi-elliptic non-interacting DOS, $n_c = 0.15$, and $J_K/D = 0.75$. μ_0 and μ_L indicate the chemical potential corresponding to a small and a large Fermi surface, respectively. Inset: focus around μ_L. (From Ref. [21].)

scales. For $T \approx T_K$, the mean-field result, depicted by Fig. 4, looks like a Fermi distribution with a thermal window around the non-interacting chemical potential, μ_0, corresponding to n_c electrons (small Fermi surface). For $T \ll T_K$, in the Fermi liquid regime, the distribution is spread and forms a step around μ_L (large Fermi surface). Figure 4 only describes the occupation of the c-electrons, which is fixed to n_c. At $T = 0$, there are states of given momentum $\mathbf{k}$ which are not fully occupied by c-electrons. The Fermi liquid picture is recovered because the quasiparticles are not pure c-states, but a linear combination of c and f-states. The quasiparticle occupation is complete, i.e., equal to one, for states with an energy $\varepsilon_{\mathbf{k}} < \mu_L$, and it vanishes for states with higher energy. This behavior is consistent with the Luttinger theorem which predicts that, at $T = 0$, the Fermi surface contains both c-electrons and f-fermions. It is not surprising that the Kondo lattice satisfies the Luttinger theorem within the mean-field approximation. The reason is that the effective mean-field model (2) does not contain an explicit many body interaction term. We expect this result to survive beyond the mean-field in the Fermi liquid phase. The enlargement

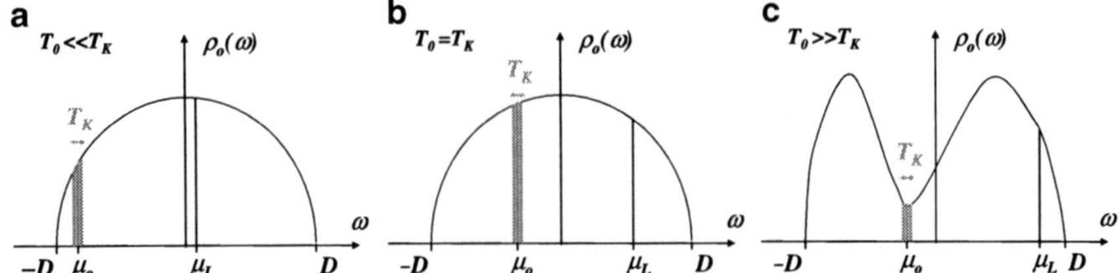

Fig. 5 Schematic plot of the non-interacting DOS. (**a**) For a regular DOS and far from the electronic half-filling $T_0 \ll T_K$. Close to the half-filling, the shape of the DOS around μ is crucial for determining T_0/T_K: (**b**) $\rho_0(\omega)$ is nearly constant around $\omega = \mu$ and $T_0 \sim T_K$. (**c**) μ is close to a minimum of ρ_0 and $T_0 \gg T_K$. Here, $\mu_0 \approx \mu$ indicates the chemical potential corresponding to n_c non-interacting c-electrons (small Fermi surface), and μ_L is the chemical potential corresponding to $n_c + 1$ non-interacting c-electrons (large Fermi surface). (From Ref. [4].)

of the Fermi surface might be a key point in the origin of the difference between T_0 and T_K: the Kondo temperature is associated with the incoherent scattering of the conduction electrons which are in the Fermi window of width T_K around μ_0 (see Fig. 5). The coherence temperature, T_0, results from the Kondo effect, but, unlike T_K, characterizes a Fermi liquid with a large Fermi surface.

3.2 Shape of the Non-interacting Density of States

Here we consider the effects due to the shape variation of the non-interacting DOS, $\rho_0(\omega)$ [4]. In order to separate this effect from the electronic filling effects, we assume that n_c is close to 1, but not exactly half full, so that the system is metallic. Equations (6) and (7) are thus simplified to

$$\frac{T_0}{T_K} \approx F_{shape} \approx \exp\left[\int_{-(D+\mu_0)}^{D-\mu_0} \frac{\rho_0(\mu_0+\omega)-\rho_0(\mu_0)}{2|\omega|\rho_0(\mu_0)} d\omega\right]. \tag{8}$$

A constant ρ_0 gives $T_0 \sim T_K$, which explains the T/T_K scaling observed in some compounds. If μ_0 is close to a local maximum of $\rho_0(\omega)$, the integrand in Eq. (8) is negative in the main part of the integration range, and $T_0 \ll T_K$. In the opposite situation, when μ_0 is close to a local minimum (see Fig. 5c), we find $T_0 \gg T_K$, which can be understood by the following argument. The incoherent Kondo screening which begins at $T \approx T_K$ involves a small number of conduction electrons which are in the Fermi thermal window of width T_K, around μ_0. At lower temperature, $T < T_K$, the Fermi surface is enlarged, due to the contribution of the f-electrons (see Fig. 4). This results from a non-zero hybridization, $r \neq 0$, in the Kondo phase. The formation of a coherent Fermi liquid ground state thus involves all the states of the large Fermi surface. In the situation described by Fig. 5c, where μ_0 is close to a local minimum of the DOS, the further we are from μ_0, the more states are available for the formation of the coherent Fermi liquid. In this case, there is a kind of self-amplification resulting in $T_0 \gg T_K$. With this picture, the magnetic screening of the Kondo impurities involves not only the c-conduction electrons but also, in a dynamical way, "growing" quasiparticles which contain some f-components.

3.3 Substitution of Magnetic Ions

Here, our analysis focuses on the concentration of Kondo impurities, x. In rare-earth based Kondo systems, the latter can be varied, e.g., by Ce–La or Yb–Lu substitution. For this purpose, we have introduced a Kondo alloy model, which is a generalization of the Hamiltonian (1) where each site of the lattice can randomly either contain a Kondo impurity, with a probability x, or not [37, 38]. The Kondo interaction is then treated within the mean-field approximation, and the different configurations of impurity distributions are averaged using a generalization of the coherent potential approximation [39]. Results described here remain valid beyond the coherent potential approximation [38].

Figure 6 illustrates the evolution of T_0/T_K as a function of the impurity concentration, x, for different values of electronic filling, n_c. The dilute limit of the model, $x \ll 1$, reproduces the universal behavior of the single impurity model, with $T_0 = T_K$. The crossover to the dense Kondo-lattice regime, with $T_0 \neq T_K$ occurs for $x \approx n_c$.

The data presented in Fig. 7 are identical to those in Fig. 6, with a rescaled axis, $x \to n_c/x$. From this rescaling, an exhaustion regime is identified, for $n_c < x$, where T_0/T_K is a universal function of n_c/x. Note that this numerical result has been obtained here for a non-interacting semi-elliptic DOS which mimics, at low n_c, the band edge of a three-dimensional system. We expect the universal exhaustion regime to depend on the dimension via the exponent characterizing the vanishing of the DOS at the band edge. Apart from the possibility of magnetic ordering, which is not considered here, the exhaustion regime might be difficult to access experimentally for the following reason: the maximal concentration of Kondo impurities is

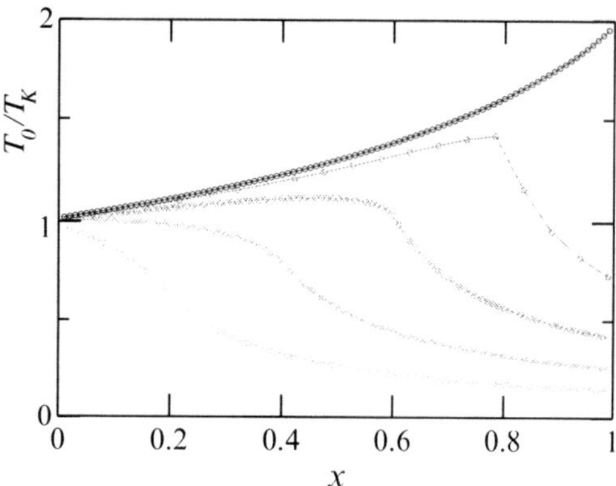

Fig. 6 T_0/T_K as a function of the impurity concentration. From *top* to *bottom*, $n_c = 1; 0.8; 0.6; 0.4; 0.2$. The *solid lines* are included as a guide for the eye. The *curves* have been computed for a semi-elliptic non-interacting DOS and $J_K/D = 0.75$. (From Ref. [37].)

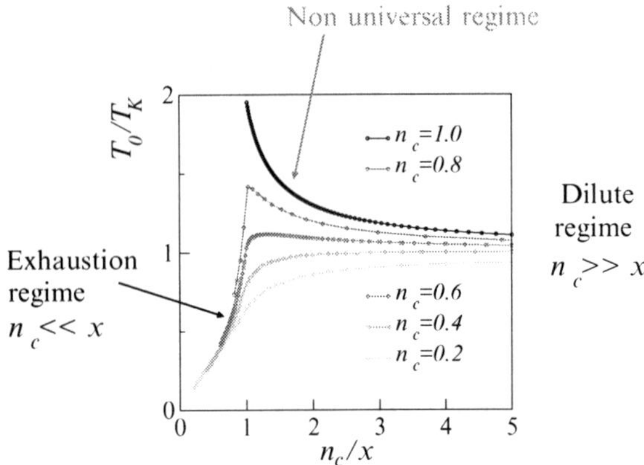

Fig. 7 Same data as Fig. 6, with a rescaled horizontal axis which represents n_c/x here. Each curve corresponds to a given electronic filling, n_c, as indicated by the legend. The parameter n_c/x has been tuned by varying x, and the curves are cut-off by the finite physical minimal value $n_c/x = n_c$.

$x = 1$. The only way to reach $n_c \ll x$ thus involves decreasing the electronic filling. Yet, in rare-earth compounds, conduction electrons usually involve more than one band. With a single band description, the exhaustion regime would be observable by increasing the concentration of magnetic ions, x. A multi-band system, however, allows for another scenario to materialize as x increases: instead of exhausting one conduction band, the system might energetically prefer transferring electrons from other bands, preventing the filling from accessing the exhaustion regime, $n_c \ll x$. In this case, a single band Kondo lattice model would not be appropriate anymore. Nevertheless, we expect that other Kondo systems can be realized experimentally, in which the universal regime $n_c \ll x$ could be observed. For most of dense compounds, $x \approx 1$, the electronic filling is $n_c \approx x$. This is a non-universal regime, where all the lattice structure becomes relevant.

4 Destabilizing the Kondo Phase

Microscopically, the Kondo effect is characterized by the formation of local singlets. The Kondo phase can be destabilized by an external magnetic field, or by the fluctuations of the internal Weiss field induced by RKKY interactions. Here, these two situations are analyzed from the mean-field approximation: the breakdown of the Kondo effect is identified to a continuous vanishing of the $f-c$ effective hybridization, r, which occurs at $T = 0$.

4.1 Effect of a Magnetic Field

We consider an external magnetic field, h, applied to the Kondo spins in the longitudinal direction z. This mimics a situation where the Landé factor of the magnetic

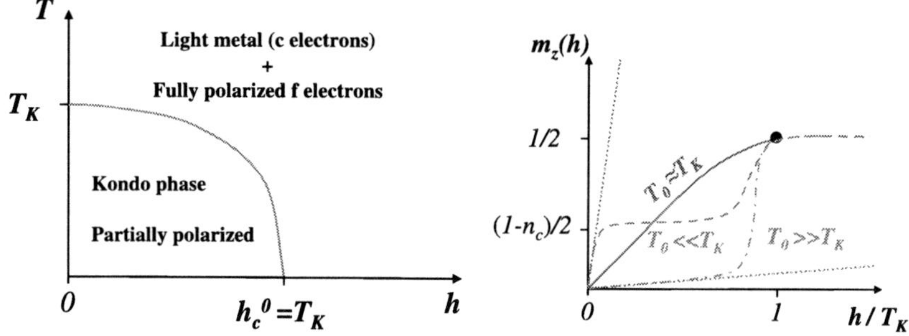

Fig. 8 *Left*: Schematic phase diagram of the Kondo lattice as a function of a magnetic field h. The red solid line indicates $h_c(T)$ which separates the Kondo phase ($r \neq 0$) from the decoupled phase ($r = 0$). *Right*: Schematic plot of the magnetization $m_z(h/T_K)$. For $T_0 \gg T_K$ (*dash dotted line*), $T_0 \approx T_K$ (*solid line*), and $T_0 \ll T_K$ (*dashed line*). The *dotted line* indicates the initial slope in the linear response regime, where $m_z(h) = h/T_0$. The *black dot* refers to the complete polarization of the local Kondo spins, $m_z = 1/2$, which occurs at the critical field $h_c^0 = T_K$. (From Ref. [4].)

ions is much larger than the one of c-electrons. A supplementary contribution is thus added to the Hamiltonian (1), $H \mapsto H - h\sum_i S_i^z$. The mean-field approximation described in Section 2 is generalized [4], resulting in an effective Hamiltonian formally similar to Eq. (2), with spin-dependent $f-$fermion potentials, $\lambda \mapsto \lambda + \sigma h$. The effective hybridization, r, remains spin-independent.

For a sufficiently small magnetic field, a low temperature solution with $r \neq 0$ is obtained. This situation, depicted by Fig. 8, characterizes a phase where the Kondo effect coexists with a partial polarization of both Kondo spins and c-electrons. A finite critical field $h_c(T)$ is obtained, above which c-electrons decouple from fully polarized local moments. At $T = 0$, in the weak coupling limit, $J_K \ll D$, the mean-field approximation yields the universal relation $h_c^0 \equiv h_c(T = 0) = T_K/\alpha$. Since $\alpha = 1.13$, we have $h_c^0 \approx T_K$. This result is not surprising if we consider the Kondo temperature as the energy scale characterizing the local singlet formation: the Kondo effect is destroyed when the Zeeman energy becomes larger than T_K. At finite temperature, assuming a constant $c-$DOS yields the critical line $[h_c(T)/h_c^0]^2 + [T/T_K]^2 = 1$.

The $T = 0$ magnetization $m_z(h) \equiv \frac{1}{2N}\sum_i(\langle f_{i\uparrow}^\dagger f_{i\uparrow}\rangle - \langle f_{i\downarrow}^\dagger f_{i\downarrow}\rangle)$ obtained from the mean-field solution is plotted in Fig. 8 as a function of the reduced magnetic field h/T_K. The low field regime is given by the linear response, $m_z(h) = h\chi_{loc}(T = 0)$, which, by definition, yields $m_z(h) = h/T_0$. Since the critical field characterizing a full polarization of the local moments is of the order of the Kondo temperature, $h_c^0 \approx T_K$, we distinguish three typical cases. $T_0 \approx T_K$ is a standard situation, where m_z increases linearly with h, until saturation. For $T_0 \gg T_K$, the linear regime allows only a small magnetization when $h < h_c^0$. Thus, at about $h \approx h_c^0$, such systems exhibit a meta-magnetic transition from an unpolarized Fermi-liquid to the polarized spin lattice. In the opposite case, $T_0 \ll T_K$, the linear regime saturates around $h \sim T_0 \ll h_c^0$. The intermediate regime $T_0 < h < T_K$ is expected to be non-universal. Eventually, a magnetization plateau can occur, at $m_z \approx (1 - n_c)/2$, similar to the one obtained for the entropy (see Fig. 2).

It is well known that the transition $r = 0$, defining T_K within the mean-field approximation, becomes a crossover when more accurate methods are used. There are several examples of systems where a finite temperature crossover ends up to a quantum critical point, i.e. a transition at zero temperature. Whether the transition at the critical field h_c^0 would survive beyond the mean-field is an open issue.

4.2 Extra-RKKY Interaction

The possibility of destabilizing a Kondo phase in favor of a magnetically ordered ground state was first discussed by Doniach [8], by comparison of the Kondo temperature, $T_K \sim De^{-1/J_K\rho_0(\mu_0)}$, with the RKKY energy, $J_{RKKY} \sim \rho_0(\mu_0)J_K^2$.

The mean-field approximation for the Kondo lattice Hamiltonian (1) can hardly provide a correct description of a transition from a Kondo ground state to a phase with magnetic ordering. This results from the two following reasons:
(i) *Difficulty in driving the system to a second order magnetic transition:* This would require a mechanism leading to a continuous vanishing of the zero temperature effective hybridization, $r(T = 0)$. However, from the Kondo lattice Hamiltonian (1), the only possibility of vanishing $r(T = 0)$ involves the non-interacting DOS, $\rho_0(\mu_0)$ vanishing. This does not correspond to the physical situation of a magnetic transition induced by the RKKY interaction.
(ii) *Difficulty in describing the criticality of transport properties:* Within the mean-field approach, the system is either in a strong coupling regime, with $r \neq 0$ and a heavy Fermi liquid ground state, or in a fully decoupled regime, with $r = 0$. Since the transport properties are governed by the conduction electrons, no universal non-Fermi-liquid behavior can be predicted at the mean-field level.

These two difficulties can be understood from the large-N formulation of the mean-field approximation: Coleman [33], and Read et al. [34] have shown that magnetic ordering involves at least processes of order $1/N$, i.e., fluctuations around the mean-field. Describing the criticality obtained from calculations taking into account the fluctuations and the emerging compact gauge field theory is beyond the scope of this presentation. We simply refer to the works of Senthil et al. [40, 41], who analysed the possibility of a quantum phase transition between a Fermi liquid phase with a large Fermi surface, and a (partially) fractionalized Fermi liquid phase. A Kondo breakdown quantum critical point has also been identified by Pepin [42, 43], who obtained, for example, a specific exponent diverging logarithmically in temperature, as observed in a number of heavy fermion metals.

Whilst *Difficulty* (ii) cannot be easily cured without the fluctuations, we describe here how *Difficulty* (i) can be fixed at the mean-field level. The Kondo lattice Hamiltonian (1) is generalized to a so-called Kondo–Heisenberg lattice, by adding a supplementary superexchange term, as follows: $H \mapsto H + \sum_{ij} J_{ij}\mathbf{S}_i\mathbf{S}_j$. This general model has been introduced first by Sengupta and Georges [44], and studied later within various methods. In terms of auxiliary fermions, introduced in Section 2, the superexchange term is rewritten as $J_{ij}\mathbf{S}_i\mathbf{S}_j = \frac{J_{ij}}{2}\sum_{\sigma\sigma'} f^{\dagger}_{i\sigma}f_{i\sigma'}f^{\dagger}_{j\sigma'}f_{j\sigma}$, which describes the spin-flip processes between two Kondo impurities, on sites i and j. Up to now, this description is very general and holds for any magnetic coupling J_{ij}, which can be periodic (ferro or antiferromagnetic), or randomly distributed (disorder case). We will now describe two complementary mean-field approaches for the superexchange.

The first one was introduced by Coqblin et al. [45] in the case of an antiferromagnetic nearest neighbor exchange $J_{ij} = J_{AF} > 0$. In the paramagnetic Kondo phase, the intersite exchange is approximated within a Resonant Valence Bound decoupling: $J_{ij}\mathbf{S}_i\mathbf{S}_j \mapsto \Gamma_{ij}\sum_\sigma f^\dagger_{i\sigma}f_{j\sigma}$, with $\Gamma_{ij} = \frac{J_{AF}}{2}\sum_\sigma\langle f_{i\sigma}f^\dagger_{j\sigma}\rangle$. The magnetic, Heisenberg-like, interaction generates an effective self-consistent dispersion for the f-fermions. Using this method, Coqblin et al. show that the Kondo effect disappears abruptly for low band filling and/or strong intersite coupling [22, 23].

We have developed another mean-field method [46], based on the DMFT, which we first applied to a disordered system where J_{ij} are given by a Gaussian distribution with an average $[J_{ij}] = 0$, and a variance $[J^2_{ij}] \equiv J^2_d$. In this case, the superexchange term generates a local energy-dependent self-energy for the f-fermions, which is determined by self-consistent relations similar to the ones obtained by Sachdev and Ye for a pure spin disordered system [47]. We have also applied this method to a model with constant nearest-neighbor antiferromagnetic exchange [48], where the self-consistent equation for the f-fermions self-energy depends on the lattice structure. In both cases, disordered or periodic model, we obtained a quantum critical point corresponding to the breakdown of Kondo effect, characterized by a vanishing of the effective hybridization, $r(T = 0) = 0$. Figure 9 depicts the phase diagram that we obtained for the disordered Kondo–Heisenberg model [46]. We also have obtained a similar phase diagram for the periodic, i.e., non-disordered, case [48]. Here, one relevant point is that the quantum critical point emerging from our mean-field

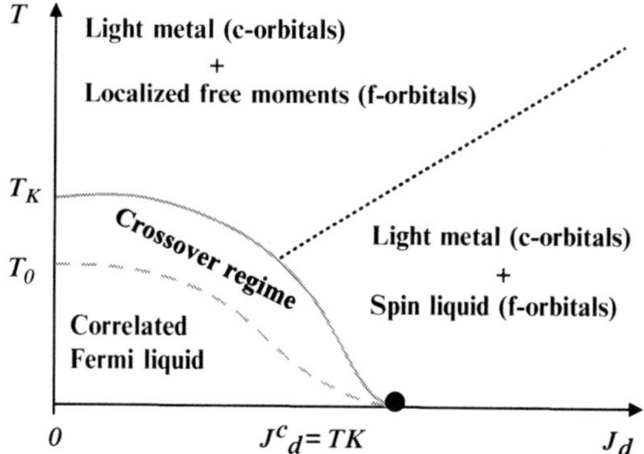

Fig. 9 Schematic phase diagram of the disordered Kondo–Heisenberg model in the $J_d - T$ plane. Kondo temperature (*solid line*) and coherence temperature (*dashed line*) as function of J_d are shown for fixed values of J_K and n_c in the case of $T_0 < T_K$. The system is a heavy Fermi liquid below $T_0(J_d)$. Above the line $T_K(J_d)$, the localized spins are essentially free for $J_d < T$, whilst forming a highly correlated spin liquid for $J_d > T$. All the lines represent crossovers. (From the analysis of Ref. [46].)

approach does not necessarily correspond to the onset of magnetic ordering. This suggests an interpretation in terms of topological transition, without breaking of symmetry, but with a violation of Luttinger theorem, as discussed in Refs. [40,41]. Furthermore, $1/N$ corrections can reveal a magnetic instability at a value of the coupling, J_d which is smaller than $J_d^c = T_K$. In this latter case, a symmetry breaking is expected. Whether criticality in heavy fermions is governed by a Kondo breakdown critical point or by a magnetic transition, what is the nature of the non-Kondo phase, and whether the Kondo breakdown coincides or not with magnetic ordering are still open questions. The answers probably depend on the system.

The quantum critical transition obtained here results from the competition between the Kondo effect and the fluctuations of the f-fermions. These fluctuations are precisely the microscopical mechanism which is required to fix *Difficulty* (i), i.e., the difficulty in driving the system to a second order magnetic transition within the mean-field approximation. In the approach introduced by Coqblin et al., the f-fermion fluctuations are generated by an effective dispersion, which is connected to the Resonant Valence Bound decoupling. In our approach, the fluctuations are included within a self-consistent local self-energy. In both approaches, fluctuations are characterized by an energy, $J_{RKKY} = J_{AF}$ or J_d, and the Kondo effect disappears above the critical value $J_{RKKY} \approx T_K$.

5 Conclusions

Important properties of Kondo systems can be obtained from the mean-field approximation. Some of them have not been presented here, like, for example, the effect of a pseudo-gap [49,50]. Also, for the sake of clarity, the description was restricted to the 'standard' Kondo model. Of course, the mean-field approximation has been generalized and applied to more realistic models, with, for example, a momentum-dependent hybridization between conduction electrons and f ions [51].

Here we have focused on the low energy scales: the Kondo temperature, T_K, characterizing the temperature crossover below which conduction electrons and local moments are strongly coupled; the coherence energy, T_0, characterizing the Fermi liquid ground state; and J_{RKKY}, the intersite magnetic correlation energy. At the mean-field level, J_{RKKY} is negligible or neglected, and any study of magnetic criticality requires the inclusion of fluctuations around the mean-field. Nevertheless, the Kondo effect can break down at the mean-field level if the RKKY interaction is added 'by hand', generalizing the Kondo lattice to a Kondo–Heisenberg model.

For the 'pure' Kondo lattice, the mean-field approximation provides explicit expressions for T_0 and T_K. Both quantities depend on the Kondo coupling with the same exponential factor, $e^{-1/J_K\rho_0(\mu_0)}$. This explains the strong sensitivity of these energy scales with respect to changes in the system (e.g. doping, or pressure). The ratio T_0/T_K, which does not depend on J_K, appears to be a promising quantity for analyzing universal behaviors of heavy fermion compounds. Whilst it is equal to one in dilute systems, T_0/T_K depends in fact on the electronic filling, n_c, the band shape, and the impurity concentration, x. A universal regime with $T_0 \ll T_K$ is expected in the exhaustion limit, $n_c \ll x$, which might be difficult to access experimentally. More typical situations correspond either to the universal dilute regime, $n_c \gg x$, or to the non-universal dense regime $x \sim n_c$. In the latter case, the shape of the non-interacting DOS becomes relevant, and can lead to $T_0 \ll T_K$ if the chemical potential, μ_0, is close to a local maximum of the DOS, or to $T_0 \gg T_K$ if μ_0 is close to a local minimum.

The experimental determination of T_0 and T_K is straightforward for systems which are 'deeply' in a Fermi liquid phase: for example, one can determine T_0 from the specific heat Sommerfeld coefficient, and T_K from the temperature dependence of the magnetic part of the entropy. The determination of T_K might become more tricky in the vicinity of a magnetic transition, where the freezing of the entropy is no longer due to Kondo singlet formation, but to magnetic intersite correlations instead. In this case, one should find another physical observable which would enable an unambiguous determination of T_K. In this case, a systematic experimental analysis of the ratio T_0/T_K might reveal interesting universal behavior.

Some of the results obtained at the mean-field level have been confirmed by exact numerical methods. For example, DMFT calculations have shown that T_0/T_K does not depend on the Kondo coupling. One limitation of the mean-field is also well known: its weakness in describing criticality, where fluctuations become important. Nevertheless, some predictions of the mean-field have not been checked yet with more accurate methods (DMFT or $1/N$ corrections). This is, for example, the case of the band shape effect.

Acknowledgements I thank the organizers of the NATO Advanced Research Workshop on Properties and Applications of Thermoelectric Materials. I acknowledge my collaborators on the work presented here: P. Fulde, A. Georges, M. Grilli, and V. Zlatic. I am also grateful to A. Rosch, M. Vojta, A. Klopper and H. Weber for useful discussions regarding this manuscript, and to P. Nozières, B. Coqblin, and C. Lacroix for fruitful advises. A part of this work was funded by the DFG through SFB 680, SFB/TR 12, and FG 960.

References

1. P. Fulde, P. Thalmeier, and G. Zwicknagl, *Strongly Correlated Electrons*, Solid State Physics Vol. 60 (Elsevier, New York, 2006).
2. H. v. Löhneysen, A. Rosch, M. Vojta, and P. Wölfle, Rev. Mod. Phys. **79**, 1015 (2007).
3. A.C. Hewson, *The Kondo Problem to Heavy Fermions* (Cambridge University Press, Cambridge, England, 1993).
4. S. Burdin, and V. Zlatiĉ, arXiv:0812.1137, to appear in Phys. Rev. B.
5. V. Zlatiĉ, R. Monnier, J. Freericks, and K.W. Becker, Phys. Rev. B **76**, 085122 (2007).
6. V. Zlatiĉ, R. Monnier, and J. Freericks, Phys. Rev. B **78**, 045113 (2008).
7. Y.F. Yang, Z. Fisk, H.O. Lee, J.D. Thompson, and D. Pines, Nature **454**, 611 (2008).
8. S. Doniach, Physica B & C **91**, 231 (1977).
9. J.J. Joyce, A.J. Arko, J. Lawrence, P.C. Canfield, Z. Fisk, R.J. Bartlett, and J.D. Thompson, Phys Rev. Lett. **68**, 236 (1992).
10. J.J. Joyce, A.J. Arko, A.B. Andrews, and R.I.R. Blyth, Phys. Rev. Lett. **72**, 1774 (1994).
11. L.H. Tjeng, S.-J. Oh, C.T. Chen, J.W. Allen, and D.L. Cox, Phys. Rev. Lett. **72**, 1775 (1994).

12. D. Malterre, M. Grioni, and Y. Baer, Adv. Phys. **45**, 299 (1996).
13. R. Mallik, E.V. Sampathkumaran, P.L. Paulose, J. Dumschat, and G. Wortmann, Phys. Rev. B **55**, 3627 (1997).
14. E.D. Mun, Y.S. Kwon, and M.H. Jung, Phys. Rev. B **67**, 033103 (2003).
15. J.M. Lawrence, P.S. Riseborough, C.H. Booth, J.L. Sarrao, J.D. Thompson, and R. Osborn Phys. Rev. B **63**, 054427 (2001).
16. E.D. Bauer, C.H. Booth, J.M. Lawrence, M.F. Hundley, J.L. Sarrao, J.D. Thompson, P.S. Riseborough, and T. Ebihara, Phys. Rev. B **69**, 125102 (2004).
17. P. Nozières, Ann. Phys. (Paris) **10**, 19 (1985).
18. P. Nozières, Eur. Phys. J. B **6**, 447 (1998).
19. C. Lacroix, Solid State Commun. **54**, 991 (1985).
20. C. Lacroix, J. Magn. Magn. Mat. **60**, 145 (1986).
21. S. Burdin, A. Georges, and D.R. Grempel, Phys. Rev. Lett. **85**, 1048 (2000).
22. B. Coqblin, M.A. Gusmao, J.R. Iglesias, C. Lacroix, A. Ruppenthal, and A.S.D. Simoes, Physica B **281**, 50 (2000).
23. B. Coqblin, C. Lacroix, M.A. Gusmao, and J.R. Iglesias, Phys. Rev. B **67**, 064417 (2003).
24. M. Jarrell, H. Akhlaghpour, and Th. Pruschke, Phys. Rev. Lett. **70**, 1670 (1993).
25. M. Jarrell, Phys. Rev. B **51**, 7429 (1995).
26. A.N. Tahvildar-Zadeh, M. Jarrell, and J. K. Freericks, Phys. Rev. B **55**, R3332 (1997).
27. A.N. Tahvildar-Zadeh, M. Jarrell, and J. K. Freericks, Phys. Rev. Lett. **80**, 5168 (1998).
28. A.N. Tahvildar-Zadeh, M. Jarrell, Th. Pruschke, and J.K. Freericks, Phys. Rev. B **60**, 10782 (1999).
29. A. Georges, G. Kotliar, W. Krauth, and M.J. Rozenberg, Rev. Mod. Phys. **68**, 13 (1996).
30. W. Metzner, and D. Vollhardt, Phys. Rev. Lett. **62**, 324 (1989).
31. P. Nozières, J. Phys. Soc. Jpn. **74**, 4 (2005).
32. C. Lacroix, and M. Cyrot, Phys. Rev. B **20**, 1969 (1979).
33. P. Coleman, Phys. Rev. B **28**, 5255 (1983).
34. N. Read, D.M. Newns, and S. Doniach, Phys. Rev. B **30**, 3841 (1984).
35. B. Coqblin, and J.R. Schrieffer, Phys. Rev. **185**, 847 (1969).
36. T.A. Costi, and N. Manini, J. Low Temp. Phys. **126**, 835 (2002).
37. S. Burdin, and P. Fulde, Phys. Rev. B **76**, 104425 (2007).
38. R.K. Kaul, and M. Vojta, Phys. Rev. B **75**, 132407 (2007).
39. J.A. Blackman, D.M. Esterling, and N.F. Berk, Phys. Rev. B **4**, 2412 (1971).
40. T. Senthil, S. Sachdev, and M. Vojta, Phys. Rev. Lett. **90**, 216403 (2003).
41. T. Senthil, M. Vojta, and S. Sachdev, Phys. Rev. B **69**, 035111 (2004).
42. C. Pepin, Phys. Rev. Lett. **98**, 206401 (2007).
43. I. Paul, C. Pepin, and M.R. Norman, Phys. Rev. Lett. **98**, 026402 (2007).
44. A.M. Sengupta, and A. Georges, Phys. Rev. B **52**, 10295 (1995).
45. J.R. Iglesias, C. Lacroix, and B. Coqblin, Phys. Rev. B **56**, 11820 (1997).
46. S. Burdin, D.R. Grempel, and A. Georges, Phys. Rev. B **66**, 045111 (2002).
47. S. Sachdev, and J.W. Ye, Phys. Rev. Lett. **70**, 3339 (1993).
48. S. Burdin, M. Grilli, and D.R. Grempel, Phys. Rev. B **67**, 121104 (2003).
49. D. Withoff, and E. Fradkin, Phys. Rev. Lett. **64**, 1835 (1990).
50. L. Fritz, and M. Vojta, Phys. Rev. B **70**, 214427 (2004).
51. H. Weber, and M. Vojta, Phys. Rev. B **77**, 125118 (2008).

Lightning Source UK Ltd.
Milton Keynes UK
17 August 2009

142719UK00001B/19/P

9 789048 128907